# MEMS AND NEMS

## Systems, Devices, and Structures

## *Nano- and Microscience, Engineering, Technology, and Medicine Series*

*Series Editor*
Sergey Edward Lyshevski

*Titles in the Series*

**MEMS and NEMS: Systems, Devices, and Structures**
Sergey Edward Lyshevski

*Forthcoming*

**Microelectrofluidic Systems: Modeling and Simulation**
Tianhao Zhang, Krishendu Chakrabarty, and Richard B. Fair

**Nanodynamics in Engineering and Biology**
Michael Pycraft Hughes

# MEMS AND NEMS

## Systems, Devices, and Structures


Sergey Edward Lyshevski
Department of Electrical and Computer Engineering
Purdue University at indianapolis

CRC PRESS
Boca Raton London New York Washington, D.C.

**Library of Congress Cataloging-in-Publication Data**

Lyshevski, Sergey Edward.
MEMS and NEMS : systems, devices, and structures / Sergey Edward Lyshevski.
p. cm. — (Nano- and microscience, engineering, technology, and medicine series)
Includes bibliographical references and index.
ISBN 0-8493-1262-0 (alk. paper)
1. Microelectromechanical systems. I. Title. II. Series.

TK7875 .L9597 2001
621.381—dc21 2001049944

This book contains information obtained from authentic and highly regarded sources. Reprinted material is quoted with permission, and sources are indicated. A wide variety of references are listed. Reasonable efforts have been made to publish reliable data and information, but the author and the publisher cannot assume responsibility for the validity of all materials or for the consequences of their use.

Neither this book nor any part may be reproduced or transmitted in any form or by any means, electronic or mechanical, including photocopying, microfilming, and recording, or by any information storage or retrieval system, without prior permission in writing from the publisher.

The consent of CRC Press LLC does not extend to copying for general distribution, for promotion, for creating new works, or for resale. Specific permission must be obtained in writing from CRC Press LLC for such copying.

Direct all inquiries to CRC Press LLC, 2000 N.W. Corporate Blvd., Boca Raton, Florida 33431.

**Trademark Notice:** Product or corporate names may be trademarks or registered trademarks, and are used only for identification and explanation, without intent to infringe.

**Visit the CRC Press Web site at www.crcpress.com**

© 2002 by CRC Press LLC

No claim to original U.S. Government works
International Standard Book Number 0-8493-1262-0
Library of Congress Card Number 2001049944
Printed in the United States of America 2 3 4 5 6 7 8 9 0
Printed on acid-free paper

621.381
L993

# PREFACE

The intent of this book is to introduce micro- and nano-electromechanical systems, devices, and structures to a wide audience. This book is written for a one-semester senior undegraduate or first-year graduate course on nano- and microelectromechanical systems. Therefore, a typical background needed includes calculus and physics. The purpose of this book is to bring together the various concepts, methods, techniques, and technologies needed to solve a wide array of problems including synthesis, modeling, simulation, analysis, design and optimization of high-performance microelectromechanical and nanoelectromechanical systems (MEMS and NEMS), devices, and structures. Microfabrication aspects and technologies are also covered to assist the readers. The availability of advanced fabrication technologies has been the considerable motivation for further developments. The emphasis of this book is on the fundamental principles of MEMS and NEMS and practical applications of the basic theory in engineering practice and technology.

This book is written in the textbook style, with the goal to reach the widest possible range of readers who have an interest in the subject. Specifically, the objective is to satisfy the existing growing demands of undergraduate and graduate students, engineers, professionals, researchers, and instructors in the fields of micro- and nanoengineering, science, and technologies. With these goals, the structure of the book was devised. Efforts were made to bring together fundamental (basic theory), applied, and technology (fabrication) aspects in different areas which are viable to study, understand, and research advanced topics in MEMS and NEMS in a unified and consistent manner.

Recently, there has been an accelerating level of interest in micro- and nanoengineering and technologies. A 21$^{st}$ century nano- and microtechnology revolution will lead to fundamental breakthroughs in the way materials, devices, and systems are understood, designed, manufactured, and used. Nano- and microengineering will change the nature of the majority of human-made structures, devices, and systems.

Current technological needs and trends include leading-edge fundamental, applied, and experimental research as well as technology developments, in order to model, analyze, characterize, design, optimize, simulate and fabricate nano- and microscale systems, devices, and structures. Current developments have been focused on analysis and design of molecular structures and devices which will lead to revolutionary breakthroughs in the data processing, computing, data storage, imaging, molecular intelligent automata, etc. Specifically, molecular computers, logic gates and switches, actuators and sensors, digital and analog nanocircuits have been devised and studied. Micro- and nanoengineering and science lead to fundamental breakthroughs in the way materials, devices and systems are devised, designed, fabricated, and used. High-performance MEMS and NEMS, micro- and nanoscale structures and devices will be widely used in nanocomputers, medicine (nanosurgery and nanotherapy, design and implants of nonrejectable artificial organs, drug delivery and diagnosis), biotechnology (genome synthesis), etc.

New phenomena in nano- and microelectromechanics, physics and chemistry, novel nanomanufacturing technologies, benchmarking control of complex molecular structures, design of large-scale architectures, and optimization, among other problems, must be addressed and studied. The major objective of this book is the development of basic theory (through multidisciplinary fundamental and applied research) to achieve the highest degree of understanding complex phenomena and effects, as well as development of novel paradigms and methods in optimization, analysis, and control of NEMS and MEMS properties and behavior. This will lead to new advances and will allow the designer to comprehensively solve a number of long-standing problems in synthesis, analysis, control, modeling, simulation, virtual prototyping, fabrication, implementation, and commercialization of novel NEMS and MEMS.

In addition to technological developments, the ability to synthesize and optimize NEMS and MEMS depends on analytical and numerical methods. Novel paradigms and concepts should be devised and applied to analyze and study complex phenomena and effects. Advanced interdisciplinary research must be carried out to design, develop, and implement high-performance NEMS and MEMS. The objectives are to expand the frontiers of the NEMS- and MEMS-based research through pioneering fundamental and applied multidisciplinary studies and developments.

The fundamental goal of this book is to develop the basic theoretical foundations in order to design and develop, analyze, and prototype high-performance NEMS and MEMS. This book is focused on the development of fundamental theory of NEMS and MEMS, as well as their components (subsystems and devices) and structures, using advanced multidisciplinary basic and applied developments. In particular, how to perform the comprehensive studies will be illustrated with analysis of the processes, phenomena, and relevant properties at nano- and micro-scales, development of MEMS and NEMS architectures, physical representations, structural synthesis and optimization, etc. It is the author's goal to substantially contribute to these basic issues, efficiently deliver the rigorous theory to the reader, and integrate the challenging problems in the context of well-defined applications addressing specific issues. The primary emphasis will be on the development of basic theory to attain fundamental understanding of MEMS and NEMS, processes in nano- and microscale structures, devising novel devises and structures, as well as the application of the developed theory.

It should be acknowledged that no matter how many times the material is reviewed and efforts are spent to guarantee the highest quality, the authors cannot guarantee that the manuscript is free from minor errors. If you find something that you feel needs correction, clarification, or modification, please notify me at lyshevs@engr.iupui.edu. Your help and assistance are greatly appreciated and deeply acknowledged.

**Acknowledgments**

Many people contributed to this book. First thanks go to my beloved family. I would like to express my sincere acknowledgments and gratitude to many colleagues and students. It gives me great pleasure to acknowledge the help I received from many people in the preparation of this book. The outstanding CRC Press team, especially Nora Konopka (Acquisitions Editor, Electrical Engineering) and Samar Haddad (Project Editor), tremendously helped and assisted me by providing valuable and deeply treasured feedback. Many thanks to all of you.

Sergey Edward Lyshevski
Department of Electrical and Computer Engineering
Purdue University Indianapolis
Indianapolis, Indiana 46202-5132
E-mail: lyshevs@engr.iupui.edu

**CONTENTS**

# CHAPTER 1

# OVERVIEW AND INTRODUCTION

Nano- and microelectromechanics, optoelectronics, as well as nano- and microfabrication have accomplished phenomenal growth over the past few years due to rapid advances in theoretical developments, experimental results (using state-of-the-art measurement and instrumentation hardware), high-performance computer-aided design software and computationally efficient environments. Recent fundamental and applied research and developments in nano- and microelectromechanics, informatics, and nanotechnology have notably contributed to the current progress. Leading edge basic research, novel technologies, software, and hardware are integrated to devise new systems and study existing systems. These prominent trends provide researchers, engineers, and students with the needed concurrent design of integrated complex microelectromechanical and nanoelectromechanical systems (MEMS and NEMS). The synergy of engineering, science, and technology is essential to attaining the goals and objectives. It becomes increasingly difficult to perform analysis and design of micro- and nanoscale systems, subsystems, devices and structures without the unified theme and multidisciplinary synergy due to integration of new phenomena, complex processes, and compatibility. Therefore, micro- and nanoelectromechanics theories must be further developed and applied. Micro- and nanoelectromechanics are based upon fundamental theory, engineering practice, and leading edge technologies in fabrication of micro- and nanoscale systems, subsystems, devices, and structures which have dimensions of micrometers and nanometers.

The development and deployment of MEMS and NEMS are critical to the U.S. economy and society because nano- and microengineering and science will lead to major breakthroughs in information technology, computers, medicine, health, manufacturing, transportation, energy, avionics, security, etc. For example, MEMS and NEMS have an important impact on medicine and bioengineering (DNA and genetic code analysis and synthesis, drug delivery, diagnostics and imaging), bio and information technologies, avionics and aerospace (nano- and microscale actuators and sensors, smart reconfigurable geometry wings and blades, space-based flexible structures, microgyroscopes), automotive systems and transportation (transducers and accelerometers), manufacturing and fabrication (micro- and nanoscale smart robots), safety, etc. Therefore, MEMS and NEMS will have an tremendous positive direct and indirect social and economic impacts. New MEMS and NEMS will be devised and used in a variety of applications (electronics, medicine, metrology, etc.). Micro- and nanoscale structures, devices, and systems frequently have the specific applications such as high-frequency resonators, electromagnetic field and stress sensors, etc. Nanoscale devices and systems will allow one to access the atomic scale phenomena, where quantum effects are predominant. Though, depending upon the size and architecture, MEMS can be designed using scaling paradigm, as the microstructures dimension is within 0.1 μm, the secondary effects cannot be neglected for nanostructures. Hence, advanced high-fidelity modeling and analysis concepts are used for various MEMS. NEMS usually are not scalable which leads to numerous challenges. To support the nano- and microengineering, basic, applied, and experimental research as well as engineering developments must be performed.

Nanoengineering studies nanoscale structures, devices and systems, whose structures and components exhibit novel physical (electromagnetic, electromechanical, optical, etc.), chemical, electrochemical, and biological properties, phenomena, and effects. The dimension of NEMS and their components (nanostructures) is from $10^{-10}$ m (molecule size) to $10^{-7}$ m; that is, from 0.1 to 100 nanometers. To manufacture nanostructures, nanotechnology is applied. In contrast, the dimension of MEMS and their components (devises and microstructures) is from 100 nanometers to the centimeter range.

Conventional microelectronics technologies are usually used to fabricate MEMS. Studying nanoscale systems, one concentrates on the atomic and molecular levels in manufacturing, fabrication, design, analysis, optimization, integration, synthesis, etc. Reducing the dimension of systems leads to the application of novel materials (carbon nanotubes, molecular wires, etc.) and new fabrication technologies. The problems to be solved range from high-yield mass production, assembling, and self-organization to devising novel high-performance MEMS and NEMS. For example, micro- and nanoscale switches, logic gates, actuators, and sensors should be devised, studied, optimized, and fabricated.

All living biological systems function as a result of atomic and molecular interactions. Different biological and organic systems, subsystems, and structures are built. The molecular building blocks (proteins, nucleic acids, lipids, carbohydrates, DNA and RNA) are applied to perform biomimicking and prototyping to devise novel MEMS and NEMS. The devised MEMS and NEMS must have the desired properties, characteristics, and functionality. Analytical and numerical methods must be developed in order to analyze the dynamics, three-dimensional geometry, bonding, and other features of atoms and molecules. In order to accomplish this analysis, complex electro-magnetic, mechanical, and other physical and chemical properties must be studied.

Micro- and nanosystems will be widely used in medicine and health. Among possible applications are drug synthesis, drug delivery, nanosurgery, nanotherapy, genome synthesis, diagnostics, novel actuators and sensors, disease diagnosis and prevention, nonrejectable artificial organs design and implant, biocompatible materials and structures, etc. For example, with drug delivery, the therapeutic potential will be enormously enhanced due to direct effective delivery of new types of drugs to the specified body sites. The molecular building blocks of DNA-based structures, proteins, nucleic acids, lipids, as well as carbohydrates and their non-biological mimics are examples of materials that possess unique properties determined by their size, folding, and patterns at the nano- and microscale. Significant progress has been made, and new analytical tools must be developed capable of characterizing the chemical, electrical, and mechanical properties of cells including processes such as cell division, locomotion, and propulsion.

Microelectromechanical systems, which integrate motion microdevices (actuators and sensors), radiating energy microdevices (antennas, microstructures with windings), microscale driving/sensing circuitry, and controlling/processing integrated circuits (ICs), are widely used.

Figure 1.1 illustrates the functional block-diagram of MEMS.

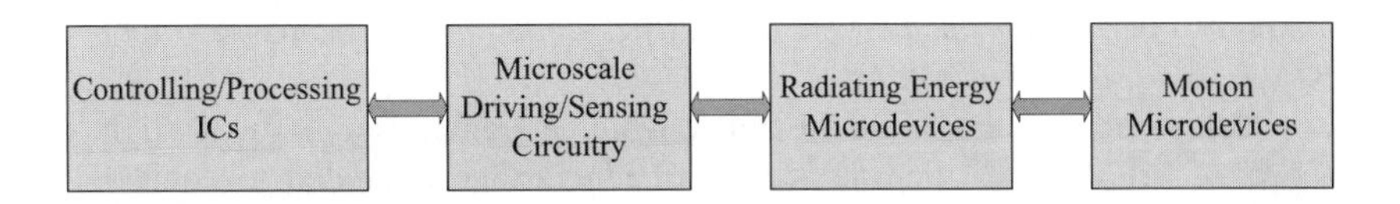


Figure 1.1 Functional block-diagram of MEMS.

MEMS is defined as follows:

*The MEMS is the batch-fabricated integrated microscale system (motion, electromagnetic, radiating energy and optical microdevices/microstructures – driving/sensing circuitry – controlling/processing ICs) that:*

1. *Converts physical stimuli, events, and parameters to electrical, mechanical, and optical signals and vice versa;*
2. *Performs actuation, sensing, and other functions;*
3. *Comprises control (intelligence, decision-making, evolutionary learning, adaptation, self-organization, etc.), diagnostics, signal processing, and data acquisition features,*

*and microscale features of electromechanical, electronic, optical, and biological components (structures, devices, and subsystems), architectures, and operating principles are basics of the MEMS operation, design, analysis, and fabrication.*

MEMS are comprised and built using microscale subsystems, devices, and structures.

The following classification defines the microdevice:

*The microdevice is the batch-fabricated integrated microscale motion, electromagnetic, radiating energy, or optical microscale device that*

1. *Converts physical stimuli, events, and parameters to electrical, mechanical, and optical signals and vice versa;*
2. *Performs actuation, sensing, and other functions,*

*and microscale features of electromechanical, electronic, optical, and biological structures, topologies, and operating principles are basics of the microdevice operation, design, analysis, and fabrication.*

Finally, the definition is given for the microstructure.

*The microstructure is the batch-fabricated microscale electromechanical, electromagnetic, mechanical, or optical composite microstructure (that can integrate simple microstructures) that is a functional component of the microdevice and serve to attain the desired microdevice's operating features (e.g., convert physical stimuli, events, and parameters to electrical, mechanical, and optical signals and vice versa, as well as perform actuation, sensing, and other functions), and microscale features of electromechanical, electromagnetic, mechanical, optical, or biological structure (geometry, topology, operating principles, etc.) are basics of the microstructure operation, design, analysis, and fabrication.*

Compared with the microstructures, which are elementary structures, the scope of MEMS and microdevices has been further expanded to devising (synthesis) new systems and devices, developing novel paradigms and theories, system-level integration, high-fidelity modeling, data-intensive analysis, control, optimization, computer-aided design, fabrication, and implementation. These have been done through biomimicking and prototyping that allow one to expand the existing devised and man-made systems and devices.

It is evident that chemical and bioengineered transducers (actuators and sensors) can be designed, used, and classified. For example, inorganic actuators, gas sensors, etc. Correspondingly, chemical and biological microelectromechanical systems, devices, and structures exist. Therefore, the above given definitions can be expanded adding chemical and biological/ bioengineered components. Furthermore, biomimicking and prototyping are very important paradigms to devise novel MEMS and NEMS.

Many species of bacterium move around their aqueous environment using flagella which is the protruding helical filament driven by the rotating molecular motor. This bionanomotor–flagella complex provides the propulsion thrust for cells to swim. Let us examine the *Escherichia coli* (*E. coli*) bacteria. Biological (bacterial) nanomotors convert chemical energy into electrical energy, and electrical energy into mechanical energy. The bionanomotor uses the proton or sodium gradient, maintained across the cell's inner membrane as the energy source. The motion is due the downhill transport of ions. The research in complex chemo-electro-mechanical energy conversion allows one to understand the torque generation, energy conversion, bearing, and sensing–feedback–control mechanisms. With the ultimate goal to devise novel organic and inorganic micro- and nanomachines through biomimicking and prototyping, one can invent:

- Unique radial and axial micro- and nanomachine topologies;
- Electrostatic- and electromagnetic-based actuation mechanisms;
- Noncontact electrostatic bearing;
- Novel sensing–feedback–control mechanisms;
- Advanced excitation concepts;
- New micro- and nanomachine configurations.

Biomimetic systems are the man-made systems which are based on biological principles, or on biologically inspired building blocks integrated as the systems structures, devices, and subsystems. These developments benefit greatly from adopting strategies and architectures from the biological world. Based on biological principles, bio-inspired systems and materials are currently being formed by self-assembling and other patterning methods. Artificial inorganic and organic nanomaterials are introduced into cells with diagnostics features as well as active (smart) structures. This research provides the enabling capabilities to achieve potential breakthroughs that guarantee major broad-based economic benefits and dramatically improve the quality of life.

Let us study bionanomotors in order to devise, design, and synthesize new high-performance micro- and nanomachines with fundamentally new organization, topologies, operating principles, enhanced functionality, superior capabilities, and enhanced operating envelopes. The *E. coli* bionanomotor and the nanomotor–flagella complex are shown in Figure 1.2.

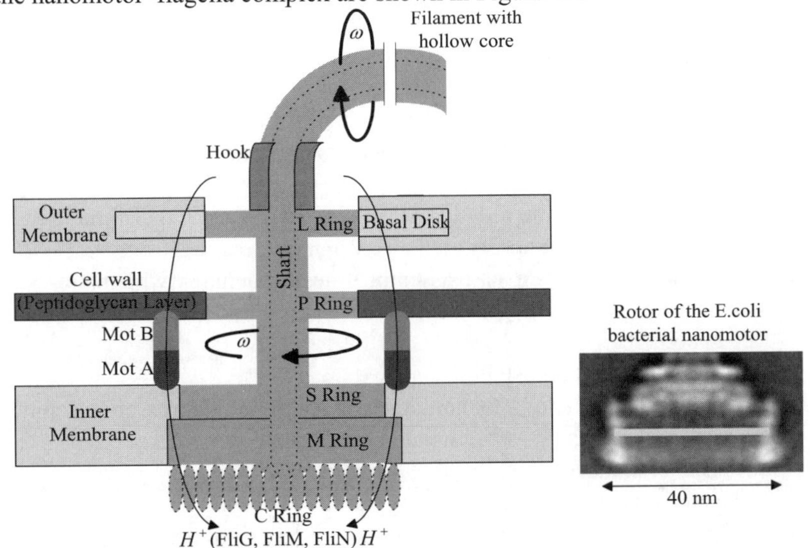


Figure 1.2 *E. coli* bacterial bionanomotor–flagella complex and rotor image.

The protonomotive force in the *E. coli* bionanomotors is axial. It should be emphasized that the protonomotive force can be radial as well. Through biomimicking, two machine (rotational motion transducer) topologies are radial and axial. In particular, the magnetic flux in the radial (or axial) direction interacts with the time-varying axial (or radial) electromagnetic field, and the electromagnetic torque (to actuate the motor) can be produced.

Using the radial topology, the cylindrical machine with permanent-magnet poles on the rotor is shown in Figure 1.3. The electrostatic noncontact bearings allow one to maximize efficiency and reliability, improve ruggedness and robustness, minimize cost and maintenance, decrease size and weight, optimize packaging and integrity, etc. Figure 1.3 illustrates a machine with noncontact bearing.

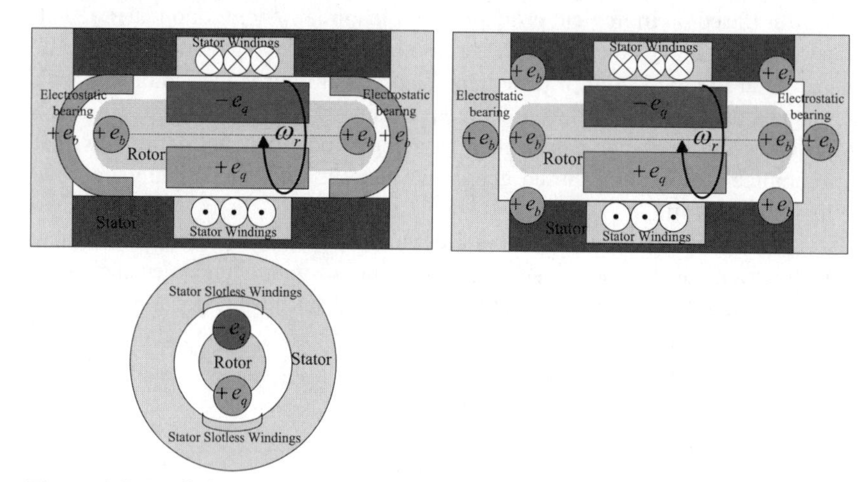


Figure 1.3 Radial topology machine with electrostatic noncontact bearings (poles are $+e_q$ and $-e_q$, and electrostatic bearing formed by $+e_b$).

The advantages of the radial topology is that the net radial force on the rotor is zero. The disadvantages are:

- Difficulty to fabricate and assemble these machines with nano- and microstructures (stator with deposited windings and rotor with deposited magnets),
- The air gap is not adjustable.

Analyzing the *E. coli* bionanomotor, the nano- and microscale machines (rotational transducers) with axial flux topology are devised. The synthesized axial machine is illustrated in Figure 1.4.

The advantages of the axial topology are:

- Affordability to manufacture and assemble machines because permanent magnets are flat (permanent-magnet thin films can be deposited using the technologies, processes, materials, and chemicals reported in Chapters 3 and 8 of this book),

- There are no strict shape-geometry and sizing requirements imposed on the magnets,
- There is no rotor back ferromagnetic material required (silicon or silicon-carbide techniques and processes can be straightforwardly applied),
- The air gap can be adjusted,
- It is easy to lay out the implanted or deposited microwindings (even as molecular wires) on the flat stator.

The disadvantages are:

- Lower torque, force, and power densities,
- Decreased winding utilization,
- Lower torque density compared with the radial topology.

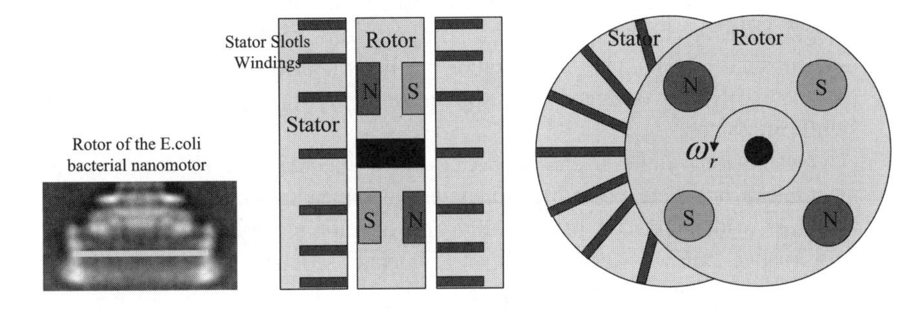


Figure 1.4 Axial topology machine with permanent magnets.

The stationary magnetic field is established by the permanent magnets, and stators and rotors can be fabricated using surface miromachining and high-aspect-ratio technologies. Slotless stator windings can be laid out as the deposited microwindings or as implanted molecular wires.

The following is step-by-step procedure in the design microdevices, with emphasis on the discussed micro- and nanoscale machines (transducers):

1. Devise machines researching operating principles, topologies, configurations, geometry, electromagnetic systems (closed-ended, open-ended, or integrated), interior or exterior rotor, etc.;
2. Define application and environmental requirements as well as specify performance specifications;
3. Perform electromagnetic, energy conversion, mechanical, and sizing-dimension estimates;
4. Define technologies, techniques, and processes to fabricate micro- and nanostructures (e.g., stator with deposited windings, rotor with permanent magnets, and bearing) and assemble them;

5. Based upon data-intensive heterogeneous analysis, determine air gap, select permanent magnets and materials, as well as perform thorough electromagnetic, mechanical, vibroacoustic and thermodynamic design with performance analysis and outcome prediction;
6. Modify and refine the design optimizing machine performance;
7. Design control laws to control micromachines and implement these controllers using ICs (this task itself can be broken down to many subtasks and problems related to control laws design, optimization, analysis, simulation, ICs topologies, ICs fabrication, micromachine-ICs integration, etc.).

The reported results clearly illustrate that before being engaged in the fabrication, it is a critical need to address fundamental synthesis, design, analysis, and optimization issues that are common to all micro- and nanomachines. In fact, machine performance and its applicability directly depend upon synthesized topologies, configurations, electromagnetic systems, mathematical models, etc. Optimization methods applied, simulation software used, control laws designed, ICs applied, and machine-ICs integration also significant factors.

Systematic synthesis and modeling allow the designer to devise novel phenomena and new operating principles guaranteeing synthesis of superior machines with enhanced functionality and operationability. To design high-performance MEMS and NEMS, fundamental, applied, and experimental research must be performed further developing the synergetic micro- and nanoelectromechanical theories.

Several fundamental electromagnetic and mechanical laws are: quantum mechanics, Maxwell's equations, nonlinear mechanics, and energy conversion. It was shown that the design of MEMS and NEMS is not a simple task because complex electromagnetic, mechanical, thermodynamic, and vibroacoustic problems must be examined in the time domain solving partial differential equations. Advanced interactive software with application-specific toolboxes, robust methods, and novel computational algorithms must be used.

Computer-aided design of MEMS and NEMS is valuable due to:

1. Calculation and thorough evaluation of a large number of options with heterogeneous performance analysis and outcome prediction;
2. Knowledge-based intelligent synthesis and evolutionary design which allow one to define optimal solution with minimal effort, time, cost, reliability, confidence, and accuracy;
3. Concurrent nonlinear electromagnetic and mechanical analysis to attain superior performance of MEMS and NEMS while avoiding costly and time-consuming fabrication and testing;
4. Possibility to solve complex partial differential equations in the time domain integrating systems patterns with nonlinear material characteristics;
5. Development of robust, accurate, and efficient rapid design and prototyping environments and tools which have innumerable features to assist the user to set the problem up and to obtain the engineering parameters.

Through the structural synthesis, the designer devises MEMS and NEMS that must be modeled, analyzed, simulated, and optimized. Synthesis, design, and optimization guarantee the superior performance capabilities. As shown in Figure 1.5, devising and developing novel micro- and nanomachines one maximizes efficiency, reliability, power and torque densities, ruggedness, robustness, durability, survivability, compactness, simplicity, controllability, and accuracy. The application of the structural synthesis paradigm leads to the minimization of cost, maintenance, size, weight, volume, and losses. Packaging and integrity are optimized using the structural synthesis concept.

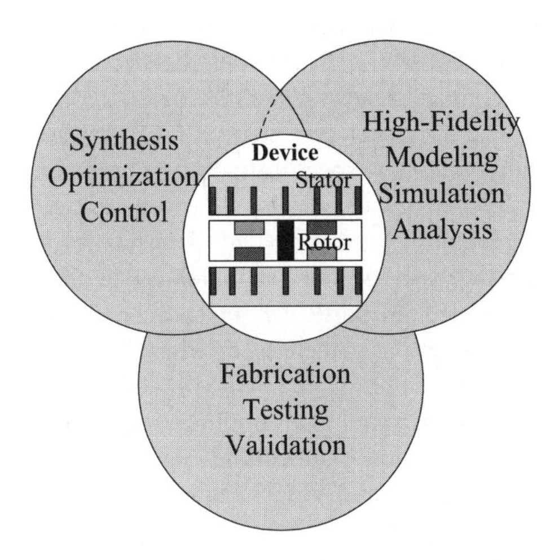


Figure 1.5 Design of high-performance nano- or microscale machines.

Micro- and nanotechnologies drastically change the fabrication and manufacturing of structures, devices, and systems through:

- Predictable properties and characteristics of micro- and nanocomposites, compounds, and materials (e.g., light weight and high strength, thermal stability, low volume and size, extremely high power, torque, force, charge and current densities, specified thermal conductivity and resistivity, etc.);
- Virtual prototyping (design cycle, cost, and maintenance reduction);
- Improved accuracy, precision, reliability and durability;
- Higher efficiency, safety, capability, flexibility, integrity, supportability, affordability, survivability and redundancy;
- Improved stability and robustness;
- Environmental compatibility, integrity and flexibility.

For MEMS and NEMS, many engineering problems can be formulated, attacked, and solved using the micro- and nonoelectromechanic paradigms. Electromechanics deals with benchmarking and emerging problems in integrated electrical–mechanical–computer engineering, science, and technology. Many of these problems have not been addressed and solved, and sometimes, the existing solutions cannot be treated as the optimal one. This reflects obvious trends in fundamental, applied, and experimental research in response to long-standing unsolved problems, as well as engineering and technological enterprise and entreaties of steady evolutionary demands.

Micro- and nanoelectromechanics focuses on the multidisciplinary synergy, integrated design, analysis, optimization, biomimicking and virtual prototyping of high-performance MEMS and NEMS, system intelligence, learning, adaptation, decision-making, and control through the use of advanced hardware devised and computationally efficient heterogeneous software developed. Integrated multidisciplinary features and the need for synergetic paradigms approach quickly. The structural complexity of MEMS and NEMS has been increased drastically due to specifications imposed on systems-devices, hardware and software advancements, as well as stringent *achievable* performance requirements. Answering the demands of the rising systems complexity, performance specifications and intelligence, the fundamental theory must be further expanded. In particular, in addition to devising subsystems, devices and structures, there are other issues which must be addressed in view of constantly evolving nature of the MEMS and NEMS (e.g., synthesis, analysis, design, modeling, simulation, optimization, complexity, intelligence, decision making, diagnostics, fabrication, packaging, etc.). Competitive *optimum-performance* MEMS and NEMS can be designed only by applying the advanced hardware and software concepts.

One of the most challenging problems in systems design is the topology–architecture–configuration synthesis, system integration, optimization, as well as selection of hardware and software (analytic and numerical methods, computation algorithms, tools and environments to perform control, sensing, execution, emulation, information flow, data acquisition, simulation, visualization, virtual prototyping and evaluation). As was emphasized, the attempts to design state-of-the-art high-performance MEMS and NEMS and to guarantee the integrated design can be pursued through analysis of complex patterns and paradigms of evolutionary developed biological systems. Even at the device level, micro- and nanoscale devices must be devised first, and structural synthesis must be performed integrating modeling, analysis, optimization, and design problems.

# CHAPTER 2

# NEW TRENDS IN ENGINEERING AND SCIENCE: MICRO- AND NANOSCALE SYSTEMS

## 2.1 INTRODUCTION TO DESIGN OF MEMS AND NEMS

Recent trends in engineering and science have increased the emphasis on integrated synthesis, analysis, design and control of advanced MEMS and NEMS. The synthesis, design and optimization processes are evolutionary in nature. They start with a given set of requirements and specifications. High-level functional and physics-based synthesis is performed first in order to start design at the system-, subsystem-, component-, device- and structure- levels. Using the advanced high-performance subsystems, components, devices, and structures synthesized (devised), the initial design is performed, and NEMS/MEMS are tested (through modeling-simulation or experiments) against the requirements. If requirements and specifications are not met, the designer revises and refines the system architecture, synthesizes and integrates new components, and performs optimization. Other alternative solutions are sought and examined.

At each level of the design hierarchy, the system performance in the behavioral domain is used to evaluate, optimize and refine the synthesis and optimization processes as well as to devise new solutions. Each level of the design hierarchy corresponds to a particular abstraction level and has the specified set of evolutionary activities and design tools that support the design at this level. The comprehensive heterogeneous synthesis and design require the application of a wide spectrum of paradigms, methods, computational environments, fabrication technologies, etc. It is the author's goal to cover manageable topics, advanced methods, and promising leading-edge technologies. Specifically, synthesis, modeling, analysis, simulation, design, optimization, computer-aided-design, and fabrications are reported and examined in this book. The conceptual view of the MEMS/NEMS synthesis-design must be introduced at the beginning in order to set the objectives, goals, problems encountered, as well as to illustrate the need for synergetic multidisciplinary developments. Therefore, the generic synthesis and design issues should be introduced first, and then, synthesis and design should be covered and studied in detail. This book concludes with the casestudies (Chapter 8) to cover the synthesis, design, and fabrication topics using the results reported in the previous chapters.

The definition for MEMS was given in Chapter 1. The stand-alone microtransducer can be viewed as the microdevice or microsystem (if the microtransducer is integrated with controlling ICs). Therefore, let us examine the possible requirements and specifications within the scope of the consequent synthesis, design, and fabrication problems. Different criteria are used to synthesize and design microtransducers (microscale actuators and sensors) with ICs due to different behavior, physical properties, operating principles, and performance criteria imposed for these microdevices. It should be emphasized that the level of hierarchy must be defined. For example, there may be no need to study the behavior of millions of transistors on each IC chip because the end-to-end ICs behavior rather should be evaluated. That is, the end-to-end ICs transient dynamics can be sufficient from the system-level viewpoints. Thus, ICs can be designed as the stand-alone MEMS microelectronic components. However, ICs must guarantee MEMS operating features, e.g., control of electromagnetic-based electromechanical devices and structures, input-output interface, analog-to-digital and digital-to-analog conversion, filtering, data acquisition, etc. The design flow is illustrated in Figure 2.1.1.

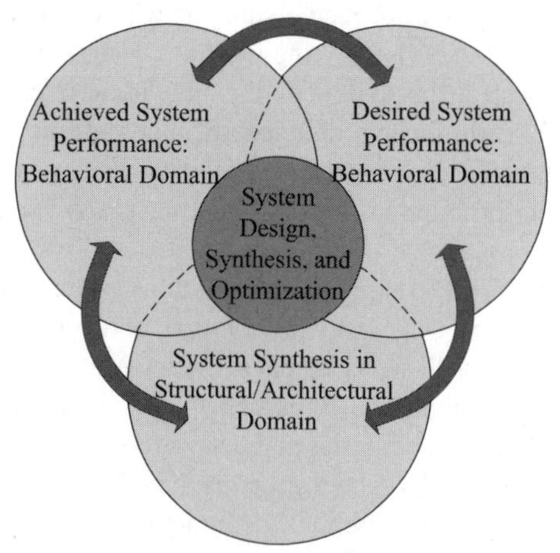


Figure 2.1.1 Design flow in synthesis of MEMS and NEMS.

The automated synthesis can be applied to implement the design flow introduced. The design of systems is a process that starts from the specification of requirements and progressively proceeds to perform a functional design and optimization. These are gradually refined through a series of sequential synthesis steps. Specifications typically include the performance requirements derived from desired systems functionality, operating envelope, affordability, reliability, and other requirements. Both *top-down* and *bottom-up* approaches should be combined to design high-performance MEMS and NEMS augmenting hierarchy, integrity, regularity, modularity, compliance, and completeness in the synthesis process.

Even though the basic foundations have been developed, some urgent areas have not been emphasized and researched thoroughly. The systems synthesis must guarantee an eventual consensus between behavioral and structural domains, as well as ensure descriptive and integrative features in the design. These can be achieved by applying the evolutionary micro- and nanoelectromechanics which allows one to extend and augment the results of quantum and classical mechanics, electromagnetics, microelectronics, informatics, and control theories, as well as to apply advanced integrated hardware and software.

To acquire and expand the engineering-science-technology core, there is the need to augment interdisciplinary areas as well as to link and place the synergetic perspectives integrating hardware (actuators–sensors–ICs) with system intelligence, control, decision-making, signal processing, data acquisition, etc. New multidisciplinary developments are needed, and the electromechanical theory should be considered as the unified cornerstone. Micro- and nanoelectromechanics, as the breakthrough paradigms in the design and analysis of MEMS and NEMS, were introduced to attack, integrate and solve a great variety of emerging problems. Electromechanics integrates systems synthesis, design, optimization, modeling, simulation, analysis, software-hardware developments and co-design, intelligence, decision making, advanced control (including self-adaptive, robust, and intelligent control), signals and image processing, biomimicking, virtual prototyping, etc. Fundamentals of electrical, mechanical, and computer engineering as well as science and technology are utilized with the ultimate objective of guaranteeing the synergistic combination of fundamental theory, precision engineering, microelectronics, microfabrication, and informatics in design, analysis, and optimization.

In Chapter 1 it was emphasized that micro- and nanoscale transducers (actuators and sensors) must be designed and integrated with the corresponding radiating energy devises, controlling and signal processing ICs, input-output devices, etc. The principles of matching and compliance are the general design principles which require that the system architectures should be synthesized integrating all subsystems and components. The matching conditions, functionality and systems compliance have to be determined and guaranteed. For example, the actuators–sensors–radiating energy devices–ICs compliance and operating functionality must be satisfied. It is evident that MEMS and NEMS devised must be controlled, and controllers should be designed. These robust control laws must be examined and verified through modeling, analysis, and simulation. Most importantly, these controllers must be implemented using ICs. Research in how to control of MEMS and NEMS aims to find methods for designing intelligent controllers, system architecture synthesis (studying and optimizing sensing-control-actuation), deriving feedback maps, obtaining gains, etc. Other important co-design problems include developing and verifying control algorithms, as well as examining execution, emulation and evaluation software.

The design of high-performance MEMS and NEMS implies the subsystems, components, devices and structures synthesis, design, and developments. Among a large variety of issues, the following problems must be resolved:

- Synthesis, characterization and design of micro- and nanoscale transducers, actuators and sensors according to their applications and overall systems requirements by means of specific computer-aided design software;
- Design of high-performance radiating energy, microelectronic and optical devices;
- Integration of actuators with sensors and ICs;
- Control and diagnostic;
- Wireless communication;
- Affordable and high-yield fabrication technologies and techniques.

Synthesis, modeling, analysis and simulation are the sequential activities. The synthesis starts with the discovery of new or application of existing physical operating principles, examining novel phenomena and effects, analysis of specifications imposed on the behavior, study of the system performance, preliminary modeling and simulation, and the assessment of the available experimental results. Heterogeneous simulation and analysis start with the model developments (based upon MEMS and NEMS devised). The designer mimics, studies, analyzes, evaluates, and assesses the systems behavior using state, performance, control, events, disturbance, decision-making, and other variables. Thus, fundamental, applied and experimental research and engineering developments are applied.

Modeling, simulation, analysis, virtual prototyping, and visualization are critical and urgently important aspects for developing and prototyping advanced MEMS and NEMS. As a flexible high-performance modeling and design environment, MATLAB[R] has become a standard cost-effective tool. Competition has prompted cost and product cycle reductions. To speed up design with assessment analysis, facilitate enormous gains in productivity and creativity, attain intelligence and evolutionary learning, integrate control and signal processing, accelerate prototyping features, visualize the results, perform data acquisition and data-intensive analysis with outcome prediction, the MATLAB environment is used. In MATLAB, the following commonly used toolboxes can be applied: SIMULINK[R], Control System, Optimization, Signal Processing, Symbolic Math, Partial Differential Equations, Neural Networks, and other application-specific toolboxes. The MATLAB environment offers a rich set of capabilities to efficiently solve a variety of complex analysis, modeling, simulation, control, and optimization problems. A wide array of MEMS and NEMS can be modeled, simulated, analyzed, and optimized. The micro- and nanoscale components, devices, and structures can be designed and simulated using other environments. For example, the VHDL (Very High-Speed Integrated Circuit Hardware Description Language) and SPICE (Simulation Program with Integrated Circuit Emphasis) environments are used to design, simulate, and analyze ICs.

## 2.2 BIOLOGICAL AND BIOSYSTEMS ANALOGIES

In Chapter 1, the *E. coli* bionanomotor (the nanomotor–flagella complex is illustrated in Figure 1.1) was used to devise micro- and nanoscale electromagnetic-based axial topology machines. In this section we consider other example of biological systems.

In general, the coordinated behavior, motion, visualization, sensing, motoring (actuation), decision making, memory, learning and other functions performed by living organisms are the results of the electromagnetic and electrochemical transmission of information by neurons, energy conversion, and other mechanisms and phenomena.

One cubic centimeter of the brain contains millions of nerve cells, and these cells communicate with thousands of neurons creating data processing (communication) networks by means of data transmission channels. The information from the brain to the muscles is transmitted within milliseconds, and the baseball, football, basketball and tennis players calculate the speed and velocity of the ball, analyze the situation, make the decision based upon the prediction, and respond (e.g., run or jump, throw or hit the ball, etc.). The human central nervous system, which includes the brain and spinal cord, serves as the link between the sensors (sensor receptors) and motors peripheral nervous system (effector, muscle, and gland cells). It should be emphasized that the nervous system has the following major functions: sensing, integration, decision-making (computing and adaptation), and motoring (actuation). The human brain consists of the hindbrain (controls homeostasis and coordinates movement), midbrain (receiving, integration, and processing the sensory information), and forebrain (neural processing and integration of information, image processing, short- and long-term memories, learning functions, adaptation, decision making and motor command development). The peripheral nervous system consists of the sensory system. Sensory neurons transmit information from internal and external environment to the central nervous system, and motor neurons carry information from the brain or spinal cord to effectors. The sensory system supplies information from sensory receptors to the central nervous system, and the motor nervous system feeds signals (commands) from the central nervous system to muscles (effectors) and glands. The spinal cord mediates reflexes that integrate sensor inputs and motor outputs, and, through the spinal cord, the neurons carry information to and from the brain. The transmission of electrical signals along neurons is a very complex phenomenon. The membrane potential for a nontransmitting neuron is due to the unequal distribution of ions (sodium and potassium) across the membrane. The resting potential is maintained due to the differential ion permeability and the so-called $Na^+$–$K^+$ pump. The stimulus changes the membrane permeability, and ions can depolarize or hyperpolarize the membrane's resting potential. This potential (voltage) change is proportional to the strength of the stimulus. The stimulus is transmitted due to the axon mechanism. The nervous system is illustrated in Figure 2.2.1.

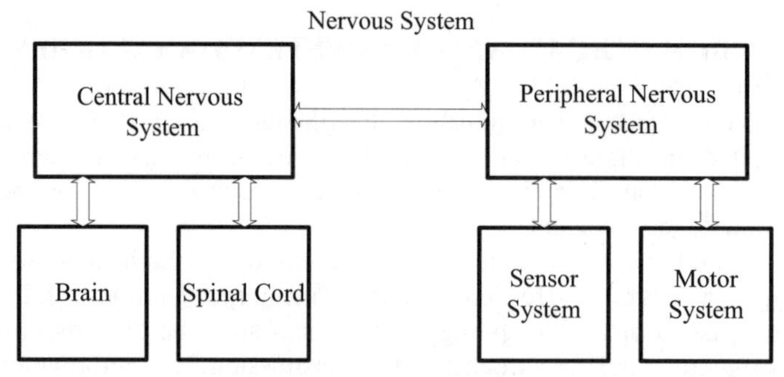


Figure 2.2.1 Vertebrate nervous system: high-level functional diagram.

All biological systems have biosensors as documented in Figure 2.2.1. Biosensors utilize biochemical reactions and sense chemicals, electromagnetic fields, specific compound, materials, radiation, etc. Many biosensors have been developed to measure and sense various compounds, chemicals, or stimuli. For example, a biosensor can be an immobilized enzyme or cell that monitors specific changes in the microenvironment. Microbial sensors are applied to the industrial process to measure organic compounds. These microbial sensors consist of immobilized whole cells and an oxygen probe used for determination of substrates and products (concentration of compounds is determined from microbial respiration activity which can be directly measured by an oxygen probe). Immobilized microorganisms and an electrode, which are the major components of the microbial sensors, sense organic compounds (concentration of compounds is indirectly determined from electroactive metabolites such as proton, carbon dioxide, hydrogen, formic acid, and other measured by the electrode).

There is a great diversity in the organization of different nervous systems. The cnidarian (*hydra*) nerve net is an organized system of simple nerves (with no central control) which performs elementary tasks (jellyfishes swim). Echinoderms have a central nerve ring with radial nerves (for example, sea stars have central and radial nerves with nerve net). Planarians have small brains that send information through two or more nerve trunks, as illustrated in Figure 2.2.2.

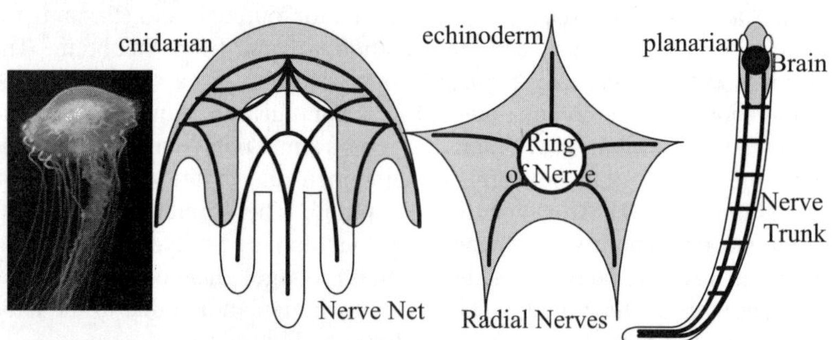


Figure 2.2.2 Overview of invertebrate nervous systems.

Let us very briefly examine the jellyfish. Jellyfishes (made up of 95% water, 3-4% salts, and 1-2% protein) have been on the earth for over 650 million years, and they have no heart, bones, brain, or eyes. A network of nerve cells allows them to move and react to food, danger, light, cold, etc. Sensors around the bell rim provide information as to whether they are heading up or down, to the light or away from it. Using the jet propulsion system, jellyfishes can swim up and down. Jellyfish are very efficient predators killing the pray by stinging it. Thus, actuation (propulsion using pulsing muscle within the bell, and stinging cells–cnidocytes which contain tiny harpoons called nematocyst triggered by the contact) and sensing mechanisms are in place even in the simplest invertebrate.

Living organisms, which consist of atoms, molecules, molecular structures and molecular systems, must be studied at the nanoscale using the corresponding theories. In fact, sugars, amino acids, hormones, and DNA are nanometers in size. In general, membranes that separate one cell from another or one subcellular organelle from another are larger structures. However, they are also built using atoms and molecules. These nanoscale molecular structures and systems perform different tasks and functions. For example, potassium and sodium ions generate nerve impulses.

The ability of organisms to function in a particular way depends on the presence, absence, concentration, location, interaction, and architectures (configurations) of these nanoscale structures and systems. Biotechnology is the synergy of nanoscience, engineering and technology (nanofabrication) to compliment and enhance fundamental research and applied developments. For example, one can apply complex biological processes to devise and fabricate nanosystems. In particular, fabrication and biomimicking can be performed researching structures, architectures and biological materials of biomolecules, cells, tissues, membranes, biomotors, etc.

Living systems perform

- Precise control of the atomic composition and architecture in very large complex and multifunctional structures (protein complexes, which integrate millions of atoms, become dysfunctional by the change of a single atom);
- Actuation–sensing–communication–control (sense the environment and deliver the measured information to the control and actuation subsystems for the appropriate response and action through decision-making and adaptation);
- Self-diagnostics, health-monitoring, reconfiguration, and repairing (sense and identify damage, perform adaptation-reconfiguration, and repair-reconfigurate to the functional state and operational level);
- Adaptation, reconfiguration, optimization, tuning, and switching based upon evolutionary learning in response to short, long, reversible, and irreversible environmental changes;

- Self-assembling and self-organization (thousands of individual components and compounds are precisely defined in complex functional and operational structures);
- Building highly hierarchical multifunctional structures (e.g., collagen, which is a very strong and robust fiber, is built through aggregation of single amino acid strands into triple helices, triple helices into micro-fibrils, microfibrils into fibrils, fibrils into fibers);
- Patterning and prototyping (templates of DNA or RNA are used as blueprints for structures);
- Design of low entropy structures increasing the entire system entropy (protein folding, which is a highly ordered arrangement of the constituent amino acids, is driven by the maximization of the entropy of the surrounding water);
- Reproducing and mimicking the similar structures and systems.

Important conclusions can be made. In general, living organisms and their systems provide a proof-of-concept principle for highly integrated multifunctional man-made nanoscale structures and systems. Nano-science, engineering, and technology will provide paradigms, concepts, methods, and tools to understand how living systems perform their functions, tasks, roles, and missions.

## 2.3 OVERVIEW OF NANO- AND MICROELECTROMECHANICAL SYSTEMS

Foreseen by Richard Feyman, the term nanotechnology was first used by N. Taniguchi in his 1974 paper "On the Basic Concept of Nanotechnology." In the last two decades, nanoengineering, nanoscience, nanotechnology, and nanomanufacturing have been popularized by many scientists and engineers. Advancing miniaturization towards the molecular level with the ultimate goal to design and manufacture nanocomputers and nanomanipulators (nanoassemblers), as well as intelligent MEMS and NEMS (which have nanocomputers as the core components), the designer faces a great number of unsolved problems and challenges.

Possible basic concepts in the development of nanocomputers are listed below. Mechanical "computers" have the richest history traced thousand years back. While very creative theories and machines have been developed and demonstrated, the feasibility of mechanical nanocomputers is questioned by some researchers due to the number of mechanical components (which are needed to be controlled), as well as due to unsolved manufacturing (assembling) and technological difficulties. Chemical nanocomputational structures can be designed based upon the processing information by making or breaking chemical bonds, and storing the information in the resulting chemicals. In contrast, in quantum nanocomputers, the information can be represented by a quantum state (e.g., the spin of the atom can be controlled by the electromagnetic field).

Electronic nanocomputers can be designed using conventional concepts tested and used for the last thirty years. In particular, molecular transistors, quantum dots, molecular logics, and other nanoelectronic devices can be used as the basic elements. The nanoswitches (memoryless processing elements), logic gates and registers can be manufactured on the scale of a single molecule. The so-called quantum dots are metal "boxes" that hold the discrete number of electrons which is changed applying the electromagnetic field. The quantum dots are arranged in the quantum dot cells. Consider the quantum dot cells which have five dots and two quantum dots with electrons. Two different states are illustrated in Figure 2.3.1 (the shaded dots contain the electron, while the white dots do not contain the electron). It is obvious that the quantum dots can be used to synthesize the logic nanodevices.

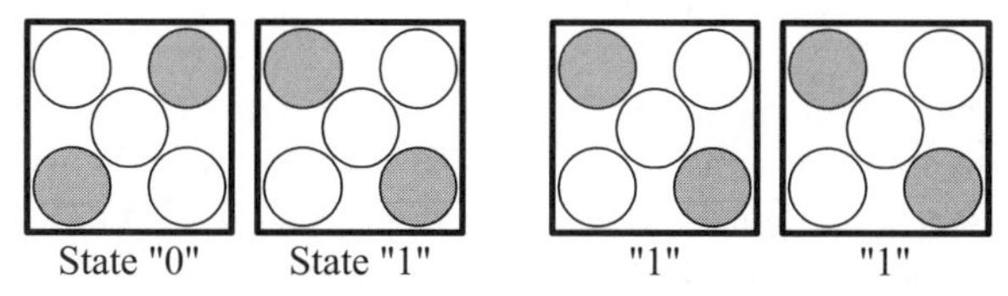


Figure 2.3.1 Quantum dots with states "0" and "1", and "1 1" configuration.

Through biosystems analogy and biomimicking, a great variety of man-made electromechanical systems have been designed, made, and implemented. As conventional electromechanical systems, many MEMS and NEMS (micro- and nanotransducers, actuators, sensors, and other devices) are controlled by changing the electromagnetic field. To analyze, design, develop, and deploy novel MEMS and NEMS, the designer must synthesize advanced architectures, apply micro- and nanoelectromechanics, integrate the latest advances in micro- and nanoscale transducers and structures, ICs and molecular electronics, materials and fabrications, structural design and optimization, modeling and simulation, etc. It is evident that novel optimized MEMS and NEMS architectures (with multiprocessors, memory hierarchies and multiple parallelism in order to guarantee high-performance computing and real-time decision-making), new structures and transducers, ICs and radiating energy devices (antennas), as well as other subsystems and components play a critical role in advancing the research, developments, and implementation. In this book we report synthesis, modeling, analysis, simulation, design, optimization, and fabrication aspects.

Electromechanical systems, as shown in Figure 2.3.2, can be classified as (1) conventional electromechanical systems, (2) microelectromechanical systems (MEMS), and (3) nanoelectromechanical systems (NEMS).

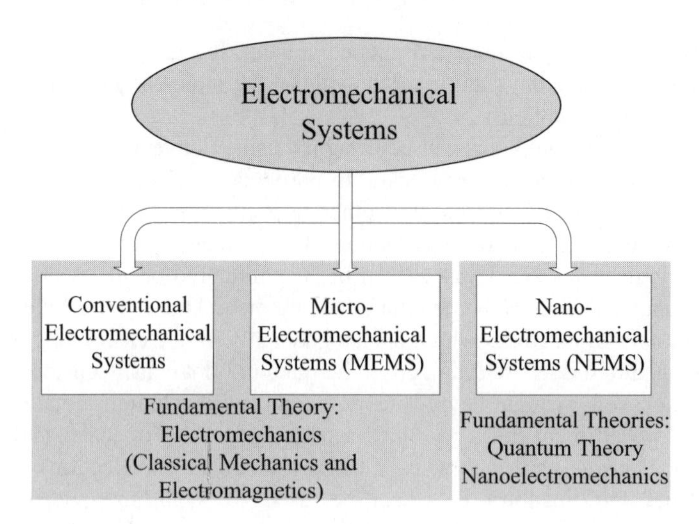


Figure 2.3.2 Classification of electromechanical systems.

The operational principles and basic foundations of conventional electromechanical systems and MEMS are the same (assuming that nanostructures which must be studied at the molecular-level are not used as the MEMS components), while NEMS are studied using different concepts and theories. Frequently, the designer applies the classical Lagrangian and Newtonian mechanics as well as electromagnetics (Maxwell's equations) to study conventional electromechanical systems and MEMS. In contrast, NEMS are examined using quantum theory and nanoelectromechanics. Figure 2.3.2 documents the fundamental theories used to study the processes and phenomena in conventional, micro-, and nanoelectromechanical systems. Figure 2.3.2 also illustrates that in general, electromechanical systems must be classified and analyzed distinguishing MEMS and NEMS.

As complex multifunctional systems, MEMS and NEMS can integrate different structures, devices and subsystems as their components. The research in integration and optimization (optimized architectures and structural optimization) of these subsystems and devices has not been performed. In particular, the end-to-end (processors–networks–input/output devices–ICs/antennas–actuators/sensors) performance and behavior must be examine. Through this book we will study different MEMS architectures, and fundamental and applied theoretical concepts will be developed and documented in order to devise and design the next generation of superior high-performance systems.

The large-scale MEMS and NEMS, which can integrate processor (multiprocessor) and memories, high-performance networks and input-output (IO) subsystems, are of far greater complexity than stand-alone single-mode MEMS commonly used today. In particular, the large-scale MEMS and NEMS can integrate:

- Thousands of nodes of high-performance stand-alone transducers (actuators/sensors and smart structures) controlled by ICs and antennas (radiating energy devices can be on-chip and out-of-chip to control transducers);
- High-performance processors or superscalar multiprocessors;
- Multi-level memory and storage hierarchies with different latencies (thousands of secondary and tertiary storage devices supporting data archives);
- Interconnected, distributed, heterogeneous databases;
- High-performance wireless communication networks (robust, adaptive intelligent networks).

It must be emphasized that even the simplest microdevice (for example, stand-alone microactuator) usually cannot function alone. For example, at least the ICs is needed to control microactuator, or microsensor must be used to measure the actuator performance variables (e.g., position, displacement, velocity, torque, force, current, voltage, etc.).

The complexity of phenomena and effects in large-scale MEMS and NEMS requires one to pursue new fundamental and applied research as well as engineering and technological developments. In addition, there is a critical need for coordination across a broad range of hardware and software. For example, design of advanced micro- and nanotransducers, synthesis of optimized (balanced) architectures, development of new programming languages and compilers, performance and debugging tools, operating system and resource management, high-fidelity visualization and data-representation systems, design of high-performance networks, etc. New algorithms and data structures, advanced system software and distributed access to very large data archives, sophisticated data mining and visualization techniques, as well as advanced data-intensive analysis are needed. In addition, advanced processor and multiprocessors are required to achieve sustained capability with required functionally and adaptability of *usable* large-scale MEMS and NEMS.

The fundamental and applied research in MEMS and NEMS has been dramatically affected by the emergence of high-performance computing. Analysis and simulation of MEMS and NEMS have significant effects. However, the problems in analysis, modeling, and simulation of large-scale MEMS and NEMS that involves the complete analysis of molecular dynamics cannot be solved because the classical quantum theory cannot be feasibly applied to moderate difficulty molecules or simplest nanostructures (1 nm cube of smart nanostructure can have thousands of molecules). There are a number of very challenging research problems in which advanced theory and high-end computing are required to advance the basic theory and engineering practice. The multidisciplinary fundamentals of micro- and nanoelectromechanics must be developed to guarantee the possibility to synthesize, analyze, and fabricate high-performance MEMS and NEMS with desired (specified) performance characteristics and desired level of confidence in the accuracy of the results.

Synergetic fundamental and applied research will dramatically shorten the time and cost of MEMS/NEMS development for medical, biomedical, aerospace, automotive, electronic and manufacturing applications.

The importance of mathematical model developments and numerical analysis has been emphasized. Numerical simulation enhances and compliments, but does not substitute fundamental research. Furthermore, meaningful and explicit simulations are based on reliable fundamental studies (mathematical model developments, data-intensive analysis, efficient computational methods and algorithms, etc.). These simulations must be validated through experiments. It is evident that heterogeneous simulations lead to understanding of performance of complex MEMS and NEMS (micro- and nanoscale structures, devices, and subsystems), reduce the time and cost of deriving and leveraging the MEMS/NEMS from concept to systems, and from systems to market. Fundamental and applied research is the core of the modeling and simulation. Correspondingly, focused efforts must be concentrated on comprehensive modeling and efficient computing.

To comprehensively study MEMS and NEMS, advanced modeling and computational tools are required primarily for 3D+ (three-dimensional geometry dynamics in time domain) data-intensive analysis used to study the end-to-end dynamic behavior of actuators, sensors, molecules, micro- and nanocircuitry, etc. The mathematical models of NEMS, MEMS, and their components (structures, devices, and subsystems) must be developed. These mathematical models (augmented with efficient computational algorithms, terascale computers, and advanced software) will play the major role in stimulating the synthesis and design of MEMS and NEMS from biomimicking and virtual prototyping standpoints.

There are three broad categories of problems for which new algorithms and computational methods are critical:

1. Problems for which basic fundamental theories are developed, but the complexity of solutions is beyond the range of current and near-future computing technologies. For example, the conceptually straightforward quantum mechanics and molecular dynamics cannot be straightforwardly applied even for simple nanoscale smart structures. In contrast, it will be illustrated in Chapter 6 that it is possible to perform robust predictive simulations of molecular-scale behavior for micro- and nanoscale structures which contain millions of molecules using the functional charge density paradigm.
2. Problems for which fundamental theories are not completely developed to justify direct simulations, but basic research can be advanced or developed through numerical results.
3. Problems for which the developed advanced modeling and simulation methods will produce major advances and will have a major impact. For example, 3D+ data-intensive analysis and study of end-to-end behavior of micro- and nanostructures and devices.

For MEMS and NEMS, as well as for devices and structures, high-fidelity modeling and massive data-intensive computational simulations (mathematical models development within contemporary intelligent libraries and databases/archives, intelligent experimental data manipulation and storage, heterogeneous data grouping and correlation, visualization, data mining and interpretation, etc.) offer the promise of developing and understanding the mechanisms, phenomena and processes in order to improve efficiency and design high-performance systems. Predictive model-based simulations require terascale computing and an unprecedented level of integration between engineering and science. These modeling and simulation developments will lead to new fundamental results. To model and simulate MEMS and NEMS, in Chapter 6 we augment modern quantum mechanics, electromagnetics, and electromechanics at the nanoscale. Our goal is to further enhance the micro- and nanoelectromechanical theories.

One can perform the steady-state and dynamic (transient) analysis. While steady-state analysis is important, and the structural optimization to comprehend the actuators/sensors–antennas–ICs design can be performed, MEMS and NEMS must be analyzed in the time domain. The long-standing goal of micro- and nanoelectromechanics is to develop the basic fundamental conceptual theories in order to study and examine the interactions between actuation and sensing, computing and communication, signal processing and hierarchical data storage (memories), as well as other processes and phenomena in MEMS and NEMS. Using the concept of electromagnetic-electromechanical interactions, the fundamental electromechanical theory can be enhanced and applied to nanostructures in order to predict the performance through analytic analysis and numerical simulations. Mathematical models of nodes can be developed, and both single molecules and groups of molecules can be studied and examined. It is critical to perform this research to determine a number of the parameters to make accurate performance evaluation and to analyze the phenomena performing simulations and comparing experimental, modeling and simulation results.

Current advances and developments in mathematical modeling and simulation of complex phenomena in MEMS and NEMS are increasingly dependent upon new approaches to heterogeneously model, compute, visualize, and validate the results. This research is needed to clarify, correlate, define, and describe the limits between the numerical results and the qualitative-quantitative analytic analysis in order to comprehend, understand and grasp the basic features. Simulations of MEMS and NEMS require terascale computing that will be available within a few years. The computational limitations and inability to develop explicit mathematical models (some nonlinear phenomena cannot be comprehended, fitted, described and precisely mapped) focus advanced studies on the basic research in robust modeling and simulation under uncertainties. Modeling and design are critical to advance and foster the theoretical and engineering enterprises. We focus our research on developments and enhancements of the micro- and nanoelectromechanical theories in order to model, simulate and design novel MEMS and NEMS. At the subsystem level, for example, micro- and nanoscale actuators and sensors should be modeled and analyzed in 3D+ (three-dimensional geometry dynamics in time domain). Rigorous methods for quantifying uncertainties for robust analysis also should be developed (however, these problems beyond the scope of this book). Uncertainties result due to the fact that it is impossible to explicitly comprehend the complex interacted processes in MEMS and NEMS (actuators/sensors and smart structures, antennas, digital and analog ICs, data movement, storage and management across multilevel memory hierarchies, archives, networks and periphery), accurately solve high-order nonlinear partial differential equations, precisely model structural and environmental changes, measure all performance variables and states, etc.

To analyze and design MEMS and NEMS, we will develop tractable analytical mathematical models. There are a number of areas where the advances must be made in order to realize the promises and benefits of modern theoretical developments. For example, to perform 3D+ modeling and data intensive analysis of actuators/sensors and smart structures, advanced analytical and numerical methods and algorithms must be used. Novel algorithms in geometry and mesh generation, data assimilation, dynamic adaptive mesh refinement, computationally efficient techniques and robust methods can be implemented in the MATLAB environment. There are a number of fundamental and computational problems that have not been addressed, formulated and solved due to the complexity of MEMS and NEMS (e.g., large-scale hybrid models, limited ability to generate and visualize the massive amount of data, computational complexity of the existing methods, etc.). Other problems include nonlinearities and uncertainties which imply fundamental limits to our ability to accurately formulate, set up, and solve analysis and design problems with the desired accuracy. Therefore, one should develop rigorous methods and algorithms for quantifying and modeling uncertainties, 3D+ geometry and mesh generation techniques, as well as methods for adaptive robust modeling and simulations under uncertainties. A broad class of fundamental and applied problems ranging from fundamental theories (quantum mechanics, electromagnetics, electromechanics, thermodynamics, structural synthesis, optimization, optimized architecture design, control, modeling, analysis, etc.) and numerical computing (to enable the major progress in design and virtual prototyping through the heterogeneous simulations, data-intensive analysis and visualization) will be addressed and studied in this book. Due to the obvious limitations and the limited scope of this book, a great number of problems and phenomena cannot be addressed and discussed (among them, robust modeling, analysis of large-scale MEMS and NEMS, etc.). However, using the results reported, these problems can be approached, examined, and advanced.

## 2.4 APPLICATIONS OF MICRO- AND NANOELECTROMECHANICAL SYSTEMS

Different MEMS and NEMS must be designed depending upon the specifications, requirements, objectives, and applications. Electromechanical, electro-opto-mechanical, and electro-chemo-opto-mechanical MEMS and NEMS have been developed. Usually, optical systems are faster, simpler, more efficient, reliable, survivable and robust compared with electromechanical systems. However, these optical systems are designed for different applications (e.g., wireless communication, computing, switching, etc.), have different configurations, and, in general, it is very difficult to make the matching comparison. For example, optical systems cannot be used as actuators. The application requirements must be counted, and the electromagnetic interference, temperature, vibration, or radiation can be the factors used to make the preferred selection. As an other example, consider micro- and nanoscale actuators. The actuator size is determined by the force or torque densities which are the functions of the materials used and size (volume). That is, the size is determined by the force or torque requirements and materials. As one uses NEMS or MEMS as the logic or sensor devices, the output electric signal (voltage or current) or electromagnetic field intensity (or density) must have the specified values. Micro- and nanoelectronic devices are needed to guarantee the interface. Definitely, nanoelectronics and molecular electronics have tremendous advantages and can be used in wireless communication, logics, memories, etc. However, to control micro- and nanoscale transducers, the power stage is determined by the rated voltage and current. Thus, FETs and MOSFETs are used as the power stage transistors, and the microscale ICs are applied.

Although MEMS and NEMS have some common features, the differences were emphasized in Chapter 1 and previous sections. In addition to different fundamental theories used to study MEMS and NEMS, distinct physical phenomena and effects, and different fabrication technologies are applied. Currently, the research and developments in NEMS and molecular nanotechnology are primarily concentrated on design, modeling, simulation, and fabrication of molecular-scale devices. In contrast, MEMS are usually fabricated using complementary metal oxide semiconductor (CMOS) technology, surface micromachining, and high-aspect-ratio (LIGA) technologies and processes. These technologies are covered in Chapters 3 and 8. Therefore, MEMS leverage conventional microelectronics techniques, processes, and materials.

It was emphasized that electromagnetic MEMS integrate motion microstructures or microtransducers controlled by ICs using radiating energy microdevices. Thus, microstructures/microtransducers, radiating energy devices, and ICs must be integrated. The direct chip attaching technique was developed and widely deployed. In particular, flip-chip MEMS assembly replaces wire banding to connect ICs with micro- and nanoscale actuators and sensors. The use of the flip-chip technique allows one to eliminate parasitic resistance, capacitance, and inductance. This results in improvements of performance characteristics. In addition, the flip-chip assembly offers advantages in the implementation of advanced flexible packaging, improving reliability and survivability, reduces weight and size, etc. The flip-chip assembly involves attaching microtransducers (actuators and sensors) directly to ICs. For example, the microtransducers can be mounted face down with bumps on the pads that form electrical and mechanical joints to the ICs substrate. Figure 2.4.1 illustrates flip-chip MEMS.

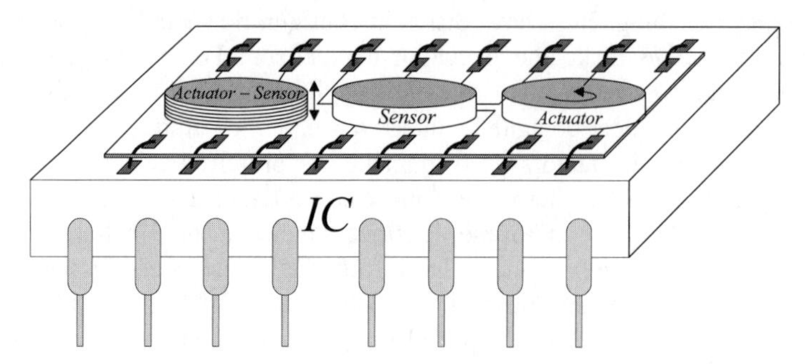


Figure 2.4.1 Flip-chip monolithic MEMS with actuators and sensors.

The large-scale integrated MEMS (a single chip that can be mass-produced at low-cost using the CMOS, micromachining, LIGA, and other technologies) can integrate:

- N nodes of microtransducers (actuators/sensors and smart structures),
- ICs and radiating energy devices (antennas),
- Optical and other devices to attain the wireless communication features,
- Processor and memories,
- Interconnection networks (communication busses),
- Input-output (IO) devices, etc.

Different architectures and configurations can be synthesized, and these problems are discussed and covered in this book. One uses MEMS and NEMS to control complex systems, processes, and phenomena. In order to control systems, many performance and decision-making variables (states, outputs, events, etc.) must be measured. Thus, in addition to actuation and sensing (performed by microtransducers integrated with ICs and radiating energy devices), computational, communication, networking, signal processing, and other functions must be performed. One can represent a high-level functional block diagram of the dynamic system–MEMS configuration as illustrated in Figure 2.4.2.

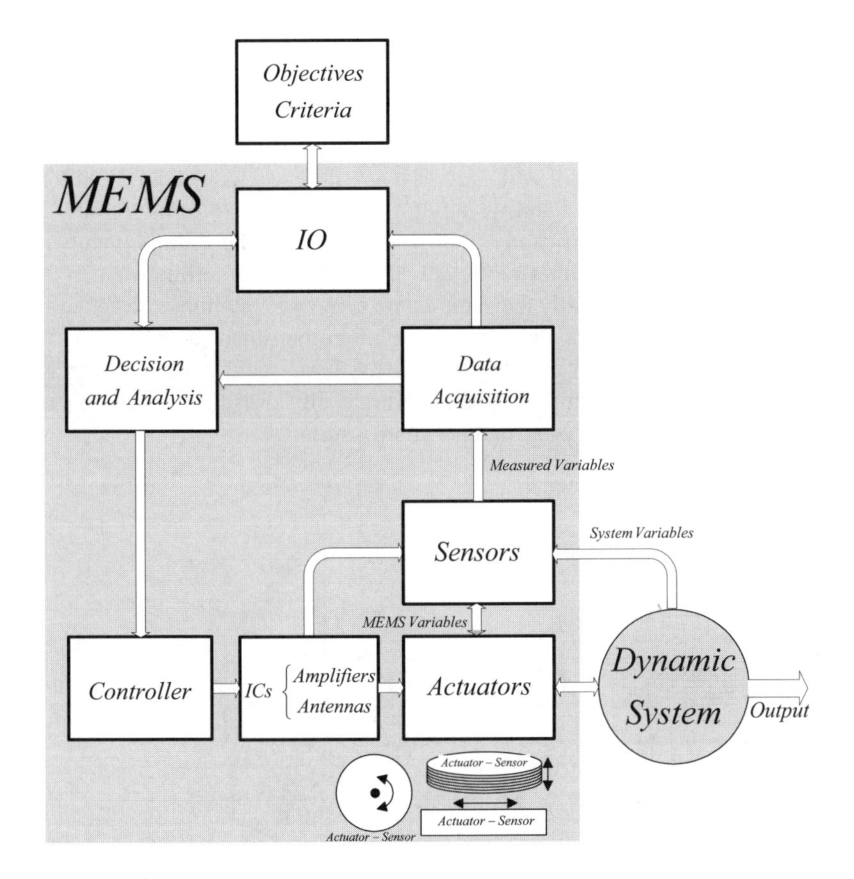


Figure 2.4.2 High-level functional block diagram of the large-scale MEMS with rotational and translational actuators, sensors and ICs.

Actuators actuate dynamic systems. These actuators respond to command stimulus (control signals) and develop torque and force. There is a great number of biological (e.g., nanobiomotor, jellyfish, human eye, locomotion system) and man-made actuators. Biological actuators are based upon electromagnetic-mechanical-optical-chemical phenomena and processes. Man-made actuators (electromagnetic, electrostatic, hydraulic, thermo, acoustic, and other motors) are devices that receive signals or stimulus (electromagnetic field, stress or pressure, thermo or acoustic, etc.) and respond with torque or force.

Consider the flight vehicles. The aircraft, spacecrafts, missiles and interceptors are controlled by displacing the control surfaces as well as by changing the control surface and wing geometry. For example, ailerons, elevators, elevons, canards, fins, flaps, rudders, stabilizers and tips of advanced aircraft can be controlled by micro- and miniscale actuators using the MEMS-based smart actuator technology. This actuator technology is uniquely suitable in the flight actuator applications.

Figure 2.4.3 illustrates the aircraft where translational and rotational actuators are used to actuate control surfaces, as well as to change the wing and control surface geometry. The application of microtransducers allows one to attain the aerodynamic flows control minimizing the drag (improving the fuel consumption and increasing the velocity). In addition, maneuverability, controllability, agility, stability and reconfigurability of flight vehicles are significantly improved expanding the flight envelope. It must be emphasized that although the force and torque density of microtransducers is usually the same as the conventional miniscale actuators, due to small dimension, the stand-alone microtransducer develops smaller force or torque. However, integrated in the large-scale multi-node arrays, microtransducers (controlled by the hierarchically distributed systems) can develop the desired force and actuate control surfaces.

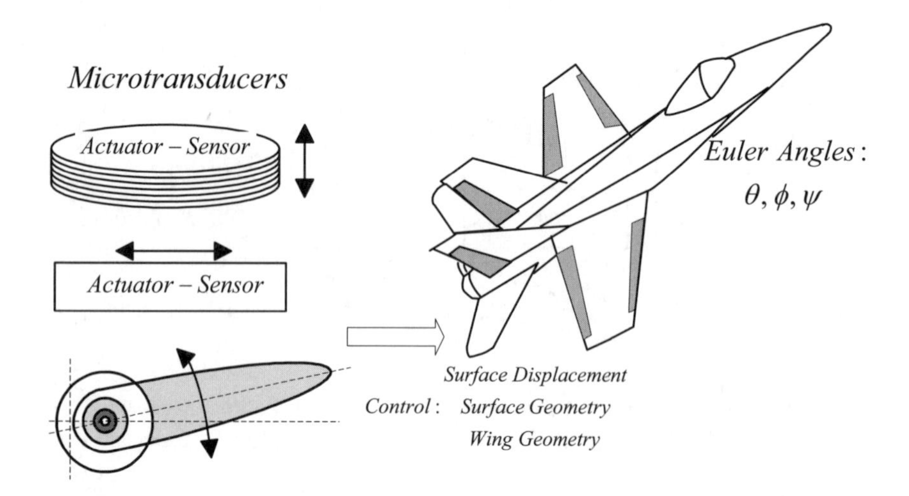


Figure 2.4.3 Aircraft with translational and rotational microtransducers.

Sensors are devices that receive and respond to signals or stimulus. For example, the aerodynamic loads (which flight vehicles experience during the flight), vibrations, temperature, pressure, velocity, acceleration, noise and radiation can be measured by micro- and nanoscale sensors, see Figure 2.4.4. It should be emphasized that there are many other sensors can be applied to measure the electromagnetic interference, displacement, orientation, position, voltages, currents, resistance and other variables of interest in power electronic devices.

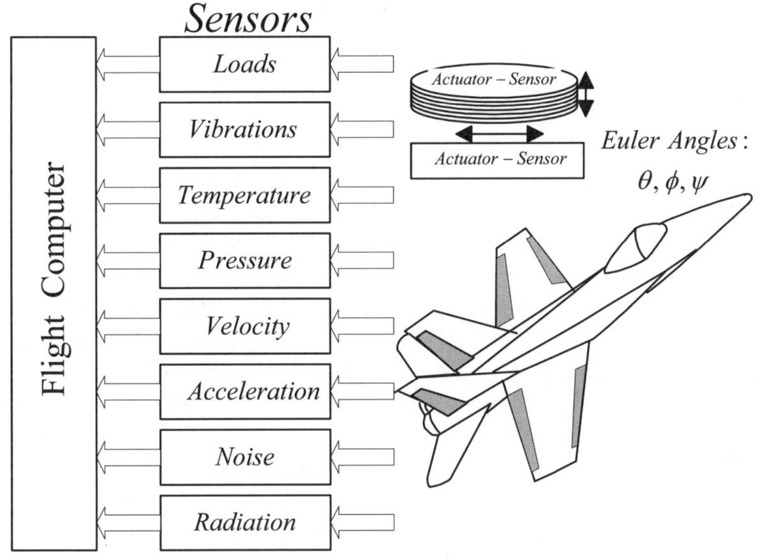


Figure 2.4.4 Application of micro- and nanoscale sensors in aircraft.

Usually, several energy conversion processes are involved to produce electric, electromagnetic, or mechanical output sensor signals. The conversion of energy is our particular interest. Using the energy-based analysis, the general theoretical fundamentals will be studied.

The major developments in MEMS and NEMS have been fabrication technology driven. The applied research has been performed mainly to manufacture structures and devices, as well as to analyze some performance characteristics for specific applications. Extensive research has been done in micro- and nanostructures. In particular, ICs and thin-film sensors have been thoroughly researched, and computer-aided design environments are available to examine them. In addition, mini- and microscale smart structures and nanocomposites have been studied, and feasible fabrication techniques, materials and processes have been developed. Recently, carbon nanotubes were discovered, and molecular wires and molecular transistors were built. However, to our best knowledge, micro- and nanodevices, MEMS and NEMS, have not been comprehensively studied at the system level. In addition, limited efforts have been concentrated to develop the fundamental theory. In this book, we will apply the quantum theory, charge density concept, advanced electromechanical theory, Maxwell's equations, as well as other cornerstone methods in order to model nanostructures, micro- and nanodevices (antennas, actuators, sensors, etc.). In particular, the micro- and nanoelectromechanical theories will be further expanded. A large variety of microtransducers with different operating features are modeled and simulated. To perform high-fidelity integrated 3D+ modeling and data-intensive analysis with post-processing and animation, the partial and ordinary nonlinear differential equations are solved.

## *Nanostructures: Giant Magneto Resistance and Multilayered Nanostructure*

Nanoscale structures and devices are devised and designed using novel phenomena and effects observed at the nanoscale. In this section we will briefly illustrate the application of nanostructures in the sensor applications. The giant magneto resistive effect was independently discovered by Peter Gruenberg (Germany) and Albert Fert (France) in the late 1980s. They experimentally observed large resistance changes in materials comprised of thin layers of various metals. Two thin films (magnetic layers with nanometer thickness) with in-plane magnetization that are antiferromagnetically coupled (magnetizations are in opposite directions in the absence of an external magnetic field) can be set into ferromagnetic alignment (magnetizations are in the same direction) applying external magnetic field. A large change in the perpendicular resistance of the multilayered nanostructures results due to these changes. This effect, known as giant magneto resistance (GMR), has enormous application features including magneto-resistive recording heads and sensors. The antiferromagnetic coupling in multilayered Fe/Cr thin films has initiated intensive studies of thin multilayered nanostructures consisting of ferromagnetic and nonferromagnetic thin films. The magnetic coupling are changed by the nonferromagnetic materials (called spacer), e.g., Cr, Cu, and Re thin films. The GMR phenomenon is associated with the spin dependent scattering of the conduction electrons and changes in the relative band structure. The electrical resistance is due to scattering of electrons within a material. Depending on the magnetic direction, a single-domain magnetic material will scatter electrons with *up* or *down* spin differently. When the magnetic layers in the multilayered GMR effect-based nanostructures are aligned anti-parallel, the resistance is high because *up* electrons that are not scattered in one layer can be scattered in the other. When the layers are aligned in parallel, all *up* electrons will not significantly contribute (not scattered) regardless of which layer they pass through, leading to a lower resistance.

Due to the fact that the saturation field of the Fe/Cr and Co/Cu thin films is large, soft magnetic multilayered nanostructures, which exhibit the GMR phenomenon, are used. For example, the *permalloy* $Ni_{80\%}Fe_{20\%}$ and copper (fabricated by sputtering in the computerized deposition systems with two independent sputtering sources) multilayered nanostructures have been studied. For different thickness of the *permalloy* and copper thin films, the saturation fields is from 0.3 to 15 A/m, and the GMR effect leads up to 20% variations of resistance. The dependencies of the GMR effect on the thin films (for example, *permalloy* and copper) thickness and geometry, results in possibility to shape the multilayered structures properties for specific requirements and different applications. Using Co in the *permalloy*-Cu multilayers, improves the relative field sensitivity and guarantees 0.1% accuracy. The dependence of the antiferromagnetic coupling on the Cu thickness is very strong, and the first and second maxima are at the 0.85 and 2 nm thickness of copper. The GMR effect is maximized (22% of resistance changes) using 39 *permalloy*-Cu multilayers with 0.8 nm and 2 nm thickness, respectively.

Multilayered nanostructures, fabricated using Co/Cu, CoCu/Cu, CoFe/Cu and other more complex thin film structures, can be applied to enhance the operating envelopes. As an example, the nanostructure which integrates Ta (5 nm) / $Ni_{80\%}Fe_{20\%}$ (2 nm) / $Ir_{20\%}Mn_{80\%}$ (10nm) / $Co_{90\%}Fe_{10\%}$ (2 nm) / Ru (1 nm) / $Co_{90\%}Fe_{10\%}$ (2 nm) / Cu (3 nm) / $Co_{90\%}Fe_{10\%}$ (1 nm) / $Ni_{80\%}Fe_{20\%}$ (5 nm) / Ta (5 nm) thin film can be fabricated, and the coefficient of the thermal expansion must be matched because the single deposition process is desired. One concludes that multilayered nanostructures, which comprise alternating ferromagnetic and nonferromagnetic thin films (each film can be within a few atomic layers thick), exhibit novel functional capabilities due to new phenomena and effects. The discussed GMR phenomena arise from quantum confinement of electrons in spin-dependent potential wells provided by the ferromagnet-spacer layer boundaries. Layers of 3d ferromagnet transition metals are indirectly magnetically coupled via spacer layers comprised of many nonferromagnetic 3d, 4d and 5d transition metals. The magnetic coupling occurs between ferromagnetic and antiferromagnetic layers as the function of thickness of the spacer layer, and the strength varies with the spacer d-band filling. The period of the coupling is related to the electronic structure of the spacer thin film and can be changed by varying the composition of the spacer layer or varying its crystallographic orientation. The resistance of metallic multilayered nanostructures depends on the magnetic arrangement of the magnetic moments of thin layers. Thus, the resistance varies as the function of the external magnetic field. The multilayered nanostructures (sensors) display much larger magneto-resistance than any metals or alloys. It was emphasized that the origin of the GMR phenomena derives from spin-dependent scattering of the conduction carriers within the magnetic layers or at the boundaries of the magnetic layers (spin-dependent scattering at the ferromagnet-spacer layer interfaces is dominant). Giant magneto resistance can be used to detect magnetic fields, and therefore, is used to read the state of magnetic bits in advanced magnetic disk drives.

Figure 2.4.5 illustrates the normalized resistance $R_N$ in the multilayered nanostructure with hysteresis (dashed line) and without hysteresis (solid line) as functions of the applied magnetic field.

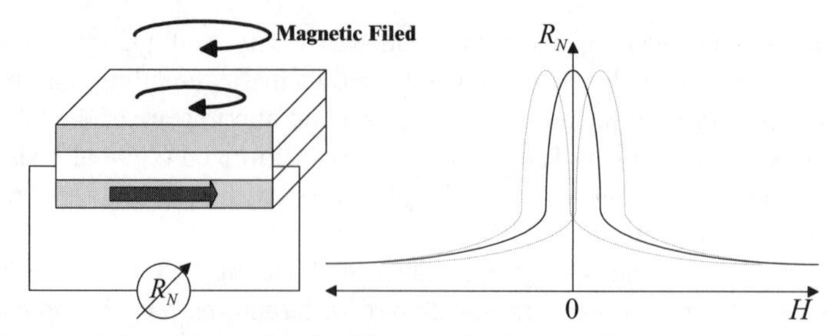


Figure 2.4.5 Normalized resistance in the multilayered nanostructure as functions of the applied magnetic field.

## 2.5 MICRO- AND NANOELECTROMECHANICAL SYSTEMS

In general, monolithic MEMS integrate microassembled devices (electromechanical and electronic microsystems and microdevices on a single chip) that have electrical, electronic and mechanical components. To fabricate MEMS, modified microelectronics technologies, techniques, processes and materials are used. Actuation and sensing cannot be viewed as the peripheral function in many applications. Integrated sensors-actuators (usually motion microstructures and microtransducers) with ICs compose the major class of MEMS. Frequently, wireless communication capabilities are the desired or required feature. Due to the use of the CMOS techniques and processes in the fabrication of actuators and sensors, MEMS leverage microelectronics in important additional areas that revolutionize the application capabilities. In fact, MEMS have considerably leveraged the microelectronics industry beyond ICs. Dual power operational amplifiers (e.g., Motorola TCA0372, DW Suffix plastic package case 751G, DP2 Suffix plastic package case 648 or DP1 Suffix plastic package case 626) as monolithic ICs can be used to control DC micromotors or microstructures, as shown in Figure 2.5.1.

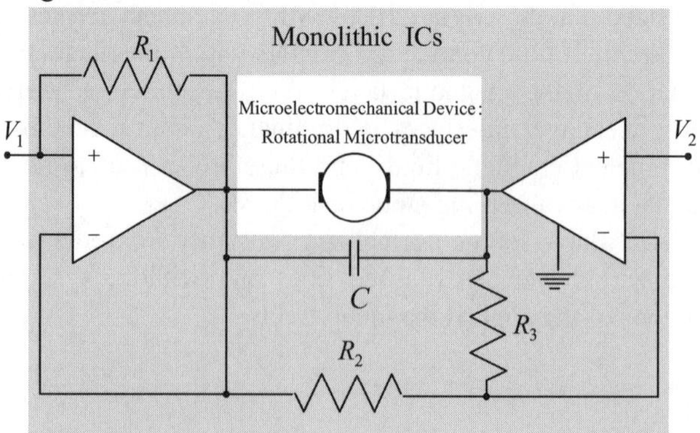


Figure 2.5.1 Application of monolithic IC to control micromotor.

Only recently has it become possible to develop high-yield fabrication techniques and fabricate MEMS at low cost. However, there is a critical demand for continuous fundamental, applied and technological improvements. Therefore, synergetic multidisciplinary activities are required. The general lack of theory to augment nonlinear electromagnetics, mechanics, optics, signal processing and control is known. These problems must be addressed and solved through focused efforts.

The need for integrated motion microstructures, microtransducers and ICs has been widely recognized. Simply scaling conventional electromechanical motion devices and augmenting them with available ICs has not met the need, and fundamental theory and fabrication processes have been developed beyond component replacement.

The set of long-range goals that challenge the analysis, design, development, fabrication, commercialization and deployment of high-performance MEMS are:

- Devising novel microscale transducers, sensors, actuators, and motion microstructures,
- Sensing, actuation, communication, and networking mechanisms,
- Sensors-actuators-ICs-radiating energy devices integration and compliance,
- Advanced high-performance MEMS architectures and configurations,
- high-fidelity modeling, data-intensive analysis, heterogeneous design, and optimization,
- Computer-aided design and development of computationally-efficient environments,
- MEMS applications, commercialization, and deployment,
- Advanced materials, processes, techniques, and technologies,
- Fabrication, packaging, microassembly, and testing.

Significant progress in the application and enhancing of CMOS (including biCMOS) technology enables the industry to fabricate microscale actuators and sensors with the corresponding ICs. This guarantees significant breakthrough and confidence. The field of MEMS has been driven by the rapid global progress in ICs, VLSI, solid-state devices, materials, microprocessors, memories and DSPs that has revolutionized instrumentation, control and systems design philosophy. In addition, this progress has facilitated explosive growth in data processing and communications in high-performance systems.

In microelectronics, many emerging problems deal with nonelectric effects, phenomena and processes (thermal and structural analysis and optimization, stress, ruggedness and packaging). It was emphasized that ICs are the necessary components to perform control, conversion, data acquisition, networking, and interfacing. For example, control signals (voltage or currents) are computed, converted, modulated, and fed to actuators and radiating energy devices by ICs. It is evident that MEMS have found applications in a wide array of microscale devices (accelerometers, pressure sensors, gyroscopes, pumps, valves and optical interconnects) due to extremely high level of integration between electromechanical, microelectronic, optical and mechanical components with low cost and maintenance, accuracy, efficiency, reliability, robustness, ruggedness, and survivability.

### 2.5.1 Microelectromechanical Systems, Devices, and Structures Definitions

Microengineering aims to integrate motion microdevices (actuators and sensors) with radiating energy microdevices (antennas, microstructures with windings), microscale driving/sensing circuitry, controlling/processing ICs, and optoelectronic devices. Affordable batch-fabricated fabrication processes were developed, and the following definition of MEMS was given in [1]:

"MEMS are batch-fabricated microscale devices (ICs and motion microstructures) that convert physical parameters to electrical signals and vice versa, and in addition, microscale features of mechanical and electrical components, architectures, structures, and parameters are important elements of their operation and design."

As was illustrated, MEMS integrate motion microstructures and devices (microtransducers) as well as ICs, radiating energy and communication devices on a single chip or on a hybrid chip. The scope of MEMS has been further expanded by devising novel paradigms, synthesizing new architectures, performing system-level synergetic integration, high-fidelity modeling, heterogeneous simulation, data-intensive analysis, control, optimization, fabrication, and implementation.

In chapter 1, we defined MEMS as:

*The MEMS is the batch-fabricated integrated microscale system (motion, electromagnetic, radiating energy and optical microdevices/microstructures – driving/sensing circuitry – controlling/processing ICs) that:*

1. *Converts physical stimuli, events, and parameters to electrical, mechanical, and optical signals and vice versa;*
2. *Performs actuation, sensing, and other functions;*
3. *Comprises control (intelligence, decision-making, evolutionary learning, adaptation, self-organization, etc.), diagnostics, signal processing, and data acquisition features,*

*and microscale features of electromechanical, electronic, optical, and biological components (structures, devices, and subsystems), architectures, and operating principles are basics of the MEMS operation, design, analysis, and fabrication.*

The MEMS are comprised and built using microscale subsystems, devices, and structures. Recall that in Chapter 1, the following important definition was made for microdevice.

*The microdevice is the batch-fabricated integrated microscale motion, electromagnetic, radiating energy, or optical microscale device that*

1. *Converts physical stimuli, events, and parameters to electrical, mechanical, and optical signals and vice versa;*
2. *Performs actuation, sensing, and other functions,*

*and microscale features of electromechanical, electronic, optical, and biological structures, topologies, and operating principles are basics of the microdevice operation, design, analysis, and fabrication.*

Finally, the microstructure is defined as given below.

*The microstructure is the batch-fabricated microscale electromechanical, electromagnetic, mechanical, or optical composite microstructure (that can integrate simple microstructures) that is a functional component of the microdevice and serve to attain the desired microdevice's operating features (e.g., convert physical stimuli, events, and parameters to electrical, mechanical, and optical signals and vice versa, as well as perform actuation, sensing, and other functions), and microscale features of electromechanical, electromagnetic, mechanical, optical, or biological structure (geometry, topology, operating principles, etc.) are basics of the microstructure operation, design, analysis, and fabrication.*

## 2.5.2 Microfabrication: Introduction

The manufacturability issues in MEMS and NEMS must be addressed. One can design and manufacture individually fabricated microscale subsystems, devices and structures. However, these individually fabricated microdevices and microstructures are impractical due to very high cost. Therefore, high-yield CMOS and other fabrication technologies (e.g., surface micromachining and LIGA) are used. Integrated MEMS at least have mechanical microstructures and ICs. Microfabricated smart multifunctional materials are used to made microscale actuators, sensors, pumps, valves, optical switches and other devices. Hundreds of millions of transistors on a chip are currently fabricated by the microelectronic industry, and enormous progress in achieving nanoscale transistor dimensions (20 nm features) was made. However, MEMS operational capabilities are measured by the system-on-chip integration, cost, performance, efficiency, size, reliability, intelligence and other criteria. There are a number of challenges in MEMS fabrication because conventional CMOS technology must be modified and integration strategies (to integrate motion structures and ICs) need to be developed. What should be fabricated first: ICs or mechanical micromachined structure? The fabrication of ICs first faces challenges because to reduce stress in silicon thin films (multifunctional material to build motion microstructures), a high-temperature annealing (usually at $1000^0$C) is needed for several hours. The aluminum ICs interconnect will be destroyed (melted), and tungsten can be used instead for interconnected metallization. This process leads to difficulties for commercially manufactured MEMS due to high cost and low reproducibility. Let us illustrate the basic steps making use the Analog Devices fabrication process. The ICs are fabricated up to metallization step. Then, mechanical structures are made using high-temperature annealing (micromachines are fabricated before metallization). Finally, ICs are interconnected. This allows the manufacturer to use low-cost conventional aluminum interconnects. The third option is to fabricate mechanical structures and then ICs. To overcome step coverage, stringer, and topography problems, motion mechanical microstructures can be fabricated in the bottoms of the etched shallow trenches (packaged directly) of the silicon substrate. These trenches are filled with a sacrificial silicon dioxide, and the silicon is planarized through chemical-mechanical polishing.

The motion mechanical microstructures can be protected (sensor applications, e.g., accelerometers and gyroscopes) and unprotected (actuator and interactive environment sensor applications). Therefore, MEMS can be packaged (encased) in a clean, hermetically sealed package, or some elements can be unprotected to allow interaction and active interfacing with the environment. This creates challenges in packaging. It is extremely important to develop novel electromechanical motion microstructures and microdevices (sticky multilayers, thin films, magnetoelectronic, electrostatic, quantum-effect-based devices, etc.) and sense their properties. Thus, microfabrication of very large scale integrated circuits (VLSI), microscale structures and devices, as well as optoelectronics must be addressed. Fabrication processes include lithography, film growth, diffusion, implantation, deposition, etching, metallization, planarization, etc. Complete microfabrication processes with integrated sequential processes are of great importance. These issues are covered in Chapters 3 and 8.

It was emphasized that microsensors sense physical variables, and microactuators control (actuate) real-world systems. These microactuators are regulated by ICs. In addition, ICs also perform computations (computing), input-output interfacing, signal conditioning, signal processing, filtering, decision-making, and other functions. For example, in microaccelerometers, the motion microstructure displaces. Using the capacitance difference, the acceleration is calculated. In microaccelerometers, computing, signal conditioning, filtering, input/output interfacing and data acquisition are performed by ICs. In the ADL-series accelerometers, these ICs are built using thousands of transistors.

Microelectromechanical systems contain microscale subsystems, devices, and structures designed and manufactured using different technologies, techniques, processes, and materials. A single silicon substrate can be used to fabricate microscale actuators, sensors, and ICs as the monolithic MEMS using the modified CMOS-based microfabrication technologies, techniques, and processes. These MEMS must be assembled, connected, and packaged. Different microfabrication techniques for MEMS components and subsystems exist. Usually, monolithic MEMS are compact, efficient, reliable, and guarantee superior performance.

As was illustrated, typically, MEMS integrate the following components: microscale actuators (actuate real-world systems), microscale sensors (detect and measure changes of the physical variables and stimuli), radiating energy and communication microdevices, microelectronics/ICs (signal processing, data acquisition, signal conversion, interfacing, communication, control, etc.).

Microactuators are needed to develop force or torque (mechanical variable). Typical examples are microscale drives, moving mirrors, pumps, servos, valves, etc. Actuation can be achieved using electromagnetic (electrostatic, magnetic, piezoelectric), hydraulic, vibration, and thermal physics-based effects. This book covers electromagnetic microactuators. The so-called comb drives (surface micromachined motion microstructures) have been widely used. These drives have movable and stationary plates (fingers). When voltage is applied, an attractive force is developed between two plates, and the motion results. A wide variety of microscale actuators have been fabricated and tested. The difficulties associated with coil fabrication are a common problem. The choice of magnetic core materials and permanent magnets are also critical issues. Electromagnetic actuators typically can be fabricated through the micromachining technologies using copper to fabricate microcoils, and nickel, nickel-iron and other thin film alloys as magnetic core materials (see Chapters 3 and 8 for details).

Piezoelectric microactuators have found wide application due to their simplicity and ruggedness (force is generated if one applies the voltage across a film of piezoelectric material). If the voltage is applied, the silicon membrane with PZT thin film deforms. Thus, the silicon-PZT membranes can be used as pumps. Chapter 8 covers PZT-based microtransducers.

Microsensors are devices that convert one physical variable (quantity) to another, e.g., electromagnetic phenomena can be converted to mechanical or optic effects. There are a number of different types of microscale sensors used in MEMS. For example, microscale thermosensors are designed and built using the thermoelectric effect (resistivity varies with temperature). Extremely low cost thermoresistors (thermistors) are fabricated on the silicon wafer, and ICs are built on the same substrate. The thermistor resistivity is a highly nonlinear function of the temperature, and compensating circuitry is used to take into account the nonlinear effects. Microelectromagnetic sensors measure electromagnetic fields, e.g., the Hall-effect sensors. Optical sensors can be fabricated on crystals that exhibit a magneto-optic effect, e.g., optical fibers. In contrast, the quantum effect sensors can sense extremely weak electromagnetic fields. Silicon-fabricated piezoresistors (silicon doped with impurities to make it n- or p-type) belong to the class of electromechanical sensors. When the force is applied to the piezoelectric resistor, the charge (voltage) induced is proportional to the applied force. Zinc oxide and lead zirconate titanate (PZT, $PbZrTiO_3$), which can be easily deposited applying conventional well-developed techniques, are used as piezoelectric crystals.

In this book, the microscale accelerometers and gyroscopes, as well as microtransducers are studied. Accelerometers and gyroscopes are based upon the capacitive sensing mechanism. In two parallel conducting plates, separated by an insulating material, the capacitance between the plates is a function of distance between plates (capacitance is inversely proportional to the distance). Thus, measuring the capacitance, the distance can be easily calculated. In accelerometers and gyroscopes, the proof mass (rotor) is suspended on silicon springs. Another important class of sensors is the thin film pressure sensors. Thin film membranes are the basic components of pressure sensors. The deformation of the membrane is usually sensed by piezoresistors or capacitive microsensors.

We have illustrated the critical need for physical- and system-level synergetic concepts in MEMS and NEMS synthesis, analysis, design, and optimization. Advances in physical-level multidisciplinary research have tremendously expanded the horizon of MEMS and NEMS. For example, magnetic-based (magnetoelectronic) memories have been thoroughly studied (magnetoelectronic devices can be grouped in three categories based upon the physics of their operation: all-metal spin transistors and valves, hybrid ferromagnetic semiconductor structures, and magnetic tunnel junctions). Writing and reading the cell data are based on different physical mechanisms. It was demonstrated that low cost, high densities, low power, high reliability and speed (write/read cycle) memories result. As the physical-level synthesis, analysis, and design are performed at the microstructure/ microdevice-level, the system-level synthesis, analysis, design and optimization must be accomplished because the design of the integrated large-scale MEMS and NEMS is required.

This section introduces the reader to the microfabrication techniques and technologies reporting the basic processes and steps to make microstructures. Different MEMS fabrication technologies are developed and applied [1-3]. In particular, micromachining and high-aspect-ratio technologies will be reported in Chapters 3 and 8. However, let us focus our attention on the basic overview to facilitate the highest level of understanding the microfabrication basics. The MEMS microfabrication technology has been adopted from microelectronics. For example, many microscale structures, devices and systems are fabricated on silicon wafers, and the basic processes, steps, materials, and equipment are similar to the ICs fabrication. In addition, the microstructures are usually made from thin films patterned using photolithographic processes and deposited applying methods developed for ICs. It is evident that some processes and materials used in MEMS fabrication are different compared with conventional microelectronic technologies. In fact, it is important to attain the compliance, compatibility and integrity in fabrication of mechanical and microelectronic components in order to make MEMS. The basic processes, which should be emphasized introducing the MEMS microfabrication technology, are deposition (deposit thin films of materials on substrates), photolithography (apply patterned masks on top of the thin films), and etching (etch thin films selectively to the photomask). The importance of these processes is due to the fact that the MEMS fabrication is based on sequential sequence of these processes to form microscale structures, devices, and systems.

## *Thin Film Deposition*

Deposition can be based on chemical reaction (chemical vapor deposition, electrodeposition, epitaxy, and thermal oxidation) as well as physical reaction (physical vapor deposition and casting). The chemical deposition processes make solid materials directly from chemical reactions in gas and/or liquid or with the substrate material. In contrast, the materials are positioned on the substrate using the physical deposition processes.

In the case of the chemical vapor deposition (CVD), the substrate is placed inside a reactor to which gases are supplied. Chemical reactions occur between the source gases. The product of these reactions is a solid material with condenses on the surfaces inside the reactor.

Two widely applied CVD technologies are the low pressure chemical vapor deposition (LPCVD) and plasma enhanced chemical vapor deposition (PECVD). Through LPCVD, thin film layers with excellent uniformity of thickness and material characteristics can be made. The major drawbacks of the LPCVD are high deposition temperature ($600^0$C or higher) and slow deposition rate. The PECVD allows one to perform the deposition at low temperature ($300^0$C) due to the energy supplied to the gas molecules by the plasma in the reactor. However, the uniformity and characteristics of thin films fabricated through PECVD do not match thin films quality fabricated using LPCVD. Furthermore, the PECVD systems usually deposit materials on one side of wafers, while the LPCVD systems deposit thin films on both sides of wafers.

The electrodeposition (electroplating and electroless plating) processes are used to electrically deposit conductive materials. To perform electroplating, the substrate is placed in a liquid solution (electrolyte), the electric potential is formed between a substrate and an electrode (external power supply is used), and due to a chemical redox process, a thin layer of material on the substrate is formed (deposited). Thus, electrodeposition is applied to fabricate thin films (thickness from 0.1 μm to more than 100 μm) of different metals (copper, gold, iron, nickel and other). In case of the electroless plating process, complex chemical solutions are used. The deposition occurs spontaneously on any surface which forms a sufficiently high electrochemical potential with the solution. Electroless plating does not require electric potential and contact to the substrate. However, the electroless plating process is difficult to control to attain the uniformity and other desired deposit characteristics.

If the substrate is an ordered semiconductor crystal (silicon), using the epitaxy process, one fabricates the microstructures with the same crystallographic orientation (for amorphous-polycrystalline substrates, the amorphous or polycrystalline thin films can be made). To attain epitaxial growth, a number of gases are introduced in a heated reactor where only the substrate is heated, and high deposition rate is achieved making films with thickness grater than 100 μm. Usually, epitaxy is used to fabricate silicon structures on the insulator substrates. Therefore, this technique primarily used for deposition of silicon to make ICs and insulated structures.

Thermal oxidation is the deposition process usually applied to form thin films for electrical insulation. Thermal oxidation is based upon oxidation of the substrate surface in an oxygen rich atmosphere at high temperature ($800^0$C to $1100^0$C). Thermal oxidation is limited to materials that can be oxidized, and this process can only form films that are oxides of the material, e.g., to form silicon dioxide on a silicon substrate.

Using physical vapor deposition (PVD) processes, materials are released from sources and transferred to the substrates. Usually, by using PVD it is impossible to attain good thin film characteristics. However, PVD allows one to fabricate affordable microstructures. The basic PVD methods are evaporation and sputtering. Applying evaporation techniques, one places the substrate in a vacuum chamber, in which a source of the material to be deposited is located. The source material is heated to the point where it starts to evaporate. Due to the vacuum, the molecules evaporate freely in the chamber, and they subsequently condense on all surfaces. The heating methods applied are electron-beam (electron beam is directed at the source material causing local heating and evaporation) and resistive (material source is heated electrically using high current to make the material evaporate) evaporation. Different methods should be used. For example, aluminum is difficult to evaporate using resistive heating, while tungsten deposition can be straightforwardly performed through resistive evaporation. In contrast to evaporation, sputtering is a technique in which the material is released from the source at lower temperature than evaporation. The substrate is placed in a vacuum chamber with the source material, and an inert gas (argon) is introduced at low pressure. A gas plasma is struck using a power source, causing the gas to become ionized. The ions are accelerated toward the surface of the source material vaporizing atoms of this material. These atoms are condensed on all surfaces including the substrate.

Applying casting, the material (which should be deposited) is dissolved in a solvent. The deposition process is performed through spraying or spinning. Once the solvent is evaporated, a thin film of the material remains on the substrate. Polymer materials can be easily dissolved in organic solvents, and casting is an efficient technique to apply photoresist to substrates (photolithography).

## *Lithography*

Lithography means the pattern transfer to a photosensitive material by selective exposure of a substrate to a radiation source. Therefore, the MEMS geometry and topography defined (pattern) through lithographic processes. Lithography is performed as sequential steps, e.g., substrate surface preparation, photoresist deposition, alignment of the mask and wafer, and exposure.

Different photosensitive materials, affected when exposed to a radiation source, are used. In particular, when photoresist is exposed to a radiation source with a specific wavelength, the chemical resistance of the photoresist changes. When the photoresist is placed in a developer solution after selective exposure, exposed or unexposed regions can be etched away. If the exposed material is etched away by the developer and the unexposed region is resilient, the material is called a positive photoresist. If the exposed material is resilient to the developer and the unexposed region is etched away, it is called a negative photoresist. To fabricate MEMS, the patterns for different sequential lithographic steps, associated to a particular microstructure, must be aligned. Therefore, the first pattern transferred to a wafer usually includes a set of alignment marks with high precision features that are needed as the reference when positioning subsequent patterns with respect to the first pattern. The alignment marks can be integrated and used in other patterns. Each pattern layer must have a precise alignment feature so that it may be registered to the rest of the layers. The alignment is needed to fabricate the desired MEMS. The alignment location, specifications, and size vary with the type of the alignment and lithographic equipment. Two alignment marks are used to align the mask and wafer, one alignment mark is sufficient to align the mask and wafer in the x- and y-axis, but two spaced marks are needed for rotational alignment.

The exposure characteristics must be examined to achieve accurate pattern transfer from the mask to the photosensitive layer taking into account the wavelength of the radiation source and the dose required to achieve the needed photoresist properties change based upon the photoresist sensitivity. The exposure dose required per unit volume of photoresist is available from the photoresists data library. However, highly reflective layer under the photoresist may result in higher exposure dose, and the photoresist thickness must be also counted. Thus, the secondary effects (reflectiveness, thickness, interference, nonuniformity, flatness, roughness, etc.) affect the pattern transfer and should be examined.

The sequential lithography steps are: prepare the wafer surface (dehydrate the substrate to attain photoresist adhesion); photoresist deposit; substrate coating; soft baking; alignment (align pattern on mask to the specified microstructure topography and geometry); exposure with the optimal dosage; post exposure bake; photoresist develop (selective removal of photoresist after exposure); hard bake (drive off remaining solvent form the photoresist); descum (remove photoresist scum).

### *Etching*

To fabricate microstructures with the ultimate goal to make MEMS, it is necessary to etch thin films deposited as well as etch the substrate itself. Two classes of etching processes are wet etching (material can be dissolved as it immersed in a chemical solution) and dry etching (material can be sputtered or dissolved using reactive ions or vapor phase etchant). Though wet etching is a very simple etching technique, in general, complex topographies and geometries cannot be made. In addition, materials exhibit *anisotropic* and *isotropic* phenomena. This leads to the application of *anisotropic* etching with different etch rates in different material directions. The *anisotropic* and *isotropic* etching effects can be beneficial to fabricate the desired microstructures. However, complex topography, accurate sizing, processes flexibility, and other requirements usually lead to the application of dry etching.

Reactive ion etching (RIE), deep reactive ion etching (DRIE), sputter etching, and vapor phase etching are the techniques used to perform dry etching.

Using the RIE, the substrate is placed inside a reactor in which several gases are introduced. A plasma is struck in the gas mixture using a power source, resulting in releasing ions from the gas molecules. These ions are accelerated towards the surface of the material being etched forming gaseous material (chemical part of reactive ion etching). The physical processes of the RIE are similar to the sputtering deposition processes. In particular, if the ions have high energy, they can knock atoms out of the material to be etched without a chemical reaction. It is a very complex task to develop the processes that balance chemical and physical etching because many parameters must be adjusted and regulated to attain the desired etching characteristics. Furthermore, the secondary effects must be examined.

Sputter etching is performed through the ion bombardment. To perform the etching through the vapor phase etching, the substrate to be etched is placed inside a chamber, in which one or more gases are introduced. The material to be etched is dissolved at the surface in a chemical reaction with the gas molecules. The most common vapor phase etching technique are silicon dioxide etching (using hydrogen fluoride) and silicon etching (using xenon diflouride). Different etching techniques are discussed in Chapter 3.

## 2.6 SYNERGETIC PARADIGMS IN MEMS

As was emphasized and illustrated, synergetic multidisciplinary research must be carried out. Rotational and translational MEMS, which integrate motion microdevices (transducers), radiating energy microdevices (antennas), communication microdevices, driving/sensing ICs, and controlling/ processing ICs, are widely used. High-performance MEMS must be devised, designed, analyzed, and optimized. The need for innovative methods to perform structural synthesis, comprehensive design, high-fidelity modeling, data-intensive analysis, heterogeneous simulation, and optimization has facilitated theoretical developments within the overall scope of engineering and science. To attack and solve the problems mentioned, far-reaching multidisciplinary research must be performed. A MEMS synthesis paradigm reported in this book (see Chapter 4) allows one to synthesize, classify and optimize rotational and translational motion microdevices based upon electromagnetic and geometric design. Microelectromechanical systems can be synthesized using a number of different operating principles, topologies, configurations, and operating features. Optimal electromagnetic, mechanical, optical, vibroacoustic and thermal design must be performed to attain superior performance. In MEMS, the issues of operational variability, viability, controllability, topology synthesis, packaging, and electromagnetic design have consistently been the most basic problems. One of the main goals of this book is to present a general conceptual framework with practical examples which expand up to the systems design and fabrication. Therefore, innovative research and manageable developments in synthesis and design of rotational and translational MEMS are reported. The MEMS classifier and synthesis concepts, which are based upon the classification paradigm reported in Chapter 4, play a central role. These benchmarking results are applied to design high-performance MEMS.

In addition, explicit distinctions must be made between various MEMS configurations, possible operating principles, and topologies. High-fidelity modeling, lumped-parameters modeling, and data-intensive analysis can be straightforwardly performed using the developed nonlinear mathematical models which allow the designer to analyze, simulate, and asses the MEMS performance in the time domain. It must be emphasized that mathematical modeling, analysis, design, and fabrication can be performed only after the designer devises (synthesizes) MEMS. For example, as the MEMS is synthesized based upon the electromagnetic features, its performance and controllability are researched, control algorithms are developed, and functionality is assessed. Microelectromechanical systems are regulated using radiating energy microdevices. Furthermore, control laws are implemented using ICs. Thus, the designer derives the MEMS configuration integrating microdevices, and then performs modeling, simulation, optimization, and assessment analysis with outcome prediction.

Chapters 3 and 8 report the fabrication aspects. However, the designer realizes that before being engaged in the fabrication (processes developments, sequential steps integration, materials and chemical selection) the MEMS must be devised. Electromagnetic-based MEMS, which are widely used in various sensing and actuation applications, must be devised, designed, modeled, analyzed, simulated, optimized, and verified. Furthermore, these rotational and translational motion microdevices need to be controlled. Chapter 4 introduces the synthesis paradigm to perform the structural synthesis of MEMS based upon electromagnetic features. As motion microdevices are devised, modeling, analysis, simulation, control, optimization, and validation are emphasized. Chapter 5 reports the results in modeling, analysis, and simulation. Analysis and modeling of nanoscale systems and structures are documented in Chapter 6. Control topics for MEMS are covered in Chapter 7. Chapter 8 documents the case-studies in design, analysis, and fabrication of MEMS.

The need for innovative integrated methods to perform data-intensive analysis, high-fidelity modeling, and design of MEMS has facilitated theoretical developments within the overall spectrum of engineering and science. This book provides one with viable tools to perform synthesis, modeling, analysis, optimization, and control of MEMS. The integrated design, analysis, optimization and virtual prototyping of high-performance MEMS can be addressed, researched, and solved through the use of the advanced electromechanical theory, state-of-the-art hardware, novel technologies, and leading-edge software.

Many problems in MEMS can be formulated, attacked, and solved using the microelectromechanics. In particular, microelectromechanics deals with benchmarking and emerging problems in integrated electrical–mechanical–computer engineering, science, and technologies. Microelectromechanics is the integrated paradigm in synthesis, design, analysis, optimization and virtual prototyping of high-performance MEMS. In addition, sensing–communication–computing–control–actuation through the use of advanced hardware, leading-edge software, and novel technologies and processes must be studied. Integrated multidisciplinary features approach quickly, and the synergetic microelectromechanics takes place.

The computer-aided design tools and environments are required to support MEMS analysis, simulation, design, optimization, and fabrication. Much effort has been devoted to attain the specified steady-state and dynamic performance of MEMS to meet the criteria and requirements imposed. Currently, MEMS are designed, optimized, and analyzed using available software packages based on the linear and steady-state (finite element) analysis. However, highly detailed heterogeneous nonlinear electromagnetic and mechanical modeling must be performed to design high-performance MEMS. Therefore, the research is concentrated on high-fidelity mathematical modeling, data-intensive analysis, nonlinear simulations, and control (design of control algorithms to attain the desired performance). The reported synthesis, modeling, analysis, simulation, optimization, and control concepts, tools, and paradigms ensure a cost-effective solution and can be used to attain rapid prototyping of high-performance state-of-the-art MEMS.

It is often very difficult, and sometimes impossible, to solve a large variety of nonlinear analysis and design problems for MEMS using conventional methods. Innovative concepts, methods, and tools that fully support the analysis, modeling, simulation, control, design, and optimization are needed. This book addresses and solves a number of long-standing problems for electromagnetic-based MEMS.

## 2.7 MEMS AND NEMS ARCHITECTURES

A large variety of micro- and nanoscale structures, devices and systems have been widely used, and a worldwide market for MEMS and NEMS and their applications will drastically increase in the near future. The differences in MEMS and NEMS were emphasized, and NEMS are much smaller than MEMS. For example, carbon nanotubes (nanostructure) can be used as the nanowires, sensors, or devices in MEMS. Different specifications are imposed on MEMS and NEMS depending upon their applications and operating principles. For example, using carbon nanotubes as the nanowires, the current density is defined by the media properties (e.g., resistivity and thermal conductivity). The maximum current is defined by the diameter and the number of the carbon nanotube layers.

Different molecular-scale nanotechnologies are applied to manufacture NEMS, and the properties of nanostructures can be controlled and changed. In contrast, monolithic and hybrid MEMS have been mainly manufactured using surface micromachining (silicon-based technology as modification of conventional microelectronics CMOS), LIGA, and LIGA-like technologies.

To deploy and commercialize MEMS and NEMS, a spectrum of problems must be solved, and a portfolio of software design tools needs to be developed using multidisciplinary concepts. In recent years, much attention has been given to MEMS fabrication, synthesis, modeling, analysis, and optimization. It is evident that MEMS and NEMS can be studied with different levels of detail and comprehensiveness, and different application-specific architectures should be synthesized and optimized. The majority of research papers study either micro- and nanoscale structures, actuators-sensors, or ICs. That is, MEMS and NEMS components are examined.

A great number of publications have been devoted to the carbon nanotubes (nanostructures). While, the results for different MEMS and NEMS components are extremely important and manageable, the systems-level research must be performed because the specifications are imposed on the systems, not on the individual elements, structures, and subsystems. Thus, MEMS and NEMS must be developed and studied to attain the comprehensiveness of the analysis and design.

The actuators are controlled by changing the voltage or current (using ICs) or by the electromagnetic field (using radiating energy devices). The ICs and antennas (MEMS components) are regulated using controllers which can include central processor and memories (as core), IO devices, etc. Micro- and nanoscale sensors are also integrated as components of MEMS and NEMS. For example, using molecular wires, carbon nanotubes or optical devices one feeds the information to the IO devices of the nanoprocessor. That is, MEMS and NEMS integrate a large number of structures, devices and subsystems which must be studied. As a result, the designer cannot consider MEMS and NEMS as the six-degree-of-freedom structures using conventional mechanics (the linear or angular displacement is a function of the applied or developed force or torque) completely ignoring the problems of how these forces or torques are generated and regulated, how components interact, how communication is performed, how energy and signal conversions are achieved, how communication is performed, etc.

In this book, we illustrate how to integrate and study the basic components of MEMS and NEMS. The synthesis, design, modeling, simulation, analysis, optimization, and prototyping of MEMS and NEMS must be attacked using advanced theories. Even though a wide range of nanoscale structures and devices (e.g., molecular diodes and transistors, transducers, switches, logics) can be fabricated with atomic precision, comprehensive systems analysis must be performed before the designer embarks in costly fabrication. Through optimization of architecture, synthesis, structural optimization of components (microtransducers, ICs and antennas), modeling, simulation, analysis as well as visualization, rapid evaluation and prototyping with assessment analysis can be performed. This facilitates cost-effective solutions reducing the design cycle as well as guaranteeing design of high-performance MEMS and NEMS which satisfy the requirements and specifications.

The large-scale integrated MEMS (a single chip that guarantees the desired operating features and performance specifications) may integrate:

- N nodes of microtransducers (actuators/sensors, smart structures and other motion microdevices),
- Radiating energy devices,
- Optical devices,
- Communication devices,
- Processors and memories,
- Interconnected networks (communication busses),
- Driving/sensing ICs,
- Controlling/processing ICs,
- Input-output (IO) devices.

Different MEMS configurations can be synthesized, and diverse architectures can be implemented based upon requirements and specifications imposed. For example, linear, star, ring, and hypercube large-scale MEMS architectures are illustrated in Figure 2.7.1.

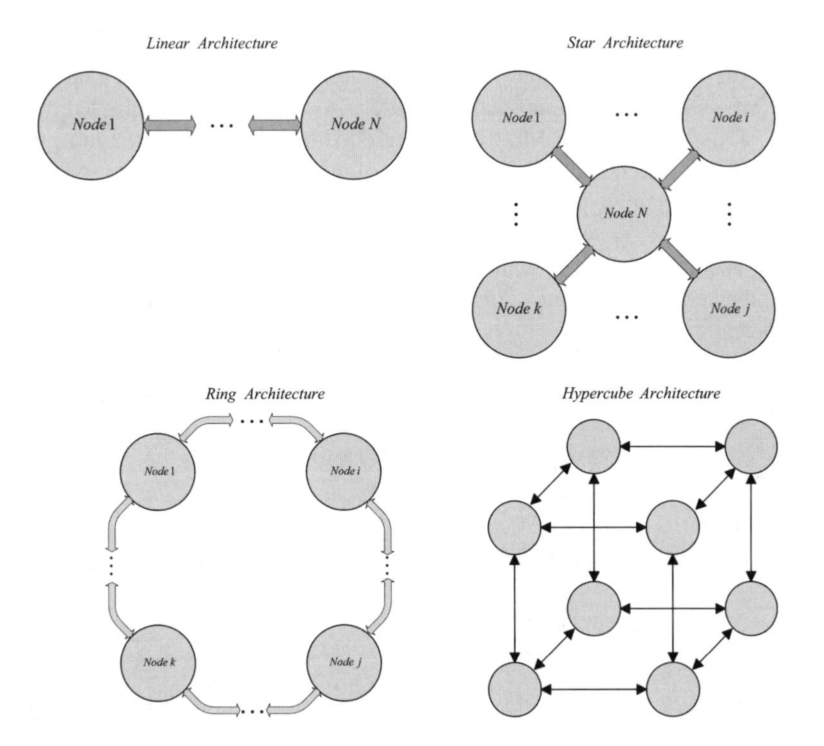


Figure 2.7.1 Linear, star, ring, and hypercube architectures.

More complex architectures can be designed, and the hypercube-connected-cycle node configuration is illustrated in Figure 2.7.2.

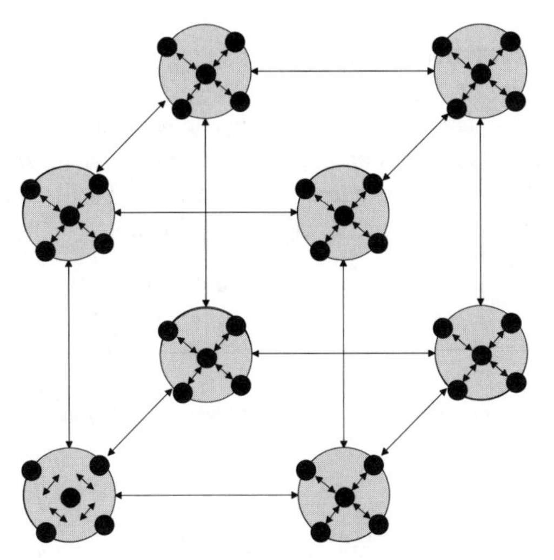

Figure 2.7.2 Hypercube-connected-cycle node architecture.

The nodes (stand-alone MEMS) can be synthesized, and the elementary MEMS can be simply pure smart structure, actuator, or sensor controlled by the ICs. The elementary MEMS can be controlled by the external electromagnetic field (that is, ICs or antenna are not a part of the microstructure or microdevice). As an alternative, MEMS node can integrate actuators and sensors, microstructures, antennas, ICs, processor (with controlling, signal processing, data acquisition and desicoin-making capabilities), memories, IO devices, communication buses, etc. Figure 2.7.3 illustrates exclusive and elementary nodes.

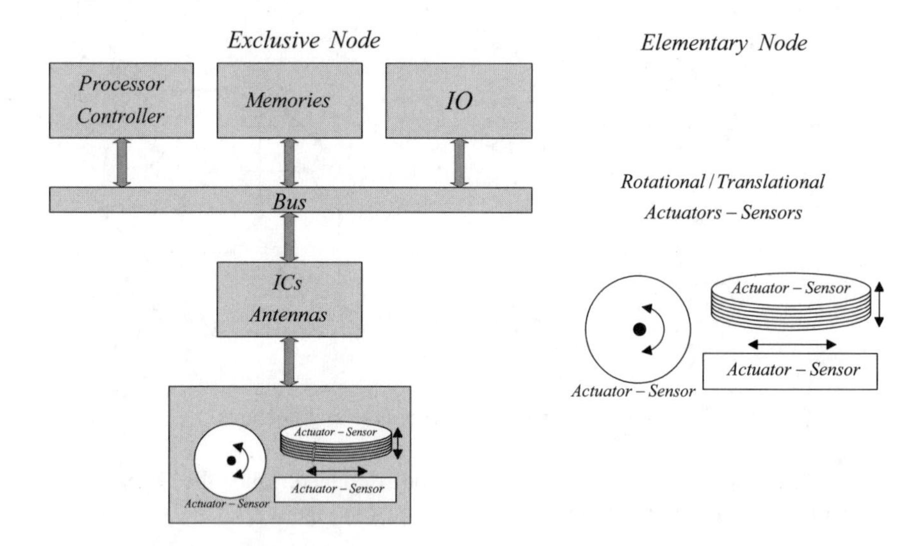


Figure 2.7.3 MEMS nodes.

MEMS and NEMS are used to control, operate, and guarantee functionality for different physical systems. For example, immune system, drug delivery, propeller, wing, relay, lock, car, aircraft, missiles, etc. To illustrate the basic components and their integration, a high-level functional block diagram is shown in Figure 2.7.4.

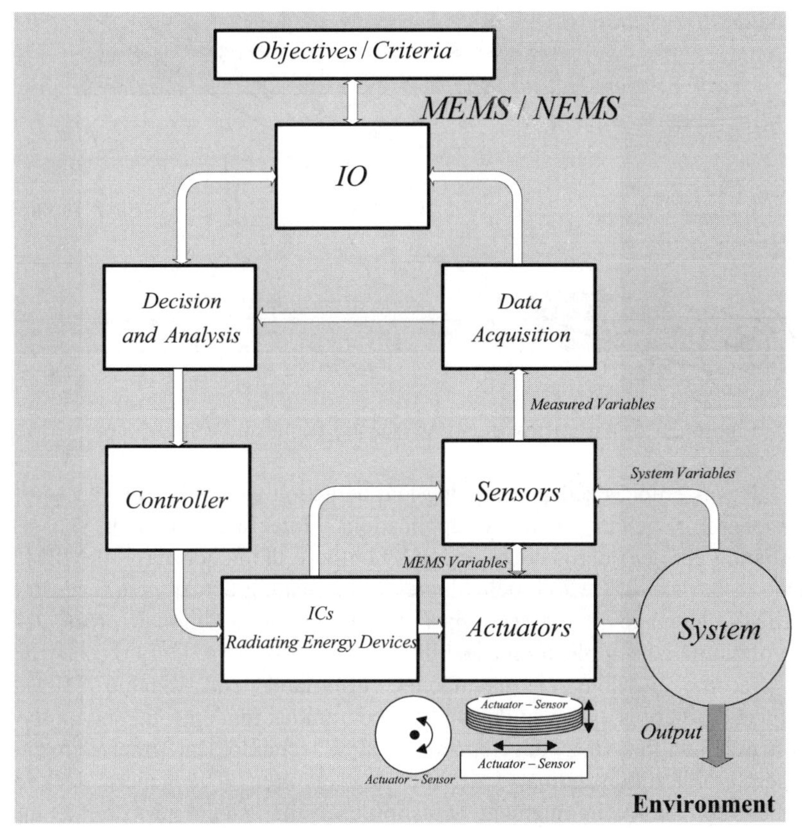


Figure 2.7.4 High-level functional block diagram of NEMS/MEMS and systems.

For example, the desired flight path of aircraft (maneuvering and landing) is maintained by displacing the control surfaces (ailerons, elevators, canards, flaps, rudders, fins and stabilizers) and/or changing the control surface and wing geometry. Figure 2.7.5 documents the application of the MEMS to actuate the control surfaces. It should be emphasized that the digital signal-level signals are generated by the flight computer, and these digital signals are converted into the desired voltages or currents fed to the microactuators by ICs. These signal-level control signals can be converted in the electromagnetic flux intensity to displace the actuators by radiating energy devices. It is also important that microtransducers can be used as sensors. As an example, the loads on the aircraft structures (airframe, wings, etc) during the flight can be sensed.

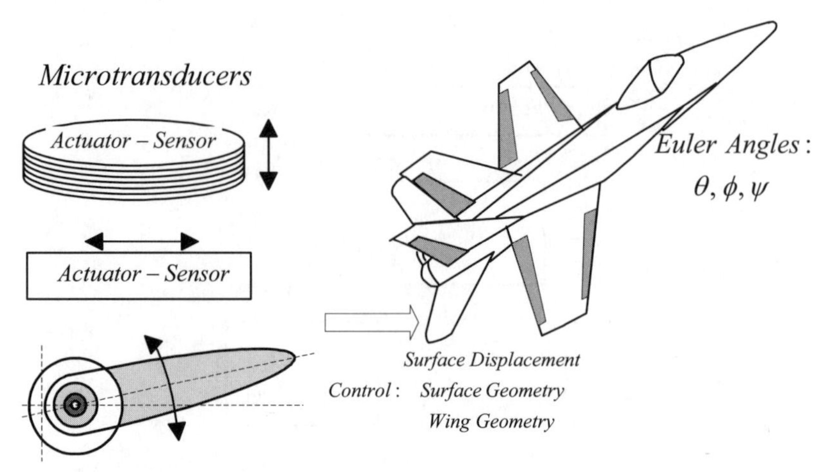


Figure 2.7.5 Aircraft with MEMS-based flight microtransducers.

It was emphasized that sensing and actuation cannot be viewed as the peripheral function in many applications. Integrated actuators/sensors, radiating energy microdevices and ICs compose the major class of MEMS. Due to the use of CMOS technologies in fabricating actuators and sensors, MEMS leverage microelectronics in important additional areas that revolutionize the application capabilities of microsystems. Only recently has it become possible to manufacture affordable and reliable MEMS. However, there is a critical demand for continuous fundamental and applied research, engineering developments, and technological improvements. Therefore, multidisciplinary activities are required. The general lack of synergetic theory to augment actuation, sensing, signal processing, and control is known. Therefore, these issues must be addressed through focused efforts. The set of long-range goals has been emphasized, and the challenges facing the development of MEMS are:

- MEMS synthesis, modeling, analysis, optimization, and design,
- Novel high-performance microtransducers (actuators and sensors),
- Devising new actuation and sensing mechanisms,
- Sensors-actuators-ICs integration and MEMS configurations,
- Sensing-communication-computing-control-actuation in MEMS,
- Advanced technologies, techniques, processes, and materials,
- Packaging, microassembly, and testing,
- MEMS implementations and applications.

Significant progress in the application of CMOS technology enables the industry to fabricate microscale actuators and sensors with the corresponding ICs. This guarantees significant breakthroughs. The field of MEMS has been driven by the rapid global progress in ICs, VLSI, solid-state devices, microprocessors, memories, and DSPs that have revolutionized instrumentation and control. In addition, this progress has facilitated explosive growth in data processing and communications in high-performance systems. In microelectronics, many emerging problems deal with non-electric phenomena and processes (optics, thermal and structural analysis and optimization, vibroacoustic, packaging, etc.). However, these non-electric phenomena frequently result due to electromagnetic effects.

It has been emphasized that ICs are the necessary component to perform control, signal processing, computing, data acquisition, and decision-making. For example, control signals (voltage or currents) are computed, converted, modulated, and fed to actuators, antennas, or microwindings. It is evident that MEMS have found application due to extremely high integration level of electromechanical components with low cost and maintenance, high accuracy, reliability, and ruggedness. The manufacturability issues were addressed. It was shown that one can design and manufacture individually-fabricated devices and subsystems. However, these devices and subsystems are impractical due to very high cost.

Piezoactuators and permanent-magnet technology has been used widely, and rotating and linear microtransducers (actuators and sensors) were designed. For example, piezoactive materials are used in ultrasonic motors. Frequently, conventional concepts of the electric machinery theory (rotational and linear direct-current, induction, and synchronous machine) are used to design and analyze MEMS-based microtransducers. The use of piezoactuators is possible as a consequence of the discovery of advanced materials in sheet and thin-film forms, especially lead zirconate titanate (PZT) and polyvinylidene fluoride. The deposition of thin films allows piezo-based micromachines to become a promising candidate for microactuation and sensing. These microtransducers can be fabricated using a deep x-ray lithography and electrodeposition processes.

To fabricate nanoscale structures, devices, and systems, molecular manufacturing methods and technologies must be developed. Self- and positional-assembly concepts are the preferable techniques compared with individually-fabricated molecular structures. To perform self- and positional-assembly, complementary pairs (CP) and molecular building blocks (MBB) should be designed. These CP and MBB, which can be built from a few to thousands of atoms, can be examined and designed using the DNA analogy. The nucleic acids consist of two major classes of molecules, e.g., DNA and RNA. Deoxyribonucleic acid (DNA) and ribonucleic acid (RNA) are the largest and most complex organic molecules which are composed of carbon, oxygen, hydrogen, nitrogen and phosphorus. The structural units of DNA and RNA are nucleotides. Each nucleotide consists of three components (nitrogen-base, pentose and phosphate) joined by dehydration synthesis. The double-helix molecular model of DNA was discovered by Watson and Crick in 1953. Deoxyribonucleic acid (long double-stranded polymer with double chain of nucleotides held together by hydrogen bonds between the bases), as the genetic material, performs two fundamental roles. It replicates (identically reproduces) itself before a cell divides, and provides pattern for protein synthesis directing the growth and development of all living organisms according to the information DNA supports. Different DNA architectures provide the mechanism for the replication of genes. Specific pairing of nitrogenous bases obeys base-pairing rules and determines the combinations of nitrogenous bases that form the rungs of the double helix. In contrast, RNA performs the protein synthesis using the DNA information. Four DNA bases are: A (adenine), G (guanine), C (cytosine) and T (thymine). The ladder-like DNA molecule is formed due to hydrogen bonds between the bases which are paired in the interior of the double helix (the base pairs are 0.34 nm apart and there are ten pairs per turn of the helix). Two backbones (sugar and phosphate molecules) form the uprights of the DNA molecule, while the joined bases form the rungs. Figure 2.7.6 illustrates that the hydrogen bonding of the bases are: A bonds to T, G bonds to C. The complementary base sequence results.

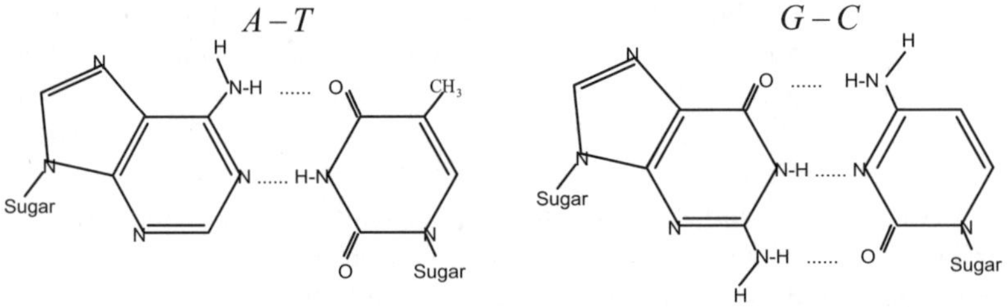

Figure 2.7.6 Deoxyribonucleic acid pairing due to hydrogen bonds.

In RNA molecules (single strands of nucleotides), the complementary bases are: A bonds to U (uracil), and G bonds to C. The complementary base bonding of DNA and RNA molecules gives one the idea of possible sticky-ended assembling (through complementary pairing) of NEMS structures and devices with the desired level of specificity, architecture, topology, organization, and reconfigurability. In structural assembling and design, the key element is the ability of CP or MBB (atoms or molecules) to associate with each other (recognize and identify other atoms or molecules by means of specific base pairing relationships). It was emphasized that in DNA, A (adenine) bonds to T (thymine), and G (guanine) bonds to C (cytosine). Using this idea, one can design the CP such as $A_1$-$A_2$, $B_1$-$B_2$, $C_1$-$C_2$, $D_1$-$D_2$, $E_1$-$E_2$, $M_1$-$M_2$, $N_1$-$N_2$, etc. That is, $A_1$ pairs with $A_2$, $B_1$ pairs with $B_2$, while $N_1$ pairs with $N_2$. This complementary pairing can be studied using electromagnetics (Coulomb law) and chemistry (chemical bonding, for example, hydrogen bonds in DNA between nitrogenous bases A and T, G and C). Figure 2.7.7 shows how two nanoscale elements with sticky ends form the complementary pair. In particular, "+" is the sticky end and "–" is its complement. Thus, the electrostatic complementary pair $A_1$-$A_2$ results.

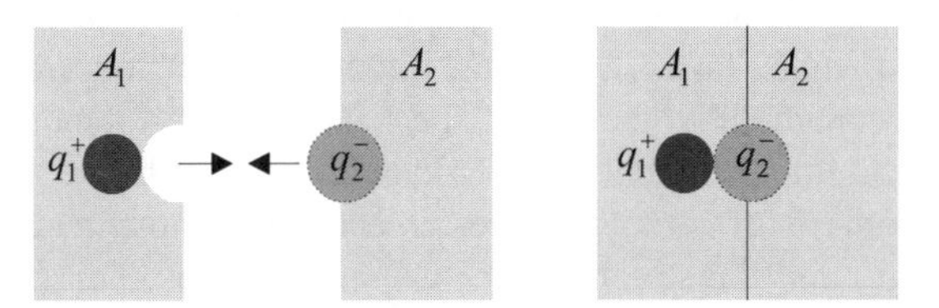


Figure 2.7.7 Sticky ended electrostatic complementary pair $A_1$-$A_2$.

An example of assembling a ring microstructure is illustrated in Figure 2.7.8. Using the sticky ended segmented (asymmetric) electrostatic CP, self-assembling of nanostructure is performed in the two-dimensional (XY) plane. It is evident that three-dimensional structures can be formed through self-assembling developing electrostatic CP.

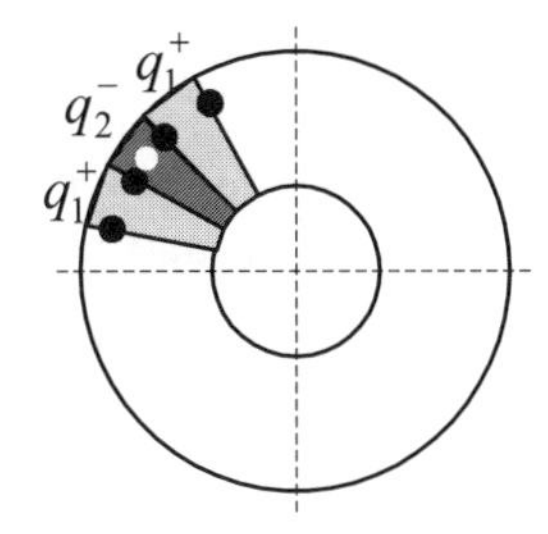


Figure 2.7.8 Two-dimensional ring self-assembling.

There are several advantages to using sticky ended electrostatic CP. In particular, the ability to recognize (identify) the complementary pair is reliably predicted, and the possibility to form stiff, strong and robust structures is attained. Self-assembled complex nanostructures can be fabricated using subsegments to form the branched junctions. This concept is well-defined electromagnetically (Coulomb law) and geometrically (branching connectivity). Using these subsegments, nanostructures with the desired geometry (e.g., cubes, octahedron, spheres, cones, etc.) can be fabricated. Furthermore, the geometry of nanostructures can be controlled sequentially introducing the CP and pairing MBB. It must be emphasized that it is possible to generate quadrilateral self-assembled nanostructures using different CP. For example, electrostatic and chemical CP can be employed. Single- and double-stranded structures can be generated and linked in the desired topological and architectural manners. The self-assembling must be controlled during the fabrication cycle, and CP and

MBB which can be paired and topologically/architecturally bonded must be added in the desired sequence. For example, polyhedral and octahedron synthesis can be made when building elements (CP or MBB) are topologically or geometrically specified. The connectivity of nanostructures determines the minimum number of linkages that flank the branched junctions. The synthesis of complex three-dimensional nanostructures is the design of topology, and the structures are characterized by their branching and linkaging.

### *Linkage Groups in Molecular Building Blocks*

Hydrogen bonds, which are weak, hold DNA and RNA strands. Strong bonds are desirable to form stiff, strong, and robust micro- and nanostructures. Using polymer chemistry, functional multimonomer groups can be designed. Polymers made from monomers with two linkage groups do not exhibit desired stiffness and strength. Tetrahedral MBB structures with four linkage groups lead to stiff and robust structures. Polymers are made from monomers, and each monomer reacts with two other monomers to form linear chains. Synthetic and organic polymers (large molecules) are nylon and dacron (synthetic), and proteins and RNA, respectively.

There are two feasible assembling techniques: self assembly and positional assembly. Self-assembling is widely used at the molecular scale, e.g., DNA and RNA. Positional assembling is used in manufacturing and microelectronic manufacturing. The current difficulties implementing positional assembly at the nanoscale with the same flexibility and integrity that is achieved in microelectronic fabrication limits the range of nanostructures which can be made. Therefore, the efforts are focused on developments of MBB, as applied to manufacture nanostructures, wit ultimate goal to guarantee:

- Affordable (low-cost) mass-production;
- High-yield and reliability;
- Simplicity, predictability, and controllability of synthesis and fabrication;
- High-performance, repeatability, and similarity of characteristics;
- Stiffness, strength, and robustness;
- Tolerance to contaminants.

It is possible to select and synthesize MBB that satisfy the requirements and specifications (nonflammability, nontoxicity, pressure, temperatures, stiffness, strength, robustness, resistivity, permiability, permittivity, etc.). Molecular building blocks are characterized by the number of linkage groups and bonds they have. The linkage groups and bonds that can be used to connect MBB are: dipolar bonds (weak), hydrogen bonds (weak), transition metal complexes bonds (weak), amide and ester linkages (weak and strong).

It must be emphasized that large MMB can be made from elementary MBB. There is a need to synthesize robust three-dimensional structures. Molecular building blocks can form planar structures which are strong, stiff, and robust in-plane, but weak and compliant in the third dimension. This problem can be resolved by forming tubular structures. It is difficult to form three-dimensional structures using MBB with two linkage groups. Molecular building blocks with three linkage groups form planar structures, which are strong, stiff, and robust in-plane but bend easily. This plane can be rolled into tubular structures to guarantee stiffness. In contrast, MBB with four, five, six, and twelve linkage groups form the strong, stiff, and robust three-dimensional structures needed to synthesize robust micro- and nanostructures. Molecular building blocks with L linkage groups are paired forming L-pair structures, and planar and nonplanar (three-dimensional) micro- and nanostructures result. These MBB can have in-plane linkage groups and out-of-plane linkage groups which are normal to the plane. For example, hexagonal sheets are formed using three in-plane linkage groups (MBB is a single carbon atom in a sheet of graphite) with adjacent sheets formed using two out-of-plane linkage groups. These structures have the hexagonal symmetry.

It is known that buckyballs ($C_{60}$), which have been studied in the literature as MBB, are formed with six functional groups. Molecular building blocks with six linkage groups can be connected together in the cubic structure. These six linkage groups corresponding to six sides of the cube or rhomb. Thus, MBB with six linkage groups form solid three-dimensional structures as cubes and rhomboids. Molecular building blocks with six in-plane linkage groups form strong planar structures. Robust, strong, and stiff cubic or hexagonal closed-packed crystal structures are formed using twelve linkage groups.

Molecular building blocks synthesized and applied should guarantee the desirable performance characteristics (stiffness, strength, robustness, resistivity, permiability, permittivity, coefficient of thermal extension, etc.), affordability, integrity and manufacturability. It is evident that stiffness, strength and robustness are predetermined by bonds (weak and strong), while resistivity, permiability and permittivity are the functions of MBB composites and media.

## References

1. S. E. Lyshevski, *Micro- and nano-Electromechanical Systems: Fundamental of Micro- and Nano- Engineering*, CRC Press, Boca Raton, FL, 2000.
2. M. Madou, *Fundamentals of Microfabrication*, CRC Press, Boca Raton, FL, 1997.
3. S. A. Campbell, *The Science and Engineering of Microelectronic Fabrication*, Oxford University Press, New York, 2001.

# CHAPTER 3

# FUNDAMENTALS OF MEMS FABRICATION

## 3.1 INTRODUCTION AND DESCRIPTION OF BASIC PROCESSES

The goals of this chapter are to reach the widest possible range of readers, not necessarily experts in microelectronics and MEMS fabrication, as well as to attain the versatility for students and professionals. In particular, this chapter aims to satisfy the existing growing demands in covering the fabrication technologies. Efforts have been made to cover basic fabrication technologies, techniques, processes, and materials. However, due to very broad and extensive technologies devised, specialized books and handbooks are written to cover them [1-4]. The availability of different fabrication technologies as well as the accessibility of the advanced techniques and processes (which can be applied primarily depending upon the equipment and infrastructure available) allows the author emphasize the basic concepts referencing the reader, interested in more comprehensive and specialized coverage, to other sources.

Microelectromechanical systems integrate motion microstructures (electromechanical components), sensors, actuators, radiating energy devices, and microelectronics. These MEMS can be fabricated utilizing different microfabrication technologies, e.g., micromachining or high-aspect ratio. While the microelectronic components are fabricated using CMOS and biCMOS process sequences, the microelectromechanical components can be made using compatible micromachining processes that selectively etch away parts of the silicon wafer or other materials and add (deposit) new structural and sacrificial layers of different materials to form mechanical, electromechanical, and electro-opto-mechanical devices and systems. The fabrication technologies used in MEMS were developed and reported in books [1, 2]. This chapter documents micromachining and microelectronics technologies, techniques, and processes. In addition, microfabrication processes and techniques to make microstructures and microtransducers are reported in Chapter 8.

The basic technologies in MEMS fabrication are CMOS and biCMOS (to fabricate ICs) and micromachining or high-aspect ratio (to fabricate motion and radiating energy microscale structures and devices). One of the main goals is to integrate microelectronics with micromachined electromechanical structures and devices in order to produce integrated high-performance MEMS. To guarantee high-performance, affordability, reliability, and manufacturability, well-developed CMOS-based batch-fabrication processes should be modified and enhanced.

In addition, assembling and packaging must be automated, e.g., auto- or self-alignment, self-assembly, and other processes should be developed. The MEMS must be protected from mechanical damage and contamination. Thus, MEMS are packaged to protect them from harsh environments, prevent mechanical damage, minimize stresses and vibrations, contamination, electromagnetic interference, etc. Therefore, MEMS are usually sealed. It is impossible to specify a generic MEMS package. Through input-output connections (power and communication buses) and networking, one delivers the power required, feeds control (command), tests (probes) signals, receives the output signals and data, interfaces, builds networks, etc. Robust packages must be designed to minimize electromagnetic interference, noise, vibration, and other undesirable effects. For example, heat generated by MEMS must be dissipated, and the thermal expansion problem must be solved. Conventional MEMS packages are usually ceramic and plastic. In ceramic packages, the die is bonded to a ceramic base, which includes a metal frame and pins for making electric outside connections. Plastic packages are connected in the similar way. However, the plastic package can be molded around the microdevice. Wear tolerance, electromagnetic and thermo insulation, among other problems, are very challenging issues. Different fabrication techniques, processes, and materials must be applied to attain the desired performance, reliability, and cost. Microsystems can be coated directly by thin films of silicon dioxide or silicon nitride which are deposited using plasma enhanced chemical vapor deposition. It is possible to deposit (at $700^0$C to $900^0$C) diamond thin films which have superior wear capabilities, excellent electric insulation, and superior thermal characteristics. Microelectromechanical systems are connected and interfaced (networked) with other systems and components (control surfaces of aircraft, flight management system, communication ports, power buses, sensors, etc.).

Bulk and surface micromachining, as well as high-aspect-ratio technologies (LIGA and LIGA-like), are the most developed fabrication methods. Silicon is the primary substrate material which is used by the microelectronic industry. A single crystal ingot (solid cylinder 300 mm diameter and 1000 mm length) of very high purity silicon is grown, sawed to the desired thickness, and polished using chemical and mechanical polishing techniques. Electromagnetic and mechanical wafer properties depend upon the orientation of the crystal growth, concentration and type of doped impurities. Depending on the silicon substrate, CMOS and biCMOS processes are used to manufacture ICs, and the processes are classified as *n-well*, *p-well*, or *twin-well*. The major steps are diffusion, oxidation, polysilicon gate formations, photolithography, masking, etching, metallization, wire bonding, etc.

### *Photolithography*

To fabricate motion and radiating energy microstructures and microdevices, the CMOS technology must be modified, e.g., new processes must be developed and novel materials should be applied. High-resolution photolithography is a technology that is applied to define two- (planar) and three-dimensional shapes (geometry). For example, the microtransducers and their components (stator, rotor, bearing, coils, etc.) geometry are defined photographically. First, a mask is produced on a glass plate. The silicon wafer is then coated with a polymer which is sensitive to ultraviolet light (this photoresistive layer is called photoresist). Ultraviolet light is shone through the mask onto the photoresist. The positive photoresist becomes softened, and the exposed layer can be removed. In general, there are two types of photoresist, e.g., positive and negative. Where the ultraviolet light strikes the positive photoresist, it weakens the polymer. Hence, when the image is developed, the photoresist is rinsed where the light struck it. A high-resolution positive image is needed, and different photolithography processes are developed based upon the basic lithography fundamentals. In contrast, if the ultraviolet light strikes negative photoresist, it strengthens the polymer. Therefore, a negative image of the mask results. Different chemical processes are involved to remove the oxide where it is exposed through the openings in the photoresist. When the photoresist is removed, the patterned oxide appears. Alternatively, electron beam lithography can be used.

Photolithography requires the design of photolithography masks, and computer-aided-design (CAD) software is available and widely applied to support the photolithography. The photolithography process and PC-based Photolithography System are illustrated in Figure 3.1.1.

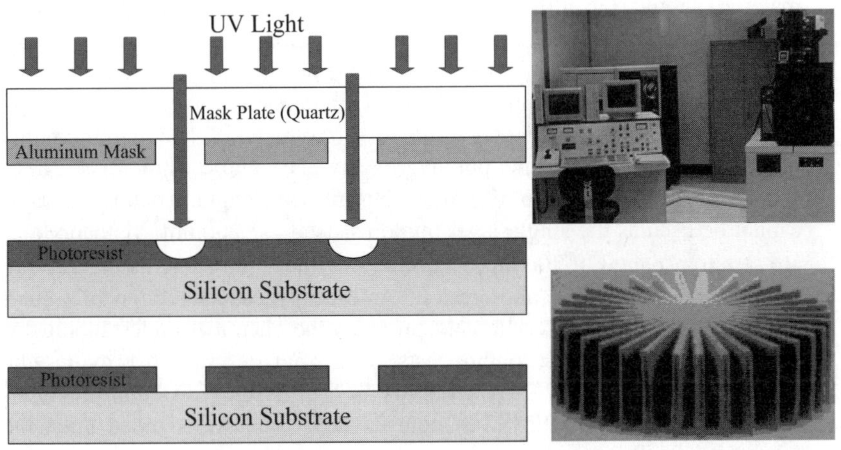


Figure 3.1.1 Photolithography process, computerized photolithography system, and fabricated microstructure.

Deep UV lithography processes were developed to decrease the feature sizes of microstructures to 0.1 μm. Different exposure wavelengths $\lambda$ are used (for example, $\lambda$ can be 435, 365, 248, 200, 150 or 100 nm). Using the Rayleigh model for image resolution, one finds the expressions for image resolution $i_R$ the depth of focus $d_F$ as given by the following formulas

$$i_R = k_i \frac{\lambda}{N_A},$$

$$d_F = k_d \frac{\lambda}{N_A^2},$$

where $k_i$ and $k_d$ are the lithographic process constants; $\lambda$ is the exposure wavelength; $N_A$ is the numerical aperture coefficient (for high-numerical-aperture we have $N_A$=0.5 – 0.6).

These formulas indicate that to pattern microstructures with the decreased feature size, the photoresist exposure wavelengths must be decreased and numerical aperture coefficient should be increased.

The so-called g- and i-line IBM lithography processes (with wavelengths of 435 nm and 365 nm, respectively) allow one to attain 0.35 μm features. The deep ultraviolet light sources (mercury source or excimer lasers) with 248-nm wavelength enables the industry to achieve 0.25 μm resolution.

The changes to short exposure wavelength possess challenges and present new highly desired possibilities. Specifically, transparent optical materials with weak absorption as well as organic transparent polymers can serve as the single-layer photoresists. High-purity synthetic fused silica and crystalline calcium fluoride are the practical choices. The ideal optical material should be fully transparent and must remain unaffected after billions of pulses. Fused silica has the absorption coefficient in the range from 0.005 to 0.10 $cm^{-1}$. Another challenge in the application of 200-nm lithography is the development of robust photoresist processes. The resins photoresists, which are widely used for 365-nm and 248-nm lithography processes, are novolac and polyhydroxystyrene. These photoresists have absorption depths of 30 to 50 nm at 200 nm wavelength. Therefore, resins cannot be used as the single-layer photoresists at $\lambda$=200 nm. Methacrylates are semitransparent at 200 nm, and these polymers can serve as the 200-nm lithography single-layer photoresists. Acid-catalyzed conversion of t-butyl methacrylate into methacrylic acid provides the chemical underpinning for several versions of these photoresists. This brief description provides the reader with an overview. It is necessary to deeply study different processes in order to fabricate MEMS. Of course, these processes depend upon the fabrication facilities available.

Lithography is the process used in ICs and MEMS fabrication to create the patterns defining the ICs', microstructures', and microdevices' features. Different lithography processes are: photolithography (as was reported), screen printing, electron-beam lithography, x-ray lithography (high-aspect ratio technology), etc. Extensions of the currently widely used optical lithography using shorter wavelength radiation result to the 100 nm minimum feature size. For microstructures and ICs that require less than 100 nm resolution features, the next-generation-lithography techniques must be applied. For example, photons and charged particles, which have short wavelengths, can be used. In particular, the high-throughput electron-beam lithography is under development. Unlike optical lithography, electron beams are not diffraction-limited, and thus, the ultimate resolution attainable is expanded. The electron-beam lithography has evolved from the early scanning-electron-microscope-type Gaussian beam systems (which expose ICs patterns as a one pixel at a time) to the massive parallel projection of pixels in electron-projection-lithography. This allows one to attain millions of pixels per shot. Figure 3.1.2 documents a scanning-electron-beam lithography system (IBM VS-2A). The digital pattern generator is based on commercial high-performance RISC processors. The system is capable of creating large area patterns. The system is computerized, and the hardware-software co-design should be accomplished (computer-aided-design and control software must be integrated within the lithography systems).

Figure 3.1.2 Scanning-electron-beam lithography system.

For last thirty years optical lithography has been widely used, and the ICs features were decreased from 5 μm to 0.1 μm (100 nm). Electron-beam and x-ray lithographies are considered as an alternative solution. The wafer throughput with electron-beam lithography is slow, and the electron-beam lithography is considered as complementary to optical lithography. However, optical lithography depends on electron-beam lithography to generate the masks. Because of its intrinsic high resolution, the electron-beam lithography is the primary process for 0.1 μm (or smaller resolution) microstructures.

It has been emphasized that the photolithography processes are embedded in ICs and MEMS fabrication. The computerized (PC-controlled) integrated stepper platforms (or the so-called step-and-repeat projection aligners) transfer the image of the microstructure or ICs from a master photomask image to a specified area on the wafer surface. The substrate is then moved (stepped), and the image can be exposed once again to another area of the wafer. This process is repeated until the entire wafer is exposed. Different steppers are applied with different capabilities and features, e.g., 100-, 157-nm or other application-specific lithography processes, contact or noncontact lithography, single- or multi-layer, different wafers (materials, size, 50-, 75-, 100-, 150-, 200- or 300-mm diameter, thickness, etc.), single- or dual-side alignment (off-axis microscope capable of viewing images on the back of the wafer while exposing features on the front side), pre-alignment, imaging, versatility, magnetic levitation, etc. The 1600DSA Ultratech computerized stepper system is illustrated in Figure 3.1.3.

Figure 3.1.3 Ultratech computerized stepper system.

In photolithography, mask aligners are used to transfer a pattern from a mask to a photoresist on the substrate. In particular, the Karl Suss MJB-3 and MA-6 mask aligners (for three and six inches wafers, respectively) are commonly used to expose photoresist-coated substrates to ultraviolet light through photo masks, see Figure 3.1.4. It was emphasized that different photoresists are sensitive to light at different wavelengths. It is important to select the mask aligner with a wavelength optimized with the respect to the photoresists used.

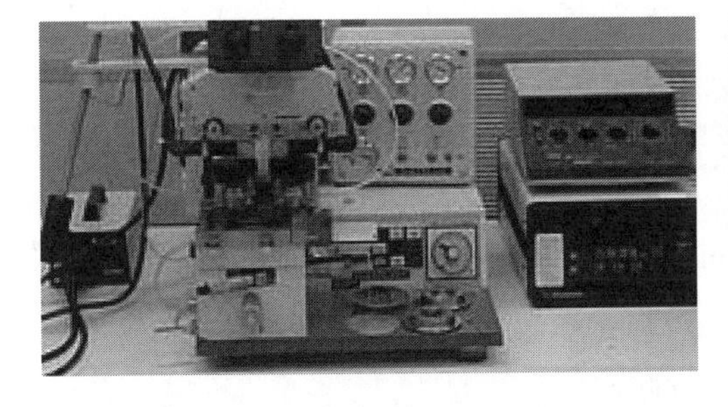

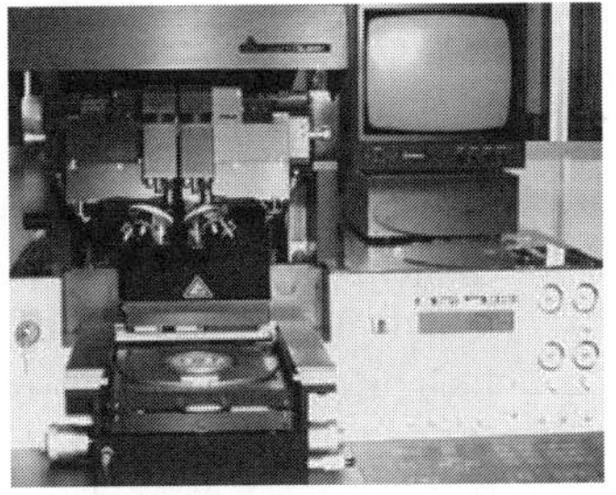

Figure 3.1.4 Karl Suss MJB-3 and MA-6 mask aligners.

The GCA Corporation manufactures a line of photolithography stepper cameras (GCA 6000, 8000, 8500 and others) controlled by IBM or Digital Equipment Corporation (LSI-11/23, -11/53, -11/73) computers with different operating systems. The GCA Mann pattern generators, which produce the patterns for ICs and microstructures, consist of a computerized controller, electronics, and cameras. Figure 3.1.5 illustrates the GCA Mann 3000 photomask pattern generator and GCA 3696 stepper (step-and-repeat) camera which allow one to achieve a micron resolution features.

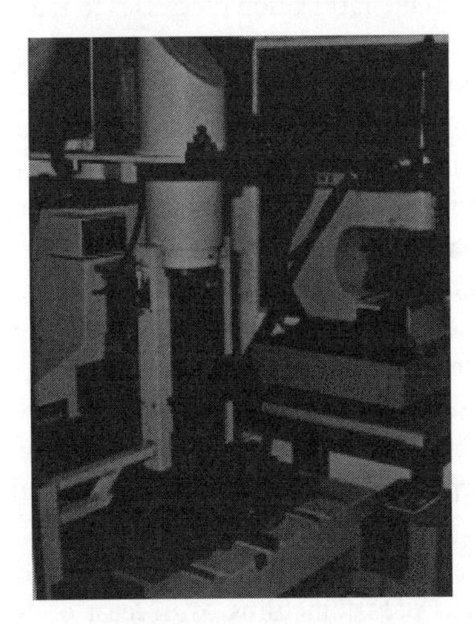

Figure 3.1.5 GCA Mann 3000 photomask pattern generator and GCA 3696 stepper camera.

There are a number of basic surface silicon micromachining techniques that can be used in order to deposit and pattern thin films (deposited on a silicon wafer) as well as to shape the silicon wafer itself, forming a set of basic microstructures.

The basic steps of the silicon micromachining are:

- Lithography;
- Deposition of thin films and materials (electroplating, chemical vapor deposition, plasma enhanced chemical vapor deposition, evaporation, sputtering, spaying, screen printing, etc.);
- Removal of material (patterning) by wet or dry techniques;
- Etching (plasma etching, reactive ion etching, laser etching, etc.);
- Doping;
- Bonding (fusion, anodic, and other);
- Planarization.

### *Etching*

Different microelectromechanical motion devices and microstructures can be designed, and silicon wafers with different crystal orientations can be used to fabricate MEMS. Reactive ion etching (dry etching process) is commonly applied. Ions are accelerated towards the material to be etched, and the etching reaction is enhanced in the direction of ion traveling. Deep trenches and pits of desired shapes can be etched in a variety of materials including silicon, silicon oxide, and silicon nitride. A combination of dry and wet etching can be sequentially integrated in the fabrication processes to make the desired MEMS.

Metal and alloy thin films can be patterned using the lift-off stenciling technique. As an example, let us document the following simple procedure. A thin film of the assisting material (silicon oxide) is deposited first, and a layer of photoresist is deposited over and patterned. The silicon oxide is then etched to undercut the photoresist. The metal (or any other material) thin film is then deposited on the silicon wafer through the evaporation process. The metal pattern is stenciled through the gaps in the photoresist, which is then removed, lifting off the unwanted metal. The sacrificial layer is then stripped off, leaving the desired metal film pattern.

The *isotropic* and *anisotropic* wet etching, as well as the concentration dependent etching, are used in bulk silicon micromachining because the microstructures are formed by etching away the bulk of silicon wafer.

Surface micromachining usually forms the structure in layers of thin films on the surface of the silicon wafer or other substrate. Hence, the surface micromachining process uses thin films of at least of two different materials, e.g., structural (usually polysilicon, metals, alloys, etc.) and sacrificial (silicon oxide) material layers. Sacrificial (silicon oxide is deposited on the wafer surface) and structural layers are deposited. Then, the sacrificial material is etched away to release the structure. A variety of different complex motion microstructures with different geometry have been fabricated using the surface micromachining technology [1, 2]. Different etching processes and materials are covered in detail in Section 3.2.

Different etching systems are widely used. As an example, the xenon difluoride (gas) etching system is illustrated in Figure 3.1.6.

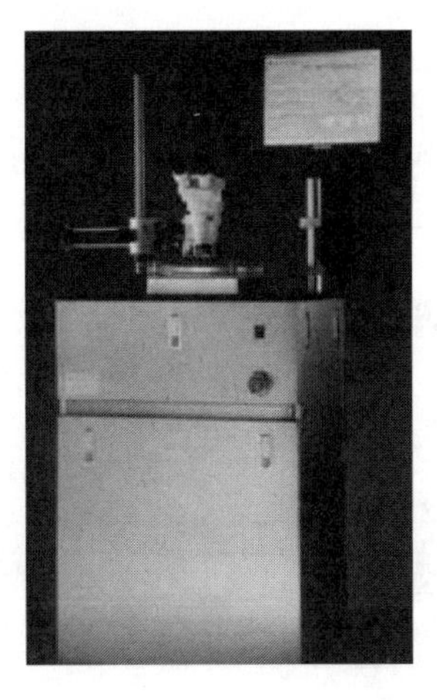

Figure 3.1.6 Xenon difluoride etching system.

### *Bonding*

Micromachined silicon wafers must be bonded together. The anodic (electrostatic) bonding technique is used to bond silicon wafer and glass substrate. The silicon wafer and glass substrate are attached, heated, and electric field is applied across the joint. This results in extremely strong bonds between the silicon wafer and glass substrate. In contrast, the direct silicon bonding is based upon applying pressure to bond silicon wafer and glass substrate. It must be emphasized that to guarantee strong bonds, the silicon wafer and glass substrate surfaces must be flat and clean.

The MEMCAD™ software (current version is 4.6), developed by Microcosm, is widely used to design, model, simulate, characterize, and package MEMS. Using the built-in Microcosm Catapult™ layout editor, augmented with a materials database and components library, three-dimensional solid models of motion microstructures can be developed. Furthermore, customizable packaging is fully supported.

### *Introduction to MEMS Fabrication and Web site Resources*

The basic fabrication techniques, processes, and materials are covered in this book. The major technologies (micromachining, which is based on the CMOS technology and LIGA) used to fabricate MEMS were emphasized. In particular, the surface micromachining is the well-developed affordable and high-yield technology which likely will be the dominant method. A simplified surface micromachining process is illustrated in Figure 3.1.7.

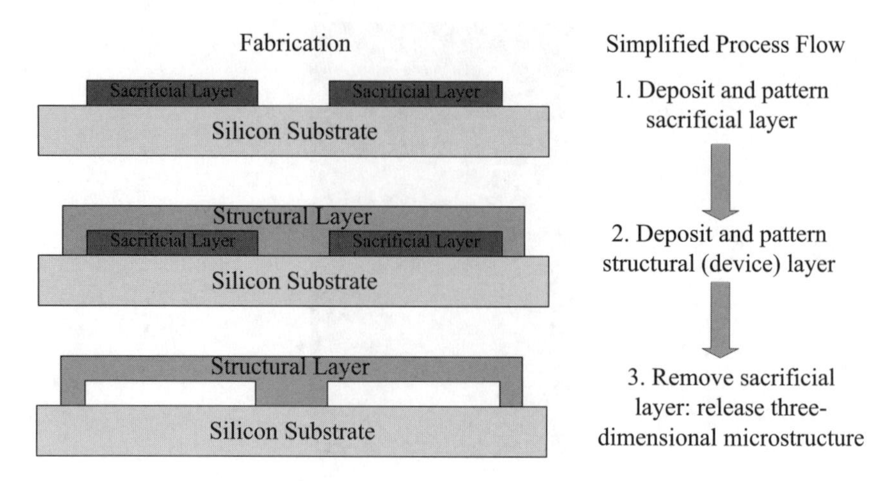


Figure 3.1.7 Simplified surface micromachining process.

Figure 3.1.7 demonstrates the major sequential steps of the simplified fabrication of the microstructure through surface micromachining. The description of processes, materials used, and techniques applied are covered in detail in this chapter and Chapter 8. The attractive feature of surface micromachining is the compatibility and compliance of three-dimensional microstructures and ICs fabrication. In fact, in addition to the microstructures and microdevices fabrication, these structures-devices must be controlled. Therefore the ICs are the important subsystems (components) of MEMS.

The following process hierarchy in fabrication of MEMS with the links to the manufacturer cites is reported on the MEMS Clearinghouse Web site (see http://mems.isi.edu/):

- Bonding
- Clean
- Lapping/polishing
- Deposition
- Etch
- Ion implant
- Miscellaneous
- Pattern transfer
- Back-end processing
- LIGA
- Mask-making
- Metrology
- Testing
- Thermal

Throughout this book, the author cannot assume responsibility for the validity and fitness of the techniques, processes, sequential steps, data, chemicals, materials, as well as for the consequences of their use. Furthermore, the microelectronic (semiconductor) and MEMS manufacturers have different proprietary techniques, processes, and materials. However, for introduction and educational purposes, the references to the MEMS Clearinghouse and university Web sites are quite appropriate. By clicking the corresponding process, the information appears on the screen, see Figure 3.1.8. In particular, using the Web site http://fab.mems-exchange.org/catalog/ and clicking on the anodic bonding, we have:

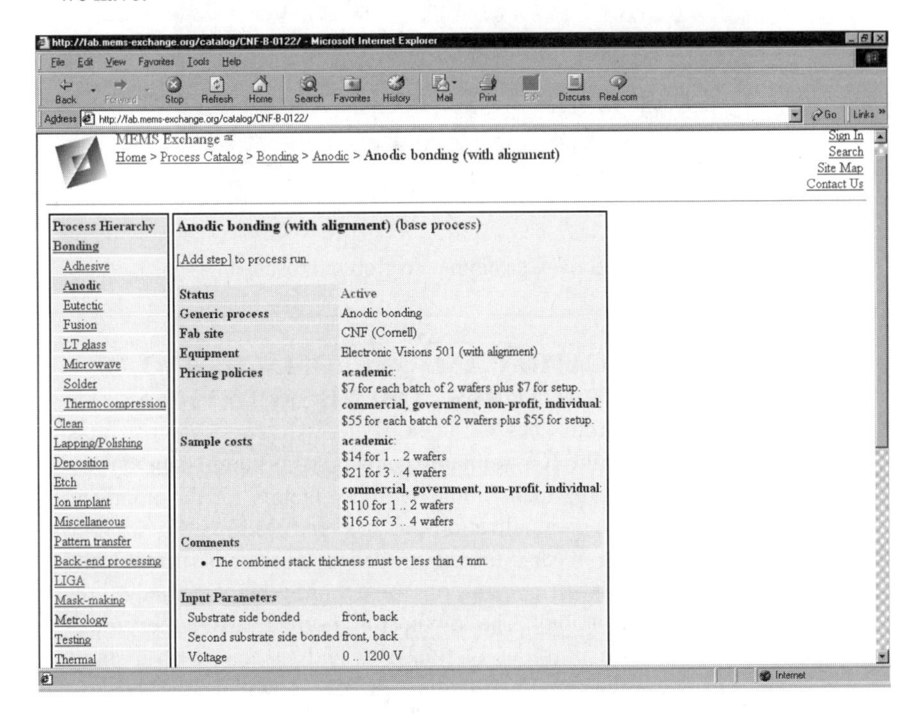


Figure 3.1.8 Anodic bonding description in the MEMS Clearinghouse Web site http://fab.mems-exchange.org/catalog/

Different computerized microscopes are used to attain the visualization, imaging and examining features. For example, scanning electron microscopes (e.g., Hitachi S-2400, 3000 and 3500 series) are widely used. The Hitachi S-2400 is the general purpose, 25 kV, diffusion-pumped, tungsten filament-based Scanning Electron Microscope. This Scanning Electron Microscope allows 300,000 magnification. After the sample loading, high voltage (from 4 to 25 kV) is applied. Focusing is performed manually or automatically, and images can be displayed and zoomed. The 4 nm resolution is achieved which can be suitable for ICs and microstructures with tens of nanometers features. For nanostructures, other microscopes are used. Figure 3.1.9 documents the Hitachi S-2400 Scanning Electron Microscope.

Figure 3.1.9 Hitachi S-2400 scanning electron microscope.

## 3.2 MICROFABRICATION AND MICROMACHINING OF ICs, MICROSTRUCTURES, AND MICRODEVICES

Complementary metal oxide semiconductor (CMOS), high-aspect-ratio, and surface micromachining technologies are key factors for development, implementation, and commercialization of ICs and MEMS [1, 2]. For MEMS, micromachining means fabrication of microscale structures and devices controlled by ICs. In general, micromachining has emerged as the extension of CMOS technology. The low-cost high-yield CMOS technology enables the fabrication of millions of transistor and capacitors on a single chip, and currently, the minimal feature sizing is within the tens of nanometer range. High-performance signal processing, signal conditioning, interfacing, and control are performed by ICs. Microstructures and microdevises, augmented with ICs, comprise MEMS which have been widely used in the actuator, sensor and communication applications.

Microfabrication technologies can be categorized as bulk, surface, and high-aspect-ratio (LIGA and LIGA-like) micromachining. The major processes in fabrication of ICs and microstructures/microdevices are: oxidation, diffusion, deposition, patterning, lithography, etching, metallization, planarization, assembling, and packaging.

Complementary metal oxide semiconductor is the well-developed technology for fabrication of ICs and thin films. Thin film fabrication processes were developed and used for polysilicon, silicon dioxide, silicon nitride, and other different materials, e.g., metals, alloys, insulators, etc. For ICs, these thin films are used to build the active and passive circuitry components, as well as attain interconnection. Doping modifies the properties of the media. It was documented that the lithography processes are used to transfer the pattern from the mask (which defines the surface topography and geometry for ICs and microstructures) to the film surface, which is then selectively etched away to remove unwanted thin films, media, and regions to complete the pattern transfer. The number of masks depends on the design complexity, fabrication technology, processes applied, and the desired ICs and microstructures' geometry. After testing, the wafers are diced, and chips are encapsulated (packaged) as final ICs, microdevices, or MEMS.

The description of the silicon-based fabrication technologies and processes is give below. Crystal growth and slicing are the processes that produce the silicon with the desired (specified) chemical, electromechanical, and thermal characteristics and properties (dimension, orientation, bow, taper, edge contour, surface flatness and scratches, minority carrier lifetime, doping type and concentration, heavy metal impurity content, electrical resistivity, Young's module, elasticity, Poison ratio, stress, thermal conductivity, thermal expansion, etc.). For silicon, the following data is useful:

- Atomic weight is 28,
- Density is 2.33 g/cm$^3$,
- Melting point is 1415$^0$C,
- Specific heat is 0.7 J/g-K,
- Thermal conductivity is 1.6 W/cm-K at 300 $^0$K,
- Coefficient of linear thermal expansion is 0.0000026,
- Intrinsic resistivity is 230,000 ohm-cm,
- Dc dielectric constant is 11.9.

To design and fabricate MEMS, the material compatibility issues must be addressed and analyzed. For example, the thermal conductivity and thermal expansion phenomena must be studied, and the appropriate materials with the closest possible coefficients of thermal conductivity and expansion must be chosen. However, the desired electromagnetic and mechanical properties (permeability, resistivity, strength, elasticity, etc.) must be attained also (see Chapter 8 for details). In fact, electromagnetic and mechanical characteristics significantly influence the MEMS performance. Therefore, the trade-offs must be examined with the ultimate goal to optimize the MEMS overall performance. Thus, it is evident that even simplest microstructures (silicon with deposited thin films) must be the topic of mechanical, electromagnetic and thermal analysis. Furthermore, though the designer tries to match the mechanical and thermal phenomena and effects, and attempts to use magnetic materials with good electromagnetic characteristics, the feasibility and affordability of the fabrication processes are also studies. The mechanical properties of different materials are documented in Table 3.2.1.

Table 3.2.1 Mechanical properties of materials.

| | Strength $10^9 \frac{N}{m^2}$ | Hardness kg/mm$^2$ | Young's Modulus $10^9$ Pa | Density g/cm$^3$ | Thermal Conductivity W/cm-K | Thermal Expansion $10^{-6}$ 1/K |
|---|---|---|---|---|---|---|
| Si | 7 | 850 | 190 | 2.3 | 1.6 | 2.6 |
| SiC | 21 | 2500 | 710 | 3.1 | 3.5 | 3.3 |
| Diamond | 54 | 6700 | 980 | 3.5 | 19 | 1 |
| $SiO_2$ | 8.3 | 800 | 70 | 2.5 | 0.015 | 0.54 |
| $Si_3N_4$ | 15 | 3500 | 380 | 3.1 | 0.2 | 0.8 |
| Fe | 13 | 420 | 200 | 7.8 | 0.8 | 12 |
| Al | 0.2 | 135 | 70 | 2.8 | 2.4 | 25 |
| Mo | 2 | 280 | 340 | 10 | 1.4 | 5.2 |
| W | 3.9 | 500 | 400 | 19 | 1.8 | 4.6 |

The mechanical properties of the following bulk and thin film materials, which are commonly used in ICs and MEMS fabrication, are given in the MEMS Clearinghouse Web site. In particular, one can consult the Material Database on *http://www.memsnet.org/material/*. The materials are:
Aluminum (Al), Amorphous Carbon (a-C:H), Amorphous Hydrogenated Silicon (a-Si:H), Amorphous Silicon (a-Si), Amorphous Silicon Dioxide (a-$SiO_2$), Antimony (Sb), Arsenic (As), Barium Titanate ($BaTiO_3$), Beryllium Oxide (BeO), Bismuth (Bi), Boron (B), Boron Carbide ($B_4C$), Boron Nitride (BN), Cadmium (Cd), Carbon (C), Carbon Nitride ($CN_x$), Chromium Boride ($CrB_2$), Chromium Carbide ($Cr_3C_2$), Chromium (Cr), Chromium Nitride, Chromium Oxide ($Cr_2O_3$), Cobalt (Co), Copper (Cu), Copper Molybdenum (CuMo), Diamond (C), Diamond-like Carbon (DLC), Fullerite ($C_{60}$ and $C_{70}$), Gallium Arsenate (GaAs), Gallium (Ga), Germanium (Ge), Glass (7059 and $SiO_2$), Gold (au), Graphite (C), Indium (In), Iron (Fe), Lead (Pb), Lead Zirconate Titanate (PZT), Lithium (Li), Magnesium (Mg), Molybdenum (mo), Molybdenum Silicide ($MoSi_2$), Mullite ($3Al_2O_32SiO_2$), Nickel (Ni), Niobium Oxide ($Nb_2O_5$), Nitride Coatings, Oxide Induced Layer in Polysilicon, Palladium (Pd), Phosphorous Bronze Metal, Phosphorous (P), Dupont Polymide (PI 2611D), Piezoelectric Sheet, Piezoresistors (diffused), Plastic, Platinum (Pt), Polyimide, Polyimide Hinges, Polysilicon, Poly Vinylidene Fluoride, Porosilicon, Sapphire, Silicon Carbide (SiC), Silicon Dioxide ($SiO_2$), Silicon Nitride Hydrogen ($SiN_xH_y$), Silicon Nitride ($Si_xH_y$), Silicon Oxide ($SiO_x$), Silicon (Si), Sillimanite ($Al_2O_3$ $SiO_2$), Silver (Ag), Sodium Silicate ($Na_2O$:$SiO_2$), Stainless Steel, Tellurium (Te), Thallium (Tl), Tin (Sn), Titanium Aluminum ($Ti_3Al$ and TiAl), Titanium Boride ($TiB_2$), Titanium Carbide (TiC), Titanium Nickel (NiTi), Titanium Nitride (TiN and $TiN_x$), Titanium Oxide ($TiO_2$), Titanium (Ti), Tungsten Carbide (WC), Tungsten Silicide ($WSi_2$), Tungsten (W), Zinc Oxide (ZnO), Zinc (Zn), Zirconium Oxide ($ZrO_2$), and Zircon ($SiO_2$ $ZrO_2$).

Emphasizing the mechanical and thermal matching (compliance) and the need for optimization of electromagnetic properties (covered in Chapter 8), we describe the major fabrication processes.

## *Oxidation*

Oxidation of silicon wafers is used for passivation of the silicon surface (the formation of a chemically, electronically, and electromechanically stable surface), diffusion, ion implantation, making dielectric films, and interfacing substrate and other materials (for example, chemical material and biosensors).

Silicon, exposed to the air at $25^0$C, is covered by the 20 Å (2 nm) layer of the silicon oxide. Thicker silicon oxide ($SiO_2$) layers can be grown at elevated temperatures in dry or wet oxygen environments. The wet and dry reactions are

$Si + 2H_2O \rightarrow SiO_2 + H_2$

and $Si + O_2 \rightarrow SiO_2$.

The oxide growth rate is expressed by the following formula

$$\frac{dx}{dt} = \frac{1}{N} k_o c_o ,$$

where $N$ is the number of molecules of oxidant per unit volume of oxide; $k_o$ and $c_o$ are the oxidation rate constant and concentration of oxidant, which are nonlinear functions of the oxide thickness, temperature, oxidant used, crystal orientation, diffusivity, pressure, etc.

If the temperature is constant, the relationship between the thickness of oxide and time is parabolic. The rate of growth is a nonlinear function of the oxygen pressure and the crystal orientation.

During the oxidation process, the dopant concentration profile in the Si-$SiO_2$ structure is redistributed due to nonuniformity of the equilibrium concentration of the impurity in silicon and silicon dioxide, and the impurity segregation coefficient (the ratio of the equilibrium concentration of the impurity in Si to that in $SiO_2$) is used. For boron doped silicon, the boron in Si near the interface is depleted, while for phosphorous, there is a buildup at the interface. High concentrations of the standard dopant atoms will increase the oxidation rate within the certain temperature range [1-4].

## *Photolithography*

Photolithography (lithography) is the process used to transfer the mask pattern (desired pattern, surface topography and geometry) to a layer of radiation- or light-sensitive material (photoresist) which is used to transfer the pattern to the films or substrates through etching processes.

Ultraviolet and radiation (optical, x-ray, electron beam, and ion beam) lithography processes are used in fabrication of ICs, motion and radiating energy microstructures and microdevices. The major steps in lithography are the fabrication of masks (pattern/topography generation) and transfer of the pattern to the wafer, see Figure 3.1.1.

The description of photolithography was documented in Section 3.1. Positive and negative photoresists are applied. The pattern on the positive photoresist after development is the same as that on the mask, while on negative photoresist it is reversed. Important photoresist characteristics are resolution, sensitivity, etch resistance, thermal stability, adhesion, viscosity, flash point, toxicity rate, etc. The photoresist processing includes dehydration baking and priming, coating, soft baking, exposure, development, inspection, post bake (UV hardening), etc. Then, the specified pattern is transferred to the wafer through etching, after which, the photoresist is stripped by strong acid solutions ($H_2SO_4$), acid-oxidant solutions ($H_2SO_4$+$Cr_2O_3$), organic solvents, alkaline strippers, oxygen plasma, or gaseous chemical reactants. Using wet and dry stripping, the photoresist must be removed without damaging silicon structures. The photolithographic cycle is completed.

In surface micromachining, an alternative solution to etching (in order to transfer the patterns from photoresist to thin films) is the lift-off process, which is an additive process. In lift-off, the photoresist is first patterned, the thin film to be patterned is deposited, and then the photoresist is dissolved. The photoresist acts as a sacrificial material under thin film regions to be removed.

### *Etching*

Etching is used to delineate patterns, remove surface damage, clean the surface, remove contaminations, and fabricate two- (planar) and three-dimensional microstructures. Wet chemical etching and dry etching (sputtering, ion beam milling, reactive ion etching, plasma etching, etc.) are used to etch semiconductors, conductors (metals), alloys, and insulators (silicon oxide, silicon nitride, etc.). Different etchants are used in micromachining and microelectronics.

Wet and dry etchants, commonly used in micromachining and ICs, are listed in Tables 3.2.2 and 3.2.3 [3, 4].

Table 3.2.2 Wet etchants.

| Material | Etchant and Etch Rate |
|---|---|
| Polysilicon | 6 ml HF, 100 ml $HNO_3$, 40 ml $H_2O$, 8000 Å/min, smooth edges<br>1 ml HF, 26 ml $HNO_3$, 33 ml $CH_3COOH$, 1500 Å/min |
| Phosphorous-doped silicon dioxide (PSG) | Buffered hydrofluoric acid (BHF)<br>28 ml HF, 170 ml $H_2O$, and 113 g $NH_4F$, 5000 Å/min<br>1 ml BHF and 7 ml $H_2O$, 800 Å/min |
| Silicon nitride ($Si_3N_4$) | Hydrofluoric acid (HF)<br>140 Å/min CVD at 1100°C<br>750 Å/min CVD at 900°C<br>1000 Å/min, CVD at 800°C |
| Silicon dioxide ($SiO_2$) | Buffered hydrofluoric acid (BHF)<br>28 ml HF, 170 ml $H_2O$, and 113 g $NH_4F$, 1000-2500 Å/min<br>1 ml BHF and 7 ml $H_2O$, 700-900 Å/min |
| Aluminum (Al) | 4 ml $H_3PO_4$, 1 ml $HNO_3$, 4 ml $CH_3COOH$, 1 ml $H_2O$, 350 Å/min<br>16-19 ml $H_3PO_4$, 1 ml $HNO_3$, 0-4 ml $H_2O$, 1500-2400 Å/min |
| Gold (Au) | 3 ml HCl, 1 ml $HNO_3$, 25-50 µm/min.<br>4 g KI, 1 g I2, 40 ml $H_2O$, 0.5-1 µm/min |
| Chromium (Cr) | 1 ml HCl, 1 ml glycerine, 800 Å/min, (need depassivation)<br>1 ml HCl, 9 ml saturated $CeSO_4$ solution, 800 Å/min (need depassivation)<br>1 ml (1 g NaOH in 2 ml $H_2O$), 3 ml (1 g $K_3Fe(CN)_6$ in 3 ml $H_2O$), 250-100 Å/min (photoresist mask) |
| Tungsten (W) | 34 g $KH_2PO_4$, 13.4 g KOH, 33 g $K_3Fe(CN)_6$, and $H_2O$ to make 1 liter, 1600 Å/min (photoresist mask) |

Table 3.2.3 Dry etchants.

| Material | Etchant (Gas) and Etch Rate |
|---|---|
| Silicon dioxide ($SiO_2$)<br>Phosphorous-doped silicon dioxide (PSG) | $CF_4 + H_2$, $C_2F_6$, $C_3F_8$, or $CHF_3$, 500-800 Å/min |
| Silicon<br>(single-crystal and polycrystalline) | $SF_6 + Cl_2$, 1000-5000 Å/min<br>$CF_4$, $CF_4O_2$, $CF_3Cl$, $SF_6Cl$, $Cl_2$+$H_2$, $C_2ClF_5O_2$, $SF_6O_2$, $SiF_4O_2$, $NF_3$, $C_2Cl_3F_5$, or $CCl_4He$ |
| Silicon nitride<br>($Si_3N_4$) | $CF_4O_2$, $CF_4$+$H_2$, $C_2F_6$, or $C_3F_8$, $SF_6He$ |
| Polysilicon | $Cl_2$, 500-900 Å/min |
| Aluminum<br>(Al) | $BCl_3$, $CCl_4$, $SiCl_4$, $BCl_3Cl_2$, $CCl_4Cl_2$, or $SiCl_4Cl_2$ |
| Gold (Au) | $C_2Cl_2F_4$ or $Cl_2$ |
| Tungsten (W) | $CF_4$, $CF_4O_2$, $C_2F_6$, or $SF_6$ |
| Al, Al-Si, Al-Cu | $BCl_3 + Cl_2$, 500 Å/min |

A large number of dry etching processes such as physical etching (sputtering and ion milling), chemical plasma etching, and the combinations of physical and chemical etching (reactive ion etching and reactive ion beam etching) are used, and recipes are available. Dry etching processes are based on the plasmas. Plasmas are fully or partially ionized gas molecules and neutral atoms/molecules sustained by the applied electromagnetic field. Usually, less than 0.1% of the gas is ionized, and the concentration of electrons is much lower than the concentration of gas molecules. The electron temperature is greater than $10000^0$K, although the gas thermal temperature is within the range from 50 to $100^0$C. Plasma etching processes involve highly reactive particles in a relatively cold medium. Adjusting process parameters controls the particles energies and gas temperature. The gas and flow rate, excitation power, frequency, reactor configuration, and pumping determine the electron density and distribution, gas density, and residence time defining the reactivity. These and wafer parameters (temperature and surface potential) define the surface interaction and etching characteristics [1-10].

The most important properties of dry etching processes are feature size control, wall profile, selectivity (the ratio of etch rates of the layer/material to be etched and to the layer/material to be kept), controllability, in-wafer and interwafer uniformities, defects, impurity, throughput, radiation damage to dielectrics, etc.

## *Doping*

Doping processes are used to selectively dope the substrate to produce either n- or p-type regions. These doped regions are used to fabricate passive and active circuitry components, form etch-stop-layers (very important feature in buck and surface micromachining), as well as produce conductive silicon-based micromechanical devices. Diffusion is achieved by placing wafers in a high-temperature furnace and passing a carrier gas that contains the desired dopant. For silicon, boron is the most common p-type dopant (acceptors), and arsenic and phosphorous are n-type dopants (donors). The dopant sources may be solid, liquid, and gaseous. Nitrogen is usually used as the carrier gas. Two major steps in diffusion are predeposition (impurity atoms are transported from the source to the wafer surface and diffused into the wafer, the number of atoms that enter the wafer surface is limited by the solid solubility of the dopant in the wafer), and drive-in (deposited wafer is heated in a diffusion furnace with an oxidizing or inert gas to redistribute the dopant in the wafer to reach a desired doping depth and uniformity). After deposition the wafer has a thin highly-doped oxide layer on the silicon, and this oxide layer is removed by a hydrofluoric acid.

The diffusion theory is based on Fick's first and second laws. Assuming constant diffusion coefficients at the process temperature, we have

$$J = -D\frac{\partial C(t,x)}{\partial x}$$

and $\dfrac{\partial C(t,z)}{\partial t} = D\dfrac{\partial^2 C(t,z)}{\partial z^2} = D\nabla^2 C$ ,

where $J$ is the net flux; $D$ is the diffusion coefficient; $C$ is the impurity concentration; $z$ is the depth.

For predeposition diffusion, the surface concentration is constant, and the diffusion profile is described as a complementary error function. Using the following boundary conditions

$C(0,z) = 0$, $C(t,0) = C_0$ and $C(t,\infty) = 0$,

we have the expression for the impurity concentration

$$C(t,z) = C_0 erfc^{\frac{z}{2\sqrt{Dt}}}, t > 0.$$

For drive-in, the total dopant in the wafer is constant, and the diffusion profile obeys a Gaussian function. These descriptions are valid for low doping concentrations.

For high doping concentration, the diffusion coefficient is a nonlinear function of concentration and other parameters. Using Ohm's law, we have

$$J = -D\frac{\partial C}{\partial z} + \mu CE,\; E = -\eta\frac{kT}{qC}\frac{dC}{dz},$$

where $\mu$ is the mobility, $\mu = \frac{q}{kT}$; $E$ is the electric field; $\eta$ is the screening factor which varies from 0 to 1.

Thus, one has $J = -D(1+\eta)\frac{dC}{dz}$.

Complex mathematical models, given in form of nonlinear partial differential equations, were derived to describe the diffusion profiles for boron, arsenic, phosphorous and antimony at high concentrations. Silicon dioxide is usually used as the diffusion mask for silicon wafers. Ion implantation, as a well-developed technique for introducing impurity atoms into a wafer below the surface by bombarding it with a beam of energetic impurity ions, is widely used.

### *Metallization*

Metallization is the formation of metal films for interconnections, ohmic contacts, rectifying metal-semiconductor contacts, and protection (e.g., attenuation of electromagnetic interference and radiation, etc.). Metal thin films can be deposited on the surface by vacuum evaporation (deposition of single element conductors, resistors, and dielectrics), sputtering, chemical vapor deposition, plating, and electroplating. The electroplating will be covered in Chapter 8. Sputtering is the deposition of compound materials and refractive metals by removal of the surface atoms or molecular fragments from a solid cathode (target) by bombarding it with positive ions from an inert gas (argon), and these atoms or molecular fragments are deposited on the substrate to form a thin film.

### *Deposition*

Atmospheric pressure and low pressure chemical vapor deposition (APCVD and LPCVD) are used to deposit metals, alloys, dielectrics, silicon, polysilicon and other semiconductor, conductor and insulator materials and compounds. The chemical reactants for the desired thin film are introduced into the CVD chamber in the vapor phase. The reactant gases then pyrochemically react at the heated surface of the wafer to form the desired thin film. Epitaxial growth, as the CVD process, allows one to grow a single crystalline layer upon a single crystalline substrate. Homoepitaxy is the growth of the same type of material on the substrate (e.g., p+ silicon etch-stop-layer on an n-type substrate for a layer formation). Heteroepitaxy is the growth of one materials on a substrate which is a different type of medium. Silicon homoepitaxy is used in bulk micromachining to form the etch-stop-layers. Plasma enhanced chemical vapor deposition (PECVD) uses a RF induced plasma to provide additional energy to the reaction. The major advantage of PECVD is that it allows one to deposit thin films at lower temperatures compared with conventional CVD.

## *MEMS Assembling and Packaging*

Assembling and packaging of MEMS includes microstructure and die inspection, separation, attachment, wire bonding, and packaging or encapsulation. For robust packaging, one must match the thermal expansion coefficients for the microstructures to minimize mechanical stresses. The connections and the package must provide actuation and sensing capabilities, high-quality sealing, robustness, protection, noise immunity, high fidelity path for signals to and from the chip, reliable input-output interconnection, etc. Application specific considerations include operation in harsh adversarial environments (contaminates, electromagnetic interference, humidity, noise, radiation, shocks, temperature, vibration, etc.), hybrid multichip packages, direct exposure of portions of the MEMS to outside stimuli (i.e., light, gas, pressure, vibration, temperature, and radiation), etc.

Bonding, assembling, and packaging are processes of great importance. In MEMS, different bonding techniques are used to assemble individually micromachined (fabricated) structures to form microdevices or complex microstructures, as well as integrate microstructures/microdevices with controlling ICs bonding an entire wafer or individual dies. Silicon direct bonding is used to bond a pair of silicon wafers together directly (face to face), while anodic bonding is used to bond silicon to glass [1]. In silicon direct bonding, the polished sides of two silicon wafers are connected face to face, and the wafer pair is annealed at the high temperature. During annealing, the bonds are formed between the wafers (these bonds can be strong as bulk silicon). The process is carried out through the following steps: (1) wafers cleaning in a strong oxidizing solution (organic clean, HF dip, and ionic clean) which results in the hydrophilic wafer surface; (2) wafers rinsing in water and drying; (3) wafers squeezing (face to face) – wafers stick together due to hydrogen bonding of hydroxyl groups and van der Waals forces on the surface of the wafers. The bonded pairs are then annealed at high temperature to guarantee strong bonds in an inert environment (nitrogen) for approximately one hour.

The silicon direct bonding technique is a simple process which does not require special equipment other than a cleaning station and an oxidation furnace. The requirements imposed on wafers are smoothness, flatness, and minimal possible level of contaminants. The quality of the wafer bonding is determined by inspection and testing using infrared illumination. Silicon, silicon dioxide, and silicon nitride coated surfaces can be used for bonding. By contacting in a controlled ambient, microdevices can be sealed in a gas or vacuum environment.

Anodic (electrostatic) bonding is used to bond silicon to glass. The glass can be in the form of a plate or wafer, or as a thin film between two silicon wafers. Anodic bonding is performed at lower temperatures ($450^0$C or less), bonding metallized microdevices. In anodic bonding, the silicon wafer is placed on a heated plate, the glass plate is placed on top of the silicon wafer, and a high negative voltage is applied to the glass. As the glass is heated, positive sodium ions become mobile and drift towards the negative electrode. A depletion region is formed in the glass at the silicon interface, resulting in a high electric field at the silicon-glass interface. This field forces the silicon and glass into intimate contact and bonds oxygen atoms from the glass with silicon in the silicon wafer leading to permanent hermetically sealed bonds. Anodic bonding of two silicon wafers can be formed by coating two wafer surfaces using sputtered glasses. Anodic bonding requires smooth bonding surfaces. This requirement is not so critical compared with the fusion bonding process because the high electrostatic forces pull small gaps into contact. It is important that glass must be selected with the thermal expansion coefficient that matches silicon. The difference in the thermal expansion coefficients of the glass and silicon will result in stress between the bonded pair after cooling to room temperature. Corning 1729 and Pyrex 7740 are widely used.

## 3.3 MEMS FABRICATION TECHNOLOGIES

Microelectromechanics integrates fundamental theories (electromagnetics, micromechanics and microelectronics), engineering practice, and fabrication technologies. Using fundamental research, MEMS can be devised, designed, and optimized. In addition to theoretical fundamentals, affordable (low-cost) high-yield fabrication technologies are required in order to make three-dimensional microscale structures, devices, and MEMS. Micromachining and high-aspect-ratio are key fabrication technologies for MEMS. Microelectromechanical systems fabrication technologies fall into three broad categories. In particular: bulk micromachining, surface micromachining and LIGA (LIGA-like) technologies.

Different fabrication techniques and processes are available. In general, the fabrication processes, techniques and materials depend upon the available facilities and equipment. High-yield proprietary integrated CMOS-based MEMS-oriented industrial fabrication technologies are developed by the leading MEMS manufacturers. It is the author's goal for this book to be used for the MEMS, micro- and nanoengineering courses which may integrate laboratories and experiments. However, due to different facilities, equipment, infrastructure, distinct course structures and different number of credit-hours allocated, it is difficult to focus and provide detailed comprehensive coverage of all possible MEMS fabrication processes including packaging, calibration, and testing. There exist an infinite number of different developments, experiments, demonstrations, and laboratories in MEMS, microdevices, microstructures and microelectronics. Excellent Web sites (Case Western University, Georgia Institute of Technology, Massachusetts Institute of Technology, Stanford University, Universities of

California at Berkeley, Los Angeles and Santa Barbara, University of Wisconsin Madison, etc.) support the possible fabrication technologies, equipment, facilities, and MEMS-NEMS infrastructure developments.

The general information, specific fabrication technologies, techniques and processes are available at http://argon.eecs.berkeley.edu:8080/ (Microlab Home Page of the University of California at Berkeley). This information can be found to be very useful and applicable. In particular, the following topics are covered:

- Equipment wish list,
- Process and equipment operating manual,
- Equipment qualification,
- Process monitoring tests,
- Sensor data plots,
- CMOS baseline information,
- Mask-making request form.

Clicking the Process and Equipment Operating Manual (available at the Web site http://argon.eecs.berkeley.edu:8080/text/labmanual.html), one receives very meaningful and detailed information. Specifically, the following topics (by chapters) are available:

*Chapter 1 General*

1.1 Table of Contents
1.2 Microlab Safety Rules and Procedures
1.3 Process Modules
1.4 Baseline CMOS Process
1.5 VLSI Etchants
1.6 Miscellaneous Etchants
1.7 Miscellaneous Plating Solutions
1.8 Chemical Safety Data
1.9 Substrate Specifications
1.10 Wafer Identification
1.13 Standard CMOS In-Line Test Measurements
1.14 Standard Equipment and Process Monitoring
1.15 Microlab Computers
1.16 Preventive Maintenance by Process Staff
1.17 Microlab Orientation Seminar
1.18 Microlab Members' Laboratory Guide
1.19 Microlab Computers Operational Procedures

*Chapter 2 Cleaning Procedures*

2.1 General Cleaning Procedures
2.2 Tylan and Lam Wafer Preparation and Rework
2.3 Sink 3 Operation
2.4 Sinks 1, 2, 4, 5 Operation
2.5 Old Lab Sinks

2.6 Sinks 6, 7, 8 and Spin Dryer Operation
2.7 Tousimis 815 Critical-Point Dryer (small samples)
2.8 Tousimis 915B Critical-Point Dryer
2.9 Sink 9 and Spin Dryer Operation
*Chapter 3 Mask Making*
3.1 Mask Generation Using CAD Software
Appendix B: L-Edit Mask Making Manual
3.2 GCA Wafer Stepper Alignment Key Design Guide
3.3 GCA 3600 Pattern Generator
3.4 APT Emulsion Mask Procedure
3.5 APT Chrome Mask Procedure
3.6 Iron Oxide Masks
3.7 Ultratek Mask Copier
3.8 CMOS Baseline Test Structures
3.9 New CMOS Baseline Test Structures
*Chapter 4 Photolithography*
4.1 SVG 8800 Coat and Develop Tracks
4.2 General Resist Parameters
4.3 OCG 825 Reversal Image Process (MOD 27 in Chapter 1.3)
4.4 HMDS
4.5 Microposit Lift Off Procedure
4.6 Polyimide Technology (MOD 28 in Chapter 1.3)
4.7 Headway Photoresist Spinner
4.8 Quintel Mask Aligner
4.9 Canon Mask Aligner
4.10 GCA Wafer Steppers
4.11 Pattern Transfer
4.12 ASML DUV Stepper Model 5500/90
4.13 GCAWS Focus/Exposure Tests
4.14 GCA Baseline Test Procedures
4.15 SVG Photoresist Coat Track
4.16 SVG Photoresist Development Track
4.17 Karl Suss MA6 Mask Aligner
4.18 Karl Suss BA6 Bond Aligner
4.19 UVBAKE - Fusion M150PC Photostabilizer System
*Chapter 5 Furnaces*
5.1 Tylan Atmospheric Furnaces
5.2 Tylan LPCVD Furnaces
5.3 Contamination Monitoring of MOS-Clean Furnaces
5.4 Heatpulse/Rapid Thermal Annealing
5.5 Tylan15 Boron+ Process
5.6 Tylan8 POCl3 Process
5.7 Tystar9 LPCVD Furnace (6")
5.8 Tystar10 LPCVD Furnace (6")
5.9 Tystar11 MOS LTO LPCVD Furnace

5.10 Tystar12 Non-MOS LTO LPCVD Furnace
5.11 Tystar19 Si-Ge LPCVD Furnaces (MOS)
5.12 Tylan18 and Tylan20 Furnaces
5.13 Tylan17 MOS Clean Oxidation/Annealing Furnace

*Chapter 6 Thin Film Systems*

6.1 Hummer (for SEM samples ONLY)
6.2 Ion Beam Milling with the Veeco Microetch System
6.3 NRC System
6.4 Randex System
6.5 Veeco 401
6.6 CPA Sputtering System
6.7 Gartek
6.8 Davis and Wilder Evaporation System
6.9 Exciton Cluster Tool
6.10 ULTEK E-Beam Evaporator
6.11 Electron-Beam Evaporator System (imetal)
6.12 Bigblue Sputtering System
6.13 MRC 8600
6.14 Specialty Coating Systems Parylene
6.15 MBE2 Manual
6.16 Pulsed Laser Deposition (lextra)

*Chapter 7 Etching Systems*

7.1 Lam1 Silicon Nitride Etcher
7.2 Lam2 Oxide Autoetcher
7.3 Lam3 Alluminum RIE Etcher
7.4 Lam4 Poly-Si Rainbow Etcher
7.5 Lam5 Poly-Si Rainbow TCP Etcher
7.6 Lam Etchers Overview
7.7 Semigroup Reactive Ion Etching System
7.8 Technics C
7.9 Matrix 106 Resist Removal System (Asher)
7.10 Plasma-Therm Parallel Plate Etcher
7.11 Oxford RIE System
7.12 STS Poly-Si ICP Etcher
7.13 Technics AandB
7.14 Xenon Difluoride Etching System
7.15 Matrix 106 Resist Removal System

*Chapter 8 Testing/Inspection Equipment*

8.1 Tencor Alphastep 200 Profiler
8.2 Nanospec Film Thickness Measurement Model 3000 (New Upgrade)
8.3 Four Point Probe Resistivity Measurement
8.4 Sonogage Resistivity Measurement
8.5 Vickers Micro-Measurement

8.6 Leo SEM
8.7 Ellipsometer (ellips)
8.8 Ellipsometer (sopra)
8.9 35mm Photomicrograph System
8.10 Electronic Balances
8.11 I-V Probe Station
8.12 Reichert Inspection Microscope
8.13 Memscope
8.14 JEOL Scanning Electron Microscope and Nanometer Pattern Generation System
8.15 Electroglas Autoprobe in DCL
8.16 Nanospec DUV Microspectrophotometer
8.17 Siemens D5000 X-Ray Diffractometer
8.18 Flexus Operation
8.19 New 4100 Auger Electron Spectrometer Operating Guide
8.20 4100 Auger Electron Spectrometer Operating Guide
8.21 Contact Angle Measurement System (Kruss)
8.22 Computer-Micro-Vision-System
8.23 UVScope

*Chapter 9 Packaging*

9.1 Disco Automatic Dicing Saw
9.2 Westbond Ultrasonic Bonder Model 7400B
9.3 Research Devices M8-B Flip Chip Aligner Bonder
9.4 Karl Suss SB6 Bonder

*Chapter 10 Planarization*

10.1 Chemical-Mechanical Polisher (CMP)
10.2 SSEC CMP Wafer Cleaner (CMPWC)

The Web site http://www-snf.stanford.edu/ (Stanford University) provides processes information for specific equipment. In addition, general fabrication processes and techniques are covered with useful links. In particular,

*E-Beam, SEM and Maskmaking*

- Overview of maskmaking at Stanford Nanofabrication Facility (SNF)
- Micronics laserwriter
- Laser writer request form for maskmaking
- Contract maskmaking
- Making transparency masks
- Electron beam photoresists
- Nikon reticle alignment marks
- Contact mask alignment marks
- Karl Suss vacuum chuck drawing (for backside maskmaking)

*Optical Photolithography*

- Overview of photolithography processes
- List of available photoresists

- AZ3612 photoresist
- SPR220 photoresist
- Polyimide photoresist
- Electron beam photoresists
- Lift-off processing
- Photoresist for implant processing
- Wafer backside protection w/photoresist
- Edge bead removal mask processing
- Photoresist exposure time table
- Manual coating procedures
- Manual developing procedures
- Image reversal processing
- Processing clear (glass, quartz, sapphire) substrates
- Maskmaking

  *Chemical Vapor Deposition*
- Tylan recipe index metallization and sputtering
- Comparison of sputter systems dry etching
- Overview
- Nitride plasma etch
- Oxide plasma etch
- Polysilicon plasma etch

  *Annealing, Oxidation and Doping*
- Tylan recipe index
- Oxide thickness calculator wet processing
- General chemical handling policies
- Overview of wet benches

  *Wet Oxide Etch*
- Non-metal wafers (wbnonmetal)
- Wafers with standard metals (wbmetal)
- Non-standard materials (wbgeneral)
- Typical etch rates of non-metal films in standard etchants
- Use of Triton X-100 (wbnonmetal, wbmetal and wbgeneral only

  *Wet Bench Clean Procedures*
- Photoresist strip procedures
- Non-standard metal wafer clean (wbgeneral)
- Standard prediffusion clean overview (procedures at specific wet benches: wbdiff, wbgeneral, wbsilicide)
- Standard metal clean overview (wet bench-specific procedures: wbmetal, wbgeneral)
- Standard pre-LPCVD or pre-metal clean overview (wet-bench-specific procedures: wbdiff, wbgeneral, wbsilicide)
- Tweezer clean procedures

*Other Non-Standard Wet Bench Processes (wbgeneral only)*

- KOH etching of silicon
- Decontamination following KOH etch
- Non-standard nitride strip
- TMAH etching of silicon

*In-Line Process Characterization Tools*

- Summary of tool capabilities

*Links to Process Characterization Studies*

*Links to Other Process Resources*

It is evident that a great deal of additional important information regarding the basic fabrication techniques, processes, and materials can be obtained from the above mentioned as well as other Web sites. The Web sites listed are given as the possible sources of additional data from educational standpoints because the high-technology industry has developed proprietary technique and processes to guarantee affordability and high-yield. The fitness, applicability and reliability depend upon the equipment and infrastructure, as well as MEMS needed to be made.

### 3.3.1 Bulk Micromachining

Bulk and surface micromachining are based on the modified CMOS technology with specifically designed micromachining processes. Bulk micromachining was developed more than thirty years ago to fabricate three-dimensional microstructures [11]. Bulk micromachining of silicon uses wet and dry etching techniques in conjunction with etch masks and etch-stop-layers to develop microstructures from silicon substrates. Bulk micromachining of silicon is affordable, high-yield and well-developed technology. Microstructures are fabricated by etching areas of the silicon substrates releasing the desired three-dimensional microstructures.

The *anisotropic* and *isotropic* wet etching processes, as well as concentration dependent etching techniques, are widely used in bulk micromachining. The microstructures are formed by etching away the bulk of the silicon wafer. Bulk machining with crystallographic and dopant-dependent etch processes, when combined with wafer-to-wafer bonding, produces complex three-dimensional microstructures with the desired geometry. One fabricates microstructures by etching deeply into the silicon wafer. There are several ways to etch silicon wafers.

The *anisotropic* etching uses etchants (usually potassium hydroxide KOH, sodium hydroxide NaOH, $H_2N_4$ and ethylene-diamine-pyrocatecol EDP) that etch different crystallographic directions at different etch rates. Certain crystallographic planes (stop-planes) etch very slowly. Through *anisotropic* etching, three-dimensional structures (cons, pyramids, cubes and channels into the surface of the silicon wafer) are fabricated. In contrast, the *isotropic* etching etches all directions in the silicon wafer at same (or close) etch rate. Therefore, hemisphere and cylinder structures can be made. Deep reactive ion etching uses plasma to etch straight walled structures (cubes, rectangular, triangular, etc.).

In bulk micromachining, wet and dry etching processes are widely used.

Wet etching is the process of removing material by immersing the wafer in a liquid bath of the chemical etchant. Wet etchants are categorized as *isotropic* etchants (attack the material being etched at the same rate in all directions) and *anisotropic* etchants (attack the material or silicon wafer at different rates in different directions, and therefore, shapes/geometry can be precisely controlled). In other words, the *isotropic* etching has a uniform etch rate at all orientations, while for *anisotropic* etching, the etch rate depends on crystal orientation. Some etchants attack silicon at different rates depending on the concentration of the impurities in the silicon (concentration dependent etching). *Isotropic* etchants are available for silicon, silicon oxide, silicon nitride, polysilicon, gold, aluminum, and other commonly used materials. Since *isotropic* etchants attack the material at the same rate in all directions, they remove material horizontally under the etch mask (undercutting) at the same rate as they etch through the material. The hydrofluoric acid etches the silicon oxide faster than the silicon. *Anisotropic* etchants, which etch different crystal planes at different rates, are widely used, and the most popular *anisotropic* etchant is potassium hydroxide (KOH) because it is the safest one to use. The application of the concentration dependent etching can be illustrated as explained below.

High levels of boron (p-type dopant) will reduce the rate at which the doped silicon is etched in the KOH system by several orders of magnitude, stopping the etching of the boron rich silicon (as described above, the boron impurities are doped into the silicon by diffusion). Let us illustrate the technique. A thick silicon oxide mask is formed over the silicon wafer and patterned to expose the surface of the silicon wafer where the boron is to be doped. The silicon wafer is then placed in a furnace in contact with a boron diffusion source. Over a period of time boron atoms migrate into the silicon wafer. As the boron diffusion is completed, the oxide mask is stripped off. A second mask can be deposited and patterned before the wafer is immersed in the KOH system that etches the silicon that is not protected by the mask etching around the boron doped silicon.

The available *anisotropic* etchants of silicon (ethylene-diamine-pyrocatecol, potassium hydroxide, and hydrazine) etch single-crystal silicon along the crystal planes. From CMOS technology, etch masks and etch-stop techniques are available which can be used in conjunction with silicon *anisotropic* etchants to selectively prevent regions of silicon from being etched. Therefore, microstructures are fabricated on a silicon substrate by combining etch masks and etch-stop patterns with different *anisotropic* etchants.

Wet etching of silicon is used for shaping and polishing, as well as for characterizing structural and compositional features. The fundamental etch reactions are electrochemical. Oxidation – reduction is followed by dissolution of the oxidation products. The etching process is either reaction-rate limited (etching process depends on the chemical reaction rate) or diffusion-limited (etching process depends on the transport of etchant by diffusion to or from the surface through the liquid). Diffusion controlled processes have lower activation energies than reaction-rate controlled processes. Therefore diffusion-controlled processes are robust (insensitive) to temperature variations. However, diffusion-controlled processes are affected by agitation which increases the supply of reactant material to the semiconductor surface, increasing the etch rate. Changes the etching conditions and parameters (temperature, etchant components, their molarity, and proportions) change the rate-limiting process. The supply of minority carriers to the semiconductor surface limits the dissolution rate in etching reactions that result in a depletion of electrons or holes. Creation of electron-hole pairs on the surface (by illumination or by application of electric currents) or providing generation sites increases the etch rate. Additional factors that determine the rate of etching of crystalline semiconductors include orientation, type and concentration of doping atoms, lattice defects, and surface structure.

One of the most important microfabrication features is the etching directionality. If the etch rate in the $x$ and $y$ directions is equal to that in the $z$ direction, the etch process is said to be *isotropic* (nondirectional). The etching of single-crystal silicon, polycrystalline, and amorphous silicon in HF, BHF, $HNO_3$ or $CH_3COOH$ etchants (which form the so-called HNA etchant system) results in *isotropic* process. Etch processes which are *anisotropic* or *directional* have the etch rate in the $z$ direction higher than the lateral ($x$ or $y$ direction) etch rate. An example of this etch profile is the etching of (100) single-crystal silicon in the KOH/water or ethylene-diamine-pyrocatechol/water (EDP) etchants. The *vertical anisotropic* etching is the directional etching in which the lateral etch rate is zero (for example, etching of the 110 single-crystal silicon in the KOH system, or etching the silicon substrate by ion bombardment assisted plasma etching, e.g., reactive ion etching or ion beam milling).

*Isotropic* etching in liquid reagents is the most widely used process for removal of damaged surfaces, creating structures in single-crystal slices, and patterning single-crystal or polycrystalline semiconductor films.

For *isotropic* etching of silicon, the most commonly used etchants are mixtures of hydrofluoric (HF) and nitric ($HNO_3$) acids in water or acetic acid ($CH_3COOH$). In this co-called HNA etchant system, after the hole injection and $OH^-$ attachment to the silicon to form $Si(OH)_2$, hydrogen is released to form $SiO_2$. Hydrofluoric acid is used to dissolve $SiO_2$ to form water soluble $H_2SiF_6$. The reaction is

$$Si + HNO_3 + 6HF \rightarrow H_2SiF_6 + H_2NO_2 + H_2O + H_2.$$

Water can be used as a diluent for this etchant. However, acetic acid $CH_3COOH$ is preferred because it controls the dissociation of the nitric acid and preserves the oxidizing power of $HNO_3$ for a wide range of dilution (i.e., it acts as a buffer). Thus, the oxidizing power of the etchant remains almost constant. The HF-$HNO_3$ system was examined in [12]. Figure 3.3.1 shows the results in the form of isoetch curves for various constituents by weight.

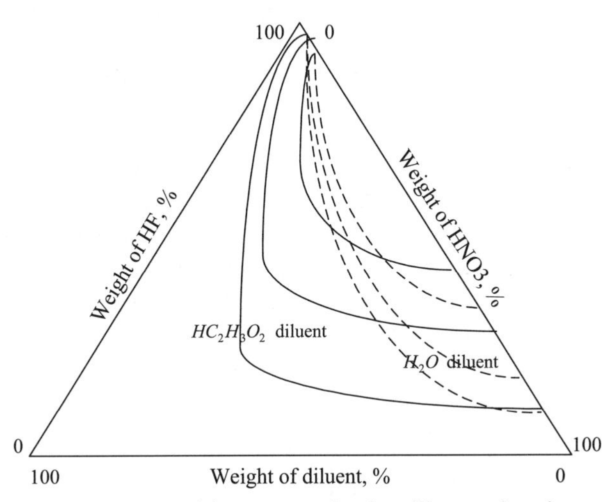


Figure 3.3.1 *Isotropic* etching curves for the silicon using the HF-$HNO_3$-diluent etchant system.

It should be noted here that normally available concentrated acids are 49% and 70% for HF and $HNO_3$, respectively. Either water (dashed line curves) or acetic acid (solid-line curves) are used as the diluent in this system. At high HF and low $HNO_3$ concentrations, the etch rate is controlled by the concentration of $HNO_3$. Etching tends to be difficult to initiate with the actual onset of etching highly variable. In addition, it results in relatively unstable silicon surfaces which proceed to slowly grow a layer of $SiO_2$ over a period of time. The etch is limited by the rate of the oxidation-reduction reaction, and therefore, it tends to be orientation dependent. At low HF and high $HNO_3$ concentrations, the etch rate is controlled by the ability of HF to remove the $SiO_2$ as it is formed. These etches are self-passivating in that the surface is covered with a 30-50 Å layer of $SiO_2$. The primary limit on the etch rate is the rate of removal of the silicon complexes by diffusion. The etching process in this region is *isotropic* and acts as the polishing etching. The etch system in the $HF:HNO_3=1:1$ range is initially insensitive to the addition of diluent when the percentage of diluent is less than 10%. From 10-30%, the etch rate decreases with the addition of diluent. If diluent is greater than 30%, even small changes of diluent cause large changes in the etch rate.

*Anisotropic* etchants of silicon, such as EDP, KOH, and hydrazine are orientation dependent. That is, they etch the different crystal orientations with different etch rates. Anisotropic etchants etch the (100) and (110) silicon crystal planes faster than the (111) crystal planes. For example, the etch rates are 500:1 for (100) versus (111) orientations, respectively [13]. Silicon dioxide, silicon nitride, and metallic thin films (chromium and gold) provide good etch masks for typical silicon anisotropic etchants. These films are used to mask areas of silicon that must be protected from etching and to define the initial geometry of the regions to be etched.

Two techniques have been widely used in conjunction with silicon *anisotropic* etching to guarantee the etch-stop. Heavily-boron-doped silicon (so-called $p^+$ etch-stop) is effective in stopping the etch. The pn-junction technique can be used to stop etching when one side of a reverse-biased junction-diode is etched away. *Anisotropic* etchants for silicon are usually alkaline solutions used at elevated temperatures. For *isotropic* etchants, two main reactions are oxidation of the silicon, followed by dissolution of the hydrated silica. The commonly used oxidant is $H_2O$ in aqueous alkaline systems (NaOH or KOH) [14-16], cesium hydroxide [17], hydrazine and EDP [18], quaternary ammonium hydroxides [19], or sodium silicates [20]. The most commonly used anisotropic etchants (etchant systems) for silicon are EDP with water, KOH with water or isopropyl, NaOH with water, and $H_2N_4$ with water or isopropyl (the etch rate of mask varies from 1 Å/min to 20 Å/min). The KOH - water etching system exhibits much higher (110) to (111) etch ratios than the EDP system. The etch rate ratios of the 100-, 110- and 111-plane in the EDP are 50, 30 and 1 [21], while in the KOH system, the rate ratios are 100, 600 and 1 [16]. Therefore, the KOH system is used for groove etching in 110-plane silicon wafers. The differences in etch ratios permit deep, high-aspect-ratio grooves with minimal undercutting of the mask. A disadvantage of the KOH system is that silicon dioxide ($SiO_2$) is etched at a rate which limits its use as a mask in many applications. For microstructures requiring long etching times, the silicon nitrade ($Si_3N_4$) is the preferred masking material for the KOH system, while if the EDP system is used, masks can be made applying a variety of materials (for example, $SiO_2$, $Si_3N_4$, Cr, or Au).

The etching process is a charge-transfer mechanism, and etch rates depend on dopant type and concentration. Highly doped materials may exhibit high etch rate due to the greater availability of mobile carriers. This occurs in the HNA etching system ($HF:HNO_3:CH_3COOH$ or $H_2O$ = 2:3:8), where typical etch rates are 1-3 μm/min at p or n concentrations [22]. The *anisotropic* etchants (EDP and KOH) exhibit a different preferential etching behavior. Silicon heavily doped with boron reduces the etch rate in the range from 5 to 100 times when etching in the KOH system, and by 250 times when etching in the EDP system. Thus, the etch rate is a function of boron concentration, and the etch-stops formed by the $p^+$ technique are less than 10 μm thick. The electrochemical each-stop process, which does not require heavy doping and guarantee the possibility to create thicker etch-stop-layers because the etch-stop-layer can be grown epitaxially, is widely used.

The widely implemented dry etching process in micromachining applications is reactive ion etching. In this process, ions are accelerated towards the material to be etched, and the etching reaction is enhanced in the direction of travel the ions. Reactive ion etching is an *anisotropic* etching process. Deep trenches and pits (up to a few tens of microns) of the specified shape with vertical walls can be etched in a variety of commonly used materials, e.g., silicon, polysilicon, silicon oxide, and silicon nitride. Compared with the *anisotropic* wet etching, dry etching is not limited by the crystal planes in the silicon. Figure 3.3.2 illustrates the *anisotropic* etched 400 μm deep and 20 μm width grooves (in 110-silicon), and three-dimensional silicon structure are made using reactive ion etching.

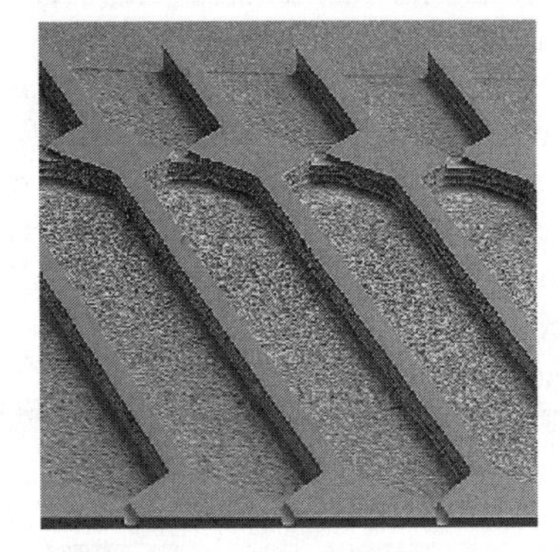

Figure 3.3.2 *Isotropic* and reactive ion etching of silicon.

### 3.3.2 Surface Micromachining

Different techniques and processes for depositing and patterning thin films are used to produce complex microstructures and microdevices on the surface of silicon wafers (surface silicon micromachining) or on the surface of other substrates. Surface micromachining technology allows one to fabricate the structure as layers of thin films. This technology guarantees the fabrication of three-dimensional microdevices with high accuracy, and the surface micromachining can be called a thin film technology. Each thin film is usually limited to thickness up to 5 μm which leads to fabrication of high-performance planar-type microscale structures and devices. The advantage of surface micromachining is the use of standard CMOS fabrication processes and facilities, as well as compliance with ICs. Therefore, this technology is widely used to manufacture microscale actuators and sensors (microdevices).

Surface micromachining has become the major fabrication technology in recent years because it allows one to fabricate complex three-dimensional microscale structures and devices. Surface micromachining with single-crystal silicon, polysilicon, silicon nitride, silicon oxide, and silicon dioxide (as structural and sacrificial materials which are deposited and etched), as well as metals and alloys, is widely used to fabricate thin micromechanical structures and devices on the surface of a silicon wafer.

This affordable low-cost high-yield technology is integrated with electromechanical microstructures – ICs fabrication processes guaranteeing the needed microstructures-IC fabrication compatibility. Surface micromachining is based on the application of sacrificial (temporary) layers that are used to maintain subsequent layers and are removed to reveal (release) fabricated microstructures. This technology was first demonstrated for ICs, and applied to fabricate motion microstructures in the 1980s. On the surface of a silicon wafer, thin layers of structural and sacrificial materials are deposited and patterned. Then, the sacrificial material is removed, and microelectromechanical structure or device is fabricated.

Figure 3.3.2 illustrates a typical process sequence of the surface micromachining fabrication technology.

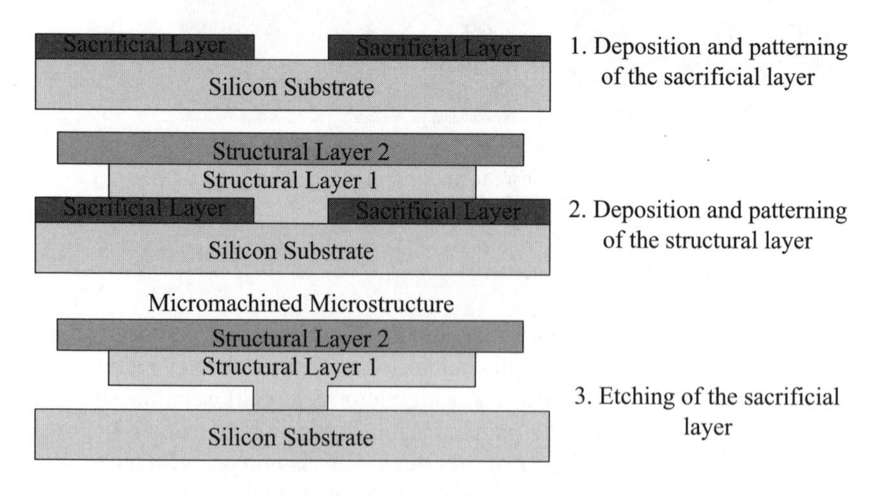


Figure 3.3.2 Surface micromachining.

Usually, the sacrificial layer is made using silicon dioxide ($SiO_2$), phosphorous-doped silicon dioxide (PSG), or silicon nitride ($Si_3N_4$). The structural layers are then typically formed with polysilicon, metals, and alloys. The sacrificial layer is removed. In particular, after fabrication of the surface microstructures and microdevices (micromachines), the silicon wafer can be wet bulk etched to form cavities below the surface components, which allows a wider range of desired motion for the device.

The wet etching can be done using:

- Hydrofluoric acid (HF),
- Buffered hydrofluoric acid (BHF),
- Potassium hydroxide (KOH),
- Ethylene-diamene-pyrocatecol (EDP),
- Tetramethylammonium hydroxide (TMAH),
- Sodium hydroxide (NaOH).

Surface micromachining has been widely used in commercial fabrication of MEMS and microdevices (microtransducers, actuators and sensors such as rotational/translational microservos, accelerometers, gyroscopes, etc.), and microstructures (gears, flip-chip electrostatic actuators, membranes, mirrors, etc.). As was emphasized, surface micromachining means the fabrication of micromechanical structures and devices by deposition and etching of structural and sacrificial layers (thin films). Simple microstructures (beams, gears, membranes, etc.) and complex microdevices (actuators, motors, and sensors) are fabricated on top of a silicon substrate. The most important attractive features of the surface micromachining technology are the small microstructure dimensions and the opportunity to integrate micromechanics, microelectronics (ICs), and optics on the same chip. Using the ICs compatible batch processing, affordable, low-cost, high-yield microstructure fabrication is achieved for high volume applications. For example, to fabricate microscale gears (microgear train), a sacrificial silicon dioxide is deposited on the wafer and patterned. Then, a structural layer of polysilicon is deposited and patterned. This polysilicon layer becomes the structural microgears element. Other layers are then deposited and patterned making the rest of the microstructure (microscale gears). Etching in the hydrofluoric or buffered hydrofluoric acids removes the sacrificial layers releasing the microgear [1].

There are three key challenges in fabrication of microstructures using surface micromachining: control and minimization of stress and stress gradient in the structural layer to avoid bending or buckling of the released microstructure; high selectivity of the sacrificial layer etchant to structural layers and silicon substrate; avoidance of stiction of the released (suspended) microstructure to the substrate. By choosing appropriate deposition and doping parameters, the stress and stress gradient in thin films can be controlled and optimized (minimized). The sacrificial layers can be etched with high selectivity against structural layer and silicon substrate using hydrofluoric or buffered hydrofluoric acids. Two methods to prevent the stiction are commonly used: (1) application of gaseous hydrofluoric acid and control of temperature using the substrate heater; (2) supercritical phase transition of carbon dioxide above the critical point (73 bar and $31^0$C). After etching the sacrificial layer and rinsing, the rinsing liquid is exchanged by liquid carbon dioxide which is subsequently transferred in the supercritical state; thus, the phase transition liquid-gaseous is avoided, and capillary forces do not occur.

The sacrificial layers are removed by the lateral etching. This selective etching can be performed using hydrofluoric acid which etches $SiO_2$ but not single-crystal and polycrystalline silicon. Alternatively, the KOH etching system with polysilicon as the sacrificial layer and silicon nitride as the cover material can be used. If 100-oriented silicon is used, substrate etching will terminate on the 111-plane. The deposition is performed using low pressure chemical vapor deposition (LPCVD) from pure silane. The requirements on the deposition rate, thickness, and stress controls lead one to analysis of mechanical properties and film morphology. The morphological range is controlled by deposition and nucleation conditions. Calculations based on single-crystal data and texture functions indicate that fine-grained, randomly oriented films are needed to attain isotropic mechanical properties. This requirement restricts the film growth conditions to the amorphous-polycrystalline boundary or for LPCVD silicon, to the temperature region from $575^0$C to $610^0$C. The optimization of mechanical properties of thin films are achieved via flow, pressure, and temperature control. This requires fast measurement and control methods.

The primary issue in the deposited thin films is the control of the built-in strain. For example, the maximum membrane deflection $d$ for a loaded plate which is subject to built-up strain is a nonlinear function of the pressure ($p$), strain ($e$), length ($l$), thickness ($h$), Young's module ($E$) and Poisson ratio ($\rho$).

In particular, one has the following formula

$$d = \frac{l^4\left(1+p^2\right)\rho}{Eh^2} f(e).$$

If the strain field is compressive ($e < 0$), bending and buckling occur. If the field is tensile, membrane deflections will be reduced for a given pressure. Compressive polysilicon can be converted to tensile polysilicon via annealing which converts thin films to the fine-grained form (this involves a volume contraction which causes the tensile field). The comprehensive database for the mechanical properties of the deposited thin films was developed in order to achieve computer-aided design and manufacturing capabilities. Polysilicon surface micromachining technology, as applied to the microscale sensors, is described in [23].

As was shown, the beam-design surface micromachining process employs thin films of two different materials (polysilicon is typically used as the structural material, and silicon oxide as the sacrificial material). These materials are deposited and patterned. The sacrificial material is etched away to release three-dimensional structure. Complex microstructures have many layers, and fabrication complexity increases.

A simple surface micromachined cantilever beam is shown in Figure 3.3.2. A sacrificial layer of silicon oxide is deposited on the surface of the silicon wafer. Two layers of polysilicon and ferromagnetic alloy are then deposited and patterned using dry etching. The wafer is wet etched to remove the silicon oxide layer under the beam releasing the beam which is attached to the wafer by the anchor.

Surface micromachining is an additive fabrication technique which uses modified CMOS technology and materials (e.g., doped and undoped single-crystal silicon and polysilicon, silicon nitride, silicon oxide, and silicon dioxide for the electrical and mechanical microstructures, and aluminum alloys for the metal connections), and involves the building of a microstructure or microdevice on top the surface of a supporting substrate. This technique complies with other CMOS technologies to fabricate ICs on a substrate.

To fabricate high-performance mechanical microstructures and microdevices using the silicon and other materials, the internal stresses of thin films must be controlled. It is desirable to grow/deposit the polysilicon, silicon nitride, silicon dioxide, metals, alloys, and insulators thin films within minimum time. However, the high deposition speed results in high internal stress in thin films, and this highly compressive internal stresses leads to the bending and buckling effects. Thus, the thin film deposition process should be controlled and optimized in order to minimize or eliminate the internal stress. For example, the stress of a polysilicon thin film can be controlled by doping it with boron, phosphorus, or arsenic. However, doped polysilicon films are rough and interfere with ICs. The stress in polysilicon can be controlled by annealing (annealing the polysilicon after deposition at elevated temperatures changes thin films to be stress free or tensile). The annealing temperature sets the films' final stress. Using this method ICs can be embedded into polysilicon films through selective doping, and hydrofluoric acid will not change the mechanical properties of the material. The stress of a silicon nitride film can be controlled by regulating the deposition temperature and the silicon/nitride ratio. The stress of a silicon dioxide thin film can be controlled and minimized by changing the deposition temperature and post annealing. It is difficult to control the stress in silicon dioxide accurately. Therefore, silicon dioxide is usually not used as the structural material. Silicon dioxide is used for electric insulation or as a sacrificial layer under the polysilicon structural layer. Sacrificial layers are temporary layers which will be selectively removed later allowing partial or complete release of the structures. Silicon nitride may also be used for electronic insulation and as a sacrificial layer.

### *Example Process*

The design and fabrication of motion microstructures start with the microstructure synthesis, identification of the microstructure functionality, specifications, and performance. Let us develop the fabrication flow to fabricate the thin membrane. The polysilicon membrane can be fabricated by oxidizing a silicon substrate, patterning the silicon dioxide, deposition and patterning of polysilicon over the silicon dioxide, and removal of the silicon dioxide. To attain the actuation features, the NiFe thin film alloy (magnetic material) should be then deposited.

The major fabrications steps and processes for thin membrane are given in Table 3.3.1.

Table 3.3.1 Major fabrication steps and processes.

| Process Steps | Description |
|---|---|
| Step 1.<br><br>Silicon dioxide grow | Silicon dioxide is grown thermally on a silicon substrate. For example, growth can be performed in a water vapor ambient at $1000^0$C for one hour. The silicon surfaces will be covered by 0.5 to 1 μm of silicon dioxide (thermal oxide thickness is limited to a few microns due to the diffusion of water vapor through silicon dioxide). Silicon dioxide can be deposited without modifying the surface of the substrate, but this process is slow to minimize the thin film stress. Silicon nitride may also be deposited, and its thickness is limited to 4-5 μm. |
| Step 2.<br><br>Photoresist | A photoresist (photo-sensitive material) is applied to the surface of the silicon dioxide. This can be done by spin coating the photoresist suspended in a solvent. The result after spinning and driving-off the solvent is a photoresist with thickness from 0.2 to 2 μm. The photoresist is then soft baked to drive off the solvents inside. |
| Step 3.<br><br>Photolithography Exposure, and Development | The photoresist is exposed to ultraviolet light patterned by a photolithography mask (photomask). This photo-mask blocks the light and defines the pattern to guarantee the desired surface topography. Photomasks are usually made using fused silica, and optical transparency at the exposure wavelength, flatness, and thermal expansion coefficient must be met. On one surface of the glass (or quartz), an opaque layer is patterned (usually hundreds of Å thick chromium layer). A photomask is generated based upon the desired form of the polysilicon membrane. The surface topography is specified by the mask. The photoresist is developed next. The exposed areas are removed in the developer. In a positive photoresist, the light will decrease the molecular weight of the photoresist, and the developer is selectively remove (etch) the lower molecular weight material. |

| | |
|---|---|
| Step 4.<br><br>Etch silicon dioxide | The silicon dioxide is etched. The remaining photoresist will be used as a *hard mask* which protects sections of the silicon dioxide. The photoresist is removed by the wet etching (hydrofluoric acid, sulfuric acid, and hydrogen peroxide) or dry etching (oxygen plasma). The result is a silicon dioxide thin film on the silicon substrate. |
| Step 5.<br><br>Deposit polysilicon | Polysilicon thin film is deposited over the silicon dioxide. For example, polysilicon can be deposited in the LPCVD system at $600^0$C in a silane ($SiH_4$) ambient. The typical deposition rate is 65-80 Å/min to minimize the internal stress and prevent bending and buckling (polysilicon thin film must be stress free or have a tensile internal stress). The thickness of the thin film is up to 4 μm. |
| Step 6.<br><br>Photoresist | Photoresist is applied to the polysilicon thin film, and the planarization must be done. The patterned silicon dioxide thin film changes the topology of the substrate surface. It is difficult to apply a uniform coat of photoresist over a surface with different heights. This results in photoresist film which different thickness, nonuniformity, and corners and edges of the patterns may not be covered. For 1 μm (or less) height, this problem is not significant; but for thicker films and multiple layers, replanarization is required. |
| Step 7.<br><br>Photolithography Exposure, and Development | A photomask containing the desired topography (form) of the polysilicon membrane is aligned to the silicon dioxide membrane. Alignment accuracy (tolerances) can be done within the nanometer range, and the accuracy depends upon the size features of the microstructure. |
| Step 8.<br><br>Etch polysilicon | The polysilicon thin film is etched with the photoresist protecting the desired polysilicon membrane form. It is difficult to find a wet etch for polysilicon which does not attack photoresist. Therefore, dry etching through plasma etching can be applied. Selectivity of the plasma between polysilicon and silicon dioxide is not concerned because the silicon dioxide will be removed later. Therefore, the polysilicon can be overetched by etching it longer than needed. This results in higher yield. |

| | |
|---|---|
| Step 9. Photoresist remove | The photoresist protecting the polysilicon membrane is removed. |
| Step 10. Deposit NiFe | NiFe thin film is deposited. |
| Step 11. Remove silicon dioxide: release the thin film membrane | The silicon dioxide is removed by wet etching (hydrofluoric or buffered hydrofluoric acids) because plasma etching cannot easily remove the silicon dioxide in the confined space under the polysilicon thin film. Hydrofluoric acid does not attack pure silicon. Hence, the polysilicon membrane and silicon substrate will not be etched. After the silicon dioxide is removed, the polysilicon membrane is formed (released). This membrane can bend down and stick to the surface of the substrate during drying after the wet etch. To avoid this, a rough polysilicon, which does not stick, can be used. Other solution is to fabricate the polysilicon membrane with the internal stress attaining that the polysilicon membrane it bended (curved) up during drying. Both solutions lead to the specific mechanical properties of the polysilicon membrane which might not be optimal from the operating requirements standpoints. Therefore, the alternatives are sought, and, in general, it possible to fabricate the polysilicon membrane with no stress. |

It is evident that the conventional CMOS processes and materials were used to develop the fabrication flow (steps) in order to fabricate thin film membrane. Therefore, CMOS fabrication facilities can be converted to fabricate microstructures, microdevices, and MEMS. Figure 3.3.3 illustrates the application of the surface micromachining technology to fabricate the polysilicon thin film membrane on the silicon substrate.

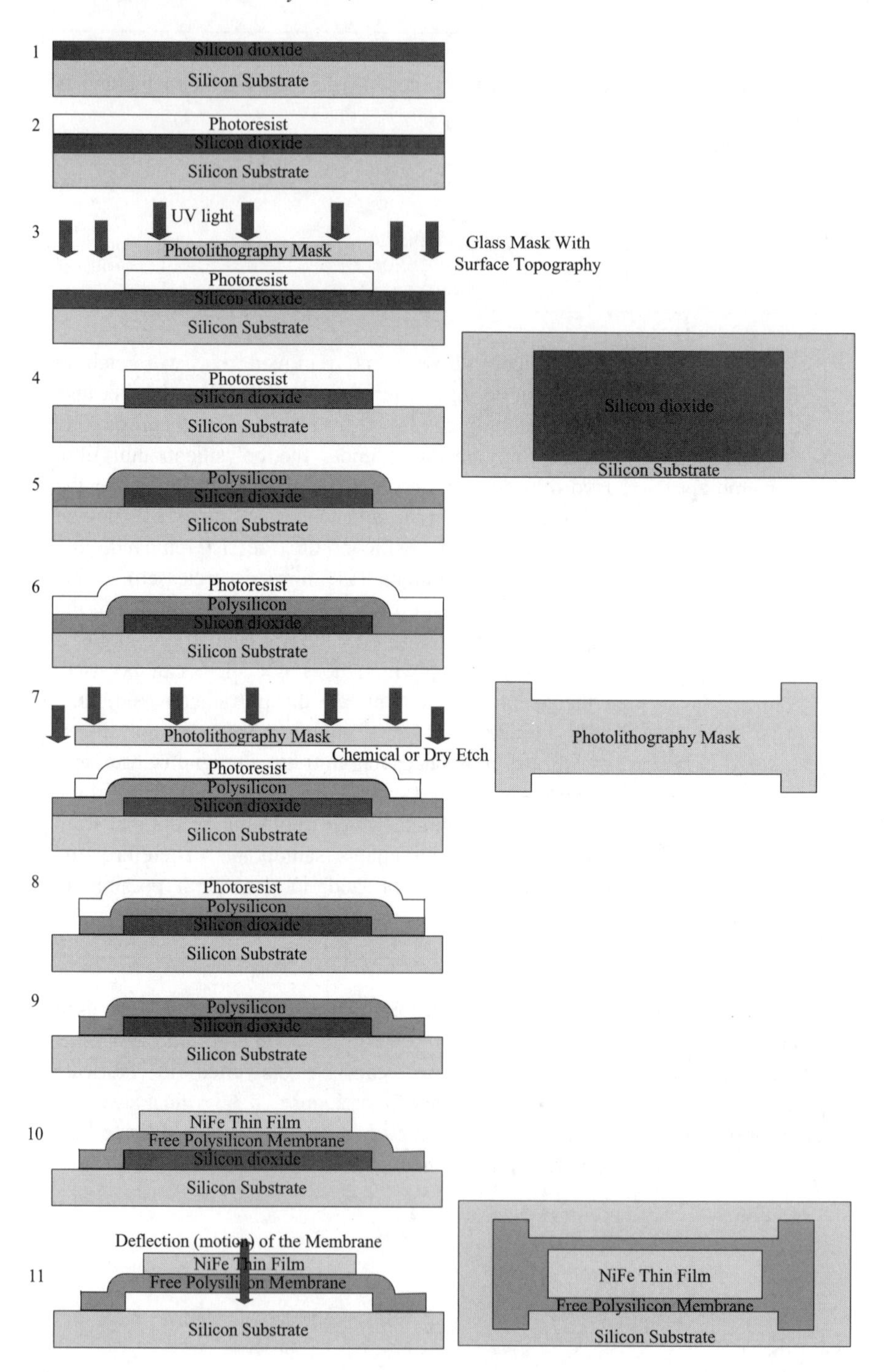


Figure 3.3.3 Micromachining fabrication of the polysilicon thin film membrane.

In electromagnetic microstructures and microdevices, metals, alloys, ferromagnetic materials, magnets, and wires (windings) must be deposited. Different processes are used [1, 24]. Using electron-beam lithography (a photographic process that uses an electron microscope to project an image of the required structures onto a silicon substrate coated with a photosensitive resist layer), the process of fabrication of micromagnets on the silicon substrate is illustrated in Figure 3.3.4. Development removes the photoresist which has been exposed to the electron beam. A ferromagnetic metal or alloy is then deposited, followed by lift-off of the unwanted material. This process allows one to make microstructures within nanomiter dimension features.

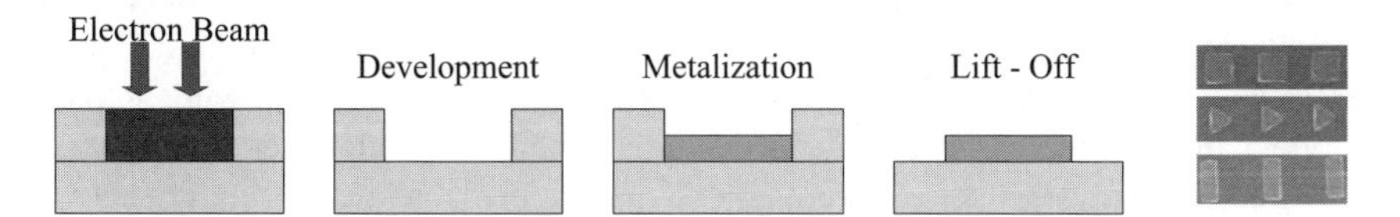


Figure 3.3.4 Electronic beam lithography in micromagnet fabrications.

The trade-offs in performance and efficiency must be studied researching different materials. For example, 95% (Ni) - 5% (Fe) to 80% (Ni) and 20% (Fe) nickel-iron alloys lead to the saturational magnetization in the 2 T range [25], and the permeability can be greater than 2000. In addition, narrow (soft) and medium B-H characteristics are achieved. The saturation magnetization and the *B-H* curves for $Ni_{x\%}Fe_{100-x\%}$ thin films are illustrated in Figure 3.3.5. As emphasized in Chapter 8, the magnetic characteristics depend upon other factors including thickness.

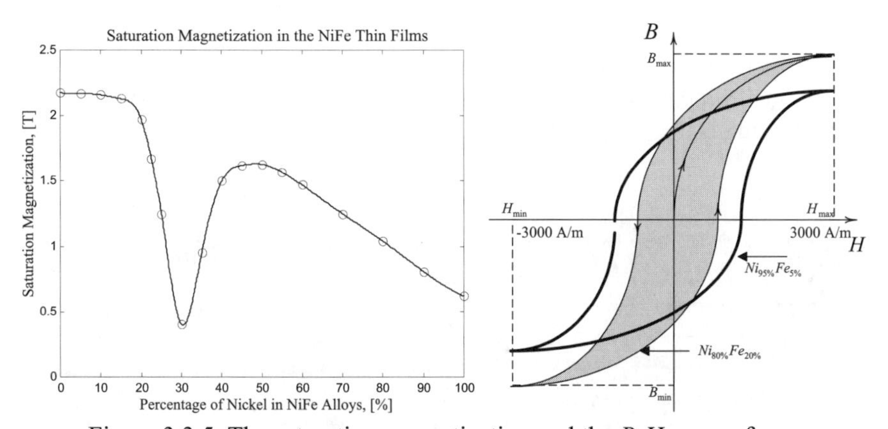


Figure 3.3.5 The saturation magnetization and the *B-H* curves for $Ni_{x\%}Fe_{100-x\%}$ thin films.

The piezoelectric effect has been used in translational (cantilever, membrane, etc.) and rotational microtransducers (actuators and sensors). The piezoelectric thin films are controlled by the electromagnetic field. For example, applying the voltage to the piezoelectric film, the film expands or contracts. Typical piezoelectric thin films being used in microactuators are zinc oxide (ZnO), lead zirconate titanate (PZT), polyvinylidene difluoride (PVDF), and lead magnesium niobate (PMN). The linear microactuators fabricated using shape memory alloys (TiNi) [26, 27] have found limited application compared with the piezoelectric.

For MEMS, there is a critical need to develop the fabrication technologies which are compatible with the silicon-based microelectronic fabrication technologies, processes, and materials used to manufacture ICs.

The electromechanical design of microtransducers (actuators and sensors), which are called microdevices, is divided into the following steps: devising operating principles based upon electromagnetic phenomena and effects, analysis of electromagnetic–mechanical features (ruggedness, elasticity, friction, vibration, thermodynamics, etc.), modeling, simulation, optimization, fabrication, test, validation, and performance analysis.

Microtransducers have stationary and rotating members (stator and rotor) and radiating energy microdevices. Different direct-current, induction and synchronous microtransducers were fabricated and tested. Many microtransducers were designed using permanent magnets. Surface micromachining technology was used to fabricate rotational micromachines with the minimal rotor outer radius 50 μm, air gap 1 μm, and bearing clearance 0.2 μm [1, 28-32]. In particular, heavily phosphorous-doped polysilicon was used to fabricate rotors and stators, and silicon nitride was used for electrical insulation. The maximal angular velocity achieved for the permanent-magnet stepper micromotor was 1000 rad/sec.

The cross-section of the slotless synchronous microtransducer (motor and generator) fabricated on the silicon substrate with polysilicon stator with deposited windings, polysilicon rotor with deposited permanent-magnets, and bearing is illustrated in Figure 3.3.6.

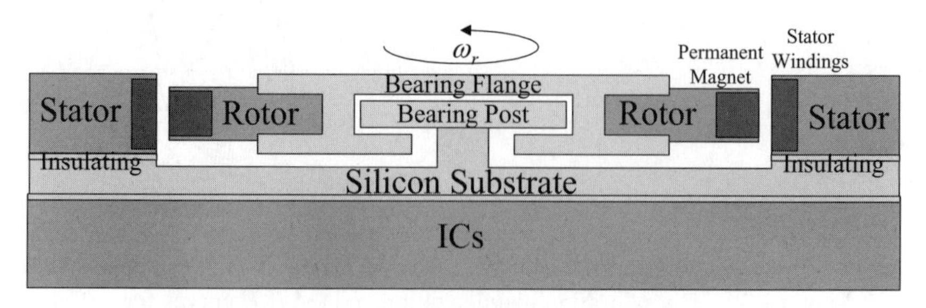


Figure 3.3.6 Cross-section schematics for slotless permanent-magnet brushless microtransducer with ICs.

Silicon dioxide, formed by thermal oxidation of silicon, can be used as the insulating layer ($SiO_2$ is commonly applied as mask and sacrificial material). The driving/sensing and controlling/processing ICs controls the brushless micromotor.

To fabricate microtransducers and ICs on a single- or double-sided chip (the application of the double-sided technique significantly enhances the performance), similar fabrication technologies and processes must be used, and the compatibility issues should be addressed and resolved. The surface micromachining processes were integrated with the CMOS technology (e.g., similar materials, lithography, etching, and other techniques). The analysis of the microtransducers feature size clearly indicates that the feature size of the microstructures and ICs is in the same order (for example, the microwindings must be deposited, microbearing and microcavities made, etc.). Furthermore, the surface micromachining processes are application-specific and strongly affected by the microtransducer devised.

To fabricate the integrated MEMS, post-, mixed-, and pre-CMOS/micromachining techniques can be applied. In a post-CMOS/micromachining technique, the ICs is passivated to protect it from the surface micromachining processes. The aluminum metallization is replaced by the tungsten (which has low resistivity and has the thermal extension coefficient matching the silicon thermal extension coefficient) metallization in order to raise the post-CMOS temperature higher than $450^0$C, but the diffusion barrier (formed by TiN or $TiSi_2$) must be used to avoid the $WSi_2$ formation at $600^0$C as well as adhesion and contact layer for the tungsten metallization. However, due to hillock formation in the tungsten during annealing, high contact resistance, and performance degradation due to the heavily doped structural and sacrificial layers, the mixed-CMOS/micromachining technique is used. In particular, performing the processes in sequence to fabricate ICs and microstructures/microdevices, the performance can be optimized, and for the standard bipolar CMOS (biCMOS) technology, the minimal modifications are required. The pre-CMOS/micromachining technique allows one to fabricate microstructures/microdevices before ICs. However, due to the vertical three-dimensional microstructure features, passivation, oxidation, step coverage, and interconnection cause difficulties [1]. Microstructures can be fabricated in trenches etched in the silicon epilayer, and then, the trenches are filled with the silicon oxide, planirized (polished), and sealed. After completing these steps, ICs are fabricated using conventional CMOS technology, and additional steps are integrated to expose and release the embedded microstructures/microdevices [1].

### 3.3.3 High-Aspect-Ratio (LIGA and LIGA-Like) Technology

There is a critical need to develop the fabrication technologies allowing one to fabricate high-aspect-ratio microstructures and microdevices. The LIGA process, which denotes Lithography–Galvanoforming–Molding (in German Lithografie–Galvanik–Abformung), is capable of producing three-dimensional microstructures of a few centimeters high with the aspect ratio (depth versus lateral dimension) of more than 100. This ratio can be achieved only through buck micromachining using wet *anisotropic* etching.

The LIGA technology is based on the x-ray lithography which guarantees shorter wavelength (from few to ten Å which lead to negligible diffraction effects) and larger depth of focus compared with optical lithography. The ability to fabricate microstructures and microdevices in the centimeter range is particularly important in the actuators applications since the specifications are imposed on the rated force and torque developed by the microdevices. Due to the limited force and torque densities, the designer faces the need to increase the actuator dimensions. The LIGA and LIGA-like processes are based on deep x-ray lithography and electroplating of metal and alloy structures, allowing one to achieve structural heights in the centimeter range [32-35]. This type of processing expands the material base significantly and allows the fabrication of new high-performance electromechanical microtransducers.

In translational (liner) and rotational microactuators, the electromagnetic force and torque depend on the change in energy which is stored in the active volume and the energy density of material [36]. In particular, the expression for the co-energy is used to derive the electromagnetic force and torque. High-performance actuators with maximized active volumes and minimized surface areas have been designed and fabricated using LIGA and LIGA-like technologies [1]. In these processes, a substrate with a plating base is covered with a thick photoresist (thickness can be in the centimeter range). The photoresist is cured and exposed by x-rays from a synchrotron source (x-ray lithography).

The photoresist strain, which is due adhesion, causes well-known difficulties. This problem is solved by combining surface micromachining, patterning the sacrificial layers under the plating base, and optimizing the processes (details are reported in Chapter 8). The achievable structural height of LIGA or LIGA-like fabricated structures is defined mainly by the photoresist processing. The photoresist procedures (based on solvent bonding of polymethyl-methacrylate PMMA and subsequent mechanical height adjustments) have been optimized and used to produce low strain photoresist layers with thickness from 50 μm to the centimeters range. Large area exposures of photoresist with thickness up to 10 cm have been achieved with x-ray masks. After electroplating, replanarization can be made through precision polishing.

Figure 3.3.7 illustrates the basic sequential processes (steps) in LIGA technology. Here, the x-ray lithography is used to produce patterns in very thick layers of photoresist. The x-rays from a synchrotron source are shone through a special mask onto a thick photoresist layer (sensitive to x-rays) which covers a conductive substrate (step 1). This photoresist is then developed (step 2). The pattern formed is electroplated with metal (step 3). The metal structures produced can be the final product, however, it is common to produce a metal mould (step 4). This mould can then be filled with a suitable media (e.g., metal, alloy, polymer, etc.) as shown in step 5. The final structure is released (step 6).

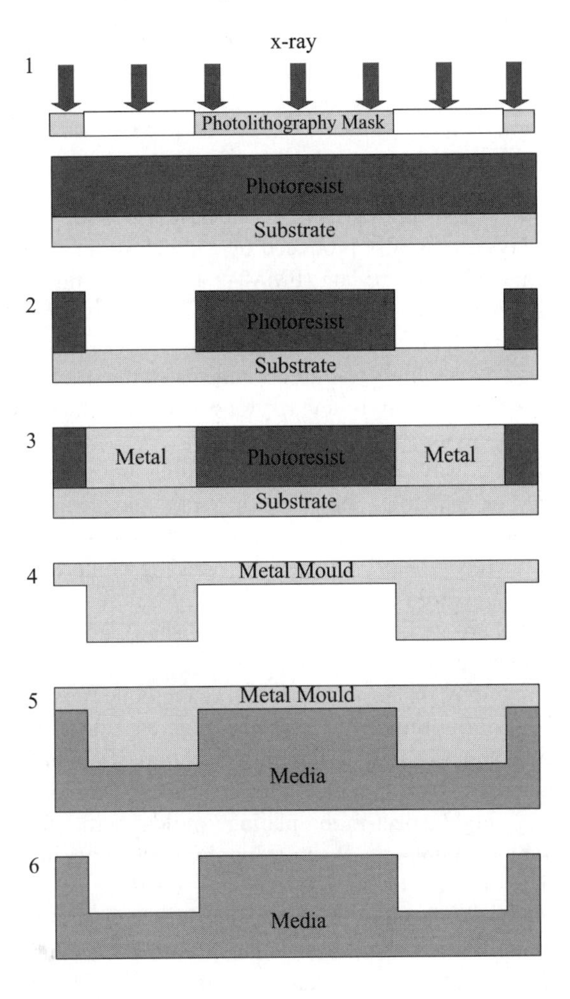


Figure 3.3.7 LIGA fabrication technology.

The described LIGA technology (frequently referred to as the high-aspect-ratio technique) allows one to fabricate microstructures with small lateral dimensions compared with thickness. Thick and narrow microstructures guarantee high ruggedness in the direction perpendicular to the substrate and compliance in the lateral directions. For actuators, high-aspect-ratio technology offers the possibility to fabricate high torque and force density microtransducers. As was emphasized, high-intensity, low-divergence, and hard x-rays are used as the exposure source for the lithography.

For exposure wavelengths $\lambda$, the image resolution and the depth of focus are (see Section 3.1)

$$i_R = k_i \frac{\lambda}{N_A},$$

$$d_F = k_d \frac{\lambda}{N_A^2}.$$

Due to short exposure wavelength, the desired features size is achieved.

These x-rays are usually produced by a synchrotron radiation source [1, 32-35]. Polymethyl-methacrylate (PMMA) and polylactides are used as the x-ray resists because PMMA (Plexiglas™ or Lucite™) and polylactides photoresists have high sensitivity to x-rays, thermal stability, desired absorption, as well as high resolution and resistance to chemical, ion, and plasma etching. Polyglycidyl-methacryl-atecoethylacrylate (PGMA) is used as the negative x-ray resist.

The exposure wavelength varies depending upon the x-ray radiation source used. For example, the 0.2 nm x-ray wavelength allows one to transfer the pattern from the high-contrast x-ray mask into the photoresist layer with a few centimeters thickness so that the photoresist relief may be fabricated with an extremely high depth to width ratio. The sidewalls of the plated structures are vertical and smooth (polished), and therefore, they can be used as optical surfaces.

Photolithography using commercially available positive photoresists and near-UV light sources can produce high-aspect-ratio plating molds. Although, in comparison to LIGA, this technique is limited in terms of thickness and aspect ratio, positive photoresists can provide a simple means of fabricating high-aspect-ratio plating molds with the conventional photolithography equipment. Positive photoresists with high transparency and high viscosity can be used to achieve coatings of 20-80 μm thick. Multiple coatings are needed to obtain the thicker layers of photoresist. If contrast printing is used, edge bead removal and good contact between the mask and the substrate are important. Conditions of softbake, exposure, and development should also be modified due to the large thickness of the photoresist. Longer softbake times are preferable to remove the solvent from the photoresist, and a high exposure is necessary. In particular, the energy density needed to exposure a 30-μm thick photoresist is 1500 $mJ/cm^2$. Long development time is required in order to completely remove the resist of exposed area. Hardbake conditions must be optimized because while hardbake improves the adhesion of the photoresist and chemical resistance to the plating solution, it also causes the distortion of the photoresist. High-aspect-ratios (10 or higher) can be obtained for 20 μm thin films. Dry etching, based upon reactive ion etching of polyimides to form high-aspect-ratio molds is used. Electromagnetically controlled dry etching of fluorinated polyimides with Ti or Al masks has been used for deep etching to attain high-aspect-ratio, good mask selectivity, and smooth sidewalls.

A critical part of the high-aspect-ratio processes is plating to form the metallic electromechanical microstructures in the mold. Using plating, metal is deposited from ions in a solution following the shape of the plating mold. This is the additive process, and the thickness of the plated metal can be large since the plating rate can be high. A variety of metals (Al, Au, Cu, Fe, Ni, and W) and alloys (NiCo, NiFe, and NiSi) can be deposited or codeposited. It is important that roughness (smoothness) of the reflective metal surfaces with the desired shape can be achieved even for optical applications. Electroplating (well-known from chemistry and covered in Chapter 8) and electroless plating (reduction of the metal ions occurs by the chemical reaction between a reducing agent and metal ion on a properly activated substrate) are the commonly used plating processes. The metal seed layer can be deposited and removed from the substrate or sacrificial layer. The plating rate and the grain size are controlled by the current density, temperature, duty cycle, etc. (see Chapter 8).

Usually, the fabrication of MEMS is done using copper, iron, nickel, alloys, and other materials through electrodeposition and electroless plating on the selected areas of the silicon substrates (for example, deposition of the copper microwindings and magnetic NiFe alloy thin film). The electroless plating can be conducted on a chemically activated silicon substrates without deposition on the photoresist mold [1]. The optimization of plating conditions (changing the current density, temperature, pH, waveforms, duty cycle, forward and reverse current, etc.) is critical to obtain smooth surfaces and practical plating rates and to avoid spontaneous plating.

## References


1. M. Madou, *Fundamentals of Microfabrication*, CRC Press, Boca Raton, FL, 1997.
2. S. A. Campbell, *The Science and Engineering of Microelectronic Fabrication*, Oxford University Press, New York, 2001.
3. S. Ghandi, *VLSI Fabrication Principles*, John Wiley, New York, 1983.
4. S. Wolfe and R. Tauber, *Silicon Processing for the VLSI Era*, Lattice Press, Sunset Beach, CA, 1986.
5. *Quick Reference Manual for Silicon IC Technology*, Editors W. Beadle, J. Tsai, and R. Plummer, John Wiley, New York, 1985.
6. R. Colclaser, *Microelectronics -- Processing and Device Design,* John Wiley, New York, 1980.
7. C. Mead and L. Conway, *Introduction to VLSI Systems*, Addison-Wesley, Reading, MA, 1980.
8. D. Reinhard, *Introduction to Integrated Circuit Engineering*, Houghton Mifflin, Boston, MA, 1987.
9. S. M. Sze, *Semiconductor Devices - Physics and Technology,* John Wiley, New York, 1985.
10. S. M. Sze, *VLSI Technology*, McGraw-Hill, New York, 1983.
11. K.E. Petersen, "Silicon as a mechanical material," *IEEE Proceedings*, pp. 420-457, 1982.
12. H. Robins and B. Schwartz, "Chemical etching of silicon, the system HF, $HNO_3$, $H_2O$ and $HC_2H_3O_2$," *J. Electrochem. Soc.*, vol. 114, pp. 108-112, 1960.
13. D. L. Kendall, "On etching very narrow grooves in silicon," *J. Appl. Phys. Letters*, vol. 26, pp. 195-201, 1975.
14. W. K. Zwicker and S. K. Kurtz, "Anisotropic etching of silicon using electrochemical displacement reactions," in *Semiconductor Silicon*, Editors H. R. Huff and R.R. Burgess, Electrochemical Society Press, Princeton, NJ, 1973.
15. J. B. Price, "Anisotropic etching of silicon with KOH-$H_2O$-isopropyl alcohol," in *Semiconductor Silicon*, Editors H. R. Huff and R.R. Burgess, Electrochemical Society Press, Princeton, NJ, pp. 338-353, 1973.
16. K. E. Bean, "Anisotropic etching of silicon," *IEEE Trans. Electron. Dev.,* vol. ED-25, pp. 1185-1993, 1978.
17. L. D. Clark, , J. L. Lund and D. J. Edell, "Cesium hydroxide (CsOH): a useful etchant for micromachining silicon," *Tech. Digest of IEEE Solid-State Sensor and Actuator Workshop*, Hilton Head, SC, pp. 5-8, 1988.
18. M. J. Declercq, J. P. DeMoor and J. P. Lambert, "A comparative study of three anisotropic etchants for silicon," *Electrochem. Soc. Ext. Abstr.,* vol. 75, no. 2, pp. 446-448, 1975.

19. M. Asano, T. Dho and H. Muraoka, "Application of choline in semiconductor technology," *Electrochem. Soc. Ext. Abstr.,* vol. 76, no. 2, pp. 911-916, 1976.
20. I. J. Pugacz-Muraszkiewicz and B.R. Hammond, "Applications of silicates to the detection of flaws in glassy passivation films deposited on silicon substrate," *J. Vac. Sci. Technol.,* vol. 14, no. 1, pp. 49-55, 1977.
21. H. Seidel, "The mechanism of anisotropic silicon etching and its relevance for micromachining," *Digest of Tech. Papers, Transducers 87, Intl. Conf. Solid-State Sensors and Actuators*, pp. 120-125, 1987.
22. H. Huraoka, T. Ohhashi and T. Sumitomo, "Controlled preferential etching technology, in Semiconductor" in *Semiconductor Silicon*, Editors H. R. Huff and R.R. Burgess, Electrochemical Society Press, Princeton, NJ, 1973.
23. H. Guckel, "Surface micromachined physical sensors," *Sensors and Materials*, vol. 4, no. 5, pp. 251-264, 1993.
24. C. H. Ahn, Y.J. Kim and M.G. Allen, "A planar variable reluctance magnetic micromotor with fully integrated stator and wrapped coil," *Proceedings, IEEE Micro Electro Mechanical Systems Workshop*, Fort Lauderdale, FL, pp. 1-6, 1993.
25. J. W. Judy, R. S. Muller and H. H. Zappe, "Magnetic microactuation of polysilicon flexible structure," *J. of Microelectromechanical Systems*, vol. 4, no. 4, pp. 162-169, 1995.
26. J. A. Walker, K.J. Gabriel and M. Mehregany, "Thin-film processing of TiNi shape memory alloy," *Sensors and Actuators*, A21-A23, 1990.
27. A. D. Johnson, "Vacuum deposited TiNi shape memory film: characterization and application in microdevices," *J. Micromechanics and Microengineering*, vol. 34, 1991.
28. H. Guckel, K. J. Skrobis, T. R. Christenson, J. Klein, S. Han, B. Choi, E. G. Lovell and T. W. Chapman, "Fabrication and testing of the planar magnetic micromotor," *J. Micromechanics and Microengineering*, vol. 1, pp. 135-138, 1991.
29. M. Mehregany and Y. C. Tai, "Surface micromachined mechanisms and micro-motors," *J. Micromechanics and Microengineering*, vol. 1, pp. 73-85, 1992.
30. M. P. Omar, M. Mehregany and R.L. Mullen, "Modeling of electric and fluid fields in silicon microactuators," *International J. of Applied Electromagnetics in Materials*, vol. 3, pp. 249-252, 1993.
31. S. F. Bart, M. Mehregany, L. S. Tavrow, J. H. Lang and S. D. Senturia, "Electric micromotor dynamics," *Trans. Electron Devices*, vol. 39, pp. 566-575, 1992.
32. E. W. Becker, W. Ehrfeld, P. Hagmann, A. Maner, D. Mynchmeyer, "Fabrication of microstructures with high aspect ratios and great

structural heights by synchrotron radiation lithography, galvanoformung, and plastic moulding (LIGA process)," *Microelectronic Engineering*, vol. 4, pp. 35-56, 1986.
33. H. Guckel, T.R. Christenson, K.J. Skrobis, J. Klein, and M. Karnowsky, "Design and testing of planar magnetic micromotors fabricated by deep x-ray lithography and electroplating," *Technical Digest of International Conference on Solid-State Sensors and Actuators, Transducers 93*, Yokohama, Japan, pp. 60-64, 1993.
34. H. Guckel, K.J. Skrobis, T.R. Christenson and J. Klein, "Micromechanics for actuators via deep x-ray lithography," *Proceedings of SPIE Symposium on Microlithography*, San Jose, CA, pp. 39-47, 1994.
35. H. Guckel, T.R. Christenson, J. Klein, T. Earles, S. Massoud-Ansari, "Microelectromagnetic actuators based on deep x-ray lithography," *Proc. International Symposium on Microsystems, Intelligent Materials and Robots*, Sendai, Japan, 1995.
36. S. E. Lyshevski, *Nano- and Microelectromechanical Systems: Fundamentals of Nano- and Microengineering*, CRC Press, Boca Raton, FL, 2000.

# CHAPTER 4

# DEVISING AND SYNTHESIS OF MEMS AND NEMS

## 4.1 MEMS MOTION MICRODEVICES CLASSIFICATION AND SYNTHESIS

New advances in microstructures, micro- and nanoscale electromechanical devices, radiating energy microdevices, driving/sensing and controlling/processing analog and digital ICs, as well as fabrication technologies, provide enabling benefits and capabilities to design and manufacture MEMS and NEMS. Critical issues are to design high-performance MEMS and NEMS which satisfy the specified criteria and requirements, e.g., systems functionality, compatibility, integrity, compliance, power and thermal management, etc. While enabling technologies have been developed to manufacture MEMS and NEMS, a spectrum of challenging problems in devising high-performance systems remains. There are several key focus areas to be studied. In particular, synthesis, optimization, fabrication, nonlinear model development, analysis, system design, and simulations.

An important problem addressed and studied in this chapter is the synthesis of motion micro- and nanostructures and devices (electromagnetic system and shape/geometry synthesis, optimization, and database developments). The proposed concept allows the designer to devise novel high-performance devices. Using the proposed concept one can devise (synthesize) and optimize different micro- and nanoscale devices and structures. The Synthesis and Classification Solver reported directly leverage high-fidelity modeling, allowing the designer to attain physical and behavioral (steady-state and transient) data-intensive analysis, heterogeneous simulations, optimization, performance assessment, outcome prediction, etc.

It was emphasized that the designer synthesizes MEMS and NEMS by devising, mimicking and prototyping new operational principles, synthesizing high-performance motion and radiating energy microdevices, microscale driving/sensing circuitry, and controlling/processing ICs.

A step-by-step procedure in the design of microdevices is as follows:

- Define application and environmental requirements;
- Specify performance specifications;
- Devise motion microstructures and microdevices, radiating energy microdevices, microscale driving/sensing circuitry, and controlling/processing ICs;
- Develop the fabrication process using micromachining and high-aspect-ratio technologies compatible with CMOS technologies;

- Perform electromagnetic, energy conversion, mechanical, thermal, vibroacoustic, and sizing/dimension estimates;
- Perform hyterogeneous electromagnetic – mechanical – thermal – vibroacoustic design with performance analysis and outcome prediction;
- Verify, modify, and refine design with ultimate goals and objectives to optimize the performance.

In this section, the design and optimization of motion microdevices is reported. To illustrate the procedure, we consider two-phase permanent-magnet synchronous slotless microtransducers as shown in Figure 4.1.1.

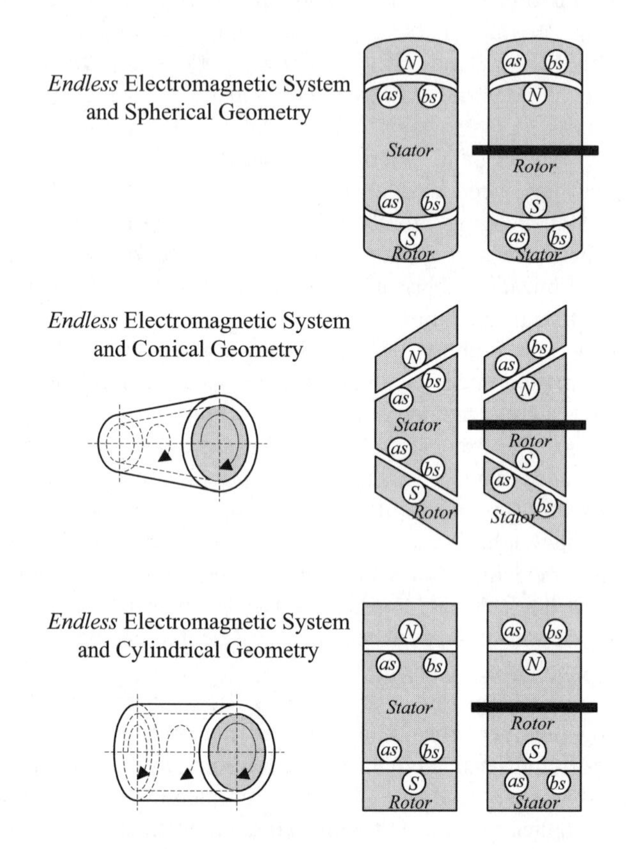


Figure 4.1.1 Permanent-magnet synchronous microtransducers with *endless* electromagnetic system and different geometry.

It is evident that the electromagnetic system is *endless*, and different geometries can be utilized as shown in Figures 4.1.1. In contrast, in translational (linear) synchronous microtransducers, the *open-ended* electromagnetic system results. The microelectromechanical motion devices were classified in [1], and qualitative and quantitative comprehensive analysis must be performed.

Motion microdevice geometry and electromagnetic systems must be integrated into the synthesis, analysis, design, and optimization patterns. Motion microdevices can have the plate, spherical, torroidal, conical, cylindrical, and asymmetrical geometry. Using these distinct geometry and diverse electromagnetic systems, we propose to classify MEMS. This idea is extremely useful in the study of existing MEMS as well as in the synthesis of an infinite number of innovative motion microdevices. In particular, using the possible geometry and electromagnetic systems (*endless*, *open-ended*, and *integrated*), novel high-performance MEMS can be synthesized.

The basic electromagnetic microtransducers (microdevices) under our consideration are: induction, synchronous, rotational, and translational (linear). That is, microdevices are classified using a type classifier

$$Y = \{y : y \in Y\}.$$

Motion microdevices are categorized using a geometric classifier (plate $P$, spherical $S$, torroidal $T$, conical $N$, cylindrical $C$, or asymmetrical $A$ geometry) and an electromagnetic system classifier (*endless* $E$, *open-ended* $O$, or *integrated* $I$). The microdevice classifier, documented in Table 4.1.1, is partitioned into 3 horizontal and 6 vertical strips, and contains 18 sections, each identified by ordered pairs of characters, such as $(E, P)$ or $(O, C)$.

In each ordered pair, the first entry is a letter chosen from the bounded electromagnetic system set

$$M=\{E, O, I\}.$$

The second entry is a letter chosen from the geometric set

$$G=\{P, S, T, N, C, A\}.$$

That is, for electromagnetic microdevises, the electromagnetic system – geometric set is

$$M \times G=\{(E, F), (E, S), (E, T), \cdots, (I, N), (I, C), (I, A)\}.$$

In general, we have

$$M \times G = \{(m, g) : m \in M \text{ and } g \in G\}.$$

Other categorization can be applied. For example, single-, two-, three-, and multi-phase microdevices are classified using a phase classifier

$$H = \{h : h \in H\}.$$

Therefore, we have

$$Y \times M \times G \times H = \{(y, m, g, h) : y \in Y, m \in M, g \in G \text{and} h \in H\}.$$

Topology (radial or axial), permanent magnets shaping (strip, arc, disk, rectangular, triangular, etc.), thin films permanent magnet characteristics (*BH* demagnetization curve, energy product, hysteresis minor loop, etc.), commutation, *emf* distribution, cooling, power, torque, size, torque-speed characteristics, packaging, as well as other distinct features are easily classified. Permanent-magnet stepper micromotors, fabricated and tested in the mid 1990s and covered in this book, are two-phase synchronous micromotors.

Table 4.1.1 Classification of electromagnetic microdevices (microtransducers) using the electromagnetic system–geometry classifier.

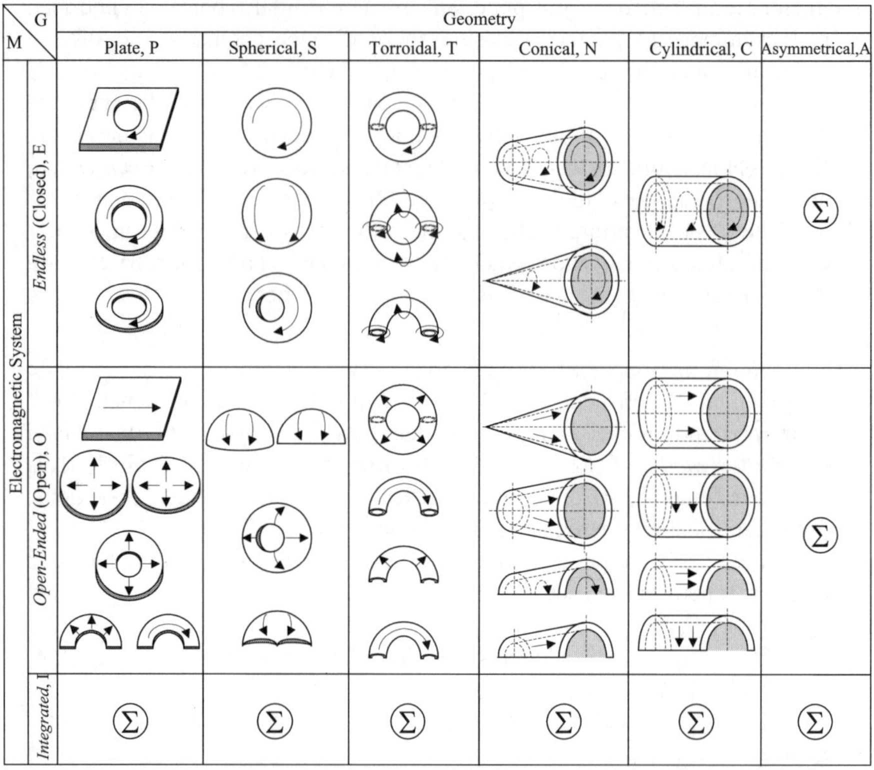

| M \ G | | Geometry | | | | | |
|---|---|---|---|---|---|---|---|
| | | Plate, P | Spherical, S | Torroidal, T | Conical, N | Cylindrical, C | Asymmetrical, A |
| Electromagnetic System | *Endless* (Closed), E | | | | | | $\Sigma$ |
| | *Open-Ended* (Open), O | | | | | | $\Sigma$ |
| | *Integrated*, I | $\Sigma$ | $\Sigma$ | $\Sigma$ | $\Sigma$ | $\Sigma$ | $\Sigma$ |

Hence, the devised electromagnetic microdevices (microtransducers) are classified by a *N*-tuple as:

{microdevice type, electromagnetic system, geometry, topology, phase, winding, connection, cooling, fabrication, materials, packaging, etc.}.

To solve a large variety of problems in modeling, analysis, performance prediction, optimization, control, and fabrication, MEMS must be devised (synthesized) first. Neural networks or generic algorithms can be efficiently used. Neural networks and generic algorithms have evolved to the mature concepts which allow the designer to perform reliable analysis, design, and optimization. Qualitative reasoning in the synthesis, classification and optimization of MEMS is based upon artificial intelligence, and the ultimate goal is to analyze, model, and optimize qualitative models of MEMS when knowledge, processes, and phenomena are not precisely known due to uncertainties. For example, micromachined motion microstructures properties and characteristics (charge density, thermal noise, mass, geometry, etc.) are not precisely known, nonuniform, and varying. It is well known that qualitative models and classifiers are more reliable compared with traditional models if there is a need to perform qualitative analysis, classification, modeling, design, optimization, and prediction. Quantitative analysis, classification, and design use a wide range of physical laws and mathematical methods to guarantee validity and robustness using partially available quantitative information.

Synthesis and performance optimization can be based on the knowledge domain. Qualitative representations and compositional (three-dimensional geometric) modeling are used to create control knowledge (existing knowledge, modeling and analysis assumptions, specific plans and requirements domains, task domain and preferences) for solving a wide range of problems through evolving decision-making. The solving architectures are based upon qualitative reusable fundamental domains (physical laws and phenomena). Qualitative reasoning must be applied to solve complex physics problems in MEMS, as well as to perform engineering analysis and design.

Emphasizing the heuristic concept for choosing the initial domain of solutions, the knowledge domain is available to efficiently and flexibly map all essential phenomena, effects, characteristics, and performances. In fact, the classification table, documented in Table 4.1.1, ensures classification, modeling, synthesis, and optimization in qualitative and quantitative knowledge domains carrying out analytical and numerical analysis of MEMS. To avoid excessive computations, high-performance (optimal) structures and devices can be found using qualitative analysis and design. That is, qualitative representations and compositional structural modeling can be used to create control knowledge in order to solve fundamental and engineering problems efficiently. The Synthesis and Classification Solver, which gives knowledge domain using compositional structural classification, modeling, analysis and synthesis, was developed applying qualitative representations. This Synthesis and Classification Solver can integrate modeling and analysis assumptions, expertise, structures, knowledge-based libraries, and preferences that are used in constraining the search (initial structural domain).

Synthesis, classification, and structural optimization are given in terms of qualitative representations and compositional modeling. This guarantees explicit domain due to the application of fundamental concepts. The Synthesis and Classification Solver can be verified solving problems analytically and numerically. Heuristic synthesis strategies and knowledge regarding physical principles must be augmented for designing micro- and nanostructures as well as micro- and nanodevices. Through qualitative analysis, classification, and design, one constrains the search domain. The solutions can be automatically generated, and the synthesized MEMS performance characteristics and end-to-end behavior can be predicted through mathematical modeling, simulations, and analysis. Existing knowledge, specific plans and requirements domains, task domain, preferences and logical relations, make it possible to reason about the modeling and analysis assumptions explicitly, which is necessary to successfully solve fundamental and engineering problems.

The Venn diagram provides a way to represent information about different MEMS topologies, configurations, and architectures. One can use regions labeled with capital letters to represent sets and use lowercase letters to represent elements. By constructing a diagram that represents some initial sets, the designer can deduce other important relations. The basic conventional form of the Venn diagram is three intersecting circles as shown in Figure 4.1.2. In this diagram, each of the circles represents a set of elements that have some common property or characteristic. Let A stand for actuators, B stand for sensors, and C stand for translational motion microstructure. Then, the region ABC represents actuators and sensors which are synthesized as translational motion microstructure, while the BC maps sensors which are the translational motion microstructures (e.g., *i*MEMS accelerometer which will be studied in this chapter).

It was illustrated that microtransducers can be designed using the *endless* electromagnetic system and conical, spherical, and conical-spherical geometry (see Figure 4.1.1). The corresponding Venn diagram is illustrated in Figure 4.1.2.

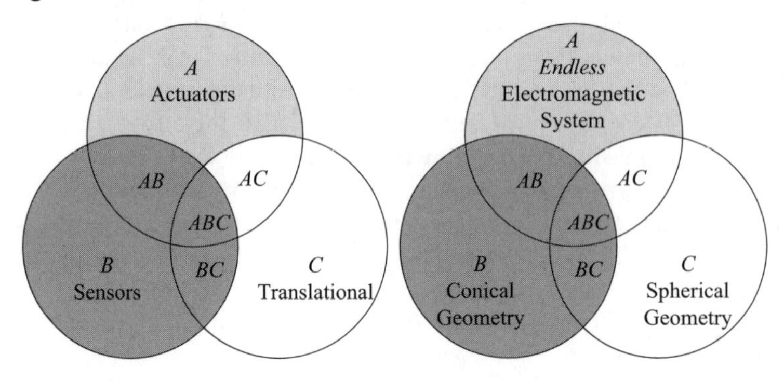


Figure 4.1.2 Venn diagram, $p = 3$: The closed curves are circles, and eight regions are labeled with the interiors that are included in each intersection. The eighth region is the outside, corresponding to the empty set.

Let $A = \{a_1, a_2, .., a_{p-1}, a_p\}$ is the collection of simple closed curves in the $XY$ plane. The collection $A$ is said to be an independent family if the intersection of $b_1, b_2, .., b_{p-1}, b_p$ is nonempty, where each $b_i$ is either $int(a_i)$ (the interior of $a_i$) or is $ext(a_i)$ (the exterior of $a_i$). If, in addition, each such intersection is connected, then $A$ is a $p$-Venn diagram, where $p$ is the number of curves in the diagram.

### *Algebra of Sets*

A set is a collection of objects (order is not significant and multiplicity is usually ignored) called the elements of the set. Symbols are used widely in the algebra of sets.

If $a$ is an element of set $A$, we have

$a \in A$.

If $a$ is not an element of set $A$, one writes

$a \notin A$.

If a set $A$ contains only the single element $a$, it is denoted as $\{a\}$.

The null set (set does not contain any elements) is denoted as $\varnothing$.

Two sets, $A$ and $B$ are equal ($A = B$) if $a \in A$ iff $a \in B$.

If $a \in A$ implies that $a \in B$, then $A$ is a subset of $B$, and $A \subset B$.

The symbols $\subset$ and $\subseteq$ describe a proper and an improper subsets.

For example, if $A \subset B$ and $B \subset A$, then $A$ is called an improper subset of $B$, $A = B$ (if there exists element $b$ in $B$ which is not in $A$, then $A$ is a proper subset of $B$).

If the set of all elements under consideration make up the universal set $U$, then $A \subset U$.

The set $A'$ is the complement of set $A$, if it is made up of all the elements of $U$ which are not elements of $A$. For each set $A$ there exists a unique set $A'$ such that $A \cup A' = U$ and $A \cap A' = \varnothing$. Furthermore, $(A')' = A$.

Two operations on sets are union $\cup$ and intersection $\cap$.

For example, an element $a \in A \cup B$ iff $a \in A$ or $a \in B$.

In contrast, an element $a \in A \cap B$ iff $a \in A$ and $a \in B$.

Using $\cup$ and $\cap$ operators, we have the following well-known algebra of sets laws:

- Closure: there is a unique set $A \cup B$ which is a subset of $U$, and there is a unique set $A \cap B$ which is a subset of $U$;
- Commutative: $A \cup B = B \cup A$ and $A \cap B = B \cap A$;
- Associative:
  $(A \cup B) \cup C = A \cup (B \cup C)$ and $(A \cap B) \cap C = A \cap (B \cap C)$;
- Distributive:
  $A \cup (B \cap C) = (A \cup B) \cap (A \cup C)$ and $A \cap (B \cup C) = (A \cap B) \cup (A \cap C)$,
  and using the index set $\Lambda$, $\lambda \in \Lambda$, one has
  $$A \cup \left( \bigcap_{\lambda \in \Lambda} B_\lambda \right) = \bigcap_{\lambda \in \Lambda} (A \cup B_\lambda) \text{ and } A \cap \left( \bigcup_{\lambda \in \Lambda} B_\lambda \right) = \bigcup_{\lambda \in \Lambda} (A \cap B_\lambda);$$
- Idempotent: $A \cup A = A$ and $A \cap A = A$;
- Identity: $A \cup \varnothing = A$ and $A \cap U = A$;

- DeMorgan's: $(A \bigcup B)' = A' \bigcap B'$ and $(A \bigcap B)' = A' \bigcup B'$;
- $U$ and $\varnothing$ laws: $U \bigcup A = U$, $U \bigcap A = A$, $\varnothing \bigcup A = A$ and $\varnothing \bigcap A = \varnothing$.

  Additional rules and properties of the complement are:
  $A \subset (A \bigcup B)$, $(A \bigcap B) \subset A$, $A \subset U$, $\varnothing \subset A$
  If $A \subset B$ then $A \bigcup B = B$, and if $B \subset A$ then $A \bigcap B = B$.

*Sets and Lattices*

A set is simply a collection of elements. For example, *a*, *b* and *c* can be grouped together as a set which is expressed as $\{a, b, c\}$ where the curly braces are used to enclose the elements that constitute a set. In addition to the set $\{a, b, c\}$, we define the sets $\{a, b\}$ and $\{d, e, g\}$.

Using the union operation, we have

$$\{a,b,c\} \bigcup \{a,b\} = \{a,b,c\} \text{ and } \{a,b,c\} \bigcup \{d,e,g\} = \{a,b,c,d,e,g\},$$

while the intersection operation leads us to

$$\{a,b,c\} \bigcap \{a,b\} = \{a,b\} \text{ and } \{a,b,c\} \bigcap \{d,e,g\} = \varnothing = \{\},$$

where $\{\}$ is the empty (or null) set.

The subset relation can be used to partially order a set of sets. If some set *A* is a subset of a set *B*, then these sets are partially ordered with respect to each other. If a set *A* is not a subset of set *B*, and *B* is not a subset of *A*, then these sets are not ordered with respect to each other. This relation can be used to partially order a set of sets in order to classify MEMS and NEMS using the Synthesis and Classification Solver introduced. Sets possess some additional structural, geometrical, as well as other properties. Additional definitions and properties can be formulated and used applying lattices.

Using a lattice, we have

- $A \subseteq A$ (reflexive law);
- If $A \subseteq B$ and $B \subseteq A$, then *A*=*B* (antisymmetric law);
- If $A \subseteq B$ and $B \subseteq C$, then $A \subseteq C$ (transitive law);
- *A* and *B* have a unique greatest bound, $A \bigcap B$. Furthermore, $G = A \bigcap B$, or *G* is the greatest lower bound of *A* and *B* if: $A \subseteq G$, $B \subseteq G$, and if *W* is any lower bound of *A* and *B*, then $G \subseteq W$;
- *A* and *B* have a unique least upper bound, $A \bigcup B$. Furthermore, $L = A \bigcup B$, or *L* is the least upper bound of *A* and *B* if: $L \subseteq A$, $L \subseteq B$, and if *P* is any upper bound of *A* and *B*, then $P \subseteq L$.

A lattice is a partially ordered set where for any pair of sets (hypotheses) there is a least upper bound and greatest lower bound. Let our current hypothesis is *H*1 and the current training example is *H*2. If *H*2 is a subset of *H*1, then no change of *H*1 is required. If *H*2 is not a subset of *H*1, then *H*1 must be changed. The minimal generalization of *H*1 is the least upper bound of *H*2 and *H*1, and the minimal specialization of *H*1 is the greatest lower bound of *H*2 and *H*1. Thus, the lattice serves as a map that allows us to locate the current hypothesis *H*1 with reference to the new information *H*2. There exists the correspondence between the algebra of propositional logic and the algebra of sets. We refer to a hypotheses as logical expressions, as rules that define a concept, or as subsets of the possible instances constructible from some set of dimensions. Furthermore, union and intersection were the important operators used to define a lattice. In addition, the propositional logic expressions can also be organized into a corresponding lattice to implement the artificial learning.

A general structure $S$ is an ordered pair formed by a set object $O$ and a set of binary relations $R$ such that

$$S=(O,R)=\bigcup_{i=1}^{n} S_i\,,$$

where $O=\{o_1, o_2,\ldots, o_{z-1}, o_z\}$, $\forall o_i \in O$; $R=\{r_1, r_2, \ldots, r_{p-1}, r_p\}$, $\forall r_i \in R$; $S_i$ is the simple structure.

In the set object $O$ we define the input $n$, output $u$, and internal $a$ variables. We have $o_i=\{q_1^i, q_2^i, \ldots, q_{g-1}^i, q_g^i\}$, $q_j^i=(n_j^i, u_j^i, a_j^i)$, $q_j^i \in O^3$. Hence, the range of $q$, as a subset of $O$, is $R(q)$. Using the input-output structural function, different MEMS and NEMS can be synthesized. The documented general theory of synthesis, classification, and structural optimization, which is built using the algebra of sets, allows the designer to derive relationships, flexibly adapt, fit, and optimize the micro- and nanoscale structures within the sets of given possible solutions.

Using the Synthesis and Classifier Solver, which is given in Table 4.1.1 in terms of electromagnetic system and geometry, the designer can classify the existing motion microdevices as well as synthesize novel high-performance microdevices. As an example, the spherical, conical, and cylindrical geometry of two-phase permanent-magnet synchronous microdevice are illustrated in Figure 4.1.3.

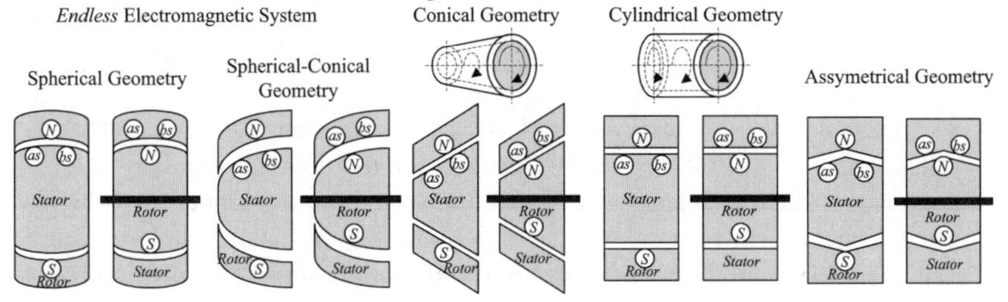


Figure 4.1.3 Two-phase permanent-magnet synchronous microdevice with *endless* electromagnetic system and distinct microtransducer geometry.

This section documents new results in the MEMS synthesis which can be used to optimize the microdevice performance. The conical (existing) and spherical-conical (devised) microdevice geometries are illustrated in Figure 4.1.3. Using the innovative spherical-conical geometry, which is different compared with the existing conical geometry, one increases the active length $L_r$ and average diameter $D_r$. For radial flux microdevices, the electromagnetic torque $T_e$ is proportional to the squared rotor diameter and axial length. In particular,

$$T_e = k_T D_r^2 L_r ,$$

where $k_T$ is the constant.

From the above relationship, it is evident that the spherical-conical micromotors develop higher electromagnetic torque compared with the conventional design. In addition, improved cooling, reduced undesirable torques components, as well as increased ruggedness and robustness contribute to the viability of the proposed solution. Thus, using the synthesis (classifier) paradigm, novel microdevices with superior performance can be devised.

The cross-section of the slotless radial-topology micromotor, fabricated on the silicon substrate with polysilicon stator (with deposited windings), polysilicon rotor (with deposited permanent-magnets), and contact bearing is illustrated in Figure 4.1.4. The fabrication of this micromotor and the processes were reported in Chapter 3.

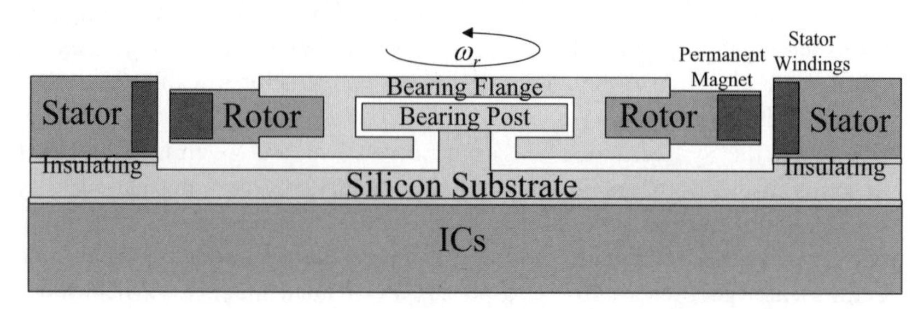


Figure 4.1.4 Cross-section schematics for slotless radial-topology permanent-magnet brushless micromotor (microtransducer) with ICs.

The analysis of the *Escherichia coli* (*E. coli*) nanobiomotor, which is illustrated in Figure 4.1.5, leads the designer to the axial-topology micro- and nanoscale transducers. As was emphasized, the designer can devise novel high-performance micro- and nanotransducers (micro- and nanomachines) through biomimicking and prototyping. The distinguished beneficial features of the devised transducers are:

- Unique radial and axial topologies;
- Electrostatic- and electromagnetic-based actuation mechanisms;
- Different electromagnetic systems;
- Noncontact electrostatic bearings.

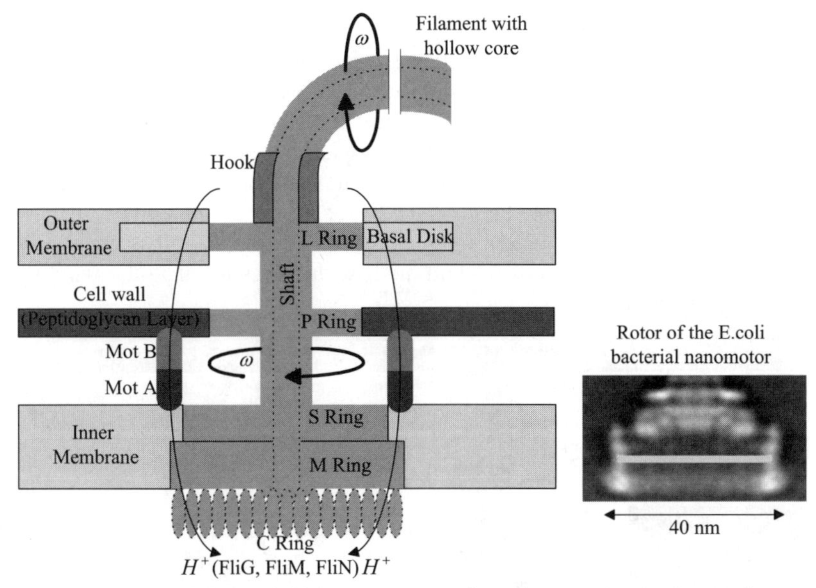


Figure 4.1.5 *E. coli* bacterial bionanomotor–flagella complex and rotor image.

Two distinct transducer topologies are radial and axial. The magnetic flux in the radial (or axial) direction interacts with the time-varying axial (or radial) electromagnetic field with ultimate goal to produce the electromagnetic torque. Using the radial flux topology, the cylindrical transducer with permanent-magnet poles on the rotor is illustrated in Figure 4.1.6.

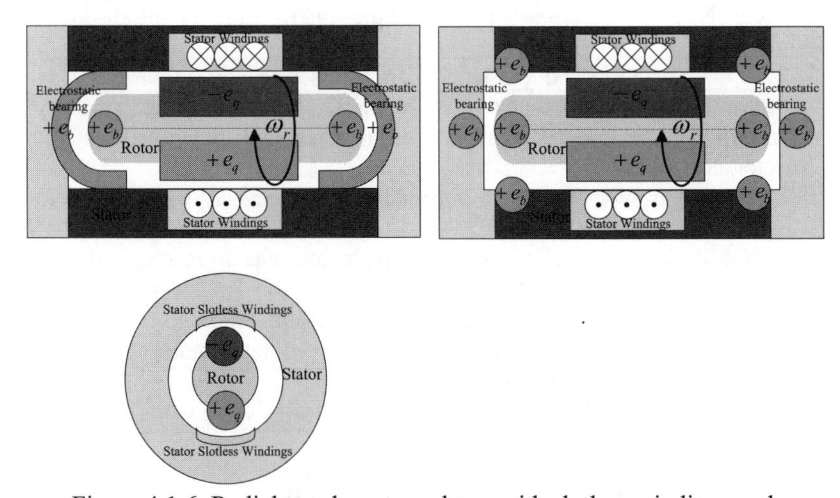


Figure 4.1.6 Radial topology transducer with slotless windings and electrostatic noncontact bearing.

The major advantages of the radial topology are that high torque and power densities can be achieved, and the net radial force on the rotor is zero. The disadvantages are that it is difficult to fabricate and assemble micro- and nanostructures (stator and rotor), and the air gap is not adjustable.

The advantages of the axial topology transducers, as shown in Figure 4.1.7, are the possibility to affordably fabricate because permanent magnets have flat surfaces (thin films permanent-magnets can be used), there are no strict shape requirements, there is no rotor back ferromagnetic material required (silicon can be used), the air gap can be adjusted, it is easy to lay out (implant and deposit) nano- and microscale wires to make the windings on the flat stator (silicon).

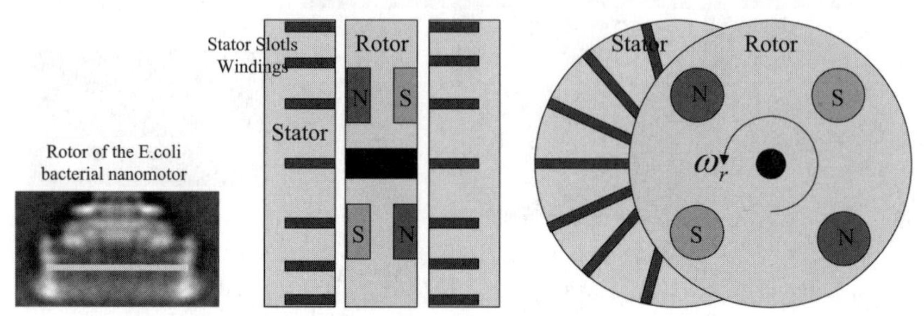

Figure 4.1.7. Axial topology transducer with permanent magnets

The stationary magnetic field is established by the permanent magnets, and stators and rotors can be fabricated using miromachining and high-aspect-ratio technologies. Slotless stator windings can be deposited on the silicon as the implanted nanowires.

Here is step-by-step procedure in the micro- and nanotransducers design:

1. Use the Synthesis and Classifier Concept: devise novel transducers by researching operational principles, topologies, configurations, geometry, electromagnetic systems (closed, open, or integrated), interior or exterior rotor, etc.
2. Define application and environmental requirements as well as specify performance specifications.
3. Perform electromagnetic, energy conversion, electro-mechanical, and sizing/dimension estimates.
4. Define and design technologies, techniques, and processes to fabricate micro- and nanostructures (e.g., stator, rotor, bearing, windings, etc.), and assemble/integrate them as the transducer (device).
5. Select materials and chemicals (substrate, insulators, conductors, permanent magnets, etc.).
6. Perform thorough electromagnetic, mechanical, thermodynamic and vibroacoustic design with performance analysis and outcome prediction.
7. Test and examine the designed micro- or nanotransducer.
8. Modify and refine the design.
9. Optimize the overall micro- or nanotransducer performance.

### 4.1.1 Microelectromechanical Microdevices

Different MEMS have been discussed, and it was emphasized that MEMS can be used as actuators, sensors, and transducers (actuators-sensors). Due to the limited torque and force densities, microdevices usually cannot develop high torque and force, and multi-node cooperative microdevices are used. In contrast, these characteristics (power, torque, and force densities) are not critical in sensor applications. Therefore, MEMS are widely used as microscale sensors. Signal-level signals, measured by sensors, are fed to analog, digital or hybrid ICs, and sensor design, signal processing, signal conditioning, and interfacing are extremely important in engineering practice.

Smart integrated sensors are the sensors in which in addition to sensing the physical variables, data acquisition, filtering, data storage, communication, interfacing, and networking are embedded. Thus, while the primary component is the sensing element (microstructure), multifunctional integration of sensors and ICs is the current demand. High-performance accelerometers, manufactured by Analog Devices using integrated microelectromechanical system technology (*i*MEMS) which is based upon CMOS processes, are studied in this section. In addition, the application of smart integrated sensors is briefly discussed.

We study the dual-axis, surface-micromachined ADXL202 accelerometer (manufactured on a single monolithic silicon chip) which combines highly accurate acceleration sensing motion microstructure (proof mass) and signal processing electronics (signal conditioning ICs). As documented in the Analog Device Catalog data, this accelerometer measures dynamic positive and negative acceleration (vibration) as well as static acceleration (force of gravity). The functional block diagram of the ADXL202 accelerometer with two digital outputs (ratio of pulse width to period is proportional to the acceleration) is illustrated in Figure 4.1.8.

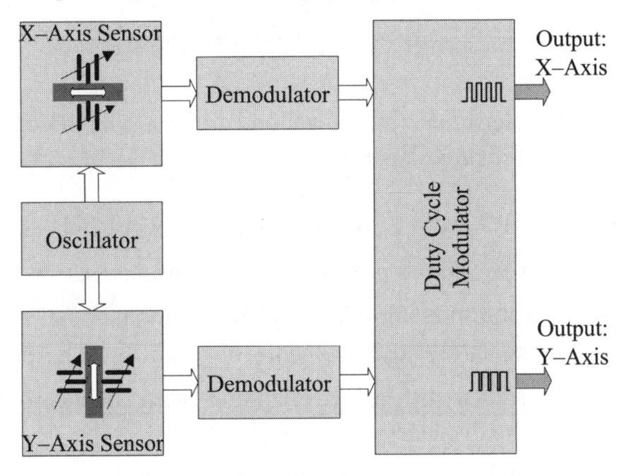


Figure 4.1.8 Functional block diagram of the ADXL202 accelerometer.

Polysilicon surface-micromachined sensor motion microstructure is fabricated on the silicon wafer by depositing polysilicon on the sacrificial oxide layer which is then etched away leaving the suspended proof mass. Polysilicon springs suspend this proof mass over the surface of the wafer. The deflection of the proof mass is measured using the capacitance difference, see Figure 4.1.9.

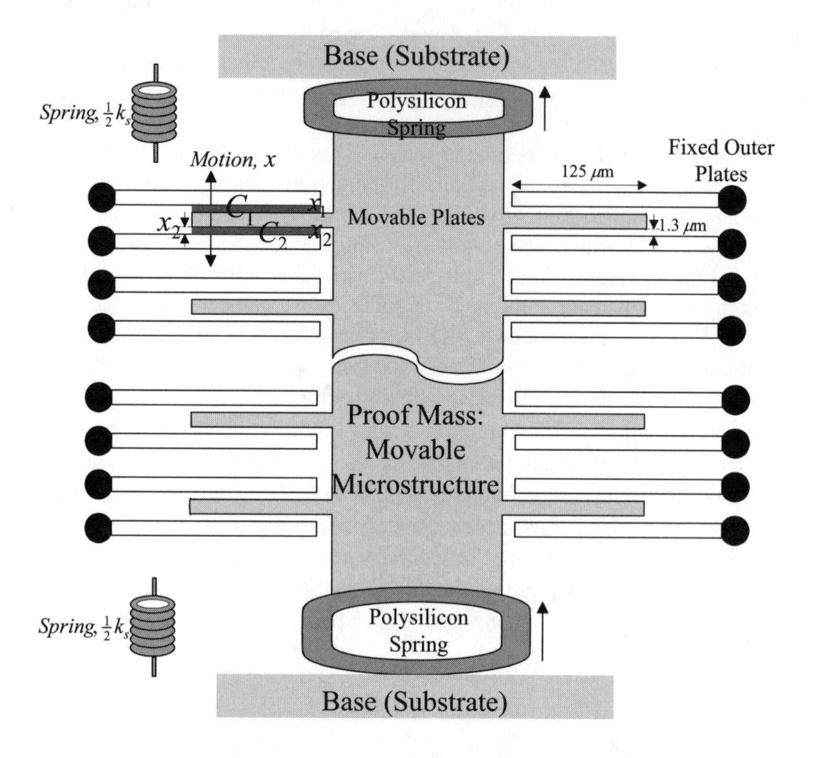


Figure 4.1.9 Accelerometer structure: proof mass, polysilicon springs, and sensing elements.

The proof mass ($1.3\ \mu m$, $2\ \mu m$ thick) has movable plates which are shown in Figure 4.1.9. The free-space (air) capacitances $C_1$ and $C_2$ (capacitances between the movable plate and two stationary outer plates) are functions of the corresponding displacements $x_1$ and $x_2$. The parallel-plate capacitance is proportional to the overlapping area between the plates ($125\ \mu m \times 2\ \mu m$) and the displacement (up to $1.3\ \mu m$).

Neglecting the fringing effects (nonuniform distribution near the edges), the parallel-plate capacitance is

$$C = \varepsilon \frac{A}{d} = \varepsilon_A \frac{1}{d},$$

where $\varepsilon$ is the permittivity; $A$ is the overlapping area; $d$ is the displacement between plates; $\varepsilon_A = \varepsilon A$

If the acceleration is zero, the capacitances $C_1$ and $C_2$ are equal because $x_1 = x_2$ (in ADXL202 accelerometer, $x_1 = x_2 = 1.3\ \mu\text{m}$ ).

Thus, one has

$$C_1 = C_2,$$

where $C_1 = \varepsilon_A \frac{1}{x_1}$ and $C_2 = \varepsilon_A \frac{1}{x_2}$.

The proof mass (movable microstructure) displacement $x$ results due to acceleration. If $x \neq 0$, we have the following expressions for capacitances

$$C_1 = \varepsilon_A \frac{1}{x_1 + x},$$

$$C_2 = \varepsilon_A \frac{1}{x_2 - x} = \varepsilon_A \frac{1}{x_1 - x}.$$

The capacitance difference is found to be

$$\Delta C = C_1 - C_2 = 2\varepsilon_A \frac{x}{x^2 - x_1^2}.$$

Measuring $\Delta C$, one finds the displacement $x$ by solving the nonlinear algebraic equation

$$\Delta C x^2 - 2\varepsilon_A x - \Delta C x_1^2 = 0.$$

This equation can be simplified.

For small displacements, the term $\Delta C x^2$ is negligible. Thus, $\Delta C x^2$ can be omitted. Then, from

$$x \approx -\frac{x_1^2}{2\varepsilon_A}\Delta C$$

one concludes that the displacement is approximately proportional to the capacitance difference $\Delta C$.

For an ideal spring, according to Hook's law, the spring exhibits a restoring force $F_s$ which is proportional to the displacement $x$. Thus,

$$F_s = k_s x,$$

where $k_s$ is the polysilicon spring constant (Figure 4.1.9 illustrates that at least two springs are used).

From Newton's second law of motion, neglecting the air friction (which is negligibly small), the following differential equation results

$$ma = m\frac{d^2 x}{dt^2} = k_s x.$$

Thus, the displacement due to the acceleration is

$$x = \frac{m}{k_s} a,$$

while the acceleration, as a function of the displacement, is

$$a = \frac{k_s}{m} x.$$

Then, making use of the measured $\Delta C$, the acceleration is found to be

$$a = -\frac{k_s x_1^2}{2m\varepsilon_A} \Delta C.$$

Making use of Newton's second law of motion, we have

$$ma = m\frac{d^2 x}{dt^2} = \underbrace{f_s(x)}_{\text{spring force}},$$

where $f_s(x)$ is the spring restoring force which is a nonlinear function of the displacement, $f_s(x) = k_{s1}x + k_{s2}x^2 + k_{s3}x^3$; $k_{s1}$, $k_{s2}$ and $k_{s3}$ are the spring constants.

Therefore, the nonlinear equation is

$$ma = k_{s1}x + k_{s2}x^2 + k_{s3}x^3.$$

The expression for the acceleration is found to be

$$a = \frac{1}{m}\left(k_{s1}x + k_{s2}x^2 + k_{s3}x^3\right),$$

where $x \approx -\frac{x_1^2}{2\varepsilon_A} \Delta C$.

This equation can be used to calculate the acceleration $a$ using the measured capacitance difference $\Delta C$.

Two proof masses (motion microstructures) can be placed orthogonally to measure the accelerations in the $x$ and $y$ axis (ADXL250), as well as the movable plates can be mounted along the sides (ADXL202). The signal conditioning, filtering, computing, and input-output interface are performed by the ICs.

Figures 4.1.10 and 4.1.11 document the ADXL202 and ADXL250 accelerometers which integrate the microscale microstructure (moving masses, springs, etc.) and ICs.

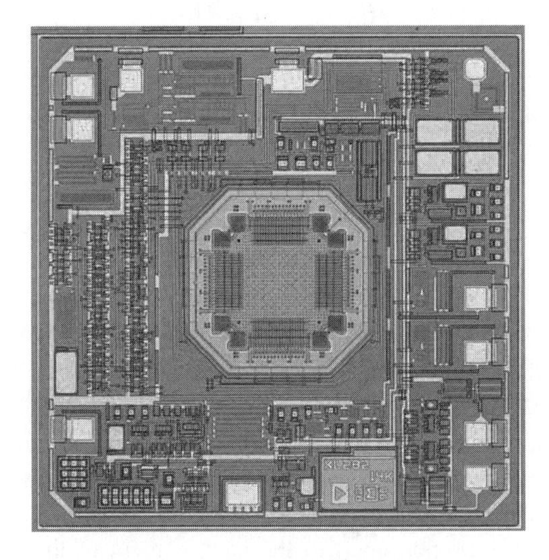

Figure 4.1.10 ADXL202 accelerometer: proof mass with fingers and ICs (courtesy of Analog Devices).

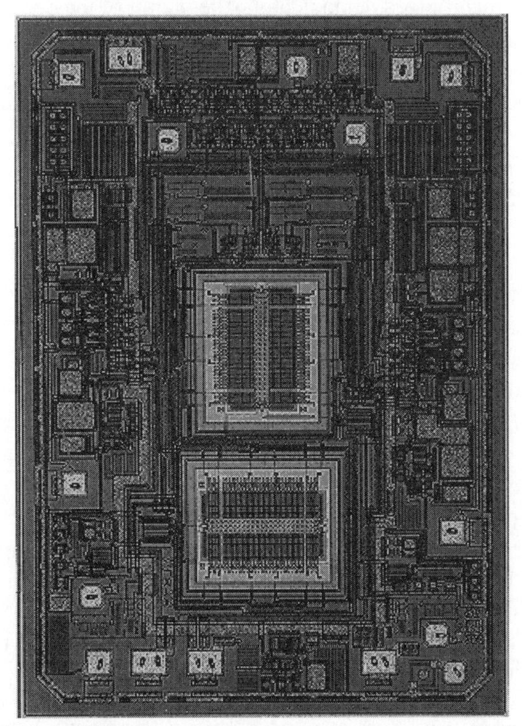

Figure 4.1.11 ADXL250 accelerometer: proof masses with fingers and ICs (courtesy of Analog Devices).

Responding to acceleration, the proof mass moves due to the mass of the movable microstructure ($m$) along the $X$ and $Y$ axes relative to the stationary member (accelerometer). The motion of the proof mass is constrained, and the polysilicon springs hold the movable microstructure (beam). Assuming that the polysilicon springs and the proof mass obey Hook's and Newton's laws, it was shown that the acceleration is calculated as

$$a = \frac{k_s}{m} x .$$

The fixed outer plates are excited by two square wave 1 MHz signals of equal magnitude that are 180 degrees out of phase from each other. When the movable plates are centered between the fixed outer plates we have $x_1 = x_2$. Thus, the capacitance difference $\Delta C$ and the output signal is zero. If the proof mass (movable microstructure) is displaced due to the acceleration, we have $\Delta C \neq 0$. Thus, the capacitance imbalance, and the amplitude of the output voltage is a function (proportional) to the displacement of the proof mass $x$. Phase demodulation is used to determine the sign (positive or negative) of acceleration. The ac signal is amplified by buffer amplifier and demodulated by a synchronous synchronized demodulator. The output of the demodulator drives the high-resolution duty cycle modulator. In particular, the filtered signal is converted to a PWM signal by the 14-bit duty cycle modulator. The zero acceleration produces 50% duty cycle. The PWM output fundamental period can be set from 0.5 to 10ms.

The Analog Devices data for differnt *i*MEMS accelerometers ADXL202/ADXL210 and ADXL150/ADXL250 are available on the Analog Devices Web site.

There is a wide range of industrial systems where smart integrated sensors are used. For example, accelerometers can be used for

- Vibroacoustic sensing, detection, and diagnostics,
- Active vibration and acoustic control,
- Situation awareness,
- Health and structural integrity monitoring,
- Internal navigation systems,
- Earthquake-actuated safety systems,
- Seismic monitoring and detection.

For example, current activities in analysis, design, and optimization of flexible structures (aircraft, missiles, manipulators, robots, spacecraft, underwater vehicles, etc.) are driven by requirements and standards which must be guaranteed. The vibration, structural integrity, and structural behavior must be addressed and studied. For example, fundamental, applied, and experimental research in aeroelasticity and structural dynamics are conducted to obtain fundamental understanding of the basic phenomena involved in flutter, force and control responses, vibration, and control. Through optimization of aeroelastic characteristics as well as applying passive and active vibration control, the designer minimizes vibration and noise, and current research integrates development of aeroelastic models and diagnostics to predict stalled/whirl flutter, force and control responses, unsteady flight, aerodynamic flow, etc. Vibration control is a very challenging problem because the designer must account complex interactive physical phenomena (elastic theory, structural and continuum mechanics, radiation and transduction, wave propagation, chaos, etc.). Thus, it is necessary to accurately measure the vibration, and the accelerometers, which allow one to measure the acceleration in the micro-g range, are used. The application of the MEMS-based accelerometers ensures small size, low cost, ruggedness, hermeticity, reliability, and flexible interfacing with microcontrollers, microprocessors, and DSPs.

High-accuracy low-noise accelerometers can be used to measure velocity and position. This provides the back-up in the case of the GPS system failures or in the dead reckoning applications (the initial coordinates and speed are assumed to be known). Measuring the acceleration, the velocity and position in the *xy* plane are found using integration. In particular,

$$v_x(t)=\int_{t_0}^{t_f} a_x(t)dt\,,\ v_y(t)=\int_{t_0}^{t_f} a_y(t)dt\,,$$

$$x_x(t)=\int_{t_0}^{t_f} v_x(t)dt\,,\ x_y(t)=\int_{t_0}^{t_f} v_y(t)dt\,.$$

Microgyroscopes have been designed, fabricated and deployed using the similar technology as *i*MEMS accelerometers. Using the difference capacitance (between the movable rotor and stationary stator plates), the angular acceleration is measured. The butterfly-shaped polysilicon rotor is suspended above the substrate, and Figure 4.1.12 shows the microgyroscope.

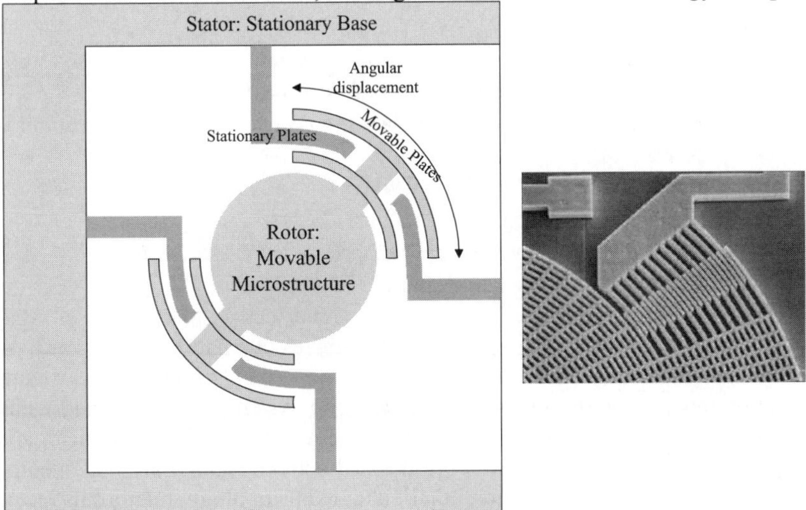


Figure 4.1.12 Angular microgyroscope structure.

### 4.1.2 Synthesis and Classification Solver

The algorithmic concept in the synthesis, classification, as well as structural and performance optimization starts by selecting an initial set of competing configurations and solutions (electromagnetic system, geometry, type, topology, etc.) for a particular problem using specifications and requirements imposed. The solutions can be generated randomly from the entire domain, however, as was emphasized earlier, available information and accessible knowledge can be readily used in order to formulate the partial (specific) domain (classifier subset). The solutions are evaluated for their efficiency and fitness. Performance and regret functionals can be designed to integrate weighted cost integrands (terms), and linear and nonlinear optimization (linear and nonlinear programming) allows one to find optimal solutions. The maxima or minima can be found using the gradient-based search. Alternatively, the evolutionary algorithms can be used, and the performance functionals are applied to compare and rank the competing solutions. The analysis and evaluation of candidate solutions are very complex problems due to infinite number of possible solutions (it is very difficult or impossible to find solutions randomly from the entire classifier domain of all possible solutions). Thus, the solutions should be examined in the partial domain (subset) of most efficient, feasible, and suitable solutions. This will allow one to define the partial classifier domain of solutions generation to efficiently solve practical problems in MEMS and NEMS design and optimization.

The following should be performed to simplify the search and optimize the algorithm to solve a wide variety of synthesis, classification, and structural optimization problems:

- Formulate and apply rules and criteria for solution sustaining based upon performance analysis, assessments, and outcomes;
- Develop and generate the partial classifier domain (subset), select solution representations;
- Initialize solutions;
- Analyze and compare solutions;
- Develop the fabrication processes to fabricate MEMS or NEMS;
- Refine the design and find the optimal solution.

Using the classifier developed, the designer can synthesize novel high-performance nano-, micro- and miniscale devices (actuators and sensors). As an example, the synthesis of a two-phase permanent-magnet synchronous microtransducer with *endless* electromagnetic system was performed using distinct geometry. The spherical, spherical-conical, conical, cylindrical, and asymmetrical geometries of the synthesized actuator/sensor are documented in Figure 4.1.3. In addition, radial- and axial-topology microtransducers were devised and analyzed.

### *Linear Programming*

*Linear Programming* is the problem that can be formulated and expressed in the so-called standard form as

minimize $cx$

subject to $Ax=b$, $x \geq 0$,

where $x$ is the vector of variables to be solved; $A$ is the matrix of known coefficients; $c$ and $b$ are the vectors of known coefficients.

The term $cx$ is called the objective function, and $Ax=b$, $x \geq 0$ are the constraints.

All these entities must have the appropriate dimensions. It should be emphasized that in general, the matrix $A$ is not square. Therefore, one cannot solve the linear programming problems by using inverse matrix $A^{-1}$. Usually $A$ has more columns than rows, and $Ax=b$ can be defined based upon the specific requirements imposed on MEMS and NEMS enabling a great spectrum in the choosing of variables $x$ in order to minimize $cx$.

Although all linear programming problems can be formulated in the standard form (all variables are non-negative), in practice it may be necessary to integrate the constraints and bounds, e.g., $x_{\min} \leq x \leq x_{\max}$ (for example, the electromagnetic field intensity, charge density, material properties, velocity, and other physical quantities are bounded). This allows one to bound the variations of variables within explicit upper or lower bounds, although this implies the limit because problems may have no finite solution. In addition, the constraints can be imposed on $Ax$. That is, $b_{\min} \leq Ax \leq b_{\max}$.

It is evident that the user needs to integrate the inequality constraints in order to solve the specific practical problems in the synthesis of MEMS and NEMS. In fact, the importance of linear programming derives by its straightforward applications and by the existence of well-developed general-purpose techniques and computationally-efficient software for finding optimal solutions. Simplex methods, introduced 50 years ago, use the *basic* solutions computed by fixing the variables at their bounds to reduce the constraints $Ax=b$ to a square system in order to solve it for unique values of the remaining variables. The *basic* solutions give extreme boundary points of the feasible region defined by $Ax=b$, $x \geq 0$. Therefore, the simplex method is based on moving from one point to another along the edges of the boundary. In contrast, barrier (interior-point) methods utilize points within the interior of the feasible region. The integer linear programming requires that some or all variables are integers. Widely used general-purpose techniques for solving integer linear programming use the solutions to a series of linear programming problems to manage the search for integer solutions and to prove the optimality.

Most linear programming problems can be straightforwardly solved using the available robust computationally efficient software. In fact, the problems with thousands variables and constraints are treated the same as the small-dimensional one. Problems having hundreds of thousands of variables and constraints are tractable and can be solved using different computational environments.

Modern linear programming and optimization software comes in two related but different tools:

1. *Algorithmic codes* which allows one to find the optimal solutions to specific linear problems using a compact listing of the variables as input as providing the compact listing of optimal solution values and related information as outputs;
2. *Modeling systems* which allows one to formulate the problems and analyze their solutions using the descriptions of linear programs in a natural and convenient form as inputs allowing the solution output to be viewed in similar terms through automatic conversion to the forms required by algorithmic codes.

The collection of statement forms for the input is often called a *modeling language*.

Large-scale linear programming algorithmic codes rely on general-structure sparse matrix techniques and other refinements developed. In additional to a variety of codes available, specialized toolboxes are applied. For example, in the MATLAB environment, the Optimization Toolbox can be very effectively used.

## *Nonlinear Programming*

In contrast to the linear programming, the *Nonlinear Programming* is a problem that can be formulated as

minimize $F(x)$

subject to $g_i(x)=0$, $h_j(x)\ge 0$ for $i=1,\ldots, n$, $n\ge 0$, $j=n+1,\ldots, m$, $m\ge n$.

Other formulations of the nonlinear programming can be set. It is evident that one minimizes the scalar-valued function $F$ of several variables ($x$ is the vector) subject to one or more other functions that serve to limit or define the values of these variables. Here, $F(x)$ is the objective function (criterion), while the other functions are called the constraints. If maximization is needed, one multiplies $F(x)$ by $-1$.

Nonlinear programming is a much more difficult problem compared with the linear programming. As a result, the special cases have been studied. The solution is found if the constraints $g_i(x)$ and $h_j(x)$ are linear (the linearly constrained optimization problem). If the objective function $F(x)$ is quadratic, the problem is called quadratic programming. One of the greatest challenges in nonlinear programming is the issues associated with the local optima where the requirements imposed on the derivatives of the functions are satisfied. Algorithms that overcome this difficulty are called the global optimization algorithms, and the corresponding techniques are available.

To solve the nonlinear programming problems, specific codes are used because, in general, globally optimal solutions are sought. The nonlinear optimization can be performed in MATLAB using the Optimization Toolbox. The MAPLE and MATHEMATICAL (Global Optimization Toolbox) environments are also available to perform nonlinear optimization.

The analysis of electromagnetic micro- and nanoscale structures and transducers will be covered in the next chapters. However, even with the incomplete background, let us illustrate the application of nonlinear programming. It was documented that the micromotors develop the electromagnetic torque

$$T_e = k_T D_r^2 L_r .$$

That is, the $T_e$ is proportional to the squared rotor diameter and axial length. Here, $k_T$ is the constant.

It is frequently desired to maximize the electromagnetic torque in order to attain high torque density within the allowed micromotor dimension. That is, using the rotor diameter and axial length as the state variables, the nonlinear programming problem results (in general, $D_r$ and $L_r$ are related due to electromagnetic features and fabrication processes). It will be illustrated that the simple formulation reported is the idealization of complex electromagnetic phenomena and effects in micro- and nanoscale transducers. In general, the nonlinear programming in design and optimization of MEMS and NEMS must be formulated using Maxwell's equations or other comprehensive mathematical models which describe complex phenomena and effects with minimum level of simplifications and assumptions.

## 4.2 NANOELECTROMECHANICAL SYSTEMS

Carbon nanotubes, discovered in 1991, are molecular structures which consist of graphene cylinders closed at either end with caps containing pentagonal rings [2]. Carbon nanotubes are produced by vaporizing carbon graphite with an electric arc under an inert atmosphere. The carbon molecules organize a perfect network of hexagonal graphite rolled up onto itself to form a hollow tube. Buckytubes are extremely strong and flexible and can be single- or multi-walled. The standard arc-evaporation method produces only multilayered tubes, and the single-layer uniform nanotubes (constant diameter) were synthesized only in the late 1990s. One can fill nanotubes with any media, including biological materials. The carbon nanotubes can be conducting or insulating medium depending upon their structure.

A single-walled carbon nanotube (one atom thick), which consists of carbon molecules, is illustrated in Figure 4.2.1. The application of these nanotubes, formed with a few carbon atoms in diameter, provides the possibility to fabricate devices on an atomic and molecular scale. The diameter of a nanotube is 100,000 times less that the diameter of the sawing needle. The carbon nanotubes, which are much stronger than steel wire (for the comparison, it is assumed that the diameter is the same), are the perfect conductor (better than silver), and have thermal conductivity better than diamond. In carbon nanotubes, carbon atoms bond together forming the pattern. Single-wall carbon nanotubes are manufactured using laser vaporization, arc technology, vapor growth, as well as other methods. Figure 4.2.2. illustrates the carbon ring with six atoms. When such a sheet rolls itself into a tube so that its edges join seamlessly together, a nanotube is formed.

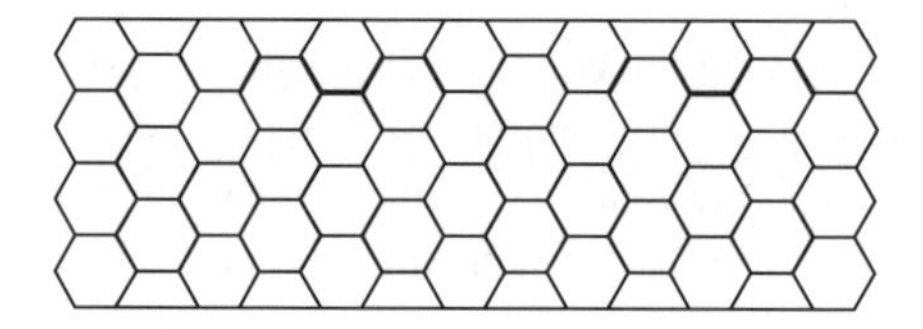

Figure 4.2.1 Single-walled carbon nanotube.

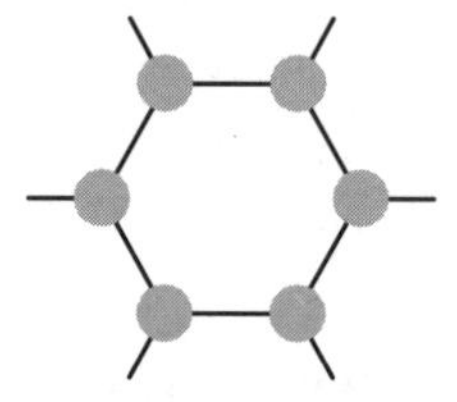

Figure 4.2.2 Single carbon nanotube ring with six atoms.

Carbon nanotubes can be widely used in MEMS and NEMS. Two slightly displaced (twisted) nanotube molecules, joined end to end, act as the diode. Molecular-scale transistors can be manufactured using different alignments. There are strong relationships between the nanotube electromagnetic properties and its diameter and degree of the molecule twist. In fact, the electromagnetic properties of the carbon nanotubes depend on the molecule's twist, and Figures 4.2.3 illustrate the possible configurations. If the graphite sheet forming the single-wall carbon nanotube is rolled up perfectly (all its hexagons line up along the molecules axis), the nanotube is a perfect conductor. If the graphite sheet rolls up at a twisted angle, the nanotube exhibits the semiconductor properties. The carbon nanotubes, which are stronger than steel wire, can be added to the plastic to make the conductive composite materials.

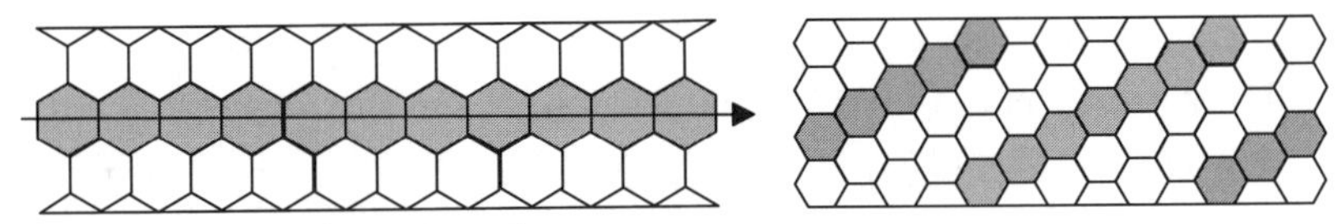

Figure 4.2.3 Carbon nanotubes.

The vapor grown carbon nanotubes with N layers are illustrated in Figure 4.2.4, and the industrially manufactured nanotubes have Angstroms diameter and nanometers length.

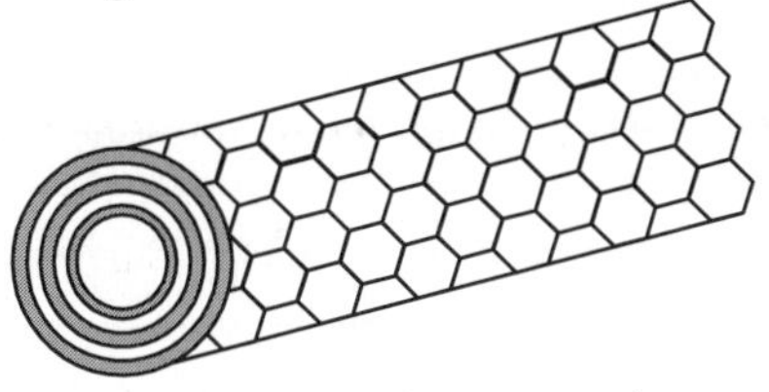

Figure 4.2.4 Three-layer carbon nanotube.

The carbon nanotubes can be organized as large-scale complex neural networks to perform computing and data storage, sensing and actuation, etc. The density of ICs designed and manufactured using the carbon nanotube technology exceeds by thousands of times the density of ICs developed using conventional currently available silicon-based technologies even if nanoscale lithography processes are applied.

Metallic solids (conductors, for example copper, silver, and iron) consist of metal atoms. These metallic solids usually have hexagonal, cubic, or body-centered cubic close-packed structures, see Figures 4.2.5. Each atom has 8 or 12 adjacent atoms. The bonding is due to valence electrons that are delocalized throughout the entire solid. The mobility of electrons is examined to study the conductivity properties.

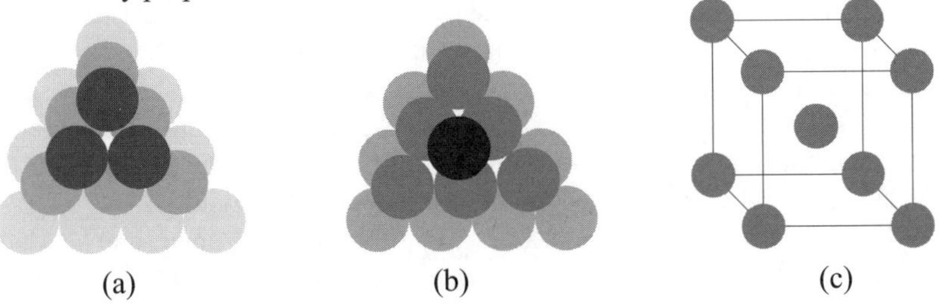


Figure 4.2.5 Close packing of metal atoms: (a) cubic packing; (b) hexagonal packing; (c) body-centered cubic packing.

More than two electrons can fit in an orbital. Furthermore, these two electrons must have two opposite spin states (spin-up and spin-down).

Therefore, the spins are said to be paired. Two opposite directions in which the electron spins (up $+\frac{1}{2}$ and down $-\frac{1}{2}$) produce oppositely directed magnetic fields. For an atom with two electrons, the spin may be either parallel (S = 1) or opposed and thus cancel (S = 0). Because of spin pairing, most molecules have no net magnetic field, and these molecules are called *diamagnetic*. In the absence of the external magnetic field, the net magnetic field produced by the magnetic fields of the orbiting electrons and the magnetic fields produced by the electron spins is zero. The external magnetic field will produce no torque on the *diamagnetic* atom as well as no realignment of the dipole fields. Accurate quantitative analysis can be performed using the quantum theory. Using the simplest atomic model, we assume that a positive nucleus is surrounded by electrons which orbit it in various circular orbits (an electron on the orbit can be studied as a current loop, and the direction of current is opposite to the direction of the electron rotation). The torque tends to align the magnetic field, produced by the orbiting electron, with the external magnetic field. The electron can have a spin magnetic moment of $\pm 9 \times 10^{-24}$ A-m$^2$. The plus and minus signs indicate that there are two possible electron alignments; in particular, aiding or opposing to the external magnetic field. The atom has many electrons, and only the spins of those electrons in shells which are not completely filed contribute to the atom's magnetic moment. The nuclear spin contributes only slightly to the atom moment. The magnetic properties of the media (diamagnetic, paramagnetic, superparamagnetic, ferromagnetic, antiferromagnetic, ferrimagnetic) result due to the combination of the listed atom moments.

Let us briefly emphasize the *paramagnetic* materials. The atom can have small magnetic moment, however, the random orientation of the atoms results that the net torque is zero. Thus, the media do not show the magnetic effect in the absence the external magnetic field. As the external magnetic field is applied, due to the atom moments, the atoms will align with the external field. If the atom has large dipole moment (due to electron spin moments), the material is called ferromagnetic. In antiferromagnetic materials, the net magnetic moment is zero, and thus the ferromagnetic media are only slightly affected by the external magnetic field.

Using carbon nanotubes, one can design electromechanical and electromagnetic nanoswitches, which are illustrated in Figure 4.2.6.

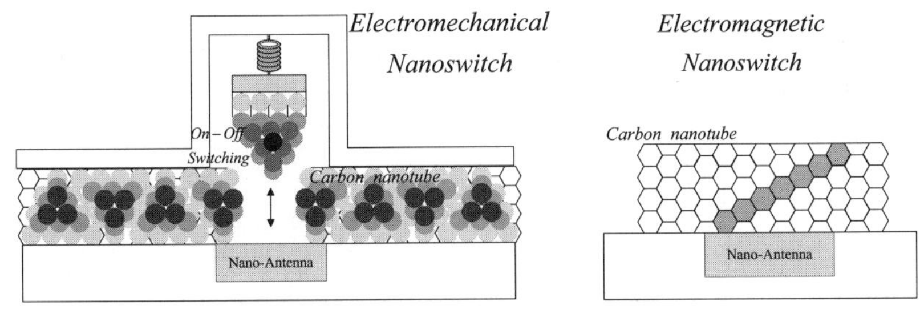


Figure 4.2.6 Application of carbon nanotubes in nanoswitches.

The carbon-based nanoelectronics is the revolutionary breakthrough in electronics [1, 2]. Fibers made using carbon nanotubes are more than 100 times stronger and weight 5 times less than steel (for the same diameter), have conductivity 5 times greater than silver, and transmit heat better than diamond. The current density of carbon nanotubes, 1,4-dithiol benzene (molecular wire) and copper are $10^{11}$, $10^{12}$ and $10^{6}$ electroncs/sec-nm$^2$, respectively. The current technologies and processes allow one to fill carbon nanotubes with other media (metals, as well as other inorganic and organic materials). The electromagnetic properties of the carbon nanotubes depend on the molecules' twist. If the graphite sheet forming the single-wall carbon nanotube is rolled up perfectly (all its hexagons line up along the molecules axis), the nanotube is a perfect conductor. If the graphite sheet rolls up at a twisted angle, the nanotube exhibits semiconductor properties. Carbon nanotubes can be implemented as nanostructures in high-speed, robust (high-noise immunity), high bandwidth, low-power, compatible, logic gates, switches, and analog to digital pipeline converters. These logic gates, switches, and converters can have the direct application in nanocomputers, wireless communication and networks. The carbon nanotube and molecular wires, used as building blocks to design nanodevices, and an electromagnetic nanoswitch, are shown in Figure 4.2.7.

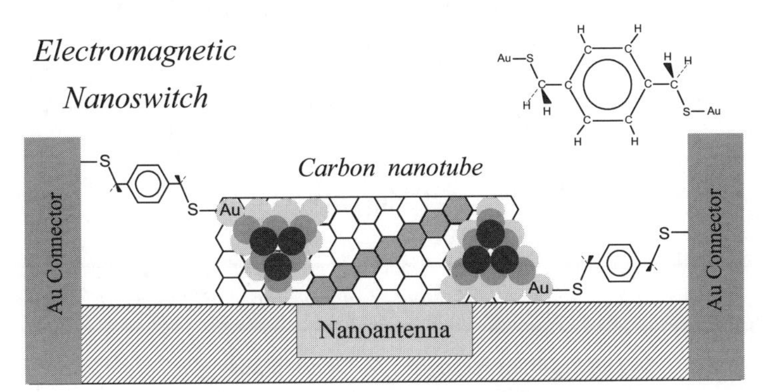


Figure 4.2.7 Nanoswitch with carbon nanotube, molecular wire (1,4-dithiol benzene), and nanoantenna.

In molecular wires, the current $i_m$ is a function of the applied voltage $u_m$, and Landauer's formula is

$$i_m = \frac{2e}{h}\int_{-\infty}^{+\infty} T(E_m, u_m)\left(\frac{1}{e^{\frac{E_m-\mu_{p1}}{k_BT}}+1} - \frac{1}{e^{\frac{E_m-\mu_{p2}}{k_BT}}+1}\right)dE_m\,,$$

where $\mu_{p1}$ and $\mu_{p2}$ are the electrochemical potentials, $\mu_{p1} = E_F + \frac{1}{2}eu_m$ and $\mu_{p2} = E_F - \frac{1}{2}eu_m$; $E_F$ is the equilibrium Fermi energy of the source; $T(E_m, u_m)$ is the transmission function obtained using the molecular energy levels and coupling.

Using the results reported in [3], we have

$$i_m = \frac{2e}{h}\int_{\mu_{p1}}^{\mu_{p2}} T(E_m, u_m)\frac{1}{4k_BT}\operatorname{sech}^2\left(\frac{E_m}{2k_BT}\right)dE_m\,,\ k_BT = 26\text{meV}.$$

The molecular wire conductance is found as

$$c_m = \frac{\partial i_m}{\partial u_m} \approx \frac{e^2}{h}\left[T(\mu_{p1}) + T(\mu_{p2})\right].$$

Micro- and nanoelectromechanical systems performance directly depends upon topologies, fabrication technologies, materials and processes, optimization, mathematical models, simulation software, as well as design and analysis capabilities. Synthesis, classification, and structural optimization allow the designer to devise novel phenomena and new operational principles guaranteeing synthesis of superior MEMS and NEMS with enhanced functionality through optimization. To design high-performance MEMS and NEMS, fundamental, applied, and experimental research must be performed devising new and further developing existing MEMS and NEMS through micro- and nanoelectromechanics. Several fundamental electromagnetic and mechanical laws are: quantum mechanics, Maxwell's equations, nonlinear electromechanics, and energy conversion. Therefore, next chapters cover mathematical model development, analysis, simulation, and control issues.

## References

1. S. E. Lyshevski, *Micro- and nano-Electromechanical Systems: Fundamental of Micro- and Nano- Engineering*, CRC Press, Boca Raton, FL, 2000.
2. R. Saito, G. Dresselhaus and M. S. Dresselhaus, *Physical Properties of Carbon Nanotubes*, Imperial College Press, London, 1999.
3. W. T. Tian, S. Datta, S. Hong, R. Reifenberger, J. I. Henderson and C. P. Kubiak, "Conductance spectra of molecular wires," *Int. Journal Chemical Physics*, vol. 109, no. 7, pp. 2874-2882, 1998.

# CHAPTER 5

# MODELING OF MICRO- AND NANOSCALE ELECTROMECHANICAL SYSTEMS, DEVICES, AND STRUCTURES

## 5.1 INTRODUCTION TO MODELING, ANALYSIS, AND SIMULATION

The design and analysis of micro- and nanoscale electromechanical systems, devices, and structures is not a simple task because electromagnetic, mechanical, thermodynamic, vibroacoustic and other problems must be studied in the time domain solving partial differential equations.

Computer-aided design of systems, devices, and structures is valuable due to

- Calculation and thorough evaluation of a large number of options with performance analysis and outcome prediction;
- Knowledge-based intelligent synthesis and evolutionary design which allow one to define optimal solution with minimal effort, time, cost, reliability, and accuracy;
- Data-intensive nonlinear electromagnetic and mechanical analysis to attain superior performance of systems, devices, and structures while avoiding costly and time-consuming fabrication and testing;
- Possibility to solve complex partial differential equations in the time domain integrating nonlinear media characteristics;
- Feasibility to develop robust and accurate rapid design software tools which have innumerable features to assist the user to set the problem up and to obtain the engineering parameters;
- Synergy between high-performance software and hardware developments.

Through the synthesis, classification and structural optimization, reported in Chapter 4, the designer devises micro- and nanoscale systems and devices which must be modeled, simulated, analyzed, and optimized. Heterogeneous synthesis, design and optimization guarantee the superior performance capabilities. Devising and developing novel MEMS and NEMS, one can maximize efficiency, reliability, power and torque densities, ruggedness, robustness, durability, survivability, compactness, simplicity, controllability, and accuracy. These features are accomplished while minimizing cost, maintenance, size, weight, volume, and losses, as well as optimizing packaging and integrity, see Figure 5.1.1.

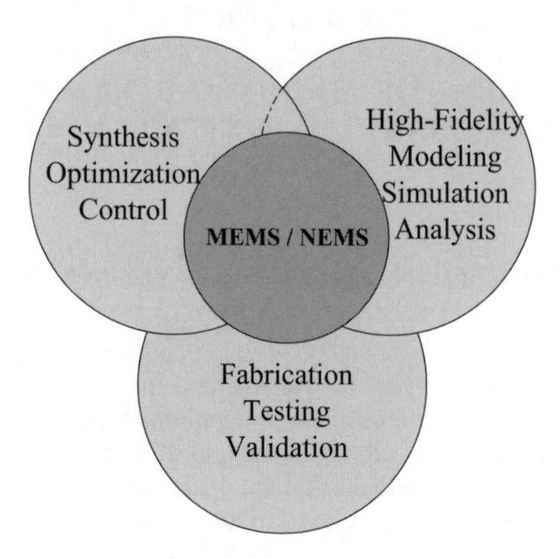


Figure 5.1.1 Design of high-performance MEMS and NEMS.

Micro- and nanotechnologies drastically change the fabrication and manufacturing of materials, devices, and systems through:

- Predictable properties of nano composites and materials (e.g., light weight, high strength, thermal stability, low volume and size, extremely high power, torque, force, charge and current densities, specified thermal conductivity and resistivity, etc.),
- Virtual prototyping (design cycle, cost, and maintenance reduction),
- Improved accuracy, precision, reliability and durability,
- High degree of efficiency and capability,
- Guaranteed flexibility, integrity and adaptability,
- Attained supportability and affordability,
- Survivability, redundancy and safety,
- Improved stability and robustness,
- Environmental competitiveness.

Using the fabrication technologies (reported in Chapters 3 and 8), MEMS and NEMS devised using the Synthesis and Classifier Solver, can be fabricated. As was reported in Chapter 4, synthesis, classification and structural optimization are based on the consideration and synthesis of the electromagnetic system, analysis of the *magnetomotive* force, geometrical and topological designs, biomimicking, and optimization. For example, different rotational (radial and axial) and translational motion microdevices (transducers) are classified using *endless* (closed), *open-ended* (open), and *integrated* electromagnetic systems. Our goal is to approach and solve a wide range of practical problems encountered in nonlinear design, modeling, analysis, control, and optimization of micro- and nanoscale electromechanical systems, devices, and structures.

Studying MEMS and NEMS, the emphasis is placed on:

- Devising and design of high-performance systems devising innovative motion and radiating energy devices, micro- and nanoscale driving/sensing circuitry, as well as controlling/signal processing ICs;
- Optimization and analysis of rotational and translation motion devices;
- Development of high-performance signal processing and controlling ICs;
- Development of high-fidelity mathematical models with minimum level of simplifications and assumptions in the time domain;
- Attaining the data-intensive analysis capabilities using complete mathematical models and robust computationally effective methods and analytical/numerical algorithms to solve nonlinear equations;
- Design of optimal robust control architectures and algorithms;
- Design of intelligent systems through self-adaptation, self-organization, evolutionary learning, decision-making, and intelligence;
- Development of advanced software and hardware to attain the highest degree of synergy, intelligence, integration, efficiency, and performance.

In this chapter, our goal is to perform nonlinear modeling, analysis, and simulation. To attain these objectives, we apply the synthesis paradigm (Synthesis and Classification Solver), develop nonlinear mathematical models to model complex electromagnetic-mechanical (electromechanical) dynamics, perform optimization, design closed-loop control systems, and perform data-intensive analysis in the time domain.

For MEMS and NEMS, devices and structures, many engineering problems can be formulated, attacked, and solved using the micro- and nanoelectromechanics. These paradigms deal with benchmarking and emerging problems in integrated electrical–mechanical–computer engineering, science, and technology. Many of these problems have not been attacked and solved, and in general, the existing solutions cannot be treated as optimal. This reflects obvious trends in synergetic fundamental, applied and experimental research in response to long-standing unsolved problems, engineering and technological enterprise and entreaties of steady evolutionary demands.

Micro- and nanoelectromechanics focus on the integrated design, data-intensive analysis, heterogeneous simulation, optimization and virtual prototyping of intelligent and high-performance MEMS and NEMS. In addition, other important problems are: system intelligence, evolutionary learning, adaptation, decision-making, and control through the use of advanced hardware devised and leading-edge software.

Integrated multidisciplinary features approach quickly. The structural complexity of micro- and nanoscale systems and devices has been increased drastically due to hardware and software advancements as well as stringent *achievable* performance requirements imposed. Answering the demands of the rising systems complexity, performance specifications, and intelligence, the fundamental theory must be further expanded. In particular, in addition to devising subsystems, devices, and structures, there are other issues which must be addressed and solved in view of constantly evolving nature of the MEMS and NEMS (e.g., analysis, design, modeling, optimization, complexity, intelligence, decision-making, diagnostics, packaging, etc.). Competitive *optimum-performance* MEMS and NEMS must be designed within the advanced hardware and software concepts.

One of the most challenging problems in MEMS and NEMS systems design is devising novel high-performance motion and radiating energy devices, architecture/configuration synthesis, system integration, optimization, as well as selection of hardware and software (environments, tools and computation algorithms to perform control, sensing, execution, emulation, information flow, data acquisition, simulation, visualization, virtual prototyping and evaluation). As was emphasized, the attempts to design state-of-the-art high-performance MEMS and NEMS and to guarantee the integrated design can be pursued through analysis of complex patterns and paradigms of evolutionary developed biological systems.

## 5.2 ELECTROMAGNETICS AND ITS APPLICATION FOR MEMS AND NEMS

To study MEMS and NEMS, micro- and nanoscale devices and structures, ICs and antennas, one applies the electromagnetic field theory and mechanics. Electric force holds atoms and molecules together. Electromagnetics plays a central role in molecular biology. For example, two DNA (deoxyribonucleic acid) chains wrap about one another in the shape of a *double helix*. These two strands are held together by electrostatic forces. Electric force is responsible for energy-transforming processes in all living organisms (*metabolism*). Electromagnetism is used to study protein synthesis and structure, nervous system, etc.

Electrostatic interaction was investigated by Charles Coulomb.

For charges $q_1$ and $q_2$, separated by a distance $x$ in free space, the magnitude of the electric force is

$$F = \frac{|q_1 q_2|}{4\pi\varepsilon_0 x^2},$$

where $\varepsilon_0$ is the permittivity of free space, $\varepsilon_0 = 8.85\times10^{-12}$ F/m or C$^2$/N-m$^2$, and $\frac{1}{4\pi\varepsilon_0} = 9\times10^9$ N-m$^2$/C.

The unit for the force is the newton [N], while the charges are given in coulombs [C].

The force is the vector. Therefore, in general, we have

$$\vec{F} = \frac{q_1 q_2}{4\pi\varepsilon_0 x^2}\vec{a}_x,$$

where $\vec{a}_x$ is the unit vector which is directed along the line joining these two charges.

The capacity, elegance and uniformity of electromagnetics arise from a sequence of fundamental laws linked together and needed to study the field quantities.

We denote the vector of electric flux density as $\vec{D}$ [F/m] and the vector of electric field intensity as $\vec{E}$ [V/m or N/C]. Using the Gauss law, the total electric flux $\Phi$ [C] through a closed surface is found to be equal to the total force charge enclosed by the surface. That is,

$$\Phi = \oint_s \vec{D} \cdot d\vec{s} = Q_s \, , \; \vec{D} = \varepsilon \vec{E} \, ,$$

where $d\vec{s}$ is the vector surface area, $d\vec{s} = ds\vec{a}_n$, $\vec{a}_n$ is the unit vector which is normal to the surface; $\varepsilon$ is the permittivity of the medium; $Q_s$ is the total charge enclosed by the surface.

Ohm's law relates the volume charge density $\vec{J}$ and electric field intensity $\vec{E}$. In particular,

$$\vec{J} = \sigma \vec{E} \, ,$$

where $\sigma$ is the conductivity [A/V-m], for copper $\sigma = 5.8 \times 10^7$, and for aluminum $\sigma = 3.5 \times 10^7$.

The current $i$ is proportional to the potential difference, and the resistivity $\rho$ of the conductor is the ratio between the electric field $\vec{E}$ and the current density $\vec{J}$. Thus,

$$\rho = \frac{\vec{E}}{\vec{J}} \, .$$

The resistance $r$ of the conductor is related to the resistivity and conductivity by the following formulas

$$r = \frac{\rho l}{A} \text{ and } r = \frac{l}{\sigma A} \, ,$$

where $l$ is the length; $A$ is the cross-sectional area.

It is important to emphasize that the parameters vary. Let us illustrate this using the molecular wire. The resistances of the wire vary due to heating. In fact, the resistivity depends on temperature $T$ [°C], and we have

$$\rho(T) = \rho_0 \left[ 1 + \alpha_{\rho 1}\left(T - T_0\right) + \alpha_{\rho 2}\left(T - T_0\right)^2 + \ldots \right] ,$$

where $\alpha_{\rho 1}$ and $\alpha_{\rho 2}$ are the coefficients.

As an example, over the small temperature range (up to 160°C) for copper at $T_0$ = 20°C, we have $\rho(T) = 1.7 \times 10^{-8} \left[ 1 + 0.0039\left(T - 20\right) \right]$.

The basic principles of electromagnetic theory should be briefly reviewed.

The total magnetic flux through the surface is given by

$$\Phi = \int \vec{B} \cdot d\vec{s},$$

where $\vec{B}$ is the magnetic flux density.

The Ampere circuital law is

$$\int_l \vec{B} \cdot d\vec{l} = \mu_0 \int_s \vec{J} \cdot d\vec{s},$$

where $\mu_o$ is the permeability of free space, $\mu_o = 4\pi \times 10^{-7}$ H/m or T-m/A.

For the filamentary current, Ampere's law connects the magnetic flux with the algebraic sum of the enclosed (linked) currents (*net current*) $i_n$, and

$$\oint_l \vec{B} \cdot d\vec{l} = \mu_o i_n.$$

The time-varying magnetic field produces the electromotive force (*emf*), denoted as $\mathscr{E}$, which induces the current in the closed circuit. Faraday's law relates the *emf* (induced voltage due to conductor motion in the magnetic field) to the rate of change of the magnetic flux $\Phi$ penetrating the loop. In approaching the analysis of electromechanical energy conversion, Lenz's law should be used to find the direction of *emf* and the current induced. In particular, the *emf* is in such a direction as to produce a current whose flux, if added to the original flux, would reduce the magnitude of the *emf*. According to Faraday's law, the induced *emf* in a closed-loop circuit is defined in terms of the rate of change of the magnetic flux $\Phi$. One has

$$\mathscr{E} = \oint_l \vec{E}(t) \cdot d\vec{l} = -\frac{d}{dt} \int_s \vec{B}(t) \cdot d\vec{s} = -N\frac{d\Phi}{dt} = -\frac{d\psi}{dt},$$

where $N$ is the number of turns; $\psi$ denotes the flux linkages.

This formula represents the Faraday law of induction, and the induced *emf* (*induced voltage*) given by $\mathscr{E} = -\frac{d\psi}{dt} = -N\frac{d\Phi}{dt}$, is our particular interest

The current flows in an opposite direction to the flux linkages. The electromotive force (*energy-per-unit-charge quantity*) represents a magnitude of the potential difference $V$ in a circuit carrying a current. We have

$$V = -ir + \mathscr{E} = -ir - \frac{d\psi}{dt}.$$

The unit for the *emf* is volts.

The Kirchhoff voltage law states that around a closed path in an electric circuit, the algebraic sum of the *emf* is equal to the algebraic sum of the voltage drop across the resistance.

Another formulation is: the algebraic sum of the voltages around any closed path in a circuit is zero.

The Kirchhoff current law states that the algebraic sum of the currents at any node in a circuit is zero.

The magnetomotive force (*mmf*) is the line integral of the time-varying magnetic field intensity $\vec{H}(t)$. That is,

$$mmf = \oint_l \vec{H}(t) \cdot d\vec{l}\ .$$

One concludes that the induced *mmf* is the sum of the induced current and the rate of change of the flux penetrating the surface bounded by the contour. To show that, we apply Stoke's theorem to find the integral form of Ampere's law (second Maxwell's equation), as given by

$$\oint_l \vec{H}(t) \cdot d\vec{l} = \int_s \vec{J}(t) \cdot d\vec{s} + \int_s \frac{d\vec{D}(t)}{dt} d\vec{s}\ ,$$

where $\vec{J}(t)$ is the time-varying current density vector.

The unit for the magnetomotive force is amperes or ampere-turns.

The duality of the *emf* and *mmf* can be observed using the following two equations given in terms of the electric and magnetic field intensity vectors

$$emf\ \mathscr{E} = \oint_l \vec{E}(t) \cdot d\vec{l}$$

$$mmf\ \ mmf = \oint_l \vec{H}(t) \cdot d\vec{l}\ .$$

The inductance (the ratio of the total flux linkages to the current which they link, $L = \frac{N\Phi}{i}$) and reluctance (the ratio of the *mmf* to the total flux, $\Re = \frac{mmf}{\Phi}$) are used to find *emf* and *mmf*.

Using the following equation for the self-inductance $L = \frac{\psi}{i}$, we have

$$\mathscr{E} = -\frac{d\psi}{dt} = -\frac{d(Li)}{dt} = -L\frac{di}{dt} - i\frac{dL}{dt}\ .$$

And, if $L = const$, one obtains $\mathscr{E} = -L\frac{di}{dt}$.

That is, the self-inductance is the magnitude of the self-induced *emf* per unit rate of change of current.

*Example 5.2.1*

Find the self-inductances of a nano-solenoid with air-core and filled-core ($\mu = 1000\mu_o$). The solenoid has 10 turns ($N$ = 10), the length is 100 nm ($l$=$1\times10^{-7}$ m), and the uniform circular cross-sectional area is $4\times10^{-17}$ $m^2$ ($A$=$4\times10^{-17}$ $m^2$).

*Solution.* The magnetic field inside a solenoid is $B = \dfrac{\mu_0 Ni}{l}$.

From $\mathscr{E} = -N\dfrac{d\Phi}{dt} = -L\dfrac{di}{dt}$ and applying $\Phi = BA = \dfrac{\mu_0 NiA}{l}$, one obtains the following expression for the self-inductance

$$L = \frac{\mu_0 N^2 A}{l}.$$

Then, for the solenoid with air-core one obtains $L = 1.6\pi\times10^{-14}$ H.

If the solenoid is filled with a magnetic material, we have

$$L = \frac{\mu N^2 A}{l}, \text{ and } L = 1.6\pi\times10^{-11} \text{ H}.$$ □

*Example 5.2.2*

Derive a formula for the self-inductance of a torroidal solenoid which has a rectangular cross section ($2a \times b$) and mean radius $r$.

*Solution.* The magnetic flux through a cross section is

$$\Phi = \int_{r-a}^{r+a} Bbdr = \int_{r-a}^{r+a} \frac{\mu Ni}{2\pi r} bdr = \frac{\mu Nib}{2\pi}\int_{r-a}^{r+a} \frac{1}{r}dr = \frac{\mu Nib}{2\pi}\ln\left(\frac{r+a}{r-a}\right).$$

Thus, $L = \dfrac{N\Phi}{i} = \dfrac{\mu N^2 b}{2\pi}\ln\left(\dfrac{r+a}{r-a}\right)$. □

By studying the electromagnetic torque $\vec{T}$ [N-m] in a current loop, one uses the following equation

$$\vec{T} = \vec{M} \times \vec{B},$$

where $\vec{M}$ denotes the magnetic moment.

Let us examine the torque-energy relations in micro- and nanoscale actuators. Our goal is to study the magnetic field energy. It is known that the energy stored in the capacitor is $\frac{1}{2}CV^2$, while the energy stored in the inductor is $\frac{1}{2}Li^2$. Observe that the energy in the capacitor is stored in the electric field between plates, while the energy in the inductor is stored in the magnetic field within the coils.

Let us find the expressions for energies stored in electrostatic and magnetic fields in terms of field quantities. The total potential energy stored in the electrostatic field is found using the potential difference $V$. We have

$$W_e = \tfrac{1}{2}\int_v \rho_v V dv \text{ [J]},$$

where $\rho_v$ is the volume charge density [C/m$^3$], $\rho_v = \vec{\nabla}\cdot\vec{D}$, $\vec{\nabla}$ is the curl operator.

Thus, the potential energy $W_e$ should be found using the amount of work which is required to position the charge in the electrostatic field. In particular, the work is found as the product of the charge and the potential.

Considering the region with a continuous charge distribution ($\rho_v = const$), each charge is replaced by $\rho_v dv$, and hence the following equation results

$$W_e = \frac{1}{2}\int_v \rho_v V dv.$$

In the Gauss form, by taking note of $\rho_v = \vec{\nabla}\cdot\vec{D}$ and making use of $\vec{E} = -\vec{\nabla}V$, one obtains the following expression for the energy stored in the electrostatic field

$$W_e = \frac{1}{2}\int_v \vec{D}\cdot\vec{E}dv,$$

and the electrostatic volume energy density is $\frac{1}{2}\vec{D}\cdot\vec{E}$ [J/m$^3$].

For a linear isotropic medium we have

$$W_e = \frac{1}{2}\int_v \varepsilon\left|\vec{E}\right|^2 dv = \frac{1}{2}\int_v \frac{1}{\varepsilon}\left|\vec{D}\right|^2 dv.$$

The electric field $\vec{E}(x,y,z)$ is found using the scalar electrostatic potential function $V(x,y,z)$ as

$$\vec{E}(x,y,z) = -\vec{\nabla}V(x,y,z).$$

In the cylindrical and spherical coordinate systems, we have

$$\vec{E}(r,\phi,z) = -\vec{\nabla}V(r,\phi,z) \text{ and } \vec{E}(r,\theta,\phi) = -\vec{\nabla}V(r,\theta,\phi).$$

Using $W_e = \frac{1}{2}\int_v \rho_v V dv$, the potential energy which is stored in the electric field between two surfaces (for example, in capacitor) is found to be

$$W_e = \frac{1}{2}QV = \frac{1}{2}CV^2.$$

Using the principle of virtual work, for the lossless conservative system, the differential change of the electrostatic energy $dW_e$ is equal to the differential change of mechanical energy $dW_{mec}$. That is

$$dW_e = dW_{mec}.$$

For translational motion, one has

$$dW_{mec} = \vec{F}_e \cdot d\vec{l},$$

where $d\vec{l}$ is the differential displacement.

One obtains $dW_e = \vec{\nabla}W_e \cdot d\vec{l}$.

Hence, the force is the gradient of the stored electrostatic energy,

$\vec{F}_e = \vec{\nabla} W_e$.

In the Cartesian coordinates, we have

$$F_{ex} = \frac{\partial W_e}{\partial x},\ F_{ey} = \frac{\partial W_e}{\partial y} \text{ and } F_{ez} = \frac{\partial W_e}{\partial z}.$$

*Example 5.2.3*

Consider the capacitor (the plates have area $A$ and they are separated by $x$), which is charged to a voltage $V$. The permittivity of the dielectric is $\varepsilon$. Find the stored electrostatic energy and the force $F_{ex}$ in the $x$ direction.

*Solution.* Neglecting the fringing effect at the edges, one concludes that the electric field is uniform, and $E = \frac{V}{x}$. Therefore,

$$W_e = \tfrac{1}{2}\int_v \varepsilon \left|\vec{E}\right|^2 dv = \tfrac{1}{2}\int_v \varepsilon \left(\frac{V}{x}\right)^2 dv = \tfrac{1}{2}\varepsilon \frac{V^2}{x^2} Ax = \tfrac{1}{2}\varepsilon \frac{A}{x} V^2 = \tfrac{1}{2} C(x) V^2.$$

Thus, the force is

$$F_{ex} = \frac{\partial W_e}{\partial x} = \frac{\partial\left(\tfrac{1}{2} C(x) V^2\right)}{\partial x} = \tfrac{1}{2} V^2 \frac{\partial C(x)}{\partial x}.$$

□

To find the stored energy in the magnetostatic field in terms of field quantities, the following formula is used

$$W_m = \tfrac{1}{2}\int_v \vec{B}\cdot\vec{H} dv.$$

The magnetic volume energy density is $\tfrac{1}{2}\vec{B}\cdot\vec{H}$ [J/m$^3$].

Using $\vec{B} = \mu\vec{H}$, one obtains two alternative formulas

$$W_m = \tfrac{1}{2}\int_v \mu\left|\vec{H}\right|^2 dv = \tfrac{1}{2}\int_v \frac{\left|\vec{B}\right|^2}{\mu} dv.$$

To show how the energy concept studied is applied to electromechanical devices, we find the energy stored in inductors. To approach this problem, we substitute

$\vec{B} = \vec{\nabla} \times \vec{A}$.

Using the following vector identity

$\vec{H}\cdot\vec{\nabla}\times\vec{A} = \vec{\nabla}\cdot\left(\vec{A}\times\vec{H}\right) + \vec{A}\cdot\vec{\nabla}\times\vec{H}$,

one obtains

$$W_m = \frac{1}{2}\int_v \vec{B}\cdot\vec{H}dv = \frac{1}{2}\int_v \vec{\nabla}\cdot\left(\vec{A}\times\vec{H}\right)dv + \frac{1}{2}\int_v \vec{A}\cdot\vec{\nabla}\times\vec{H}dv$$
$$= \frac{1}{2}\int_s \left(\vec{A}\times\vec{H}\right)\cdot d\vec{s} + \frac{1}{2}\int_v \vec{A}\cdot\vec{J}dv = \frac{1}{2}\int_v \vec{A}\cdot\vec{J}dv.$$

Using the general expression for the vector magnetic potential $\vec{A}(\vec{r})$ [Wb/m], as given by

$$\vec{A}(\vec{r}) = \frac{\mu_0}{4\pi}\int_{v_A}\frac{\vec{J}(\vec{r}_A)}{x}dv_J\,,\ \vec{\nabla}\cdot\vec{A} = 0,$$

we have

$$W_m = \frac{\mu}{8\pi}\int_v\int_{v_J}\frac{\vec{J}(\vec{r}_A)\cdot\vec{J}(\vec{r})}{x}dv_J dv\,.$$

Here, $v_J$ is the volume of the medium where $\vec{J}$ exists.

The general formula for the self-inductance $i = j$ and the mutual inductance $i \neq j$ of loops *i* and *j* is

$$L_{ij} = \frac{N_i\Phi_{ij}}{i_j} = \frac{\psi_{ij}}{i_j},$$

where $\psi_{ij}$ is the flux linkage through *i*th coil due to the current in *j*th coil; $i_j$ is the current in *j*th coil.

The *Neumann* formula is applied to find the mutual inductance. We have,

$$L_{ij} = L_{ji} = \frac{\mu}{4\pi}\oint_{l_i}\oint_{l_j}\frac{d\vec{l}_j\cdot d\vec{l}_i}{x_{ij}}, i \neq j\,.$$

Then, using $W_m = \frac{\mu}{8\pi}\int_v\int_{v_J}\frac{\vec{J}(\vec{r}_A)\cdot\vec{J}(\vec{r})}{x}dv_J dv$, one obtains

$$W_m = \frac{\mu}{8\pi}\int_{l_i}\int_{l_j}\frac{i_j d\vec{l}_j\cdot i_i d\vec{l}_i}{x_{ij}}.$$

Hence, the energy stored in the magnetic field is found to be

$$W_m = \tfrac{1}{2}i_i L_{ij} i_j\,.$$

As an example, the energy, stored in the inductor is $W_m = \frac{1}{2}Li^2$.

The differential change in the stored magnetic energy should be found. Using

$$\frac{dW_m}{dt} = \frac{1}{2}\left(L_{ij}i_j\frac{di_i}{dt} + L_{ij}i_i\frac{di_j}{dt} + i_i i_j\frac{dL_{ij}}{dt}\right),$$

we have

$$dW_m = \tfrac{1}{2}\left(L_{ij} i_j di_i + L_{ij} i_i di_j + i_i i_j dL_{ij}\right).$$

For translational motion, the differential change in the mechanical energy is expressed by

$$dW_{mec} = \vec{F}_m \cdot d\vec{l}\ .$$

Assuming that the system is conservative (for lossless systems $dW_{mec} = dW_m$), in the rectangular coordinate system we obtain the following equation

$$dW_m = \frac{\partial W_m}{\partial x} dx + \frac{\partial W_m}{\partial y} dy + \frac{\partial W_m}{\partial z} dz = \vec{\nabla} W_m \cdot d\vec{l}\ .$$

Hence, the force is the gradient of the stored magnetic energy, and

$$\vec{F}_m = \vec{\nabla} W_m .$$

In the XYZ coordinate system for the translational motion, we have

$$F_{mx} = \frac{\partial W_m}{\partial x},\ F_{my} = \frac{\partial W_m}{\partial y} \text{ and } F_{mz} = \frac{\partial W_m}{\partial z}.$$

For the rotational motion, the torque should be used. Using the differential change in the mechanical energy as a function of the angular displacement $\theta$, the following formula results if the rigid body (nano- or microactuator) is constrained to rotate about the *z*-axis

$$dW_{mec} = T_e d\theta ,$$

where $T_e$ is the *z*-component of the electromagnetic torque.

Assuming that the system is lossless, one obtains the following expression for the electromagnetic torque

$$T_e = \frac{\partial W_m}{\partial \theta}.$$

*Example 5.2.4*

Calculate the magnetic energy of the torroidal microsolenoid if the self-inductance is $L=2\times10^{-11}$ H when the current is $i=1\times10^{-6}$ A.

*Solution.* The stored field energy is

$$W_m = \tfrac{1}{2} L i^2 .$$

Hence,

$W_m=1\times10^{-23}$ J. □

*Example 5.2.5*

Derive the expression for the electromagnetic force developed by the microelectromagnet with the cross-sectional area *A*. Let the current $i_a(t)$ in *N* coils produces the constant flux $\Phi_m$, see Figure 5.2.1.

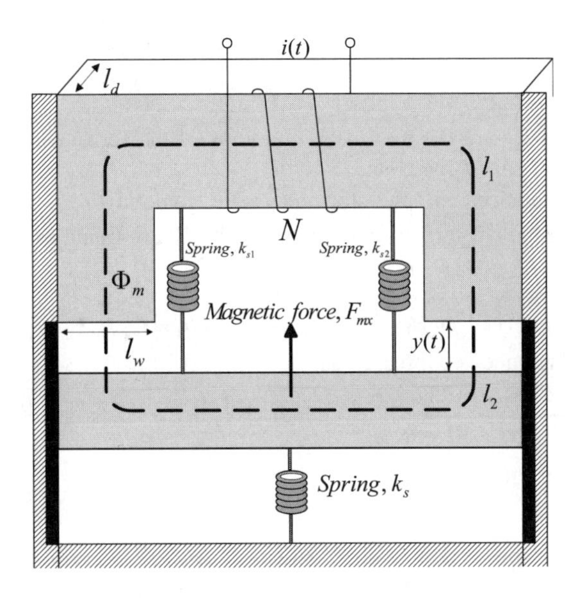


Figure 5.2.1 Microelectromagnet with springs.

*Solution.*

We assume that the flux is constant. Taking into the account the fact that the displacement (the virtual displacement is denoted as *dy*) changes only the magnetic energy stored in the air gaps, from

$$W_m = \frac{1}{2}\int_v \mu\left|\vec{H}\right|^2 dv = \frac{1}{2}\int_v \frac{\left|\vec{B}\right|^2}{\mu} dv\,,$$

we have

$$dW_m = dW_{m\,air\,gap} = 2\frac{B^2}{2\mu_0}Ady = \frac{\Phi_m^2}{\mu_0 A}dy\,.$$

Thus, if $\Phi_m$ = const (the current is constant), one concludes that the increase of the air gap (*dy*) leads to increase of the stored magnetic energy. Using $F_{mx} = \frac{\partial W_m}{\partial x}$, one finds the expression for the force in the following form

$$\vec{F}_{mx} = -\vec{a}_y \frac{\Phi_m^2}{\mu_0 A}\,.$$

The result indicates that the force tends to reduce the air-gap length, and the movable member is attached to the springs which develop three forces in addition to the electromagnetic force.

The *fringing* effect should be integrated in high-fidelity modeling. The air gap reluctance (two air gaps are in series) is expressed as

$$\Re_g = \frac{2x}{\mu_0\left(k_{g1}l_w l_d + k_{g2}x^2\right)},$$

where $k_{g1}$ and $k_{g2}$ are the nonlinear functions of the ferromagnetic material, $l_d/l_w$ ratio, $B$–$H$ curve, load, etc.

The reluctances of the ferromagnetic materials of stationary and movable members (microstructures) $\Re_1$ and $\Re_2$ are found as

$$\Re_1 = \frac{l_1}{\mu_0\mu_1 A} = \frac{l_1}{\mu_0\mu_1 l_w l_d} \text{ and } \Re_2 = \frac{l_2}{\mu_0\mu_2 A} = \frac{l_2}{\mu_0\mu_2 l_w l_d}.$$

The inductance is expressed as

$$L(x) = \frac{N^2}{\Re_g(x) + \Re_1 + \Re_2}.$$

The electromagnetic torque is

$$F_{mx} = \tfrac{1}{2}i^2\frac{dL(x)}{dx} = \tfrac{1}{2}i^2\frac{d\left(\dfrac{N^2}{\Re_g(x) + \Re_1 + \Re_2}\right)}{dx}$$

$$= \tfrac{1}{2}i^2\frac{d\left(\dfrac{N^2}{\dfrac{2}{\mu_0}\left(\dfrac{l_w l_d}{x} + k_{g1}l_w + k_{g2}l_d + k_{g3}x\right)^{-1} + \Re_1 + \Re_2}\right)}{dx}.$$

□

In micro- and nanoscale electromechanical motion devices, the coupling (magnetic interaction) between windings that are carrying currents is represented by their mutual inductances. In fact, the current in each winding causes the magnetic field in other windings. The mutually induced *emf* is characterized by the mutual inductance which is a function of the position $x$ or the angular displacement $\theta$. By applying the expressions for the coenergy $W_c\left[i, L(x)\right]$ (translational motion) or $W_c\left[i, L(\theta)\right]$ (rotational motion), the developed electromagnetic torque can be found as

$$T_e(i,x) = \frac{\partial W_c[i, L(x)]}{\partial x}$$

and

$$T_e(i,x) = \frac{\partial W_c[i, L(\theta)]}{\partial \theta}.$$

*Example 5.2.6*

Consider the microactuator (microelectromagnet) which has $N$ turns, see Figure 5.2.2. The distance between the stationary and movable members is denoted as $x(t)$. The mean lengths of the stationary and movable members are $l_1$ and $l_2$, and the cross-sectional area is $A$. Neglecting the leakage flux, find the force exerted on the movable member if the time-varying current $i_a(t)$ is supplied (one can feed sinusoidal current to the winding). The permeabilities of stationary and movable members are $\mu_1$ and $\mu_2$.

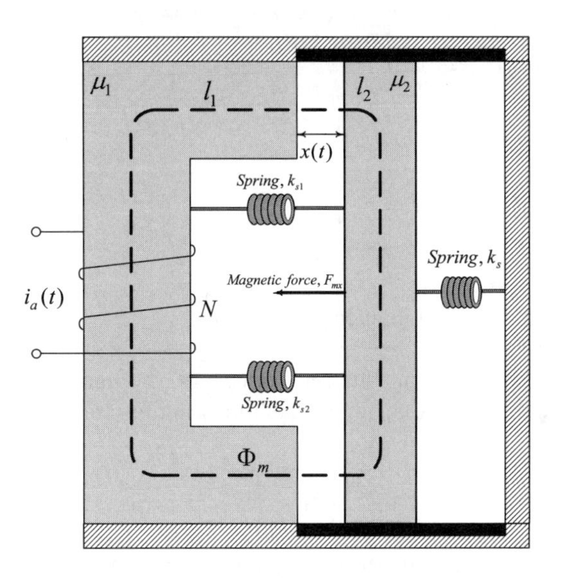


Figure 5.2.2 Schematic of a microelectromagnet.

*Solution.*

The electromagnetic force is

$$F_{mx} = \frac{\partial W_m}{\partial x},$$

where $W_m = \frac{1}{2} L i_a^2(t)$.

The magnetizing inductance is

$$L = \frac{N\Phi}{i_a(t)} = \frac{\psi}{i_a(t)},$$

where the magnetic flux is found as

$$\Phi = \frac{N i_a(t)}{\Re_1 + \Re_x + \Re_x + \Re_2}.$$

The reluctances of the ferromagnetic materials of stationary and movable members $\Re_1$ and $\Re_2$, as well as the reluctance of the air gap $\Re_x$, are found as

$$\Re_1 = \frac{l_1}{\mu_0 \mu_1 A}, \; \Re_2 = \frac{l_2}{\mu_0 \mu_2 A} \text{ and } \Re_x = \frac{x(t)}{\mu_0 A}$$

and the circuit analog with the reluctances of the various paths is illustrated in Figure 5.2.3.

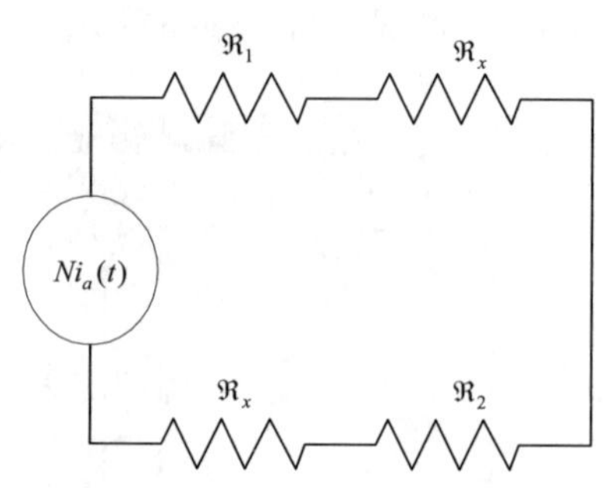


Figure 5.2.3 Circuit analog.

By making use of the reluctances in the movable and stationary members and air gap, one obtains the following formula for the flux linkages

$$\psi = N\Phi = \frac{N^2 i_a(t)}{\frac{l_1}{\mu_0 \mu_1 A} + \frac{2x(t)}{\mu_0 A} + \frac{l_2}{\mu_0 \mu_2 A}},$$

and the magnetizing inductance is a nonlinear function of the displacement.
Thus, we have

$$L(x) = \frac{N^2}{\frac{l_1}{\mu_0 \mu_1 A} + \frac{2x(t)}{\mu_0 A} + \frac{l_2}{\mu_0 \mu_2 A}} = \frac{N^2 \mu_0 \mu_1 \mu_2 A}{\mu_2 l_1 + \mu_1 \mu_2 2x(t) + \mu_1 l_2}.$$

Using

$$F_{mx} = \frac{\partial W_m}{\partial x} = \frac{1}{2}\frac{\partial\left(L(x(t)) i_a^2(t)\right)}{\partial x},$$

the force in the $x$ direction is found to be

$$F_{mx} = -\frac{N^2 \mu_0 \mu_1^2 \mu_2^2 A i_a^2}{[\mu_2 l_1 + \mu_1 \mu_2 2x(t) + \mu_1 l_2]^2}.$$

Differential equations must be developed to model the microelectromagnet studied. Using Newton's second law of motion, one obtains following nonlinear differential equations to model, analyze, and simulate the microelectromagnet dynamics

$$\frac{dx}{dt}=v,$$

$$\frac{dv}{dt}=\frac{1}{m}\left(\frac{N^2\mu_0\mu_1^2\mu_2^2Ai_a^2}{[\mu_2 l_1+\mu_1\mu_2 2x(t)+\mu_1 l_2]^2}-k_s x+k_{s1}x+k_{s2}x\right).$$

This set of two differential equations gives the lumped-parameter model if microactuator. It is evident that nonlinear differential equations are found. These equations accurately model microactuators, and therefore, can be used for analysis, simulation, and design. □

*Example 5.2.7*

Two microcoils have mutual inductance 0.00005 H ($L_{12}$=0.00005 H). The current in the first microcoil is $i_1=\sqrt{\sin 4t}$. Find the induced *emf* in the second microcoil.

*Solution.*

The induced *emf* is given as

$$\mathscr{E}_2=L_{12}\frac{di_1}{dt}.$$

By using the power rule for the time-varying current in the first coil $i_1=\sqrt{\sin 4t}$, we have

$$\frac{di_1}{dt}=\frac{2\cos 4t}{\sqrt{\sin 4t}}.$$

Hence, one obtains

$$\mathscr{E}_2=\frac{0.0001\cos 4t}{\sqrt{\sin 4t}}.$$ □

*Example 5.2.8*

Figure 5.2.4 illustrates a microactuator (microscale electromechanical device) with a stationary member and movable plunger. Our goal is to derive the differential equations to perform modeling. That is, the ultimate objective is to illustrate the lumped-parameter mathematical model development for microactuators.

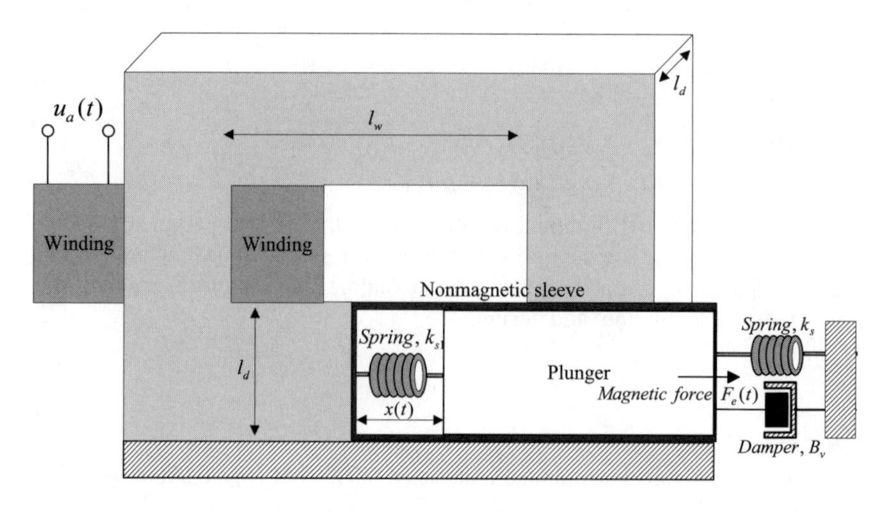


Figure 5.2.4 Schematic of a microactuator.

*Solution.*

Applying Newton's second law of motion, one finds the differential equation to model the translational motion. In particular,

$$F(t) = m\frac{d^2x}{dt^2} + B_v\frac{dx}{dt} + (k_s x - k_{s1}x) + F_e(t),$$

where $x$ denotes the displacement of a plunger (air gap); $m$ is the mass of a movable plunger; $B_v$ is the viscous friction coefficient; $k_s$ and $k_{s1}$ are the spring constants; $F_e(t)$ is the magnetic force which is given as

$$F_e(i,x) = \frac{\partial W_c(i,x)}{\partial x}.$$

The restoring/stretching forces exerted by the springs is

$$F_s = k_s x - k_{s1}x.$$

Assuming that the magnetic system is linear, the coenergy is expressed as

$$W_c(i,x) = \tfrac{1}{2}L(x)i^2.$$

The electromagnetic fore is

$$F_e(i,x) = \tfrac{1}{2}i^2\frac{dL(x)}{dx}.$$

The inductance can be found by using the following formula

$$L(x) = \frac{N^2}{\Re_f + \Re_g} = \frac{N^2\mu_f\mu_0 A_f A_g}{A_g l_f + A_f\mu_f(x+2d)},$$

where $\Re_f$ and $\Re_g$ are the reluctances of the ferromagnetic material and air gap; $A_f$ and $A_g$ are the associated cross section areas; $l_f$ and $(x + 2d)$ are the lengths of the magnetic material and the air gap.

Hence,

$$\frac{dL}{dx} = -\frac{N^2 \mu_f^2 \mu_0 A_f^{\,2} A_g}{[A_g l_f + A_f \mu_f (x+2d)]^2}.$$

Using Kirchhoff's law, the voltage equation for the electric circuit is

$$u_a = ri + \frac{d\psi}{dt},$$

where the flux linkage $\psi$ is expressed as $\psi = L(x)i$.

One obtains

$$u_a = ri + L(x)\frac{di}{dt} + i\frac{dL(x)}{dx}\frac{dx}{dt}.$$

Thus, we have

$$\frac{di}{dt} = -\frac{r}{L(x)}i + \frac{1}{L(x)}\frac{N^2 \mu_f^2 \mu_0 A_f^{\,2} A_g}{[A_g l_f + A_f \mu_f (x+2d)]^2} iv + \frac{1}{L(x)}u_a.$$

Augmenting this equation with the second-order differential equation for the mechanical system, three first-order nonlinear differential equations are found as

$$\frac{di}{dt} = -\frac{r[A_g l_f + A_f \mu_f (x+2d)]}{N^2 \mu_f \mu_0 A_f A_g} i + \frac{\mu_f A_f}{A_g l_f + A_f \mu_f (x+2d)} iv$$

$$+ \frac{A_g l_f + A_f \mu_f (x+2d)}{N^2 \mu_f \mu_0 A_f A_g} u_a,$$

$$\frac{dx}{dt} = v,$$

$$\frac{dv}{dt} = \frac{N^2 \mu_f^2 \mu_0 A_f^{\,2} A_g}{2m[A_g l_f + A_f \mu_f (x+2d)]^2} i^2 - \frac{1}{m}(k_s x - k_{s1} x) - \frac{B_v}{m} v.$$

It must be emphasized that the *fringing* effect should be integrated in high-fidelity modeling and analysis. The air gap reluctance is expressed as

$$\Re_g = \frac{x+2d}{\mu_0 \left(k_{g1} l_w l_d + k_{g2} x^2\right)},$$

where $k_{g1}$ and $k_{g2}$ are the nonlinear functions of the ferromagnetic material, $l_d/l_w$ ratio, *B–H* curve, load, etc.

Therefore, the inductance can be found as $L(x) = \dfrac{N^2}{\Re_f + \Re_g}$, and the accurate mathematical model can be straightforwardly derived using the procedure reported. □

### 5.2.1 Basic Foundations in Model Developments of Micro- and Nanoactuators in Electromagnetic Fields

Electromagnetic theory and mechanics form the basis for the development of MEMS and NEMS models.

The electrostatic and magnetostatic equations in linear isotropic media are found using the vectors of the electric field intensity $\vec{E}$, electric flux density $\vec{D}$, magnetic field intensity $\vec{H}$, and magnetic flux density $\vec{B}$. In addition, one uses the constitutive equations

$\vec{D}=\varepsilon\vec{E}$ and $\vec{B}=\mu\vec{H}$

where $\varepsilon$ is the permittivity; $\mu$ is the permeability.

The basic equations are given in Table 5.2.1.

Table 5.2.1 Fundamental equations of electrostatic and magnetostatic fields.

| | Electrostatic Model | Magnetostatic Model |
|---|---|---|
| Governing equations | $\nabla\times\vec{E}(x,y,z,t)=0$ <br> $\nabla\cdot\vec{E}(x,y,z,t)=\frac{\rho_v(x,y,z,t)}{\varepsilon}$ | $\nabla\times\vec{H}(x,y,z,t)=0$ <br> $\nabla\cdot\vec{H}(x,y,z,t)=0$ |
| Constitutive equations | $\vec{D}=\varepsilon\vec{E}$ | $\vec{B}=\mu\vec{H}$ |

In the static (time-invariant) fields, electric and magnetic field vectors form separate and independent pairs. That is, $\vec{E}$ and $\vec{D}$ are not related to $\vec{H}$ and $\vec{B}$, and vice versa. However, in reality, the electric and magnetic fields are time-varying, and the changes of magnetic field influence the electric field, and vice versa.

The partial differential equations are found by using Maxwell's equations. In particular, four Maxwell's equations in the differential form for time-varying fields are

$$\nabla\times\vec{E}(x,y,z,t)=-\mu\frac{\partial\vec{H}(x,y,z,t)}{\partial t}, \qquad \text{(Faraday's law)}$$

$$\nabla\times\vec{H}(x,y,z,t)=\sigma\vec{E}(x,y,z,t)+\vec{J}(x,y,z,t)=\sigma\vec{E}(x,y,z,t)+\varepsilon\frac{\partial\vec{E}(x,y,z,t)}{\partial t},$$

$$\nabla\cdot\vec{E}(x,y,z,t)=\frac{\rho_v(x,y,z,t)}{\varepsilon}, \qquad \text{(Gauss's law)}$$

$$\nabla\cdot\vec{H}(x,y,z,t)=0,$$

where $\vec{E}$ is the electric field intensity, and using the permittivity $\varepsilon$, the electric flux density is $\vec{D}=\varepsilon\vec{E}$; $\vec{H}$ is the magnetic field intensity, and using the permeability $\mu$, the magnetic flux density is $\vec{B}=\mu\vec{H}$; $\vec{J}$ is the current density, and using the conductivity $\sigma$, we have $\vec{J}=\sigma\vec{E}$; $\rho_v$ is the volume charge density, and the total electric flux through a closed surface is $\Phi=\oint_s \vec{D}\cdot d\vec{s}=\oint_v \rho_v dv=Q$ (Gauss's law), while the magnetic flux crossing surface is $\Phi=\oint_s \vec{B}\cdot d\vec{s}$.

The second equation

$$\nabla\times\vec{H}(x,y,z,t)=\sigma\vec{E}(x,y,z,t)+\vec{J}(x,y,z,t)=\sigma\vec{E}(x,y,z,t)+\varepsilon\frac{\partial\vec{E}(x,y,z,t)}{\partial t}$$

was derived by Maxwell adding the term $\vec{J}_d(x,y,z,t)=\varepsilon\frac{\partial\vec{E}(x,y,z,t)}{\partial t}$ in the Ampere law.

The constitutive (auxiliary) equations are given using the permittivity $\varepsilon$, permeability tensor $\mu$, and conductivity $\sigma$. In particular, one has

$$\vec{D}=\varepsilon\vec{E} \text{ or } \vec{D}=\varepsilon\vec{E}+\vec{P},$$
$$\vec{B}=\mu\vec{H} \text{ or } \vec{B}=\mu(\vec{H}+\vec{M}),$$
$$\vec{J}=\sigma\vec{E} \text{ or } \vec{J}=\rho_v\vec{v}.$$

The Maxwell's equations can be solved using the boundary conditions on the field vectors.

In two-region media, we have

$$\vec{a}_N\times(\vec{E}_2-\vec{E}_1)=0,$$
$$\vec{a}_N\times(\vec{H}_2-\vec{H}_1)=\vec{J}_s,$$
$$\vec{a}_N\cdot(\vec{D}_2-\vec{D}_1)=\rho_s,$$
$$\vec{a}_N\cdot(\vec{B}_2-\vec{B}_1)=0,$$

where $\vec{J}_s$ is the surface current density vector; $\vec{a}_N$ is the surface normal unit vector at the boundary from region 2 into region 1; $\rho_s$ is the surface charge density.

The constitutive relations that describe media can be integrated with Maxwell's equations which relate the fields in order to find two partial differential equations. Using the electric and magnetic field intensities $\vec{E}$ and $\vec{H}$ to model electromagnetic fields in MEMS, one has

$$\nabla\times(\nabla\times\vec{E})=\nabla(\nabla\cdot\vec{E})-\nabla^2\vec{E}=-\mu\frac{\partial\vec{J}}{\partial t}-\mu\frac{\partial^2\vec{D}}{\partial t^2}=-\mu\sigma\frac{\partial\vec{E}}{\partial t}-\mu\varepsilon\frac{\partial^2\vec{E}}{\partial t^2},$$

$$\nabla\times(\nabla\times\vec{H})=\nabla(\nabla\cdot\vec{H})-\nabla^2\vec{H}=-\mu\sigma\frac{\partial\vec{H}}{\partial t}-\mu\varepsilon\frac{\partial^2\vec{H}}{\partial t^2}.$$

The following pair of homogeneous and inhomogeneous wave equations

$$\nabla^2\vec{E}-\mu\sigma\frac{\partial\vec{E}}{\partial t}-\mu\varepsilon\frac{\partial^2\vec{E}}{\partial t^2}=\nabla\left(\frac{\rho_v}{\varepsilon}\right),$$

$$\nabla^2\vec{H}-\mu\sigma\frac{\partial\vec{H}}{\partial t}-\mu\varepsilon\frac{\partial^2\vec{H}}{\partial t^2}=0$$

is equivalent to four Maxwell's equations and constitutive relations. For some cases, these two equations can be solved independently. It must be emphasized that it is not always possible to use the boundary conditions using only $\vec{E}$ and $\vec{H}$, and thus, the problem cannot always be simplified to two electromagnetic field vectors. Therefore, the electric scalar and magnetic vector potentials are used. Denoting the magnetic vector potential as $\vec{A}$ and the electric scalar potential as $V$, we have

$$\nabla\times\vec{A}=\vec{B}=\mu\vec{H}\ \text{and}\ \vec{E}=-\frac{\partial\vec{A}}{\partial t}-\nabla V.$$

The electromagnetic field is derivative from the potentials. Using the Lorentz equation

$$\nabla\cdot\vec{A}=-\frac{\partial V}{\partial t},$$

the inhomogeneous vector potential wave equation to be solved is

$$-\nabla^2\vec{A}+\mu\sigma\frac{\partial\vec{A}}{\partial t}+\mu\varepsilon\frac{\partial^2\vec{A}}{\partial t^2}=-\mu\sigma\nabla V.$$

To model motion microdevices, the mechanical equations must be used, and Newton's second law is usually applied to derive the equations of motion.

Using the volume charge density $\rho_v$, the Lorenz force, which relates the electromagnetic and mechanical phenomena, is found as

$$\vec{F}=\rho_v(\vec{E}+\vec{v}\times\vec{B})=\rho_v\vec{E}+\vec{J}\times\vec{B}.$$

The Lorentz force law is

$$\vec{F}=\frac{d\vec{p}}{dt}=q(\vec{E}+\vec{v}\times\vec{B})=q\left[-\nabla V-\frac{\partial\vec{A}}{\partial t}+\vec{v}\times(\nabla\times\vec{A})\right],$$

where for the canonical momentum $\vec{p}$ we have $\frac{d\vec{p}}{dt}=-\nabla\Pi$.

That is, the Maxwell equations illustrate how charges produce the electromagnetic fields, and the Lorentz force law demonstrates how the electromagnetic fields affect charges.

The energy per unit time per unit area, transported by the electromagnetic fields is called the Pointing vector

$$\vec{S}=\frac{1}{\mu}(\vec{E}\times\vec{B}).$$

The electromagnetic force can be found by applying the Maxwell stress tensor method. This concept employs a volume integral to obtain the stored energy, and stress at all points of a bounding surface can be determined. The sum of local stresses gives the net force. In particular, the electromagnetic stress is

$$\vec{F}=\int_v \rho_v(\vec{E}+\vec{v}\times\vec{B})dv=\int_v(\rho_v\vec{E}+\vec{J}\times\vec{B})dv$$

$$=\frac{1}{\mu}\oint_s \overleftrightarrow{T}\cdot d\vec{s}-\varepsilon\mu\frac{d}{dt}\int_v \vec{S}dv.$$

The force per unit volume is

$$\vec{F}_u=\rho_v\vec{E}+\vec{J}\times\vec{B}=\nabla\cdot\overleftrightarrow{T}-\varepsilon\mu\frac{\partial\vec{S}}{\partial t}.$$

The electromagnetic stress energy tensor $\overleftrightarrow{T}$ (the second Maxwell stress tensor) is

$$(\vec{a}\cdot\overleftrightarrow{T})_j=\sum_{i=x,y,z}a_iT_{ij}.$$

We have

$$(\nabla\cdot\overleftrightarrow{T})_j=\varepsilon\left[(\nabla\cdot\vec{E})E_j+(\vec{E}\cdot\nabla)E_j-\tfrac{1}{2}\nabla_jE^2\right]$$

$$+\frac{1}{\mu}\left[(\nabla\cdot\vec{B})B_j+(\vec{B}\cdot\nabla)B_j-\tfrac{1}{2}\nabla_jB^2\right]$$

or

$$T_{ij}=\varepsilon\left(E_iE_j-\tfrac{1}{2}\delta_{ij}E^2\right)+\frac{1}{\mu}\left(B_iB_j-\tfrac{1}{2}\delta_{ij}B^2\right),$$

where $i$ and $j$ are the indexes which refer to the coordinates $x$, $y$ and $z$ (the stress tensor $\overleftrightarrow{T}$ has nine components $T_{xx}$, $T_{xy}$, $T_{xz}$, … $T_{zy}$ and $T_{zz}$); $\delta_{ij}$ is the Kronecker delta-function, which is defined to be 1 if the indexes are the same and 0 otherwise, $\delta_{xx}=\delta_{yy}=\delta_{zz}=1$ and $\delta_{xy}=\delta_{xz}=\delta_{yz}=0$.

The electromagnetic torque developed by motion microstructures is found using the electromagnetic field, and the electromagnetic stress tensor is given as

$$T_s = T_s^E + T_s^M$$

$$= \begin{bmatrix} E_1 D_1 - \frac{1}{2} E_j D_j & E_1 D_2 & E_1 D_3 \\ E_2 D_1 & E_2 D_2 - \frac{1}{2} E_j D_j & E_2 D_3 \\ E_3 D_1 & E_3 D_2 & E_3 D_3 - \frac{1}{2} E_j D_j \end{bmatrix}$$

$$+ \begin{bmatrix} B_1 H_1 - \frac{1}{2} B_j H_j & B_1 H_2 & B_1 H_3 \\ B_2 H_1 & B_2 H_2 - \frac{1}{2} B_j H_j & B_2 H_3 \\ B_3 H_1 & B_3 H_2 & B_3 H_3 - \frac{1}{2} B_j H_j \end{bmatrix}.$$

For the Cartesian, cylindrical, and spherical coordinate systems, which can be used to develop the mathematical model, we have

$$E_x = E_1,\ E_y = E_2,\ E_z = E_3,\ D_x = D_1,\ D_y = D_2,\ D_z = D_3,$$
$$H_x = H_1,\ H_y = H_2,\ H_z = H_3,\ B_x = B_1,\ B_y = B_2,\ B_z = B_3;$$
$$E_r = E_1,\ E_\theta = E_2,\ E_z = E_3,\ D_r = D_1,\ D_\theta = D_2,\ D_z = D_3,$$
$$H_r = H_1,\ H_\theta = H_2,\ H_z = H_3,\ B_r = B_1,\ B_\theta = B_2,\ B_z = B_3;$$
$$E_\rho = E_1,\ E_\theta = E_2,\ E_\phi = E_3,\ D_\rho = D_1,\ D_\theta = D_2,\ D_\phi = D_3,$$
$$H_\rho = H_1,\ H_\theta = H_2,\ H_\phi = H_3,\ B_\rho = B_1,\ B_\theta = B_2,\ B_\phi = B_3.$$

Maxwell's equations can be solved using the MATLAB environment (for example, using the Partial Differential Equations Toolbox).

The *electromotive* and *magnetomotive* forces are found as

$$emf = \oint_l \vec{E} \cdot d\vec{l} = \underbrace{\oint_l \left(\vec{v} \times \vec{B}\right) \cdot d\vec{l}}_{\text{motional induction (generation)}} - \underbrace{\oint_s \frac{\partial \vec{B}}{\partial t} d\vec{s}}_{\text{transformer induction}}$$

and $mmf = \oint_l \vec{H} \cdot d\vec{l} = \oint_s \vec{J} \cdot d\vec{s} + \oint_s \frac{\partial \vec{D}}{\partial t} d\vec{s}$.

The motional *emf* is a function of the velocity and the magnetic flux density, while the electromotive force induced in a stationary closed circuit is equal to the negative rate of increase of the magnetic flux (transformer induction).

We introduced the vector magnetic potential $\vec{A}$. Using the equation

$$\vec{B} = \nabla \times \vec{A},$$

one has the following nonhomogeneous vector wave equation

$$\nabla^2 \times \vec{A} - \mu\varepsilon \frac{\partial^2 \vec{A}}{\partial t^2} = -\mu\vec{J},$$

and the solution gives the waves traveling with the velocity $\frac{1}{\sqrt{\mu\varepsilon}}$.

### 5.2.2 Lumped-Parameter Mathematical Models of MEMS

The problems of modeling, analysis, and control of MEMS are very important in many applications. A mathematical model is a mathematical description (in the form of a function or an equation) of MEMS which integrate motion microdevices (microscale actuators and sensors), radiating energy microdevices, microscale driving/sensing circuitry, and controlling/signal processing ICs. The purpose of the model development is to understand and comprehend the phenomena, as well as to analyze the end-to-end behavior and study the system performance.

To model MEMS, advanced analysis methods are required to accurately cope with the involved highly complex physical phenomena, effects, and processes. The need for high-fidelity analysis, computationally-efficient algorithms, and simulation time reduction increases significantly for complex microdevices restricting the application of Maxwell's equations to a small number of problems possible to solve. As was illustrated in the previous section, nonlinear electromagnetic and energy conversion phenomena are described by the partial differential equations. The application of Maxwell's equations contradicts the needs for data-intensive analysis and performance evaluation capabilities with outcome prediction within overall modeling domains as particularly necessary for simulation and analysis of high-performance MEMS. Hence, other modeling and analysis methods are needed to be applied. The lumped mathematical models, described by ordinary differential equations, can be used. The process of mathematical modeling and model development is given below.

The first step is to formulate the modeling problem:

- Examine and analyze MEMS using a multi-level hierarchy concept: develop multivariable input-output subsystem (devices) pairs, e.g., motion microstructures or microdevices (microscale actuators and sensors) – radiating energy microdevices – microscale circuitry – ICs – controller – input/output devices;
- Understand and comprehend the MEMS structure and system configuration;
- Gather (collect) the data and information;
- Develop input-output variable pairs, identify the independent and dependent control, disturbance, output, reference (command), state and performance variables, as well as events;
- Making accurate assumptions, simplify the problem to make the studied MEMS mathematically tractable. Mathematical models are the idealization of physical phenomena. These mathematical models are never absolutely accurate, and comprehensive models simplify the reality to allow the designer to perform the thorough analysis and accurate predictions of the system performance. Even using Maxwell's equations, one makes assumptions in order to solve these partial differential equations.

The second step is to derive equations that relate the variables and events. In particular,

- Define and specify the basic laws (Kirchhoff, Lagrange, Maxwell, Newton, and other) to be used to obtain the equations of motion. Mathematical models of electromagnetic, electronic, and mechanical microscale subsystems can be found and augmented to derive mathematical models of MEMS using defined variables and events;
- Derive mathematical models.

The third step is the simulation, analysis, and validation:

- Identify the numerical and analytic methods to be used in analysis and simulations;
- Analytically and/or numerically solve the mathematical equations (e.g., differential or difference equations, nonlinear equations, etc.);
- Using information variables (measured or observed) and events, synthesize the fitting and mismatch functionals;
- Verify the results through the comprehensive comparison of the solution (model input-state-output-event mapping sets) with the experimental data (experimental input-state-output-event mapping sets);
- Calculate the fitting and mismatch functionals;
- Examine the analytical and numerical data against new experimental data and evidence.

If matching with the desired accuracy is not guaranteed, the mathematical model of MEMS must be refined, and the designer must start the cycle again.

Electromagnetic theory and classical mechanics form the basis for the development of mathematical models of MEMS. It was illustrated that MEMS can be modeled using Maxwell's equations and *torsional-mechanical* equations of motion. Forces and torques are found using the Maxwell stress tensor. The nonlinear partial differential equations result. However, for modeling, simulation, analysis, design, control, and optimization, the lumped-parameter mathematical models as given by ordinary differential equations can be derived and used.

Consider the rotational microstructure (bar magnet, current loop, and microsolenoid) in a uniform magnetic field, see Figure 5.2.5. The microstructure rotates if the electromagnetic torque is developed. Therefore, the electromagnetic field must be studied to find the electromagnetic torque developed.

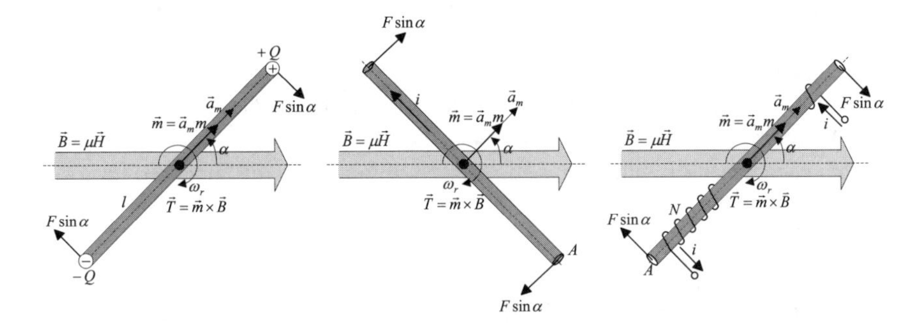


Figure 5.2.5 Clockwise rotation of a microstructure (magnetic bar, current loop, and solenoid).

The torque tends to align the magnetic moment $\vec{m}$ with $\vec{B}$, and it is well-known that

$$\vec{T} = \vec{m} \times \vec{B}.$$

For a magnetic bar with the length $l$, the pole strength is $Q$.

The magnetic moment is given as $m = Ql$, and the force is $F = QB$.

The electromagnetic torque is found to be

$$T = 2F\tfrac{1}{2}l\sin\alpha = QlB\sin\alpha = mB\sin\alpha.$$

Using the vector notations, one obtains

$$\vec{T} = \vec{m} \times \vec{B} = \vec{a}_m m \times \vec{B} = Ql\vec{a}_m \times \vec{B},$$

where $\vec{a}_m$ is the unit vector in the magnetic moment direction.

For a current loop with the cross-sectional area $A$, the torque is found as

$$\vec{T} = \vec{m} \times \vec{B} = \vec{a}_m m \times \vec{B} = iA\vec{a}_m \times \vec{B}.$$

For a solenoid with $N$ turns, one obtains

$$\vec{T} = \vec{m} \times \vec{B} = \vec{a}_m m \times \vec{B} = iAN\vec{a}_m \times \vec{B}.$$

The straightforward application of Newton's second law for the rotational motion gives

$$\sum \vec{T}_\Sigma = J\frac{d\omega_r}{dt},$$

where $\sum \vec{T}_\Sigma$ is the net torque; $\omega_r$ is the angular velocity; $J$ is the equivalent moment of inertia.

The transient evolution of the angular displacement $\theta_r$ is modeled as

$$\frac{d\theta_r}{dt} = \omega_r.$$

Augmenting the equations for the electromagnetic torque (found in terms of the magnetic field variables $\vec{m}$ and $\vec{B}$) and the *torsional-*

*mechanical* dynamics (the state variables are the angular velocity and displacement), the mathematical model of micro- and nanoscale rotational actuators results.

The energy is stored in the magnetic field, and media are classified as diamagnetic, paramagnetic, ferromagnetic, antiferromagnetic, and super-paramagnetic. Using the magnetic susceptibility $\chi_m$, the magnetization is expressed as

$$\vec{M} = \chi_m \vec{H}.$$

Magnetization curves should be studied, and the permeability is $\mu = \frac{B}{H}$. The magnetic field density $B$ lags behind the magnetic flux intensity $H$, and this phenomenon is called hysteresis. Thus, the $B$-$H$ magnetization curves must be studied.

The per-unit volume magnetic energy stored is $\oint_B H dB$. The density of the energy stored in the magnetic field is $\frac{1}{2}\vec{B} \cdot \vec{H}$. If $B$ is linearly related to $H$, we have the expression for the total energy stored in the magnetic field as $\frac{1}{2}\int_v \vec{B} \cdot \vec{H} dv$.

For translational motion, Newton's second law states that the net force acting on the object is related to its acceleration as $\sum \vec{F} = m\vec{a}$.

In the XYZ coordinate system, one obtains

$$\sum F_x = ma_x, \sum F_y = ma_y \text{ and } \sum F_z = ma_z.$$

The force is the gradient of the stored magnetic energy. That is,

$$\vec{F}_m = \vec{\nabla} W_m.$$

Hence, in the *xyz* directions, we have

$$F_{mx} = \frac{\partial W_m}{\partial x}, F_{my} = \frac{\partial W_m}{\partial y} \text{ and } F_{mz} = \frac{\partial W_m}{\partial z},$$

where the stored magnetic energy is found using the volume current density $\vec{J}$. In particular,

$$W_m = \frac{\mu}{8\pi} \int_v \int_{v_A} \frac{\vec{J}(\vec{r}_A) \cdot \vec{J}(\vec{r})}{R} dv_A dv.$$

Applying the field quantities, we have

$$W_m = \frac{1}{2}\int_v \vec{A} \cdot \vec{J} dv = \frac{1}{2}\int_v \vec{B} \cdot \vec{H} dv$$

The magnetic energy density is

$$w_m = \frac{1}{2}\vec{A} \cdot \vec{J} = \frac{1}{2}\vec{B} \cdot \vec{H}.$$

Using Newton's second law and the derived expression for stored magnetic energy, we have nine highly coupled nonlinear differential equations for the *xyz* translational motion of microactuator. In particular,

$$\frac{dF_{xyz}}{dt} = f_F\left(F_{xyz}, v_{xyz}, x_{xyz}, H\right),$$

$$\frac{dv_{xyz}}{dt} = f_v\left(F_{xyz}, v_{xyz}, x_{xyz}, F_{Lxyz}\right),$$

$$\frac{dx_{xyz}}{dt} = f_x\left(v_{xyz}, x_{xyz}\right),$$

where $F_{xyz}$ are the forces developed; $v_{xyz}$ and $x_{xyz}$ are the linear velocities and positions; $F_{Lxyz}$ are the load forces.

The expressions for energies stored in electrostatic and magnetic fields in terms of field quantities should be derived. The total potential energy stored in the electrostatic field is obtained using the potential difference $V$ as

$$W_e = \tfrac{1}{2}\int_v \rho_v V dv,$$

where the volume charge density is found as $\rho_v = \vec{\nabla}\cdot\vec{D}$, $\vec{\nabla}$ is the curl operator.

In the Gauss form, using $\rho_v = \vec{\nabla}\cdot\vec{D}$ and making use of $\vec{E} = -\vec{\nabla}V$, one obtains the following expression for the energy stored in the electrostatic field $W_e = \tfrac{1}{2}\int_v \vec{D}\cdot\vec{E}dv$, and the electrostatic volume energy density is $\tfrac{1}{2}\vec{D}\cdot\vec{E}$. For a linear isotropic medium, one finds

$$W_e = \tfrac{1}{2}\int_v \varepsilon\left|\vec{E}\right|^2 dv = \tfrac{1}{2}\int_v \frac{1}{\varepsilon}\left|\vec{D}\right|^2 dv.$$

The electric field $\vec{E}(x,y,z)$ is found using the scalar electrostatic potential function $V(x,y,z)$ as

$$\vec{E}(x,y,z) = -\vec{\nabla}V(x,y,z).$$

In the cylindrical and spherical coordinate systems, we have

$$\vec{E}(r,\phi,z) = -\vec{\nabla}V(r,\phi,z)$$

and $\vec{E}(r,\theta,\phi) = -\vec{\nabla}V(r,\theta,\phi)$.

Using the principle of virtual work, for the lossless conservative micro- and nanoelectromechanical systems, the differential change of the electrostatic energy $dW_e$ is equal to the differential change of mechanical energy $dW_{mec}$, $dW_e = dW_{mec}$. For translational motion $dW_{mec} = \vec{F}_e \cdot d\vec{l}$, where $d\vec{l}$ is the differential displacement.

One obtains

$$dW_e = \vec{\nabla} W_e \cdot d\vec{l} \, .$$

Hence, the force is the gradient of the stored electrostatic energy,

$$\vec{F}_e = \vec{\nabla} W_e \, .$$

In the Cartesian coordinates, we have

$$F_{ex} = \frac{\partial W_e}{\partial x}, \; F_{ey} = \frac{\partial W_e}{\partial y} \text{ and } F_{ez} = \frac{\partial W_e}{\partial z} \, .$$

Energy conversion takes place in micro- and nanoscale electromechanical motion devices (actuators and sensors, smart structures), antennas and ICs. We study electromechanical motion devices that convert electrical energy (more precisely electromagnetic energy) to mechanical energy and vise versa (conversion of mechanical energy to electromagnetic energy). Fundamental principles of energy conversion, applicable to micro- and nanoelectromechanical motion devices were studied to provide basic foundations. Using the principle of conservation of energy we can formulate:

*for a lossless micro- and nanoelectromechanical motion devices (in the conservative system no energy is lost through friction, heat, or other irreversible energy conversion) the sum of the instantaneous kinetic and potential energies of the system remains constant.*

The energy conversion is represented in Figure 5.2.6.

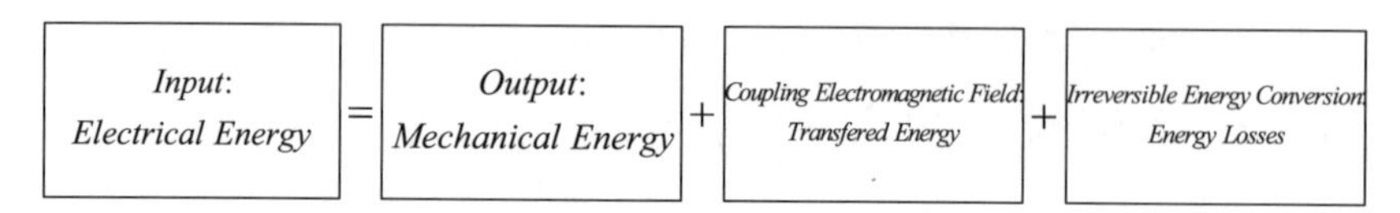


Figure 5.2.6 Energy transfer in micro- and nanoelectromechanical systems.

The general equation which describes the energy conversion is given as

$$\underbrace{\mathbf{E}_E}_{\text{Electrical Energy Input}} - \underbrace{\mathbf{L}_E}_{\text{Ohmic Losses}} - \underbrace{\mathbf{L}_M}_{\text{Magnetic Losses}} = \underbrace{\mathbf{E}_M}_{\text{Mechanical Energy}} + \underbrace{\mathbf{L}_F}_{\text{Friction Losses}} + \underbrace{\mathbf{L}_S}_{\text{Stored Energy}}$$

For conservative (lossless) energy conversion, one can write

$$\underbrace{\Delta\mathbf{W}_E}_{\text{Change in Electrical Energy Input}} = \underbrace{\Delta\mathbf{W}_M}_{\text{Change in Mechanical Energy}} + \underbrace{\Delta\mathbf{W}_m}_{\text{Change in Electromagnetic Energy}} \, .$$

The total energy stored in the magnetic field is

$$W_m = \tfrac{1}{2} \int_v \vec{B} \cdot \vec{H} dv \, ,$$

where $\vec{B}$ and $\vec{H}$ are related using the permeability $\mu$, $\vec{B} = \mu\vec{H}$.

The material becomes magnetized in response to the external field $\vec{H}$, and the dimensionless magnetic susceptibility $\chi_m$ or relative permeability $\mu_r$ are used. We have,

$$\vec{B} = \mu\vec{H} = \mu_0(1 + \chi_m)\vec{H} = \mu_0\mu_r\vec{H} = \mu\vec{H}.$$

Based upon the value of the magnetic susceptibility $\chi_m$, the materials are classified as:

- Diamagnetic, $\chi_m \approx -1 \times 10^{-5}$ ($\chi_m = -9.5 \times 10^{-6}$ for copper, $\chi_m = -3.2 \times 10^{-5}$ for gold, and $\chi_m = -2.6 \times 10^{-5}$ for silver);
- Paramagnetic, $\chi_m \approx 1 \times 10^{-4}$ ($\chi_m = 1.4 \times 10^{-3}$ for $Fe_2O_3$, and $\chi_m = 1.7 \times 10^{-3}$ for $Cr_2O_3$);
- Ferromagnetic, $|\chi_m| >> 1$ (iron, nickel, cobalt, neodymium-iron-boron and samarium-cobalt permanent magnets, etc.).

The magnetization behavior of the ferromagnetic materials is mapped by the magnetization curve, where *H* is the externally applied magnetic field, and *B* is total magnetic flux density in the medium. Typical *B-H* curves for hard and soft ferromagnetic materials are given in Figure 5.2.7.

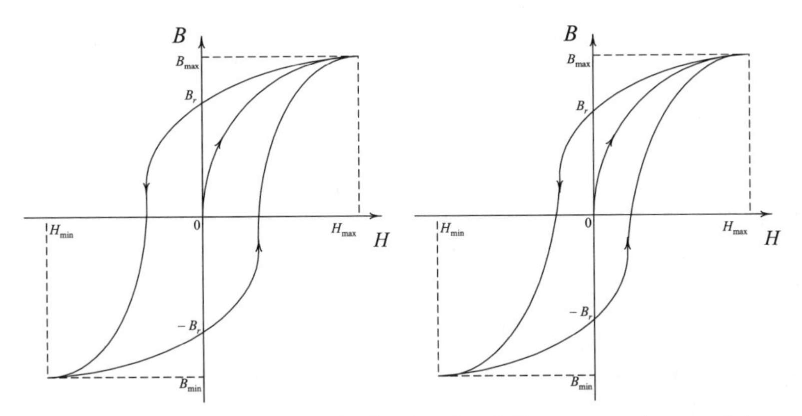


Figure 5.2.7 *B-H* curves for hard and soft ferromagnetic materials.

The *B* versus *H* curve allows one to establish the energy analysis.

Assume that initially $B_0 = 0$ and $H_0 = 0$. Let *H* increases form $H_0 = 0$ to $H_{\max}$. Then, *B* increases from $B_0 = 0$ until the maximum value of *B* (denoted as $B_{\max}$) is reached. If *H* then decreases to $H_{\min}$, *B* decreases to $B_{\min}$ through the remanent value $B_r$ (the so-called residual magnetic flux density) along the different curve, see Figure 5.2.7. For variations of *H*,

$H \in [H_{\min}\ H_{\max}]$, $B$ changes within the *hysteresis loop*, and $B \in [B_{\min}\ B_{\max}]$.

In the per-unit volume, the applied field energy is

$$W_F = \oint_B H dB,$$

while the stored energy is expressed as

$$W_c = \oint_H B dH.$$

In the volume $v$, we have the following expressions for the field and stored energy

$$W_F = v \oint_B H dB$$

and $W_c = v \oint_H B dH$.

A complete $B$ versus $H$ loop should be considered, and the equations for field and stored energy represent the areas enclosed by the corresponding curve. It should be emphasized that each point of the $B$ versus $H$ curve represents the total energy.

In ferromagnetic materials, time-varying magnetic flux produces core losses which consist of hysteresis losses (due to the hysteresis loop of the $B$-$H$ curve) and the eddy-current losses (proportional to the current frequency and lamination thickness). The area of the hysteresis loop is related to the hysteresis losses. Soft ferromagnetic materials have a narrow hysteresis loop and they are easily magnetized and demagnetized. Therefore, the lower hysteresis losses, compared with hard ferromagnetic materials, result.

Usually the flux linkages are plotted versus the current because the current and flux linkages are used rather than the flux intensity and flux density. In micro- and nanoelectromechanical motion devices almost all energy is stored in the air gap. Using the fact that the air is a conservative medium, one concludes that the coupling filed is lossless.

Figure 5.2.8 illustrates the nonlinear magnetizing characteristic (normal magnetization curve). The energy stored in the magnetic field is

$$W_F = \oint_\psi i d\psi,$$

while the coenergy is found as

$$W_c = \oint_i \psi di.$$

The total energy is

$$W_F + W_c = \oint_\psi i d\psi + \oint_i \psi di = \psi i.$$

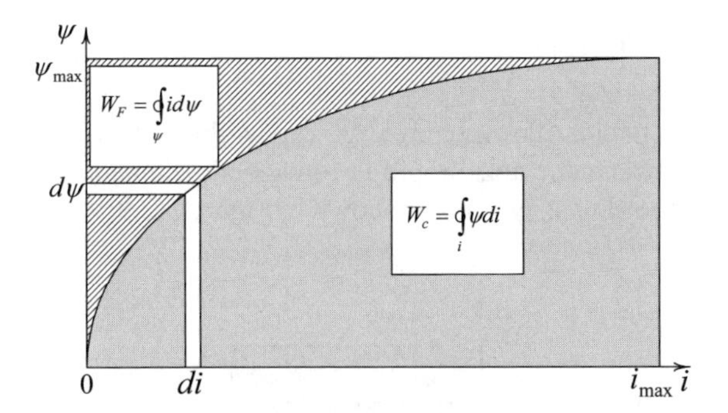


Figure 5.2.8 Magnetization curve and energies.

The flux linkage is the function of the current $i$ and position $x$ (for translational motion) or angular displacement $\theta$ (for rotational motion). That is, $\psi = f(i,x)$ or $\psi = f(i,\theta)$. The current can be found as the nonlinear function of the flux linkages and position or angular displacement. Hence, $d\psi = \frac{\partial\psi(i,x)}{\partial i}di + \frac{\partial\psi(i,x)}{\partial x}dx$, $d\psi = \frac{\partial\psi(i,\theta)}{\partial i}di + \frac{\partial\psi(i,\theta)}{\partial\theta}d\theta$, and $di = \frac{\partial i(\psi,x)}{\partial\psi}d\psi + \frac{\partial i(\psi,x)}{\partial x}dx$, $di = \frac{\partial i(\psi,\theta)}{\partial\psi}d\psi + \frac{\partial i(\psi,\theta)}{\partial\theta}d\theta$.

Therefore, we have

$$W_F = \oint_\psi i d\psi = \oint_i i\frac{\partial\psi(i,x)}{\partial i}di + \oint_x i\frac{\partial\psi(i,x)}{\partial x}dx\,,$$

$$W_F = \oint_\psi i d\psi = \oint_i i\frac{\partial\psi(i,\theta)}{\partial i}di + \oint_\theta i\frac{\partial\psi(i,\theta)}{\partial\theta}d\theta\,,$$

and $$W_c = \oint_i \psi di = \oint_\psi \psi\frac{\partial i(\psi,x)}{\partial\psi}d\psi + \oint_x \psi\frac{\partial i(\psi,x)}{\partial x}dx\,,$$

$$W_c = \oint_i \psi di = \oint_\psi \psi\frac{\partial i(\psi,\theta)}{\partial\psi}d\psi + \oint_\theta \psi\frac{\partial i(\psi,\theta)}{\partial\theta}d\theta\,.$$

Assuming that the coupling field is lossless, the differential change in the mechanical energy (which is found using the differential displacement $d\vec{l}$ as $dW_{mec} = \vec{F}_m \cdot d\vec{l}$ ) is related to the differential change of the coenergy. For displacement $dx$ at constant current, one obtains $dW_{mec} = dW_c$. Hence, the electromagnetic force is

$$F_e(i,x) = \frac{\partial W_c(i,x)}{\partial x}\,.$$

For rotational motion, the electromagnetic torque is

$$T_e(i,\theta)=\frac{\partial W_c(i,\theta)}{\partial \theta}.$$

Nano- and microscale structures, as well as thin magnetic films, exhibit anisotropy. Consider the anisotropic ferromagnetic element in the Cartesian (rectangular) coordinate systems as shown in Figure 5.2.9.

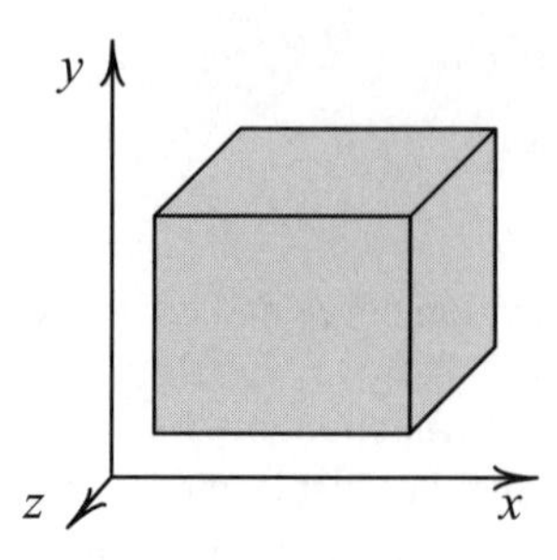


Figure 5.2.9 Material in the *xyz* coordinate system.

The permeability is $\mu(x,y,z)=\begin{bmatrix}\mu_{xx} & \mu_{xy} & \mu_{xz}\\ \mu_{yx} & \mu_{yy} & \mu_{yz}\\ \mu_{zx} & \mu_{zy} & \mu_{zz}\end{bmatrix}$, and therefore,

$$\vec{B}=\mu(x,y,z)\vec{H},\quad \begin{bmatrix}B_x\\ B_y\\ B_z\end{bmatrix}=\begin{bmatrix}\mu_{xx} & \mu_{xy} & \mu_{xz}\\ \mu_{yx} & \mu_{yy} & \mu_{yz}\\ \mu_{zx} & \mu_{zy} & \mu_{zz}\end{bmatrix}\begin{bmatrix}H_x\\ H_y\\ H_z\end{bmatrix}.$$

The analysis of anisotropic micro- and nanoscale actuators and sensors can be performed. Some actuators and sensors can be studied assuming that the media is linear, homogeneous, and isotropic. Unfortunately, this assumption is not valid in general.

Control of the microactuators/ position and linear velocity, angular displacement and angular velocity, is established by changing electromagnetic field (e.g., *E*, *D*, *H* or *B*). In fact, in this chapter it was already shown that the electromagnetic torque and force are derived in the terms of the electromagnetic field quantities. For example, the magnetic field intensity can be considered as a control. Electromagnetic fields are developed by ICs or antennas. Hence, the ICs or microantenna dynamics have to be integrated in the MEMS equations of motion. Thus, microscale antennas and ICs must be thoroughly considered.

Consider the microactuator controlled by the microantenna. Assume that the linear isotropic media has permittivity $\varepsilon_0\varepsilon_m$ and permeability $\mu_0\mu_m$.

The force is calculated using the stress energy tensor $\overleftrightarrow{T}$ which is given in terms of the electromagnetic field as

$$T_{ij} = \varepsilon_0 \varepsilon_m E_i E_j + \mu_0 \mu_m H_i H_j - \tfrac{1}{2}\delta_{ij}\left(\varepsilon_0 \varepsilon_m E^2 + \mu_0 \mu_m H^2\right),$$

where $\delta_{ij}$ is the Kronecker delta-function, defined as $\delta_{ij} = \begin{cases} 1 \text{ if } i = j \\ 0 \text{ if } i \neq j \end{cases}$.

The Maxwell's equation for electromagnetic fields can be expressed in the tensor form using the electromagnetic field stress tensor. In particular, the electromagnetic force is found as

$$\vec{F} = \int_s \vec{T} d\vec{s} .$$

The results derived can be viewed using the energy analysis, and one has

$$\sum \vec{F}(\mathbf{r}) = -\nabla \Pi(\mathbf{r}),$$

$$\Pi(\mathbf{r}) = \frac{\varepsilon_0 \varepsilon_m}{2} \int_s \vec{E} \cdot \vec{E} dv + \frac{1}{2\mu_0 \mu_m} \int_s \vec{H} \cdot \vec{H} dv .$$

Let us demonstrate how to apply the reported concept in the design of motion microtransducers (micromachines). For preliminary design, it is sufficiently accurate to apply Faraday's or Lenz's laws which give the *electromotive force* in term of the time-varying magnetic field changes. In particular,

$$emf = -\frac{d\psi}{dt} = -\frac{\partial \psi}{\partial t} - \frac{\partial \psi}{\partial \theta_r}\frac{d\theta_r}{dt} = -\frac{\partial \psi}{\partial t} - \frac{\partial \psi}{\partial \theta_r}\omega_r ,$$

where $\frac{\partial \psi}{\partial t}$ is the transformer term.

The total flux linkage is

$$\psi = \tfrac{1}{4}\pi N_S \Phi_p ,$$

where $N_S$ is the number of turns; $\Phi_p$ is the flux per pole.

For radial topology microtransducers, we have

$$\Phi_p = \frac{\mu i N_S}{P^2 g_e} R_{in\,st} L ,$$

where $i$ is the current in the phase microwinding (supplied by the IC); $R_{in\,st}$ is the inner stator radius; $L$ is the inductance; $P$ is the number of poles; $g_e$ is the equivalent gap which includes the airgap and radial thickness of the permanent magnet.

Denoting the number of turns per phase as $N_s$, the *magnetomotive force* is

$$mmf = \frac{iN_s}{P}\cos P\theta_r .$$

The simplified expression for the electromagnetic torque for radial topology brushless microtransducers is

$$T = \tfrac{1}{2} P B_{ag} i_s N_S L_r D_r ,$$

where $B_{ag}$ is the air gap flux density, $B_{ag} = \frac{\mu i N_S}{2Pg_e} \cos P\theta_r$; $i_s$ is the total current; $L_r$ is the active length (rotor axial length); $D_r$ is the outside rotor diameter.

As was illustrated, the axial topology brushless microtransducers can be designed and fabricated, and the electromagnetic torque is given as

$$T = k_{ax} B_{ag} i_s N_S D_a^2,$$

where $k_{ax}$ is the nonlinear coefficient which is found in terms of active conductors and thin-film permanent magnet length; $D_a$ is the equivalent diameter which is a function of windings and permanent-magnet topography.

As the expression for the electromagnetic torque is found, the *torsional-mechanical dynamics* used to integrate electromagnetic- and mechanical-based phenomena and derive the complete lumped-parameter mathematical model.

In particular, Newton's second law, as given by

$$\frac{d\omega_r}{dt} = \frac{1}{J} \sum \vec{T}_\Sigma,$$

$$\frac{d\theta_r}{dt} = \omega_r,$$

is used. Here, $\omega_r$ and $\theta_r$ are the angular velocity and displacement, $\sum \vec{T}_\Sigma$ is the net torque.

### 5.2.3 Direct-Current Microtransducers

It has been shown that the basic electromagnetic principles and fundamental physical laws are used to design motion micro- and nano-structures. Micro- and nanoengineering leverages from conventional theory of electromechanical motion devices, electromagnetics, integrated circuits, and quantum mechanics. The fabrication of motion microstructures is based upon CMOS- and LIGA-based technologies, and rotational and translational transducers (actuators and sensors) have been manufactured and tested. The major challenge is the difficulty in fabricating brushes and windings for direct-current microdevices (electric micromachines), reliability and ruggedness (due to bearing problems, vibration, heat and other phenomena), etc. It appears that novel fabrication technologies allow one to overcome many challenges. The most efficient class of microtransducers to be used as MEMS motion microdevices are induction and synchronous. These microtransducers do not have a collector. The stator and rotor windings, as well as permanent-magnets have been manufactured and tested for synchronous (permanent-magnet synchronous and stepper micromotors) and induction micromachines. Direct-current microtransducers are not the preferable choice. However, these micromachines will be covered first because they are simple from analysis standpoints, and, in addition, students and engineers are familiar with these transducers. Furthermore, even using the micromachining technologies, mini/micro-scale permanent-magnet DC transducers (1-2 mm in diameter and 2-6 mm long) were commercialized and manufactured for pagers, phones, cameras, etc. It must be emphasized that due to the limits imposed on the torque, force, and power densities, the micromachine sizing is defined by the required torque, force, or power needed to be produced. Therefore, though it is possible to design and fabricate micromachines, bigger machines are needed due to specifications imposed.

In this chapter we develop mathematical models of the permanent-magnet DC micromachines, discuss the operating principles, and give the basic architectures of MEMS with DC micromachines.

The list of basic variables and symbols used in this section is given below:

$i_a$ is the currents in the armature winding;

$u_a$ is the applied voltages to the armature windings;

$\omega_r$ and $\theta_r$ are the angular velocity and angular displacement of the rotor;

$E_a$ is the *electromotive force*;

$T_e$ and $T_L$ are the electromagnetic and load torques;

$r_a$ is the resistances of the armature windings;

$L_a$ is the self-inductances of the armature windings;

$B_m$ is the viscous friction coefficient;

$J$ is the equivalent moment of inertial of the rotor and attached load.

Permanent-magnet microtransducers are rotating energy-transfer electromechanical motion devices which convert energy by means of rotational motion. Transducers are the major part of MEMS, and therefore they must be thoroughly studied with the driving ICs. Micromotors (actuators) convert electrical energy to mechanical energy, while generators (sensors) convert mechanical energy to electrical energy. It is worth mentioning that the same permanent-magnet microtransducers can be used as the actuator-motor (if one applies the voltage) or as the sensor-generator (if one rotates the transducers, the voltage is induced). Hence, the energy conversion is reversible, and sensors can be operated as actuators and vice versa. That is, permanent-magnet DC microtransducers can be used as the actuators (motors) and sensors (gyroscopes, tachogenerators, etc.). Transducers have stationary and rotating members, separated by an air gap. The armature winding is placed in the rotor slots and connected to a rotating commutator, which rectifies the induced voltage, see Figure 5.2.10. One supplies the armature voltages to the armature (rotor) windings. The rotor windings and stator permanent magnets are magnetically coupled.

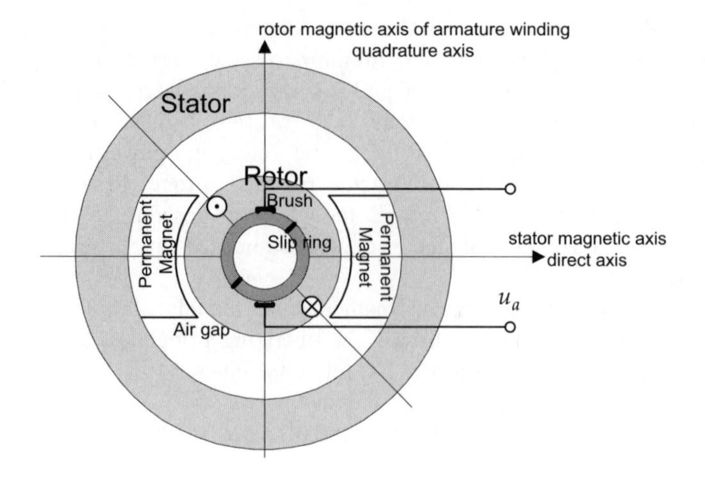


Figure 5.2.10 Permanent-magnet DC microtransducers.

The brushes, which are connected to the armature windings, ride on the commutator. The armature winding consists of identical uniformly distributed coils. The excitation magnetic field is produced by the permanent magnets. Due to the commutator (circular conducting segments), armature windings and permanent magnets produce stationary *magnetomotive forces* which are displaced by 90 electrical degrees. The armature magnetic force is along the *quadrature* (rotor) magnetic axis, while the *direct* axis stands for a permanent magnet magnetic axis. The electromagnetic torque is produced as a result of the interaction of these stationary *magnetomotive forces.*

From Kirchhoff's law, one obtains the following steady-state equation for the armature voltage for micromotors (the armature current opposes the induced *electromotive force*)

$$u_a - E_a = i_a r_a .$$

For microgenerators, the armature current is in the same direction as the generated *electromotive force*, and we have

$$u_a - E_a = -i_a r_a .$$

The difference between the applied voltage and the induced *electromotive force* is the voltage drop across the internal armature resistance $r_a$. One concludes that transducers rotate at an angular velocity at which the *electromotive force* generated in the armature winding balances the armature voltage. If a microtransducer operates as a micromotor, the induced *electromotive force* is less than the voltage applied to the windings. If a microtransducer operates as a generator, the generated (induced) *electromotive force* is greater than the terminal voltage.

The constant magnetic flux in AC and DC microtransducers are produced by permanent-magnets. Microtransducers with permanent-magnet poles are called permanent-magnet microtransducers. The permanent-magnet

DC microtransducer was illustrated in Figure 5.2.10, and a schematic diagram of permanent-magnet DC microtransducers is shown in Figure 5.2.11.

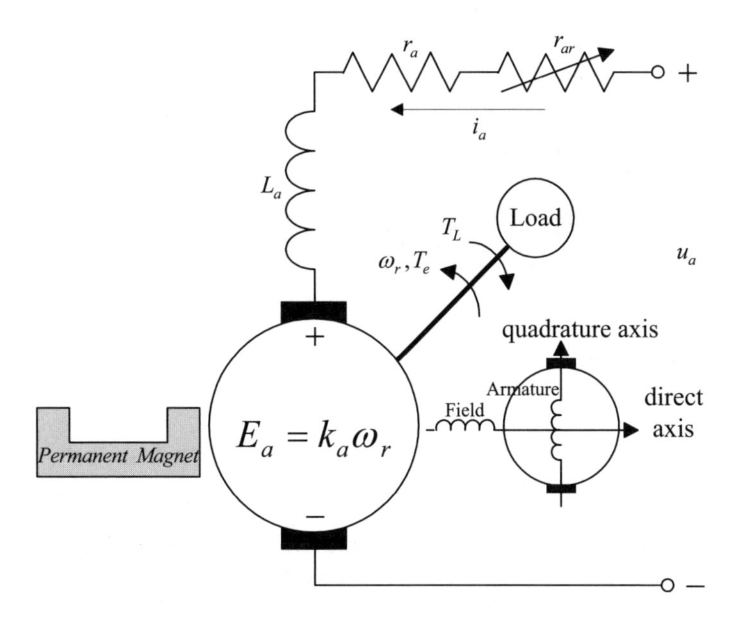


Figure 5.2.11 Schematic diagram of permanent-magnet microtransducers (current direction corresponds to the motor operation).

Using Kirchhoff's voltage law and Newton's second law of motion, the differential equations for permanent-magnet DC microtransducers can be easily derived. Assuming that the *susceptibility* is constant (in reality, Curie's constant varies as a function of temperature), one supposes that the flux, established by the permanent magnet poles, is constant. Then, denoting the *back emf* and *torque* constants as $k_a$, we have the following linear differential equations describing the transient behavior of the armature winding and *torsional-mechanical* dynamics

$$\frac{di_a}{dt} = -\frac{r_a}{L_a} i_a - \frac{k_a}{L_a}\omega_r + \frac{1}{L_a}u_a,$$

$$\frac{d\omega_r}{dt} = \frac{k_a}{J} i_a - \frac{B_m}{J}\omega_r - \frac{1}{J}T_L.$$

The lumped-parameter model in the state-space (matrix) form is found as

$$\begin{bmatrix} \frac{di_a}{dt} \\ \frac{d\omega_r}{dt} \end{bmatrix} = \begin{bmatrix} -\frac{r_a}{L_a} & -\frac{k_a}{L_a} \\ \frac{k_a}{J} & -\frac{B_m}{J} \end{bmatrix} \begin{bmatrix} i_a \\ \omega_r \end{bmatrix} + \begin{bmatrix} \frac{1}{L_a} \\ 0 \end{bmatrix} u_a - \begin{bmatrix} 0 \\ \frac{1}{J} \end{bmatrix} T_L .$$

An $s$-domain block diagram of permanent-magnet DC micromotors is illustrated in Figure 5.2.12.

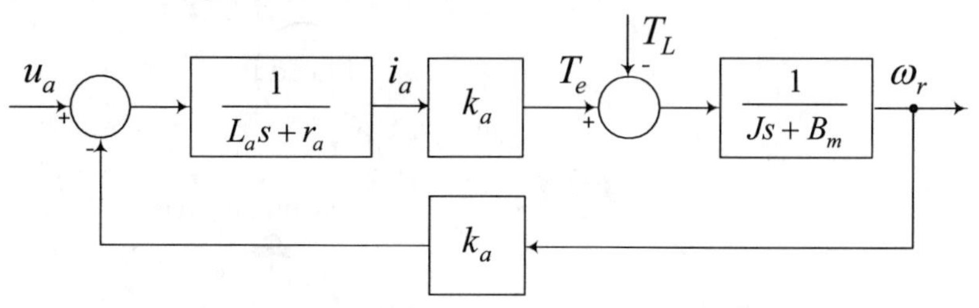


Figure 5.2.12 $s$-domain block diagram of permanent-magnet DC micromotors.

The angular velocity can be reversed if the polarity of the applied voltage is changed (the direction of the field flux cannot be changed). The steady-state torque-speed characteristic curves obey the following equation

$$\omega_r = \frac{u_a - r_a i_a}{k_a} = \frac{u_a}{k_a} - \frac{r_a}{k_a^2} T_e ,$$

and a spectrum of the torque-speed characteristic curves is illustrated in Figure 5.2.13.

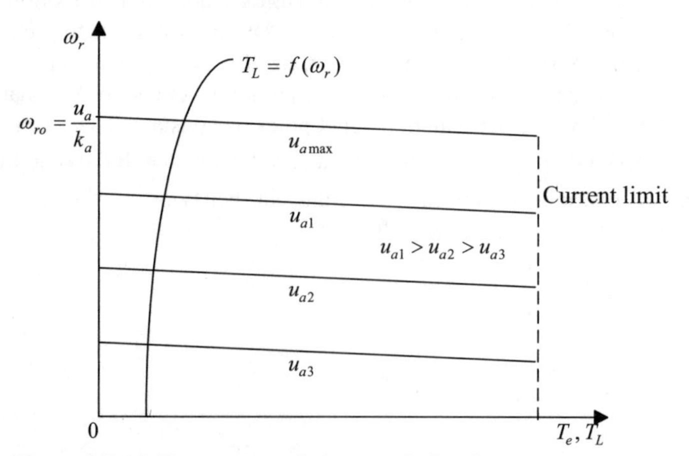


Figure 5.2.13 Torque-speed characteristics for permanent-magnet micromotors.

If the permanent-magnet DC microtransducer is used as the generator, the circuitry dynamics for the resistive load $R_L$ is given as

$$\frac{di_a}{dt} = -\frac{r_a + R_L}{L_a} i_a + \frac{k_a}{L_a} \omega_r .$$

That is, in the steady-state, the armature current is proportional to the angular velocity, and we have

$$i_a = \frac{k_a}{r_a + R_L} \omega_r .$$

Flip-chip MEMS found wide application due to low cost, and well-developed fabrication processes were applied. For example, monolithic dual power operational amplifiers, designed as a single-chip ICs, feed DC micromotor to regulate the angular velocity, see Figure 5.2.14.

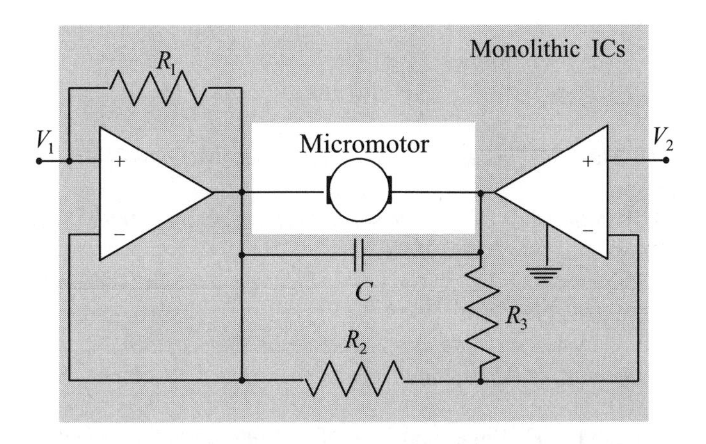


Figure 5.2.14 Application of a monolithic IC to control DC micromotors.

Motion microdevices (microtransducers - actuators and sensors) are mounted face down with bumps on the pads that form electrical and mechanical joints to the ICs driver. Figure 5.2.15 illustrates a flip-chip MEMS with permanent-magnet micromotor driven by the MC33030 monolithic ICs driver. Control algorithms are implemented to control the angular velocity of micromotors. The MC33030 integrates on-chip operational amplifier and comparator, driving and braking logic, PWM four-quadrant converter, etc. The MC33030 data and complete description are given in [4]. As in the conventional configurations, the difference between the reference (command) and actual angular velocity or displacement, linear velocity or position, is compared by the *error amplifier*, and two comparators are used. A *pnp* differential output power stage provides driving and braking capabilities, and the four-quadrant H-configured power stage guarantees high performance and efficiency. Using the error between the desired (command) and actual angular velocity or displacement, the bipolar voltage $u_a$ is applied to the armature winding. The electromagnetic torque is developed, and micromotor rotates.

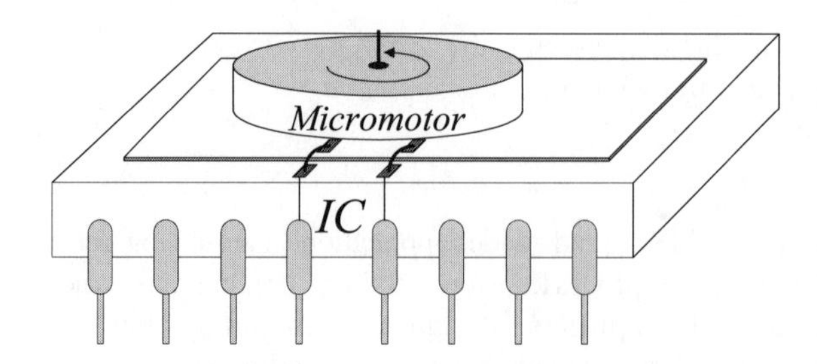


Figure 5.2.15 Flip-chip monolithic MEMS: MC33030 ICs and micromotor.

## References

1. W. H. Hayt, *Engineering Electromagnetics*, McGraw-Hill, New York, 1989.
2. J. D. Krause and D. A. Fleisch, *Electromagnetics With Applications*, McGraw-Hill, New York, 1999.
3. P. C. Krause and O. Wasynczuk, *Electromechanical Motion Devices*, McGraw-Hill, New York, 1989.
4. S. E. Lyshevski, *Micro- and nanoelectromechanical Systems, Fundamentals of Micro- and nanoengineering*, CRC Press, Boca Raton, FL, 2000.
5. C. R. Paul, K. W. Whites and S. A. Nasar, *Introduction to Electromagnetic Fields*, McGraw-Hill, New York, 1998.

## 5.3 INDUCTION MICROMACHINES

The majority of rotating micromachines and microtransducers designed and fabricated to date are the synchronous microdevices. Induction microctransducers, though have lower torque and power densities, do not require permanent magnet. Therefore, in this section we cover analysis, modeling, and control issues for induction micromotors which can be used as actuators. The following variables and symbols are used in this section:

$u_{as}, u_{bs}$ and $u_{cs}$ are the phase voltages in the stator windings *as, bs* and *cs*;

$u_{qs}, u_{ds}$ and $u_{os}$ are the *quadrature-*, *direct-*, and *zero*-axis voltages;

$i_{as}, i_{bs}$ and $i_{cs}$ are the phase currents in the stator windings *as, bs* and *cs*;

$i_{qs}, i_{ds}$ and $i_{os}$ are the *quadrature-*, *direct-*, and *zero*-axis stator currents;

$\psi_{as}, \psi_{bs}$ and $\psi_{cs}$ are the stator flux linkages;

$\psi_{qs}, \psi_{ds}$ and $\psi_{os}$ are the *quadrature-*, *direct-*, and *zero*-axis stator flux linkages;

$u_{ar}, u_{br}$ and $u_{cr}$ are the voltages in the rotor windings *ar, br* and *cr*;

$u_{qr}, u_{dr}$ and $u_{or}$ are the *quadrature-*, *direct-*, and *zero*-axis rotor voltages;

$i_{ar}, i_{br}$ and $i_{cr}$ are the currents in the rotor windings *ar, br* and *cr*;

$i_{qr}, i_{dr}$ and $i_{or}$ are the *quadrature-*, *direct-*, and *zero*-axis rotor currents;

$\psi_{ar}, \psi_{br}$ and $\psi_{cr}$ are the rotor flux linkages;

$\psi_{qr}, \psi_{dr}$ and $\psi_{or}$ are the *quadrature-*, *direct-*, and *zero*-axis rotor flux linkages;

$\omega_r$ and $\omega_{rm}$ are the electrical and mechanical angular velocities;

$\theta_r$ and $\theta_{rm}$ are the electrical and mechanical angular displacements;

$T_e$ is the electromagnetic torque developed by the micromotor;

$T_L$ is the load torque applied;

$r_s$ and $r_r$ are the resistances of the stator and rotor windings;

$L_{ss}$ and $L_{rr}$ are the self-inductances of the stator and rotor windings;

$L_{ms}$ is the stator magnetizing inductance;

$L_{ls}$ and $L_{lr}$ are the stator and rotor leakage inductances;

$N_s$ and $N_r$ are the number of turns of the stator and rotor windings;

*P* is the number of poles;

$B_m$ is the viscous friction coefficient;

$J$ is the equivalent moment of inertia;

$\omega$ and $\theta$ are the angular velocity and displacement of the reference frame.

### 5.3.1 Two-Phase Induction Micromotors

Two-phase induction micromotors (microactuators), shown in Figure 5.3.1, have two stator and rotor windings.

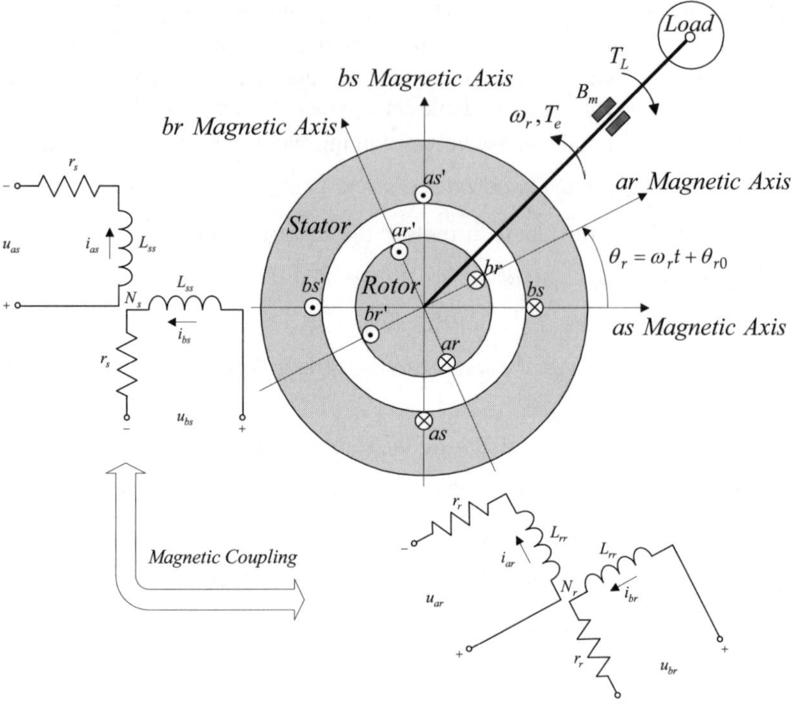


Figure 5.3.1 Two-phase symmetrical induction micromotor.

To develop lumped-parameter mathematical models of two-phase induction micromotors, we model the stator and rotor circuitry dynamics. As the control and state variables we use the voltages applied to the stator (*as* and *bs*) and rotor (*ar* and *br*) windings, as well as the stator and rotor currents and flux linkages.

Using Kirchhoff's voltage law, four differential equations are

$$u_{as} = r_s i_{as} + \frac{d\psi_{as}}{dt}, \quad u_{bs} = r_s i_{bs} + \frac{d\psi_{bs}}{dt},$$

$$u_{ar} = r_r i_{ar} + \frac{d\psi_{ar}}{dt}, \quad u_{br} = r_r i_{br} + \frac{d\psi_{br}}{dt}.$$

Hence, in the state-space form we have

$$\mathbf{u}_{abs} = \mathbf{r}_s \mathbf{i}_{abs} + \frac{d\boldsymbol{\psi}_{abs}}{dt},$$

$$\mathbf{u}_{abr} = \mathbf{r}_r \mathbf{i}_{abr} + \frac{d\boldsymbol{\psi}_{abr}}{dt}, \tag{5.3.1}$$

where $\mathbf{u}_{abs} = \begin{bmatrix} u_{as} \\ u_{bs} \end{bmatrix}$, $\mathbf{u}_{abr} = \begin{bmatrix} u_{ar} \\ u_{br} \end{bmatrix}$, $\mathbf{i}_{abs} = \begin{bmatrix} i_{as} \\ i_{bs} \end{bmatrix}$, $\mathbf{i}_{abr} = \begin{bmatrix} i_{ar} \\ i_{br} \end{bmatrix}$, $\boldsymbol{\psi}_{abs} = \begin{bmatrix} \psi_{as} \\ \psi_{bs} \end{bmatrix}$, and $\boldsymbol{\psi}_{abr} = \begin{bmatrix} \psi_{ar} \\ \psi_{br} \end{bmatrix}$ are the phase voltages, currents, and flux linkages; $\mathbf{r}_s = \begin{bmatrix} r_s & 0 \\ 0 & r_s \end{bmatrix}$ and $\mathbf{r}_r = \begin{bmatrix} r_r & 0 \\ 0 & r_r \end{bmatrix}$ are the matrices of the stator and rotor resistances.

Studying the magnetically coupled micromotor circuits, the following matrix equation for the flux linkages is found

$$\begin{bmatrix} \boldsymbol{\psi}_{abs} \\ \boldsymbol{\psi}_{abr} \end{bmatrix} = \begin{bmatrix} \mathbf{L}_s & \mathbf{L}_{sr} \\ \mathbf{L}_{sr}^T & \mathbf{L}_r \end{bmatrix} \begin{bmatrix} \mathbf{i}_{abs} \\ \mathbf{i}_{abr} \end{bmatrix},$$

where $\mathbf{L}_s$ is the matrix of the stator inductances, $\mathbf{L}_s = \begin{bmatrix} L_{ss} & 0 \\ 0 & L_{ss} \end{bmatrix}$, $L_{ss} = L_{ls} + L_{ms}$, $L_{ms} = \dfrac{N_s^2}{\Re_m}$; $\mathbf{L}_r$ is the matrix of the rotor inductances, $\mathbf{L}_r = \begin{bmatrix} L_{rr} & 0 \\ 0 & L_{rr} \end{bmatrix}$, $L_{rr} = L_{lr} + L_{mr}$, $L_{mr} = \dfrac{N_r^2}{\Re_m}$; $\mathbf{L}_{sr}$ is the matrix of the stator-rotor mutual inductances, $\mathbf{L}_{sr} = \begin{bmatrix} L_{sr}\cos\theta_r & -L_{sr}\sin\theta_r \\ L_{sr}\sin\theta_r & L_{sr}\cos\theta_r \end{bmatrix}$, $L_{sr} = \dfrac{N_s N_r}{\Re_m}$.

Using the number of turns in the stator and rotor windings, we have

$$\mathbf{i}'_{abr} = \frac{N_r}{N_s}\mathbf{i}_{abr},$$

$$\mathbf{u}'_{abr} = \frac{N_s}{N_r}\mathbf{u}_{abr},$$

$$\boldsymbol{\psi}'_{abr} = \frac{N_s}{N_r}\boldsymbol{\psi}_{abr}.$$

Then, taking note of the turn ratio, the flux linkages are written in matrix form as

$$\begin{bmatrix} \boldsymbol{\psi}_{abs} \\ \boldsymbol{\psi}'_{abr} \end{bmatrix} = \begin{bmatrix} \mathbf{L}_s & \mathbf{L}'_{sr} \\ \mathbf{L}'^{T}_{sr} & \mathbf{L}'_r \end{bmatrix} \begin{bmatrix} \mathbf{i}_{abs} \\ \mathbf{i}'_{abr} \end{bmatrix}, \tag{5.3.2}$$

where $\mathbf{L}_r' = \left(\frac{N_s}{N_r}\right)^2 \mathbf{L}_r = \begin{bmatrix} L_{rr}' & 0 \\ 0 & L_{rr}' \end{bmatrix}$, $L_{rr}' = L_{lr}' + L_{mr}'$;

$$\mathbf{L}_{sr}' = \left(\frac{N_s}{N_r}\right)\mathbf{L}_{sr} = L_{ms}\begin{bmatrix} \cos\theta_r & -\sin\theta_r \\ \sin\theta_r & \cos\theta_r \end{bmatrix},\ L_{ms} = \frac{N_s}{N_r}L_{sr},$$

$$L_{mr}' = \left(\frac{N_s}{N_r}\right)^2 L_{mr},\ L_{mr}' = L_{ms} = \frac{N_s}{N_r}L_{sr},\ L_{rr}' = L_{lr}' + L_{ms}.$$

Substituting the matrices for self- and mutual inductances $\mathbf{L}_s$, $\mathbf{L}_r'$ and $\mathbf{L}_{sr}'$ in (5.3.2), one obtains

$$\begin{bmatrix} \psi_{as} \\ \psi_{bs} \\ \psi_{ar}' \\ \psi_{br}' \end{bmatrix} = \begin{bmatrix} L_{ss} & 0 & L_{ms}\cos\theta_r & -L_{ms}\sin\theta_r \\ 0 & L_{ss} & L_{ms}\sin\theta_r & L_{ms}\cos\theta_r \\ L_{ms}\cos\theta_r & L_{ms}\sin\theta_r & L_{rr}' & 0 \\ -L_{ms}\sin\theta_r & L_{ms}\cos\theta_r & 0 & L_{rr}' \end{bmatrix} \begin{bmatrix} i_{as} \\ i_{bs} \\ i_{ar}' \\ i_{br}' \end{bmatrix}.$$

Therefore, the circuitry differential equations (5.3.1) are rewritten as

$$\mathbf{u}_{abs} = \mathbf{r}_s\mathbf{i}_{abs} + \frac{d\boldsymbol{\psi}_{abs}}{dt},\ \mathbf{u}_{abr}' = \mathbf{r}_r'\mathbf{i}_{abr}' + \frac{d\boldsymbol{\psi}_{abr}'}{dt}.$$

where $\mathbf{r}_r' = \frac{N_s^2}{N_r^2}\mathbf{r}_r = \frac{N_s^2}{N_r^2}\begin{bmatrix} r_r' & 0 \\ 0 & r_r' \end{bmatrix}$.

Assuming that the self- and mutual inductances $L_{ss}$, $L_{rr}'$, $L_{ms}$ are time-invariant and using the expressions for the flux linkages, one obtains a set of nonlinear differential equations to model the circuitry dynamics

$$L_{ss}\frac{di_{as}}{dt} + L_{ms}\frac{d\left(i_{ar}'\cos\theta_r\right)}{dt} - L_{ms}\frac{d\left(i_{br}'\sin\theta_r\right)}{dt} = -r_s i_{as} + u_{as},$$

$$L_{ss}\frac{di_{bs}}{dt} + L_{ms}\frac{d\left(i_{ar}'\sin\theta_r\right)}{dt} + L_{ms}\frac{d\left(i_{br}'\cos\theta_r\right)}{dt} = -r_s i_{bs} + u_{bs},$$

$$L_{ms}\frac{d\left(i_{as}\cos\theta_r\right)}{dt} + L_{ms}\frac{d\left(i_{bs}\sin\theta_r\right)}{dt} + L_{rr}'\frac{di_{ar}'}{dt} = -r_r' i_{ar}' + u_{ar}',$$

$$-L_{ms}\frac{d\left(i_{as}\sin\theta_r\right)}{dt} + L_{ms}\frac{d\left(i_{bs}\cos\theta_r\right)}{dt} + L_{rr}'\frac{di_{br}'}{dt} = -r_r' i_{br}' + u_{br}'.$$

Cauchy's form of these differential equations is found. In particular, we have the following nonlinear differential equations to model the stator-rotor circuitry dynamics for two-phase induction micromotors

$$\frac{di_{as}}{dt} = -\frac{L'_{rr}r_s}{L_{ss}L'_{rr} - L^2_{ms}}i_{as} + \frac{L^2_{ms}}{L_{ss}L'_{rr} - L^2_{ms}}i_{bs}\omega_r + \frac{L_{ms}L'_{rr}}{L_{ss}L'_{rr} - L^2_{ms}}i'_{ar}\left(\omega_r \sin\theta_r + \frac{r'_r}{L'_{rr}}\cos\theta_r\right)$$
$$+\frac{L_{ms}L'_{rr}}{L_{ss}L'_{rr} - L^2_{ms}}i'_{br}\left(\omega_r \cos\theta_r - \frac{r'_r}{L'_{rr}}\sin\theta_r\right) + \frac{L'_{rr}}{L_{ss}L'_{rr} - L^2_{ms}}u_{as} - \frac{L_{ms}}{L_{ss}L'_{rr} - L^2_{ms}}\cos\theta_r u'_{ar}$$
$$+\frac{L_{ms}}{L_{ss}L'_{rr} - L^2_{ms}}\sin\theta_r u'_{br},$$
$$\frac{di_{bs}}{dt} = -\frac{L'_{rr}r_s}{L_{ss}L'_{rr} - L^2_{ms}}i_{bs} - \frac{L^2_{ms}}{L_{ss}L'_{rr} - L^2_{ms}}i_{as}\omega_r - \frac{L_{ms}L'_{rr}}{L_{ss}L'_{rr} - L^2_{ms}}i'_{ar}\left(\omega_r \cos\theta_r - \frac{r'_r}{L'_{rr}}\sin\theta_r\right)$$
$$+\frac{L_{ms}L'_{rr}}{L_{ss}L'_{rr} - L^2_{ms}}i'_{br}\left(\omega_r \sin\theta_r + \frac{r'_r}{L'_{rr}}\cos\theta_r\right) + \frac{L'_{rr}}{L_{ss}L'_{rr} - L^2_{ms}}u_{bs} - \frac{L_{ms}}{L_{ss}L'_{rr} - L^2_{ms}}\sin\theta_r u'_{ar}$$
$$-\frac{L_{ms}}{L_{ss}L'_{rr} - L^2_{ms}}\cos\theta_r u'_{br},$$
$$\frac{di'_{ar}}{dt} = -\frac{L_{ss}r'_r}{L_{ss}L'_{rr} - L^2_{ms}}i'_{ar} + \frac{L_{ms}L_{ss}}{L_{ss}L'_{rr} - L^2_{ms}}i_{as}\left(\omega_r \sin\theta_r + \frac{r_s}{L_{ss}}\cos\theta_r\right) - \frac{L_{ms}L_{ss}}{L_{ss}L'_{rr} - L^2_{ms}}i_{bs}\left(\omega_r \cos\theta_r - \frac{r_s}{L_{ss}}\sin\theta_r\right)$$
$$-\frac{L^2_{ms}}{L_{ss}L'_{rr} - L^2_{ms}}i'_{br}\omega_r - \frac{L_{ms}}{L_{ss}L'_{rr} - L^2_{ms}}\cos\theta_r u_{as} - \frac{L_{ms}}{L_{ss}L'_{rr} - L^2_{ms}}\sin\theta_r u_{bs} + \frac{L_{ss}}{L_{ss}L'_{rr} - L^2_{ms}}u'_{ar},$$
$$\frac{di'_{br}}{dt} = -\frac{L_{ss}r'_r}{L_{ss}L'_{rr} - L^2_{ms}}i'_{br} + \frac{L_{ms}L_{ss}}{L_{ss}L'_{rr} - L^2_{ms}}i_{as}\left(\omega_r \cos\theta_r - \frac{r_s}{L_{ss}}\sin\theta_r\right) + \frac{L_{ms}L_{ss}}{L_{ss}L'_{rr} - L^2_{ms}}i_{bs}\left(\omega_r \sin\theta_r + \frac{r_s}{L_{ss}}\cos\theta_r\right)$$
$$+\frac{L^2_{ms}}{L_{ss}L'_{rr} - L^2_{ms}}i'_{ar}\omega_r + \frac{L_{ms}}{L_{ss}L'_{rr} - L^2_{ms}}\sin\theta_r u_{as} - \frac{L_{ms}}{L_{ss}L'_{rr} - L^2_{ms}}\cos\theta_r u_{bs} + \frac{L_{ss}}{L_{ss}L'_{rr} - L^2_{ms}}u'_{br}.$$

(5.3.3)

The electrical angular velocity $\omega_r$ and displacement $\theta_r$ are used in (5.3.3) as the state variables. Therefore, the *torsional-mechanical* equation of motion must be incorporated to describe the evolution of $\omega_r$ and $\theta_r$. From Newton's second law, we have

$$\sum T = T_e - B_m\omega_{rm} - T_L = J\frac{d\omega_{rm}}{dt},$$

$$\frac{d\theta_{rm}}{dt} = \omega_{rm}.$$

The mechanical angular velocity $\omega_{rm}$ is expressed by using the electrical angular velocity $\omega_r$ and the number of poles *P*. In particular,

$$\omega_{rm} = \frac{2}{P}\omega_r.$$

The mechanical and electrical angular displacements $\theta_{rm}$ and $\theta_r$ arc related as

$$\theta_{rm} = \frac{2}{P}\theta_r.$$

Taking note of Newton's second law of motion, one obtains two differential equations as given by

$$\frac{d\omega_r}{dt}=\frac{P}{2J}T_e-\frac{B_m}{J}\omega_r-\frac{P}{2J}T_L,$$
$$\frac{d\theta_r}{dt}=\omega_r.$$

The electromagnetic torque developed by the induction micromotors must be found. In particular, to find the expression for the electromagnetic torque developed by two-phase induction micromotors, the coenergy $W_c\left(\mathbf{i}_{abs},\mathbf{i}'_{abr},\theta_r\right)$ is used, and

$$T_e=\frac{P}{2}\frac{\partial W_c\left(\mathbf{i}_{abs},\mathbf{i}'_{abr},\theta_r\right)}{\partial\theta_r}.$$

Assuming that the magnetic system is linear, one has

$$W_c=W_f=\tfrac{1}{2}\mathbf{i}_{abs}^T\left(\mathbf{L}_s-L_{ls}\mathbf{I}\right)\mathbf{i}_{abs}+\mathbf{i}_{abs}^T\mathbf{L}'_{sr}\mathbf{i}'_{abr}+\tfrac{1}{2}\mathbf{i}'^{\,T}_{abr}\left(\mathbf{L}'_r-L'_{lr}\mathbf{I}\right)\mathbf{i}'_{abr}.$$

The self-inductances $L_{ss}$ and $L'_{rr}$, as well as the leakage inductances $L_{ls}$ and $L'_{lr}$, are not functions of the angular displacement $\theta_r$, while the following expression for the matrix of stator-rotor mutual inductances $\mathbf{L}'_{sr}$ was derived

$$\mathbf{L}'_{sr}=L_{ms}\begin{bmatrix}\cos\theta_r & -\sin\theta_r\\ \sin\theta_r & \cos\theta_r\end{bmatrix}.$$

Then, for *P*-pole two-phase induction micromotors, the electromagnetic torque is given by

$$T_e=\frac{P}{2}\frac{\partial W_c\left(\mathbf{i}_{abs},\mathbf{i}'_{abr},\theta_r\right)}{\partial\theta_r}=\frac{P}{2}\mathbf{i}_{abs}^T\frac{\partial\mathbf{L}'_{sr}(\theta_r)}{\partial\theta_r}\mathbf{i}'_{abr}=\frac{P}{2}L_{ms}\begin{bmatrix}i_{as} & i_{bs}\end{bmatrix}\begin{bmatrix}-\sin\theta_r & -\cos\theta_r\\ \cos\theta_r & -\sin\theta_r\end{bmatrix}\begin{bmatrix}i'_{ar}\\ i'_{br}\end{bmatrix}$$
$$=-\frac{P}{2}L_{ms}\left[\left(i_{as}i'_{ar}+i_{bs}i'_{br}\right)\sin\theta_r+\left(i_{as}i'_{br}-i_{bs}i'_{ar}\right)\cos\theta_r\right].\tag{5.3.4}$$

Using (5.3.4) for the electromagnetic torque $T_e$ in the *torsional-mechanical* equations of motion, one obtains

$$\frac{d\omega_r}{dt}=-\frac{P^2}{4J}L_{ms}\left[\left(i_{as}i'_{ar}+i_{bs}i'_{br}\right)\sin\theta_r+\left(i_{as}i'_{br}-i_{bs}i'_{ar}\right)\cos\theta_r\right]-\frac{B_m}{J}\omega_r-\frac{P}{2J}T_L$$
$$\frac{d\theta_r}{dt}=\omega_r.\tag{5.3.5}$$

These two differential equation must be integrated with the circuitry dynamics. In particular, augmenting differential equations (5.3.3) and (5.3.5), the following set of highly nonlinear differential equations results to model two-phase induction micromachines

$$\frac{di_{as}}{dt}=-\frac{L_{rr}^{'}r_s}{L_\Sigma}i_{as}+\frac{L_{ms}^2}{L_\Sigma}i_{bs}\omega_r+\frac{L_{ms}L_{rr}^{'}}{L_\Sigma}i_{ar}^{'}\left(\omega_r\sin\theta_r+\frac{r_r^{'}}{L_{rr}^{'}}\cos\theta_r\right)$$

$$+\frac{L_{ms}L_{rr}^{'}}{L_\Sigma}i_{br}^{'}\left(\omega_r\cos\theta_r-\frac{r_r^{'}}{L_{rr}^{'}}\sin\theta_r\right)+\frac{L_{rr}^{'}}{L_\Sigma}u_{as}-\frac{L_{ms}}{L_\Sigma}\cos\theta_r u_{ar}^{'}+\frac{L_{ms}}{L_\Sigma}\sin\theta_r u_{br}^{'},$$

$$\frac{di_{bs}}{dt}=-\frac{L_{rr}^{'}r_s}{L_\Sigma}i_{bs}-\frac{L_{ms}^2}{L_\Sigma}i_{as}\omega_r-\frac{L_{ms}L_{rr}^{'}}{L_\Sigma}i_{ar}^{'}\left(\omega_r\cos\theta_r-\frac{r_r^{'}}{L_{rr}^{'}}\sin\theta_r\right)$$

$$+\frac{L_{ms}L_{rr}^{'}}{L_\Sigma}i_{br}^{'}\left(\omega_r\sin\theta_r+\frac{r_r^{'}}{L_{rr}^{'}}\cos\theta_r\right)+\frac{L_{rr}^{'}}{L_\Sigma}u_{bs}-\frac{L_{ms}}{L_\Sigma}\sin\theta_r u_{ar}^{'}-\frac{L_{ms}}{L_\Sigma}\cos\theta_r u_{br}^{'},$$

$$\frac{di_{ar}^{'}}{dt}=-\frac{L_{ss}r_r^{'}}{L_\Sigma}i_{ar}^{'}+\frac{L_{ms}L_{ss}}{L_\Sigma}i_{as}\left(\omega_r\sin\theta_r+\frac{r_s}{L_{ss}}\cos\theta_r\right)-\frac{L_{ms}L_{ss}}{L_\Sigma}i_{bs}\left(\omega_r\cos\theta_r-\frac{r_s}{L_{ss}}\sin\theta_r\right)$$

$$-\frac{L_{ms}^2}{L_\Sigma}i_{br}^{'}\omega_r-\frac{L_{ms}}{L_\Sigma}\cos\theta_r u_{as}-\frac{L_{ms}}{L_\Sigma}\sin\theta_r u_{bs}+\frac{L_{ss}}{L_\Sigma}u_{ar}^{'},$$

$$\frac{di_{br}^{'}}{dt}=-\frac{L_{ss}r_r^{'}}{L_\Sigma}i_{br}^{'}+\frac{L_{ms}L_{ss}}{L_\Sigma}i_{as}\left(\omega_r\cos\theta_r-\frac{r_s}{L_{ss}}\sin\theta_r\right)+\frac{L_{ms}L_{ss}}{L_\Sigma}i_{bs}\left(\omega_r\sin\theta_r+\frac{r_s}{L_{ss}}\cos\theta_r\right)$$

$$+\frac{L_{ms}^2}{L_\Sigma}i_{ar}^{'}\omega_r+\frac{L_{ms}}{L_\Sigma}\sin\theta_r u_{as}-\frac{L_{ms}}{L_\Sigma}\cos\theta_r u_{bs}+\frac{L_{ss}}{L_\Sigma}u_{br}^{'},$$

$$\frac{d\omega_r}{dt}=-\frac{P^2}{4J}L_{ms}\left[\left(i_{as}i_{ar}^{'}+i_{bs}i_{br}^{'}\right)\sin\theta_r+\left(i_{as}i_{br}^{'}-i_{bs}i_{ar}^{'}\right)\cos\theta_r\right]-\frac{B_m}{J}\omega_r-\frac{P}{2J}T_L,$$

$$\frac{d\theta_r}{dt}=\omega_r, \tag{5.3.6}$$

where $L_\Sigma=L_{ss}L_{rr}^{'}-L_{ms}^2$.

These nonlinear differential equations give the lumped-parameter mathematical model of two-phase induction micromotors (microtransducers).

In the state-space (matrix) form, a set of six highly coupled nonlinear differential equations (5.3.6) is given as

$$
\begin{bmatrix} \frac{di_{as}}{dt} \\ \frac{di_{bs}}{dt} \\ \frac{di'_{ar}}{dt} \\ \frac{di'_{br}}{dt} \\ \frac{d\omega_r}{dt} \\ \frac{d\theta_r}{dt} \end{bmatrix} = \begin{bmatrix} -\frac{L'_{rr} r_s}{L_\Sigma} & 0 & 0 & 0 & 0 & 0 \\ 0 & -\frac{L'_{rr} r_s}{L_\Sigma} & 0 & 0 & 0 & 0 \\ 0 & 0 & -\frac{L_{ss} r'_r}{L_\Sigma} & 0 & 0 & 0 \\ 0 & 0 & 0 & -\frac{L_{ss} r'_r}{L_\Sigma} & 0 & 0 \\ 0 & 0 & 0 & 0 & -\frac{B_m}{J} & 0 \\ 0 & 0 & 0 & 0 & 1 & 0 \end{bmatrix} \begin{bmatrix} i_{as} \\ i_{bs} \\ i'_{ar} \\ i'_{br} \\ \omega_r \\ \theta_r \end{bmatrix}
$$

$$
+ \begin{bmatrix} \frac{L_{ms}^2}{L_\Sigma} i_{bs}\omega_r + \frac{L_{ms}L'_{rr}}{L_\Sigma} i'_{ar}\left(\omega_r \sin\theta_r + \frac{r'_r}{L'_{rr}}\cos\theta_r\right) + \frac{L_{ms}L'_{rr}}{L_\Sigma} i'_{br}\left(\omega_r \cos\theta_r - \frac{r'_r}{L'_{rr}}\sin\theta_r\right) \\ -\frac{L_{ms}^2}{L_\Sigma} i_{as}\omega_r - \frac{L_{ms}L'_{rr}}{L_\Sigma} i'_{ar}\left(\omega_r \cos\theta_r - \frac{r'_r}{L'_{rr}}\sin\theta_r\right) + \frac{L_{ms}L'_{rr}}{L_\Sigma} i'_{br}\left(\omega_r \sin\theta_r + \frac{r'_r}{L'_{rr}}\cos\theta_r\right) \\ \frac{L_{ms}L_{ss}}{L_\Sigma} i_{as}\left(\omega_r \sin\theta_r + \frac{r_s}{L_{ss}}\cos\theta_r\right) - \frac{L_{ms}L_{ss}}{L_\Sigma} i_{bs}\left(\omega_r \cos\theta_r - \frac{r_s}{L_{ss}}\sin\theta_r\right) - \frac{L_{ms}^2}{L_\Sigma} i'_{br}\omega_r \\ \frac{L_{ms}L_{ss}}{L_\Sigma} i_{as}\left(\omega_r \cos\theta_r - \frac{r_s}{L_{ss}}\sin\theta_r\right) + \frac{L_{ms}L_{ss}}{L_\Sigma} i_{bs}\left(\omega_r \sin\theta_r + \frac{r_s}{L_{ss}}\cos\theta_r\right) + \frac{L_{ms}^2}{L_\Sigma} i'_{ar}\omega_r \\ -\frac{P^2}{4J} L_{ms}\left[\left(i_{as}i'_{ar} + i_{bs}i'_{br}\right)\sin\theta_r + \left(i_{as}i'_{br} - i_{bs}i'_{ar}\right)\cos\theta_r\right] \\ 0 \end{bmatrix}
$$

$$
+ \begin{bmatrix} \frac{L'_{rr}}{L_\Sigma} & 0 & 0 & 0 \\ 0 & \frac{L'_{rr}}{L_\Sigma} & 0 & 0 \\ 0 & 0 & \frac{L_{ss}}{L_\Sigma} & 0 \\ 0 & 0 & 0 & \frac{L_{ss}}{L_\Sigma} \\ 0 & 0 & 0 & 0 \\ 0 & 0 & 0 & 0 \end{bmatrix} \begin{bmatrix} u_{as} \\ u_{bs} \\ u'_{ar} \\ u'_{br} \end{bmatrix} + \begin{bmatrix} -\frac{L_{ms}}{L_\Sigma}\cos\theta_r u'_{ar} + \frac{L_{ms}}{L_\Sigma}\sin\theta_r u'_{br} \\ -\frac{L_{ms}}{L_\Sigma}\sin\theta_r u'_{ar} - \frac{L_{ms}}{L_\Sigma}\cos\theta_r u'_{br} \\ -\frac{L_{ms}}{L_\Sigma}\cos\theta_r u_{as} - \frac{L_{ms}}{L_\Sigma}\sin\theta_r u_{bs} \\ \frac{L_{ms}}{L_\Sigma}\sin\theta_r u_{as} - \frac{L_{ms}}{L_\Sigma}\cos\theta_r u_{bs} \\ 0 \\ 0 \end{bmatrix} - \begin{bmatrix} 0 \\ 0 \\ 0 \\ 0 \\ \frac{P}{2J} \\ 0 \end{bmatrix} T_L
\tag{5.3.7}
$$

### *Modeling Two-Phase Induction Micromotors Using the Lagrange Equations*

The mathematical model can be derived using Lagrange's equations. The generalized independent coordinates and the generalized forces are

$$q_1 = \frac{i_{as}}{s},\ q_2 = \frac{i_{bs}}{s},\ q_3 = \frac{i'_{ar}}{s},\ q_4 = \frac{i'_{br}}{s},\ q_5 = \theta_r,$$

and $Q_1 = u_{as}$, $Q_2 = u_{bs}$, $Q_3 = u'_{ar}$, $Q_4 = u'_{br}$, $Q_5 = -T_L$

Five Lagrange equations are written as

$$\frac{d}{dt}\left(\frac{\partial\Gamma}{\partial\dot{q}_1}\right) - \frac{\partial\Gamma}{\partial q_1} + \frac{\partial D}{\partial\dot{q}_1} + \frac{\partial\Pi}{\partial q_1} = Q_1,$$

$$\frac{d}{dt}\left(\frac{\partial\Gamma}{\partial\dot{q}_2}\right) - \frac{\partial\Gamma}{\partial q_2} + \frac{\partial D}{\partial\dot{q}_2} + \frac{\partial\Pi}{\partial q_2} = Q_2,$$

$$\frac{d}{dt}\left(\frac{\partial\Gamma}{\partial\dot{q}_3}\right) - \frac{\partial\Gamma}{\partial q_3} + \frac{\partial D}{\partial\dot{q}_3} + \frac{\partial\Pi}{\partial q_3} = Q_3,$$

$$\frac{d}{dt}\left(\frac{\partial\Gamma}{\partial\dot{q}_4}\right) - \frac{\partial\Gamma}{\partial q_4} + \frac{\partial D}{\partial\dot{q}_4} + \frac{\partial\Pi}{\partial q_4} = Q_4,$$

$$\frac{d}{dt}\left(\frac{\partial\Gamma}{\partial\dot{q}_5}\right) - \frac{\partial\Gamma}{\partial q_5} + \frac{\partial D}{\partial\dot{q}_5} + \frac{\partial\Pi}{\partial q_5} = Q_5.$$

The total kinetic, potential, and dissipated energies are

$$\Gamma = \tfrac{1}{2}L_{ss}\dot{q}_1^2 + L_{ms}\dot{q}_1\dot{q}_3\cos q_5 - L_{ms}\dot{q}_1\dot{q}_4\sin q_5 + \tfrac{1}{2}L_{ss}\dot{q}_2^2$$
$$+ L_{ms}\dot{q}_2\dot{q}_3\sin q_5 + L_{ms}\dot{q}_2\dot{q}_4\cos q_5 + \tfrac{1}{2}L'_{rr}\dot{q}_3^2 + \tfrac{1}{2}L'_{rr}\dot{q}_4^2 + \tfrac{1}{2}J\dot{q}_5^2,$$

$$\Pi = 0,$$

$$D = \tfrac{1}{2}\left(r_s\dot{q}_1^2 + r_s\dot{q}_2^2 + r'_r\dot{q}_3^2 + r'_r\dot{q}_4^2 + B_m\dot{q}_5^2\right).$$

Thus,

$$\frac{\partial\Gamma}{\partial q_1} = 0,\ \frac{\partial\Gamma}{\partial\dot{q}_1} = L_{ss}\dot{q}_1 + L_{ms}\dot{q}_3\cos q_5 - L_{ms}\dot{q}_4\sin q_5,$$

$$\frac{\partial\Gamma}{\partial q_2} = 0,\ \frac{\partial\Gamma}{\partial\dot{q}_2} = L_{ss}\dot{q}_2 + L_{ms}\dot{q}_3\sin q_5 + L_{ms}\dot{q}_4\cos q_5,$$

$$\frac{\partial\Gamma}{\partial q_3} = 0,\ \frac{\partial\Gamma}{\partial\dot{q}_3} = L'_{rr}\dot{q}_3 + L_{ms}\dot{q}_1\cos q_5 + L_{ms}\dot{q}_2\sin q_5,$$

$$\frac{\partial\Gamma}{\partial q_4} = 0,\ \frac{\partial\Gamma}{\partial\dot{q}_4} = L'_{rr}\dot{q}_4 - L_{ms}\dot{q}_1\sin q_5 + L_{ms}\dot{q}_2\cos q_5,$$

$$\frac{\partial \Gamma}{\partial q_5} = -L_{ms}\dot{q}_1\dot{q}_3 \sin q_5 - L_{ms}\dot{q}_1\dot{q}_4 \cos q_5$$
$$+ L_{ms}\dot{q}_2\dot{q}_3 \cos q_5 - L_{ms}\dot{q}_2\dot{q}_4 \sin q_5$$
$$= -L_{ms}\left[(\dot{q}_1\dot{q}_3 + \dot{q}_2\dot{q}_4)\sin q_5 + (\dot{q}_1\dot{q}_4 - \dot{q}_2\dot{q}_3)\cos q_5\right],$$
$$\frac{\partial \Gamma}{\partial \dot{q}_5} = J\dot{q}_5,$$
$$\frac{\partial \Pi}{\partial q_1} = 0, \quad \frac{\partial \Pi}{\partial q_2} = 0, \quad \frac{\partial \Pi}{\partial q_3} = 0, \quad \frac{\partial \Pi}{\partial q_4} = 0, \quad \frac{\partial \Pi}{\partial q_5} = 0,$$
$$\frac{\partial D}{\partial \dot{q}_1} = r_s\dot{q}_1, \quad \frac{\partial D}{\partial \dot{q}_2} = r_s\dot{q}_2, \quad \frac{\partial D}{\partial \dot{q}_3} = r_r'\dot{q}_3, \quad \frac{\partial D}{\partial \dot{q}_4} = r_r'\dot{q}_4, \quad \frac{\partial D}{\partial \dot{q}_5} = B_m\dot{q}_5.$$

Taking note of $\dot{q}_1 = i_{as}$, $\dot{q}_2 = i_{bs}$, $\dot{q}_3 = i_{ar}'$, $\dot{q}_4 = i_{br}'$ and $\dot{q}_5 = \omega_r$, one obtains the following differential equations

$$L_{ss}\frac{di_{as}}{dt} + L_{ms}\frac{d\left(i_{ar}'\cos\theta_r\right)}{dt} - L_{ms}\frac{d\left(i_{br}'\sin\theta_r\right)}{dt} + r_s i_{as} = u_{as},$$

$$L_{ss}\frac{di_{bs}}{dt} + L_{ms}\frac{d\left(i_{ar}'\sin\theta_r\right)}{dt} + L_{ms}\frac{d\left(i_{br}'\cos\theta_r\right)}{dt} + r_s i_{bs} = u_{bs},$$

$$L_{ms}\frac{d\left(i_{as}\cos\theta_r\right)}{dt} + L_{ms}\frac{d\left(i_{bs}\sin\theta_r\right)}{dt} + L_{rr}'\frac{di_{ar}'}{dt} + r_r' i_{ar}' = u_{ar}',$$

$$-L_{ms}\frac{d\left(i_{as}\sin\theta_r\right)}{dt} + L_{ms}\frac{d\left(i_{bs}\cos\theta_r\right)}{dt} + L_{rr}'\frac{di_{br}'}{dt} + r_r' i_{br}' = u_{br}',$$

$$J\frac{d^2\theta_r}{dt^2} + L_{ms}\left[\left(i_{as}i_{ar}' + i_{bs}i_{br}'\right)\sin\theta_r + \left(i_{as}i_{br}' - i_{bs}i_{ar}'\right)\cos\theta_r\right] + B_m\frac{d\theta_r}{dt} = -T_L$$

For $P$-pole induction micromotors, by making use of

$$\frac{d\theta_r}{dt} = \omega_r,$$

six differential equations, as found in (5.3.6), result.

## *Control of Induction Micromotors*

The angular velocity of induction micromotors must be controlled, and the torque-speed characteristic curves should be thoroughly examined. The electromagnetic torque developed by two-phase induction micromotors is given by equation (5.3.4). To guarantee the balanced operating condition for two-phase induction micromotors, one supplies the following phase voltages to the stator windings

$$u_{as}(t) = \sqrt{2}u_M \cos(\omega_f t),$$

$$u_{bs}(t) = \sqrt{2}u_M \sin(\omega_f t),$$

and the sinusoidal steady-state phase currents are

$$i_{as}(t) = \sqrt{2}i_M \cos(\omega_f t - \varphi_i)$$

and $i_{bs}(t) = \sqrt{2}i_M \sin(\omega_f t - \varphi_i)$.

Here the following notations are used: $u_M$ is the magnitude of the voltages applied to the *as* and *bs* stator windings; $i_M$ is the magnitude of the *as* and *bs* stator currents; $\omega_f$ is the angular frequency of the applied phase voltages, $\omega_f = 2\pi f$ ; $f$ is the frequency of the supplied voltage; $\varphi_i$ is the phase difference.

The applied voltage to the micromotor windings cannot exceed the admissible voltage $u_{M\max}$. That is,

$$u_{M\min} \le u_M \le u_{M\max}.$$

The micromotor synchronous angular velocity $\omega_e$ is found using the number of poles as $\omega_e = \frac{4\pi f}{P}$. It is evident that the synchronous velocity $\omega_e$ can be regulated by changing the frequency *f*. To regulate the angular velocity, one varies the magnitude of the applied voltages as well as the frequency. The torque-speed characteristic curves of induction micromotors must be thoroughly studied. Performing the transient analysis by solving the derived differential equations (5.3.6), one can find the steady-state curves $\omega_r = \Omega_T(T_e)$ by plotting the angular velocity versus the electromagnetic torque developed.

The following principles are used to control the angular velocity of induction micromotors.

*Voltage control.*

By changing the magnitude $u_M$ of the applied phase voltages to the stator windings, the angular velocity is regulated in the stable operating region, see Figure 5.3.2a. It was emphasized that

$$u_{M\min} < u_M < u_{M\max},$$

where $u_{M\max}$ is the maximum allowed (rated) voltage.

*Frequency control.*

The magnitude of the applied phase voltages is constant $u_M^{constant}$, and the angular velocity is regulated above and below the synchronous angular velocity by changing the frequency of the supplied voltages *f*. This concept can be clearly demonstrated using the formula $\omega_e = \dfrac{4\pi f}{P}$. The torque-speed characteristics for different values of the frequency are shown in Figure 5.3.2b.

*Voltage-frequency control.*

The angular frequency $\omega_f$ is proportional to the frequency of the supplied voltages, $\omega_f = 2\pi f$. To minimize losses, the voltages applied to the stator windings should be regulated if the frequency is changed. In particular, the magnitude of phase voltages can be decreased linearly with decreasing the frequency. That is, to guarantee the *constant volts per hertz* control one maintains the following relationship

$$\frac{u_{Mi}}{f_i} = const$$

or $\dfrac{u_{Mi}}{\omega_{fi}} = const$.

The corresponding torque-speed characteristics are documented in Figure 5.3.2c. Regulating the voltage-frequency patterns, one shapes the torque-speed curves. For example, the following relation

$$\sqrt{\frac{u_{Mi}}{f_i}} = const$$

can be applied to adjust the magnitude $u_M$ and frequency $f$ of the supplied voltages. To attain the acceleration and settling time specified, overshoot and rise time needed, the general purpose (standard), soft- and high-starting torque patterns are implemented based upon the requirements and criteria imposed (see the standard, soft- and high-torque patterns as illustrated in Figure 5.3.2d).

That is, assigning

$$\omega_f = \varphi(u_M)$$

with domain

$$u_{M\min} < u_M < u_{M\max}$$

and range

$$\omega_{f\min} < \omega_f < \omega_{f\max},$$

one maintains

$$\frac{u_{Mi}}{f_i} = \text{var}$$

or $\dfrac{u_{Mi}}{\omega_{fi}} = \text{var}$.

For example, the desired torque-speed characteristics, as documented in Figure 5.3.2e, can be guaranteed.

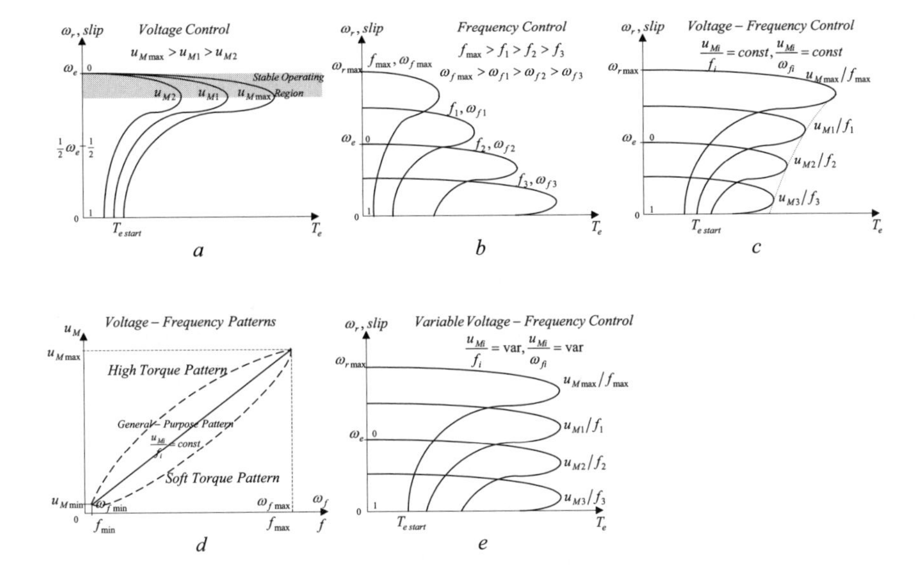


Figure 5.3.2 Torque-speed characteristic curves $\omega_r = \Omega_T(T_e)$:

(a) voltage control; (b) frequency control; (c) voltage-frequency control: *constant volts per hertz* control; (d) voltage-frequency patterns; (e) variable voltage-frequency control.

## *S-Domain Block Diagram of Two-Phase Induction Micromotors*

To perform the analysis of dynamics, to control induction machines, as well as to visualize the results, it is important to develop the *s*-domain block diagrams.

For squirrel-cage induction micromotors, the rotor windings are short-circuited, and hence

$$u'_{ar} = u'_{br} = 0 .$$

The block diagram is built using differential equations (5.3.6). The resulting *s*-domain block diagram is shown in Figure 5.3.3.

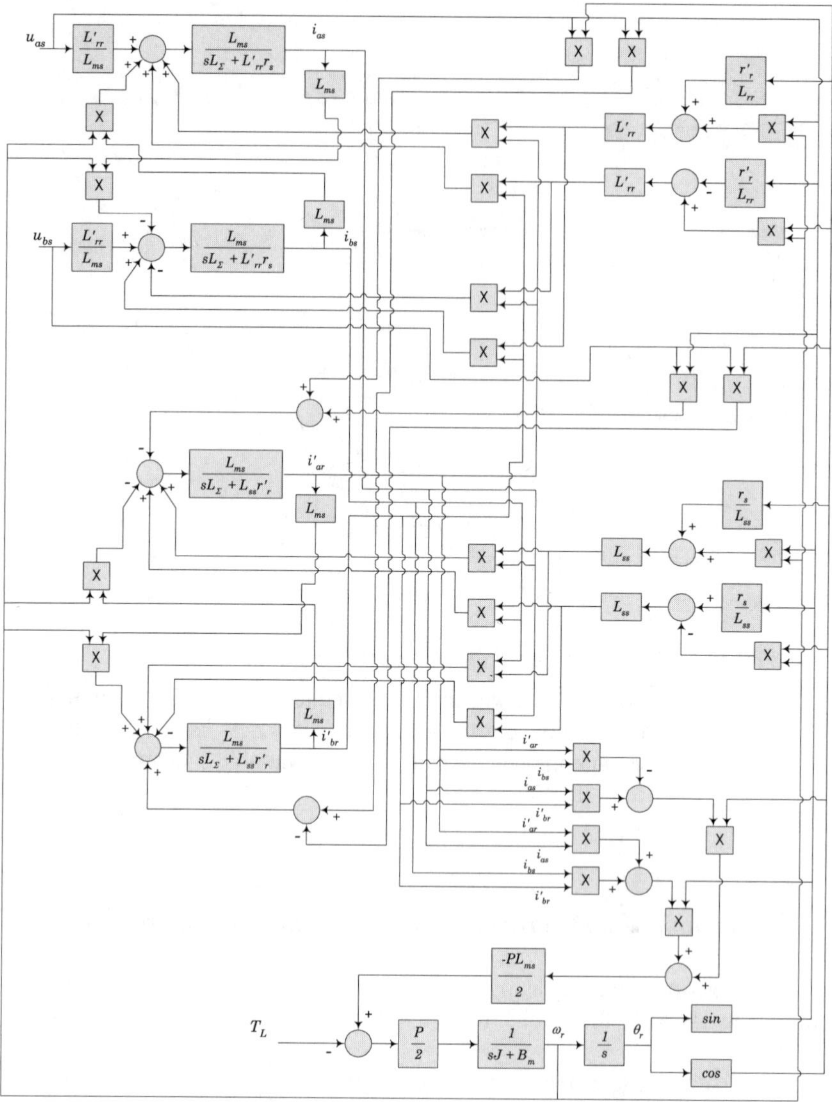


Figure 5.3.3 $s$-domain block diagram of squirrel-cage induction micromotors.

## 5.3.2 Three-Phase Induction Micromotors

### *Dynamics of Induction Micromotors in the Machine Variables*

Our goal is to develop the mathematical model of three-phase induction micromotors, as shown in Figure 5.3.4, using Kirchhoff's and Newton's second laws.

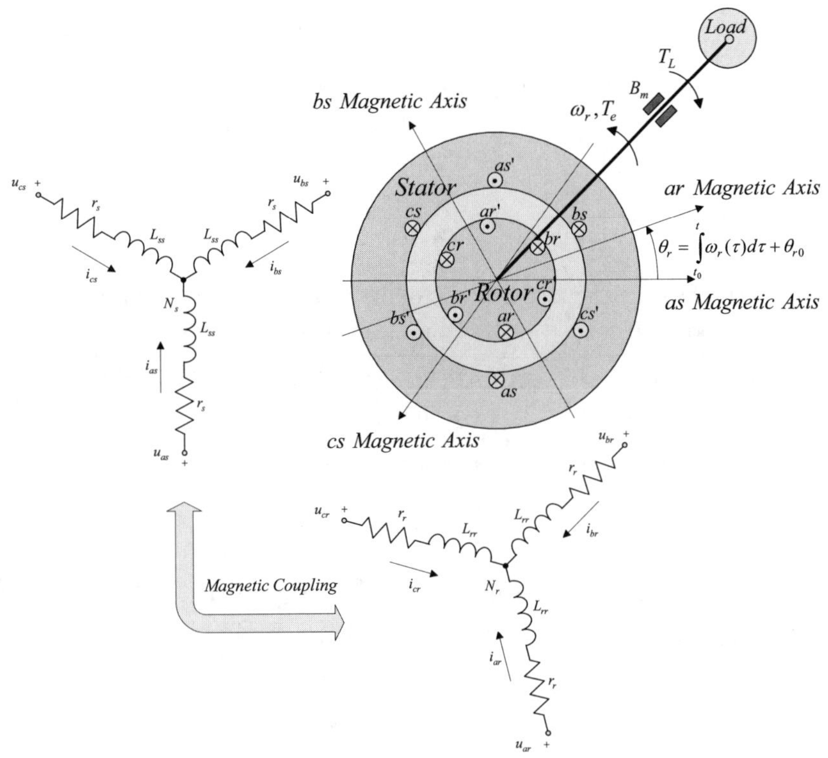


Figure 5.3.4 Three-phase symmetrical induction micromotor.

Studying the magnetically coupled stator and rotor circuitry, Kirchhoff's voltage law relates the *abc* stator and rotor phase voltages, currents, and flux linkages through the set of differential equations.

For magnetically coupled stator and rotor windings, we have

$$u_{as} = r_s i_{as} + \frac{d\psi_{as}}{dt},\ u_{bs} = r_s i_{bs} + \frac{d\psi_{bs}}{dt},\ u_{cs} = r_s i_{cs} + \frac{d\psi_{cs}}{dt},$$

$$u_{ar} = r_r i_{ar} + \frac{d\psi_{ar}}{dt},\ u_{br} = r_r i_{br} + \frac{d\psi_{br}}{dt},\ u_{cr} = r_r i_{cr} + \frac{d\psi_{cr}}{dt}. \quad (5.3.8)$$

It is clear that the *abc* stator and rotor voltages, currents, and flux linkages are used as the variables. In the state-space form, equations (5.3.8) are rewritten as

$$\mathbf{u}_{abcs} = \mathbf{r}_s \mathbf{i}_{abcs} + \frac{d\boldsymbol{\psi}_{abcs}}{dt},$$

$$\mathbf{u}_{abcr} = \mathbf{r}_r \mathbf{i}_{abcr} + \frac{d\boldsymbol{\psi}_{abcr}}{dt}, \quad (5.3.9)$$

where the *abc* stator and rotor voltages, currents and flux linkages are

$$\mathbf{u}_{abcs}=\begin{bmatrix}u_{as}\\u_{bs}\\u_{cs}\end{bmatrix},\mathbf{u}_{abcr}=\begin{bmatrix}u_{ar}\\u_{br}\\u_{cr}\end{bmatrix},\mathbf{i}_{abcs}=\begin{bmatrix}i_{as}\\i_{bs}\\i_{cs}\end{bmatrix},\mathbf{i}_{abcr}=\begin{bmatrix}i_{ar}\\i_{br}\\i_{cr}\end{bmatrix},\boldsymbol{\psi}_{abcs}=\begin{bmatrix}\psi_{as}\\\psi_{bs}\\\psi_{cs}\end{bmatrix},$$

and $\boldsymbol{\psi}_{abcr}=\begin{bmatrix}\psi_{ar}\\\psi_{br}\\\psi_{cr}\end{bmatrix}$.

In (5.3.9), the diagonal matrices of the stator and rotor resistances are

$$\mathbf{r}_s=\begin{bmatrix}r_s&0&0\\0&r_s&0\\0&0&r_s\end{bmatrix}\text{ and }\mathbf{r}_r=\begin{bmatrix}r_r&0&0\\0&r_r&0\\0&0&r_r\end{bmatrix}.$$

The flux linkages equations must be thoroughly examined, and one has

$$\begin{bmatrix}\boldsymbol{\psi}_{abcs}\\\boldsymbol{\psi}_{abcr}\end{bmatrix}=\begin{bmatrix}\mathbf{L}_s&\mathbf{L}_{sr}\\\mathbf{L}_{sr}^T&\mathbf{L}_r\end{bmatrix}\begin{bmatrix}\mathbf{i}_{abcs}\\\mathbf{i}_{abcr}\end{bmatrix},\tag{5.3.10}$$

where the matrices of self- and mutual inductances $\mathbf{L}_s$, $\mathbf{L}_r$ and $\mathbf{L}_{sr}$ are

$$\mathbf{L}_s=\begin{bmatrix}L_{ls}+L_{ms}&-\frac{1}{2}L_{ms}&-\frac{1}{2}L_{ms}\\-\frac{1}{2}L_{ms}&L_{ls}+L_{ms}&-\frac{1}{2}L_{ms}\\-\frac{1}{2}L_{ms}&-\frac{1}{2}L_{ms}&L_{ls}+L_{ms}\end{bmatrix},$$

$$\mathbf{L}_r=\begin{bmatrix}L_{lr}+L_{mr}&-\frac{1}{2}L_{mr}&-\frac{1}{2}L_{mr}\\-\frac{1}{2}L_{mr}&L_{lr}+L_{mr}&-\frac{1}{2}L_{mr}\\-\frac{1}{2}L_{mr}&-\frac{1}{2}L_{mr}&L_{lr}+L_{mr}\end{bmatrix},$$

$$\mathbf{L}_{sr}=L_{sr}\begin{bmatrix}\cos\theta_r&\cos\left(\theta_r+\frac{2}{3}\pi\right)&\cos\left(\theta_r-\frac{2}{3}\pi\right)\\\cos\left(\theta_r-\frac{2}{3}\pi\right)&\cos\theta_r&\cos\left(\theta_r+\frac{2}{3}\pi\right)\\\cos\left(\theta_r+\frac{2}{3}\pi\right)&\cos\left(\theta_r-\frac{2}{3}\pi\right)&\cos\theta_r\end{bmatrix}.$$

Using the number of turns $N_s$ and $N_r$, one finds

$$\mathbf{u}'_{abcr}=\frac{N_s}{N_r}\mathbf{u}_{abcr},\ \mathbf{i}'_{abcr}=\frac{N_r}{N_s}\mathbf{i}_{abcr}\text{ and }\boldsymbol{\psi}'_{abcr}=\frac{N_s}{N_r}\boldsymbol{\psi}_{abcr}.$$

The inductances are expressed as

$$L_{ms}=\frac{N_s}{N_r}L_{sr},\ L_{sr}=\frac{N_sN_r}{\Re_m},\text{ and }L_{ms}=\frac{N_s^2}{\Re_m}.$$

Then, we have

$$\mathbf{L}_{sr}^{'} = \frac{N_s}{N_r}\mathbf{L}_{sr} = L_{ms}\begin{bmatrix} \cos\theta_r & \cos\left(\theta_r + \frac{2}{3}\pi\right) & \cos\left(\theta_r - \frac{2}{3}\pi\right) \\ \cos\left(\theta_r - \frac{2}{3}\pi\right) & \cos\theta_r & \cos\left(\theta_r + \frac{2}{3}\pi\right) \\ \cos\left(\theta_r + \frac{2}{3}\pi\right) & \cos\left(\theta_r - \frac{2}{3}\pi\right) & \cos\theta_r \end{bmatrix},$$

and

$$\mathbf{L}_r^{'} = \frac{N_s^2}{N_r^2}\mathbf{L}_r = \begin{bmatrix} L_{lr}^{'} + L_{ms} & -\frac{1}{2}L_{ms} & -\frac{1}{2}L_{ms} \\ -\frac{1}{2}L_{ms} & L_{lr}^{'} + L_{ms} & -\frac{1}{2}L_{ms} \\ -\frac{1}{2}L_{ms} & -\frac{1}{2}L_{ms} & L_{lr}^{'} + L_{ms} \end{bmatrix}, \; L_{lr}^{'} = \frac{N_s^2}{N_r^2}L_{lr}.$$

From (5.3.10), one finds

$$\begin{bmatrix} \boldsymbol{\psi}_{abcs} \\ \boldsymbol{\psi}_{abcr}^{'} \end{bmatrix} = \begin{bmatrix} \mathbf{L}_s & \mathbf{L}_{sr}^{'} \\ \mathbf{L}_{sr}^{'T} & \mathbf{L}_r^{'} \end{bmatrix}\begin{bmatrix} \mathbf{i}_{abcs} \\ \mathbf{i}_{abcr}^{'} \end{bmatrix}. \quad (5.3.11)$$

Substituting the matrices $\mathbf{L}_s$, $\mathbf{L}_{sr}^{'}$ and $\mathbf{L}_r^{'}$, we have

$$\begin{bmatrix} \psi_{as} \\ \psi_{bs} \\ \psi_{cs} \\ \psi_{ar}^{'} \\ \psi_{br}^{'} \\ \psi_{cr}^{'} \end{bmatrix} =$$

$$\begin{bmatrix} L_{ls} + L_{ms} & -\frac{1}{2}L_{ms} & -\frac{1}{2}L_{ms} & L_{ms}\cos\theta_r & L_{ms}\cos\left(\theta_r + \frac{2}{3}\pi\right) & L_{ms}\cos\left(\theta_r - \frac{2}{3}\pi\right) \\ -\frac{1}{2}L_{ms} & L_{ls} + L_{ms} & -\frac{1}{2}L_{ms} & L_{ms}\cos\left(\theta_r - \frac{2}{3}\pi\right) & L_{ms}\cos\theta_r & L_{ms}\cos\left(\theta_r + \frac{2}{3}\pi\right) \\ -\frac{1}{2}L_{ms} & -\frac{1}{2}L_{ms} & L_{ls} + L_{ms} & L_{ms}\cos\left(\theta_r + \frac{2}{3}\pi\right) & L_{ms}\cos\left(\theta_r - \frac{2}{3}\pi\right) & L_{ms}\cos\theta_r \\ L_{ms}\cos\theta_r & L_{ms}\cos\left(\theta_r - \frac{2}{3}\pi\right) & L_{ms}\cos\left(\theta_r + \frac{2}{3}\pi\right) & L_{lr}^{'} + L_{ms} & -\frac{1}{2}L_{ms} & -\frac{1}{2}L_{ms} \\ L_{ms}\cos\left(\theta_r + \frac{2}{3}\pi\right) & L_{ms}\cos\theta_r & L_{ms}\cos\left(\theta_r - \frac{2}{3}\pi\right) & -\frac{1}{2}L_{ms} & L_{lr}^{'} + L_{ms} & -\frac{1}{2}L_{ms} \\ L_{ms}\cos\left(\theta_r - \frac{2}{3}\pi\right) & L_{ms}\cos\left(\theta_r + \frac{2}{3}\pi\right) & L_{ms}\cos\theta_r & -\frac{1}{2}L_{ms} & -\frac{1}{2}L_{ms} & L_{lr}^{'} + L_{ms} \end{bmatrix}\begin{bmatrix} i_{as} \\ i_{bs} \\ i_{cs} \\ i_{ar}^{'} \\ i_{br}^{'} \\ i_{cr}^{'} \end{bmatrix}.$$

Using (5.3.9) and (5.3.11), one obtains

$$\mathbf{u}_{abcs} = \mathbf{r}_s\mathbf{i}_{abcs} + \frac{d\boldsymbol{\psi}_{abcs}}{dt} = \mathbf{r}_s\mathbf{i}_{abcs} + \mathbf{L}_s\frac{d\mathbf{i}_{abcs}}{dt} + \frac{d(\mathbf{L}_{sr}^{'}\mathbf{i}_{abcr}^{'})}{dt},$$

$$\mathbf{u}_{abcr}^{'} = \mathbf{r}_r^{'}\mathbf{i}_{abcr}^{'} + \frac{d\boldsymbol{\psi}_{abcr}^{'}}{dt} = \mathbf{r}_r^{'}\mathbf{i}_{abcr}^{'} + \mathbf{L}_r^{'}\frac{d\mathbf{i}_{abcr}^{'}}{dt} + \frac{d(\mathbf{L}_{sr}^{'T}\mathbf{i}_{abcs})}{dt}, \quad (5.3.12)$$

where $\mathbf{r}_r^{'} = \frac{N_s^2}{N_r^2}\mathbf{r}_r$.

State-space equations (5.3.12) in expanded form using (5.3.11) are rewritten as

$$u_{as} = r_s i_{as} + (L_{ls} + L_{ms})\frac{di_{as}}{dt} - \tfrac{1}{2}L_{ms}\frac{di_{bs}}{dt} - \tfrac{1}{2}L_{ms}\frac{di_{cs}}{dt}$$
$$+ L_{ms}\frac{d(i'_{ar}\cos\theta_r)}{dt} + L_{ms}\frac{d(i'_{br}\cos(\theta_r + \frac{2\pi}{3}))}{dt} + L_{ms}\frac{d(i'_{cr}\cos(\theta_r - \frac{2\pi}{3}))}{dt},$$
$$u_{bs} = r_s i_{bs} - \tfrac{1}{2}L_{ms}\frac{di_{as}}{dt} + (L_{ls} + L_{ms})\frac{di_{bs}}{dt} - \tfrac{1}{2}L_{ms}\frac{di_{cs}}{dt}$$
$$+ L_{ms}\frac{d(i'_{ar}\cos(\theta_r - \frac{2\pi}{3}))}{dt} + L_{ms}\frac{d(i'_{br}\cos\theta_r)}{dt} + L_{ms}\frac{d(i'_{cr}\cos(\theta_r + \frac{2\pi}{3}))}{dt},$$
$$u_{cs} = r_s i_{cs} - \tfrac{1}{2}L_{ms}\frac{di_{as}}{dt} - \tfrac{1}{2}L_{ms}\frac{di_{bs}}{dt} + (L_{ls} + L_{ms})\frac{di_{cs}}{dt}$$
$$+ L_{ms}\frac{d(i'_{ar}\cos(\theta_r + \frac{2\pi}{3}))}{dt} + L_{ms}\frac{d(i'_{br}\cos(\theta_r - \frac{2\pi}{3}))}{dt} + L_{ms}\frac{d(i'_{cr}\cos\theta_r)}{dt},$$
$$u'_{ar} = r'_r i'_{ar} + L_{ms}\frac{d(i_{as}\cos\theta_r)}{dt} + L_{ms}\frac{d(i_{bs}\cos(\theta_r - \frac{2\pi}{3}))}{dt} + L_{ms}\frac{d(i_{cs}\cos(\theta_r + \frac{2\pi}{3}))}{dt}$$
$$+ (L'_{lr} + L_{ms})\frac{di'_{ar}}{dt} - \tfrac{1}{2}L_{ms}\frac{di'_{br}}{dt} - \tfrac{1}{2}L_{ms}\frac{di'_{cr}}{dt},$$
$$u'_{br} = r'_r i'_{br} + L_{ms}\frac{d(i_{as}\cos(\theta_r + \frac{2\pi}{3}))}{dt} + L_{ms}\frac{d(i_{bs}\cos\theta_r)}{dt} + L_{ms}\frac{d(i_{cs}\cos(\theta_r - \frac{2\pi}{3}))}{dt}$$
$$- \tfrac{1}{2}L_{ms}\frac{di'_{ar}}{dt} + (L'_{lr} + L_{ms})\frac{di'_{br}}{dt} - \tfrac{1}{2}L_{ms}\frac{di'_{cr}}{dt},$$
$$u'_{cr} = r'_r i'_{cr} + L_{ms}\frac{d(i_{as}\cos(\theta_r - \frac{2\pi}{3}))}{dt} + L_{ms}\frac{d(i_{bs}\cos(\theta_r + \frac{2\pi}{3}))}{dt} + L_{ms}\frac{d(i_{cs}\cos\theta_r)}{dt}$$
$$- \tfrac{1}{2}L_{ms}\frac{di'_{ar}}{dt} - \tfrac{1}{2}L_{ms}\frac{di'_{br}}{dt} + (L'_{lr} + L_{ms})\frac{di'_{cr}}{dt}.$$

Cauchy's form of differential equations, given in the state-space form, is found to be

$$
\begin{bmatrix} \frac{di_{as}}{dt} \\ \frac{di_{bs}}{dt} \\ \frac{di_{cs}}{dt} \\ \frac{di'_{ar}}{dt} \\ \frac{di'_{br}}{dt} \\ \frac{di'_{cr}}{dt} \end{bmatrix}
= \frac{1}{L_{\Sigma L}}
\begin{bmatrix}
-r_s L_{\Sigma m} & -\frac{1}{2} r_s L_{ms} & -\frac{1}{2} r_s L_{ms} & 0 & 0 & 0 \\
-\frac{1}{2} r_s L_{ms} & -r_s L_{\Sigma m} & -\frac{1}{2} r_s L_{ms} & 0 & 0 & 0 \\
-\frac{1}{2} r_s L_{ms} & -\frac{1}{2} r_s L_{ms} & -r_s L_{\Sigma m} & 0 & 0 & 0 \\
0 & 0 & 0 & -r_r L_{\Sigma m} & -\frac{1}{2} r_r L_{ms} & -\frac{1}{2} r_r L_{ms} \\
0 & 0 & 0 & -\frac{1}{2} r_r L_{ms} & -r_r L_{\Sigma m} & -\frac{1}{2} r_r L_{ms} \\
0 & 0 & 0 & -\frac{1}{2} r_r L_{ms} & -\frac{1}{2} r_r L_{ms} & -r_r L_{\Sigma m}
\end{bmatrix}
\begin{bmatrix} i_{as} \\ i_{bs} \\ i_{cs} \\ i'_{ar} \\ i'_{br} \\ i'_{cr} \end{bmatrix}
$$

$$
+\frac{1}{L_{\Sigma L}}
\begin{bmatrix}
0 & 0 & 0 & r_r L_{ms}\cos\theta_r & r_r L_{ms}\cos\left(\theta_r+\frac{2}{3}\pi\right) & r_r L_{ms}\cos\left(\theta_r-\frac{2}{3}\pi\right) \\
0 & 0 & 0 & r_r L_{ms}\cos\left(\theta_r-\frac{2}{3}\pi\right) & r_r L_{ms}\cos\theta_r & r_r L_{ms}\cos\left(\theta_r+\frac{2}{3}\pi\right) \\
0 & 0 & 0 & r_r L_{ms}\cos\left(\theta_r+\frac{2}{3}\pi\right) & r_r L_{ms}\cos\left(\theta_r-\frac{2}{3}\pi\right) & r_r L_{ms}\cos\theta_r \\
r_s L_{ms}\cos\theta_r & r_s L_{ms}\cos\left(\theta_r-\frac{2}{3}\pi\right) & r_s L_{ms}\cos\left(\theta_r+\frac{2}{3}\pi\right) & 0 & 0 & 0 \\
r_s L_{ms}\cos\left(\theta_r+\frac{2}{3}\pi\right) & r_s L_{ms}\cos\theta_r & r_s L_{ms}\cos\left(\theta_r-\frac{2}{3}\pi\right) & 0 & 0 & 0 \\
r_s L_{ms}\cos\left(\theta_r-\frac{2}{3}\pi\right) & r_s L_{ms}\cos\left(\theta_r+\frac{2}{3}\pi\right) & r_s L_{ms}\cos\theta_r & 0 & 0 & 0
\end{bmatrix}
\begin{bmatrix} i_{as} \\ i_{bs} \\ i_{cs} \\ i'_{ar} \\ i'_{br} \\ i'_{cr} \end{bmatrix}
$$

$$
+\frac{1}{L_{\Sigma L}}
\begin{bmatrix}
0 & 1.3L_{ms}^2\omega_r & -1.3L_{ms}^2\omega_r & L_{\Sigma ms}\omega_r\sin\theta_r & L_{\Sigma ms}\omega_r\sin\left(\theta_r+\frac{2}{3}\pi\right) & L_{\Sigma ms}\omega_r\sin\left(\theta_r-\frac{2}{3}\pi\right) \\
-1.3L_{ms}^2\omega_r & 0 & 1.3L_{ms}^2\omega_r & L_{\Sigma ms}\omega_r\sin\left(\theta_r-\frac{2}{3}\pi\right) & L_{\Sigma ms}\omega_r\sin\theta_r & L_{\Sigma ms}\omega_r\sin\left(\theta_r+\frac{2}{3}\pi\right) \\
1.3L_{ms}^2\omega_r & -1.3L_{ms}^2\omega_r & 0 & L_{\Sigma ms}\omega_r\sin\left(\theta_r+\frac{2}{3}\pi\right) & L_{\Sigma ms}\omega_r\sin\left(\theta_r-\frac{2}{3}\pi\right) & L_{\Sigma ms}\omega_r\sin\theta_r \\
L_{\Sigma ms}\omega_r\sin\theta_r & L_{\Sigma ms}\omega_r\sin\left(\theta_r-\frac{2}{3}\pi\right) & L_{\Sigma ms}\omega_r\sin\left(\theta_r+\frac{2}{3}\pi\right) & 0 & -1.3L_{ms}^2\omega_r & 1.3L_{ms}^2\omega_r \\
L_{\Sigma ms}\omega_r\sin\left(\theta_r+\frac{2}{3}\pi\right) & L_{\Sigma ms}\omega_r\sin\theta_r & L_{\Sigma ms}\omega_r\sin\left(\theta_r-\frac{2}{3}\pi\right) & 1.3L_{ms}^2\omega_r & 0 & -1.3L_{ms}^2\omega_r \\
L_{\Sigma ms}\omega_r\sin\left(\theta_r-\frac{2}{3}\pi\right) & L_{\Sigma ms}\omega_r\sin\left(\theta_r+\frac{2}{3}\pi\right) & L_{\Sigma ms}\omega_r\sin\theta_r & -1.3L_{ms}^2\omega_r & 1.3L_{ms}^2\omega_r & 0
\end{bmatrix}
\begin{bmatrix} i_{as} \\ i_{bs} \\ i_{cs} \\ i'_{ar} \\ i'_{br} \\ i'_{cr} \end{bmatrix}
$$

$$
+\frac{1}{L_{\Sigma L}}
\begin{bmatrix}
2L_{ms}+L'_{lr} & \frac{1}{2}L_{ms} & \frac{1}{2}L_{ms} & -L_{ms}\cos\theta_r & -L_{ms}\cos\left(\theta_r+\frac{2}{3}\pi\right) & -L_{ms}\cos\left(\theta_r-\frac{2}{3}\pi\right) \\
\frac{1}{2}L_{ms} & 2L_{ms}+L'_{lr} & \frac{1}{2}L_{ms} & -L_{ms}\cos\left(\theta_r-\frac{2}{3}\pi\right) & -L_{ms}\cos\theta_r & -L_{ms}\cos\left(\theta_r+\frac{2}{3}\pi\right) \\
\frac{1}{2}L_{ms} & \frac{1}{2}L_{ms} & 2L_{ms}+L'_{lr} & -L_{ms}\cos\left(\theta_r+\frac{2}{3}\pi\right) & -L_{ms}\cos\left(\theta_r-\frac{2}{3}\pi\right) & -L_{ms}\cos\theta_r \\
-L_{ms}\cos\theta_r & -L_{ms}\cos\left(\theta_r-\frac{2}{3}\pi\right) & -L_{ms}\cos\left(\theta_r+\frac{2}{3}\pi\right) & 2L_{ms}+L'_{lr} & \frac{1}{2}L_{ms} & \frac{1}{2}L_{ms} \\
-L_{ms}\cos\left(\theta_r+\frac{2}{3}\pi\right) & -L_{ms}\cos\theta_r & -L_{ms}\cos\left(\theta_r-\frac{2}{3}\pi\right) & \frac{1}{2}L_{ms} & 2L_{ms}+L'_{lr} & \frac{1}{2}L_{ms} \\
-L_{ms}\cos\left(\theta_r-\frac{2}{3}\pi\right) & -L_{ms}\cos\left(\theta_r+\frac{2}{3}\pi\right) & -L_{ms}\cos\theta_r & \frac{1}{2}L_{ms} & \frac{1}{2}L_{ms} & 2L_{ms}+L'_{lr}
\end{bmatrix}
\begin{bmatrix} u_{as} \\ u_{bs} \\ u_{cs} \\ u'_{ar} \\ u'_{br} \\ u'_{cr} \end{bmatrix}.
\tag{5.3.13}
$$

Here, the following notations are used

$$L_{\Sigma L} = \left(3L_{ms} + L'_{lr}\right)L'_{lr},\ L_{\Sigma m} = 2L_{ms} + L'_{lr},\ L_{\Sigma ms} = \tfrac{3}{2}L_{ms}^2 + L_{ms}L'_{lr}.$$

Newton's second law is applied to derive the *torsional-mechanical* equations, and the expression for the electromagnetic torque must be obtained.

For $P$-pole three-phase induction machines, as one finds the expression for coenergy $W_c\left(\mathbf{i}_{abcs}, \mathbf{i}'_{abcr}, \theta_r\right)$, the electromagnetic torque can be straightforwardly derived as

$$T_e = \frac{P}{2}\frac{\partial W_c\left(\mathbf{i}_{abcs}, \mathbf{i}'_{abcr}, \theta_r\right)}{\partial \theta_r}.$$

For three-phase induction micromotors we have

$$W_c = W_f = \tfrac{1}{2}\mathbf{i}_{abcs}^T\left(\mathbf{L}_s - L_{ls}\mathbf{I}\right)\mathbf{i}_{abcs} + \mathbf{i}_{abcs}^T\mathbf{L}'_{sr}(\theta_r)\mathbf{i}'_{abcr} + \tfrac{1}{2}\mathbf{i}'^{T}_{abcr}\left(\mathbf{L}'_r - \mathbf{L}'_{lr}\mathbf{I}\right)\mathbf{i}'_{abcr}$$

Matrices $\mathbf{L}_s$ and $\mathbf{L}'_r$, as well as leakage inductances $L_{ls}$ and $L'_{lr}$, are not functions of the electrical displacement $\theta_r$. Therefore, we have

$$T_e = \frac{P}{2}\mathbf{i}_{abcs}^T\frac{\partial\mathbf{L}'_{sr}(\theta_r)}{\partial\theta_r}\mathbf{i}'_{abcr}$$

$$= -\frac{P}{2}L_{ms}\begin{bmatrix} i_{as} & i_{bs} & i_{cs}\end{bmatrix}\begin{bmatrix} \sin\theta_r & \sin\left(\theta_r + \frac{2}{3}\pi\right) & \sin\left(\theta_r - \frac{2}{3}\pi\right) \\ \sin\left(\theta_r - \frac{2}{3}\pi\right) & \sin\theta_r & \sin\left(\theta_r + \frac{2}{3}\pi\right) \\ \sin\left(\theta_r + \frac{2}{3}\pi\right) & \sin\left(\theta_r - \frac{2}{3}\pi\right) & \sin\theta_r \end{bmatrix}\begin{bmatrix} i'_{ar} \\ i'_{br} \\ i'_{cr}\end{bmatrix}$$

$$= -\frac{P}{2}L_{ms}\left\{\left[i_{as}\left(i'_{ar} - \tfrac{1}{2}i'_{br} - \tfrac{1}{2}i'_{cr}\right) + i_{bs}\left(i'_{br} - \tfrac{1}{2}i'_{ar} - \tfrac{1}{2}i'_{cr}\right) + i_{cs}\left(i'_{cr} - \tfrac{1}{2}i'_{br} - \tfrac{1}{2}i'_{ar}\right)\right]\sin\theta_r\right.$$

$$\left. + \tfrac{\sqrt{3}}{2}\left[i_{as}\left(i'_{br} - i'_{cr}\right) + i_{bs}\left(i'_{cr} - i'_{ar}\right) + i_{cs}\left(i'_{ar} - i'_{br}\right)\right]\cos\theta_r\right\}. \quad (5.3.14)$$

Using Newton's second law and (5.3.14), the *torsional-mechanical* equations are found to be

$$\frac{d\omega_r}{dt} = \frac{P}{2J}T_e - \frac{B_m}{J}\omega_r - \frac{P}{2J}T_L$$

$$= -\frac{P^2}{4J}L_{ms}\left\{\left[i_{as}\left(i'_{ar} - \tfrac{1}{2}i'_{br} - \tfrac{1}{2}i'_{cr}\right) + i_{bs}\left(i'_{br} - \tfrac{1}{2}i'_{ar} - \tfrac{1}{2}i'_{cr}\right) + i_{cs}\left(i'_{cr} - \tfrac{1}{2}i'_{br} - \tfrac{1}{2}i'_{ar}\right)\right]\sin\theta_r\right.$$

$$\left. + \tfrac{\sqrt{3}}{2}\left[i_{as}\left(i'_{br} - i'_{cr}\right) + i_{bs}\left(i'_{cr} - i'_{ar}\right) + i_{cs}\left(i'_{ar} - i'_{br}\right)\right]\cos\theta_r\right\} - \frac{B_m}{J}\omega_r - \frac{P}{2J}T_L,$$

$$\frac{d\theta_r}{dt} = \omega_r. \quad (5.3.15)$$

Augmenting differential equations (5.3.13) and (5.3.15), the resulting model for three-phase induction micromotors in the *machine* variables, is found.

## *Mathematical Model of Three-Phase Induction Micromotors in the Arbitrary Reference Frame*

The *abc* stator and rotor variables must be transformed to the *quadrature*, *direct*, and *zero* quantities. To transform the *machine* (*abc*) stator voltages, currents, and flux linkages to the *quadrature-*, *direct-*, and *zero*-axis components of stator voltages, currents and flux linkages, the *direct* Park transformation is used. In particular,

$$\mathbf{u}_{qdos} = \mathbf{K}_s\mathbf{u}_{abcs},\ \mathbf{i}_{qdos} = \mathbf{K}_s\mathbf{i}_{abcs},\ \boldsymbol{\psi}_{qdos} = \mathbf{K}_s\boldsymbol{\psi}_{abcs}, \quad (5.3.16)$$

where the stator transformation matrix $\mathbf{K}_s$ is given by

$$\mathbf{K}_s = \tfrac{2}{3}\begin{bmatrix} \cos\theta & \cos\left(\theta - \tfrac{2}{3}\pi\right) & \cos\left(\theta + \tfrac{2}{3}\pi\right) \\ \sin\theta & \sin\left(\theta - \tfrac{2}{3}\pi\right) & \sin\left(\theta + \tfrac{2}{3}\pi\right) \\ \tfrac{1}{2} & \tfrac{1}{2} & \tfrac{1}{2} \end{bmatrix}. \tag{5.3.17}$$

Here, the angular displacement of the reference frame is

$$\theta = \int_{t_0}^{t} \omega(\tau)d\tau + \theta_0 .$$

Using the rotor transformations matrix $\mathbf{K}_r$, the *quadrature-*, *direct-*, and *zero*-axis components of rotor voltages, currents, and flux linkages are found by using the *abc* rotor voltages, currents, and flux linkages.

In particular,

$$\mathbf{u}'_{qdor} = \mathbf{K}_r \mathbf{u}'_{abcr},\ \mathbf{i}'_{qdor} = \mathbf{K}_r \mathbf{i}'_{abcr},\ \boldsymbol{\psi}'_{qdor} = \mathbf{K}_r \boldsymbol{\psi}'_{abcr}, \tag{5.3.18}$$

where the rotor transformation matrix is

$$\mathbf{K}_r = \tfrac{2}{3}\begin{bmatrix} \cos\left(\theta - \theta_r\right) & \cos\left(\theta - \theta_r - \tfrac{2}{3}\pi\right) & \cos\left(\theta - \theta_r + \tfrac{2}{3}\pi\right) \\ \sin\left(\theta - \theta_r\right) & \sin\left(\theta - \theta_r - \tfrac{2}{3}\pi\right) & \sin\left(\theta - \theta_r + \tfrac{2}{3}\pi\right) \\ \tfrac{1}{2} & \tfrac{1}{2} & \tfrac{1}{2} \end{bmatrix}. \tag{5.3.19}$$

From differential equations (5.3.12)

$$\mathbf{u}_{abcs} = \mathbf{r}_s \mathbf{i}_{abcs} + \frac{d\boldsymbol{\psi}_{abcs}}{dt},\ \mathbf{u}'_{abcr} = \mathbf{r}'_r \mathbf{i}'_{abcr} + \frac{d\boldsymbol{\psi}'_{abcr}}{dt},$$

by taking note of the inverse Park transformation matrices $\mathbf{K}_s^{-1}$ and $\mathbf{K}_r^{-1}$, we have

$$\mathbf{K}_s^{-1}\mathbf{u}_{qdos} = \mathbf{r}_s \mathbf{K}_s^{-1}\mathbf{i}_{qdos} + \frac{d\left(\mathbf{K}_s^{-1}\boldsymbol{\psi}_{qdos}\right)}{dt},$$

$$\mathbf{K}_r^{-1}\mathbf{u}'_{qdor} = \mathbf{r}'_r \mathbf{K}_r^{-1}\mathbf{i}'_{qdor} + \frac{d\left(\mathbf{K}_r^{-1}\boldsymbol{\psi}'_{qdor}\right)}{dt}. \tag{5.3.20}$$

Making use of (5.3.17) and (5.3.19) one finds inverse matrices $\mathbf{K}_s^{-1}$ and $\mathbf{K}_r^{-1}$. In particular,

$$\mathbf{K}_s^{-1} = \begin{bmatrix} \cos\theta & \sin\theta & 1 \\ \cos\left(\theta - \tfrac{2}{3}\pi\right) & \sin\left(\theta - \tfrac{2}{3}\pi\right) & 1 \\ \cos\left(\theta + \tfrac{2}{3}\pi\right) & \sin\left(\theta + \tfrac{2}{3}\pi\right) & 1 \end{bmatrix},$$

and $$\mathbf{K}_r^{-1} = \begin{bmatrix} \cos\left(\theta - \theta_r\right) & \sin\left(\theta - \theta_r\right) & 1 \\ \cos\left(\theta - \theta_r - \tfrac{2}{3}\pi\right) & \sin\left(\theta - \theta_r - \tfrac{2}{3}\pi\right) & 1 \\ \cos\left(\theta - \theta_r + \tfrac{2}{3}\pi\right) & \sin\left(\theta - \theta_r + \tfrac{2}{3}\pi\right) & 1 \end{bmatrix}.$$

Multiplying left and right sides of equations (5.3.20) by $\mathbf{K}_s$ and $\mathbf{K}_r$, one has

$$\mathbf{u}_{qdos} = \mathbf{K}_s \mathbf{r}_s \mathbf{K}_s^{-1} \mathbf{i}_{qdos} + \mathbf{K}_s \frac{d\mathbf{K}_s^{-1}}{dt} \psi_{qdos} + \mathbf{K}_s \mathbf{K}_s^{-1} \frac{d\psi_{qdos}}{dt},$$

$$\mathbf{u}_{qdor}' = \mathbf{K}_r \mathbf{r}_r' \mathbf{K}_r^{-1} \mathbf{i}_{qdor}' + \mathbf{K}_r \frac{d\mathbf{K}_r^{-1}}{dt} \psi_{qdor}' + \mathbf{K}_r \mathbf{K}_r^{-1} \frac{d\psi_{qdor}'}{dt}. \quad (5.3.21)$$

The matrices of the stator and rotor resistances $\mathbf{r}_s$ and $\mathbf{r}_r'$ are diagonal, and hence,

$\mathbf{K}_s \mathbf{r}_s \mathbf{K}_s^{-1} = \mathbf{r}_s$ and $\mathbf{K}_r \mathbf{r}_r' \mathbf{K}_r^{-1} = \mathbf{r}_r'$.

Performing differentiation, one finds

$$\frac{d\mathbf{K}_s^{-1}}{dt} = \omega \begin{bmatrix} -\sin\theta & \cos\theta & 0 \\ -\sin\left(\theta - \frac{2}{3}\pi\right) & \cos\left(\theta - \frac{2}{3}\pi\right) & 0 \\ -\sin\left(\theta + \frac{2}{3}\pi\right) & \cos\left(\theta + \frac{2}{3}\pi\right) & 0 \end{bmatrix},$$

$$\frac{d\mathbf{K}_r^{-1}}{dt} = \left(\omega - \omega_r\right) \begin{bmatrix} -\sin\left(\theta - \theta_r\right) & \cos\left(\theta - \theta_r\right) & 0 \\ -\sin\left(\theta - \theta_r - \frac{2}{3}\pi\right) & \cos\left(\theta - \theta_r - \frac{2}{3}\pi\right) & 0 \\ -\sin\left(\theta - \theta_r + \frac{2}{3}\pi\right) & \cos\left(\theta - \theta_r + \frac{2}{3}\pi\right) & 0 \end{bmatrix}.$$

Therefore,

$$\mathbf{K}_s \frac{d\mathbf{K}_s^{-1}}{dt} = \omega \begin{bmatrix} 0 & 1 & 0 \\ -1 & 0 & 0 \\ 0 & 0 & 0 \end{bmatrix}$$

and $\mathbf{K}_r \dfrac{d\mathbf{K}_r^{-1}}{dt} = \left(\omega - \omega_r\right) \begin{bmatrix} 0 & 1 & 0 \\ -1 & 0 & 0 \\ 0 & 0 & 0 \end{bmatrix}$.

One obtains the voltage equations for stator and rotor circuits in the *arbitrary* reference frame when the angular velocity of the reference frame $\omega$ is not specified. From (5.3.21) the following state-space differential equations result

$$\mathbf{u}_{qdos} = \mathbf{r}_s \mathbf{i}_{qdos} + \begin{bmatrix} 0 & \omega & 0 \\ -\omega & 0 & 0 \\ 0 & 0 & 0 \end{bmatrix} \psi_{qdos} + \frac{d\psi_{qdos}}{dt},$$

$$\mathbf{u}_{qdor}' = \mathbf{r}_r' \mathbf{i}_{qdor}' + \begin{bmatrix} 0 & \omega - \omega_r & 0 \\ -\omega + \omega_r & 0 & 0 \\ 0 & 0 & 0 \end{bmatrix} \psi_{qdor}' + \frac{d\psi_{qdor}'}{dt}. \quad (5.3.22)$$

From (5.3.22), six differential equations in expanded form are found to model the stator and rotor circuitry dynamics. In particular,

$$u_{qs} = r_s i_{qs} + \omega \psi_{ds} + \frac{d\psi_{qs}}{dt},$$

$$u_{ds} = r_s i_{ds} - \omega \psi_{qs} + \frac{d\psi_{ds}}{dt},$$

$$u_{os} = r_s i_{os} + \frac{d\psi_{os}}{dt},$$

$$u'_{qr} = r'_r i'_{qr} + (\omega - \omega_r)\psi'_{dr} + \frac{d\psi'_{qr}}{dt},$$

$$u'_{dr} = r'_r i'_{dr} - (\omega - \omega_r)\psi'_{qr} + \frac{d\psi'_{dr}}{dt},$$

$$u'_{or} = r'_r i'_{or} + \frac{d\psi'_{or}}{dt}. \qquad (5.3.23)$$

Using the matrix equation for flux linkages

$$\begin{bmatrix} \boldsymbol{\psi}_{abcs} \\ \boldsymbol{\psi}'_{abcr} \end{bmatrix} = \begin{bmatrix} \mathbf{L}_s & \mathbf{L}'_{sr} \\ \mathbf{L}'^{T}_{sr} & \mathbf{L}'_r \end{bmatrix} \begin{bmatrix} \mathbf{i}_{abcs} \\ \mathbf{i}'_{abcr} \end{bmatrix}$$

we have

$$\boldsymbol{\psi}_{abcs} = \mathbf{L}_s \mathbf{i}_{abcs} + \mathbf{L}'_{sr} \mathbf{i}'_{abcr},$$

$$\boldsymbol{\psi}'_{abcr} = \mathbf{L}'^{T}_{sr} \mathbf{i}_{abcs} + \mathbf{L}'_r \mathbf{i}'_{abcr}.$$

These equations can be represented using the *quadrature*, *direct*, and *zero* quantities. Employing the Park transformation matrices one has

$$\mathbf{K}_s^{-1} \boldsymbol{\psi}_{qdos} = \mathbf{L}_s \mathbf{K}_s^{-1} \mathbf{i}_{qdos} + \mathbf{L}'_{sr} \mathbf{K}_r^{-1} \mathbf{i}'_{qdor},$$

$$\mathbf{K}_r^{-1} \boldsymbol{\psi}'_{qdor} = \mathbf{L}'^{T}_{sr} \mathbf{K}_s^{-1} \mathbf{i}_{qdos} + \mathbf{L}'_r \mathbf{K}_r^{-1} \mathbf{i}'_{abcr}.$$

Thus

$$\boldsymbol{\psi}_{qdos} = \mathbf{K}_s \mathbf{L}_s \mathbf{K}_s^{-1} \mathbf{i}_{qdos} + \mathbf{K}_s \mathbf{L}'_{sr} \mathbf{K}_r^{-1} \mathbf{i}'_{qdor},$$

$$\boldsymbol{\psi}'_{qdor} = \mathbf{K}_r \mathbf{L}'^{T}_{sr} \mathbf{K}_s^{-1} \mathbf{i}_{qdos} + \mathbf{K}_r \mathbf{L}'_r \mathbf{K}_r^{-1} \mathbf{i}'_{abcr}. \qquad (5.3.24)$$

Taking note of the Park transformation matrices and applying the derived expressions for $\mathbf{L}_s$, $\mathbf{L}'_{sr}$ and $\mathbf{L}'_r$, by multiplying the matrices we obtain

$$\mathbf{K}_s \mathbf{L}_s \mathbf{K}_s^{-1} = \begin{bmatrix} L_{ls} + M & 0 & 0 \\ 0 & L_{ls} + M & 0 \\ 0 & 0 & L_{ls} \end{bmatrix},$$

$$\mathbf{K}_s\mathbf{L}'_{sr}\mathbf{K}_r^{-1} = \mathbf{K}_r\mathbf{L}'^{\,T}_{sr}\mathbf{K}_s^{-1} = \begin{bmatrix} M & 0 & 0 \\ 0 & M & 0 \\ 0 & 0 & 0 \end{bmatrix},$$

and $\mathbf{K}_r\mathbf{L}'_r\mathbf{K}_r^{-1} = \begin{bmatrix} L'_{lr} + M & 0 & 0 \\ 0 & L'_{lr} + M & 0 \\ 0 & 0 & L'_{lr} \end{bmatrix}$, $M = \frac{3}{2} L_{ms}$.

In expanded form, the flux linkage equations (5.3.24) are

$$\psi_{qs} = L_{ls}i_{qs} + Mi_{qs} + Mi'_{qr},$$

$$\psi_{ds} = L_{ls}i_{ds} + Mi_{ds} + Mi'_{dr},$$

$$\psi_{os} = L_{ls}i_{os},$$

$$\psi'_{qr} = L'_{lr}i'_{qr} + Mi_{qs} + Mi'_{qr},$$

$$\psi'_{dr} = L'_{lr}i'_{dr} + Mi_{ds} + Mi'_{dr},$$

$$\psi'_{or} = L'_{lr}i'_{or}. \tag{5.3.25}$$

Using the expressions (5.3.25) in (5.3.23), the following differential equations result

$$u_{qs} = r_s i_{qs} + \omega\left(L_{ls}i_{ds} + Mi_{ds} + Mi'_{dr}\right) + \frac{d\left(L_{ls}i_{qs} + Mi_{qs} + Mi'_{qr}\right)}{dt},$$

$$u_{ds} = r_s i_{ds} - \omega\left(L_{ls}i_{qs} + Mi_{qs} + Mi'_{qr}\right) + \frac{d\left(L_{ls}i_{ds} + Mi_{ds} + Mi'_{dr}\right)}{dt},$$

$$u_{os} = r_s i_{os} + \frac{d\left(L_{ls}i_{os}\right)}{dt},$$

$$u'_{qr} = r'_r i'_{qr} + \left(\omega - \omega_r\right)\left(L'_{lr}i'_{dr} + Mi_{ds} + Mi'_{dr}\right) + \frac{d\left(L'_{lr}i'_{qr} + Mi_{qs} + Mi'_{qr}\right)}{dt},$$

$$u'_{dr} = r'_r i'_{dr} - \left(\omega - \omega_r\right)\left(L'_{lr}i'_{qr} + Mi_{qs} + Mi'_{qr}\right) + \frac{d\left(L'_{lr}i'_{dr} + Mi_{ds} + Mi'_{dr}\right)}{dt},$$

$$u'_{or} = r'_r i'_{or} + \frac{d\left(L'_{lr}i'_{or}\right)}{dt}.$$

Cauchy's form of differential equations is

$$\frac{di_{qs}}{dt}=\frac{1}{L_{SM}L_{RM}-M^2}\left[-L_{RM}r_si_{qs}-\left(L_{SM}L_{RM}-M^2\right)\omega i_{ds}+Mr_r^{'}i_{qr}^{'}\right.$$
$$\left.-M\left(Mi_{ds}+L_{RM}i_{dr}^{'}\right)\omega_r+L_{RM}u_{qs}-Mu_{qr}^{'}\right],$$
$$\frac{di_{ds}}{dt}=\frac{1}{L_{SM}L_{RM}-M^2}\left[\left(L_{SM}L_{RM}-M^2\right)\omega i_{qs}-L_{RM}r_si_{ds}+Mr_r^{'}i_{dr}^{'}\right.$$
$$\left.+M\left(Mi_{qs}+L_{RM}i_{qr}^{'}\right)\omega_r+L_{RM}u_{ds}-Mu_{dr}^{'}\right],$$
$$\frac{di_{os}}{dt}=\frac{1}{L_{ls}}\left(-r_si_{os}+u_{os}\right),$$
$$\frac{di_{qr}^{'}}{dt}=\frac{1}{L_{SM}L_{RM}-M^2}\left[Mr_si_{qs}-L_{SM}r_r^{'}i_{qr}^{'}-\left(L_{SM}L_{RM}-M^2\right)\omega i_{dr}^{'}\right.$$
$$\left.+L_{SM}\left(Mi_{ds}+L_{RM}i_{dr}^{'}\right)\omega_r-Mu_{qs}+L_{SM}u_{qr}^{'}\right],$$
$$\frac{di_{dr}^{'}}{dt}=\frac{1}{L_{SM}L_{RM}-M^2}\left[Mr_si_{ds}+\left(L_{SM}L_{RM}-M^2\right)\omega i_{qr}^{'}-L_{SM}r_r^{'}i_{dr}^{'}\right.$$
$$\left.-L_{SM}\left(Mi_{qs}+L_{RM}i_{qr}^{'}\right)\omega_r-Mu_{ds}+L_{SM}u_{dr}^{'}\right],$$
$$\frac{di_{or}^{'}}{dt}=\frac{1}{L_{lr}^{'}}\left(-r_r^{'}i_{or}^{'}+u_{or}^{'}\right), \qquad (5.3.26)$$

where $L_{SM}=L_{ls}+M=L_{ls}+\frac{3}{2}L_{ms}$ and $L_{RM}=L_{lr}^{'}+M=L_{lr}^{'}+\frac{3}{2}L_{ms}$.

One concludes that the nonlinear differential equations are found to describe the stator-rotor circuitry transient behavior of three-phase induction micromotors. To complete the model developments, the *torsional-mechanical* equations

$$T_e-B_m\omega_{rm}-T_L=J\frac{d\omega_{rm}}{dt},$$
$$\frac{d\theta_{rm}}{dt}=\omega_{rm}, \qquad (5.3.27)$$

must be used.

The equation for the electromagnetic torque must be obtained in terms of the *quadrature*- and *direct*-axis components of stator and rotor currents.

Using the formula for coenergy

$$W_c=\tfrac{1}{2}\mathbf{i}_{abcs}^T\left(\mathbf{L}_s-L_{ls}\mathbf{I}\right)\mathbf{i}_{abcs}+\mathbf{i}_{abcs}^T\mathbf{L}_{sr}^{'}\left(\theta_r\right)\mathbf{i}_{abcr}^{'}+\tfrac{1}{2}\mathbf{i}_{abcr}^{'T}\left(\mathbf{L}_r^{'}-L_{lr}^{'}\mathbf{I}\right)\mathbf{i}_{abcr}^{'},$$

one finds $T_e=\frac{P}{2}\frac{\partial W_c\left(\mathbf{i}_{abcs},\mathbf{i}_{abcr}^{'},\theta_r\right)}{\partial\theta_r}=\frac{P}{2}\mathbf{i}_{abcs}^T\frac{\partial\mathbf{L}_{sr}^{'}(\theta_r)}{\partial\theta_r}\mathbf{i}_{abcr}^{'}$.

Hence, we have

$$T_e = \frac{P}{2}\left(\mathbf{K}_s^{-1}\mathbf{i}_{qdos}\right)^T \frac{\partial \mathbf{L}'_{sr}(\theta_r)}{\partial \theta_r}\mathbf{K}_r^{-1}\mathbf{i}'_{qdor} = \frac{P}{2}\mathbf{i}_{qdos}^T \mathbf{K}_s^{-1T} \frac{\partial \mathbf{L}'_{sr}(\theta_r)}{\partial \theta_r}\mathbf{K}_r^{-1}\mathbf{i}'_{qdor}.$$

By performing multiplication of matrices, the following formula results

$$T_e = \frac{3P}{4} M\left(i_{qs} i'_{dr} - i_{ds} i'_{qr}\right). \tag{5.3.28}$$

Thus, from (5.3.27) and (5.3.28), one has

$$\frac{d\omega_r}{dt} = \frac{3P^2}{8J} M\left(i_{qs} i'_{dr} - i_{ds} i'_{qr}\right) - \frac{B_m}{J}\omega_r - \frac{P}{2J}T_L,$$

$$\frac{d\theta_r}{dt} = \omega_r. \tag{5.3.29}$$

Augmenting the circuitry and *torsional-mechanical* dynamics, as given by differential equations (5.3.26) and (5.3.29), the model for three-phase induction micromotors in the *arbitrary* reference frame results.

We have a set of eight highly coupled nonlinear differential equations

$$\frac{di_{qs}}{dt} = \frac{1}{L_{SM}L_{RM} - M^2}\Big[-L_{RM}r_s i_{qs} - \left(L_{SM}L_{RM} - M^2\right)\omega i_{ds} + M r'_r i'_{qr} - M\left(M i_{ds} + L_{RM} i'_{dr}\right)\omega_r + L_{RM}u_{qs} - M u'_{qr}\Big],$$

$$\frac{di_{ds}}{dt} = \frac{1}{L_{SM}L_{RM} - M^2}\Big[\left(L_{SM}L_{RM} - M^2\right)\omega i_{qs} - L_{RM}r_s i_{ds} + M r'_r i'_{dr} + M\left(M i_{qs} + L_{RM} i'_{qr}\right)\omega_r + L_{RM}u_{ds} - M u'_{dr}\Big],$$

$$\frac{di_{os}}{dt} = \frac{1}{L_{ls}}\left(-r_s i_{os} + u_{os}\right),$$

$$\frac{di'_{qr}}{dt} = \frac{1}{L_{SM}L_{RM} - M^2}\Big[M r_s i_{qs} - L_{SM} r'_r i'_{qr} - \left(L_{SM}L_{RM} - M^2\right)\omega i'_{dr} + L_{SM}\left(M i_{ds} + L_{RM} i'_{dr}\right)\omega_r - M u_{qs} + L_{SM} u'_{qr}\Big],$$

$$\frac{di'_{dr}}{dt} = \frac{1}{L_{SM}L_{RM} - M^2}\Big[M r_s i_{ds} + \left(L_{SM}L_{RM} - M^2\right)\omega i'_{qr} - L_{SM} r'_r i'_{dr} - L_{SM}\left(M i_{qs} + L_{RM} i'_{qr}\right)\omega_r - M u_{ds} + L_{SM} u'_{dr}\Big],$$

$$\frac{di'_{or}}{dt} = \frac{1}{L'_{lr}}\left(-r'_r i'_{or} + u'_{or}\right),$$

$$\frac{d\omega_r}{dt} = \frac{3P^2}{8J} M\left(i_{qs} i'_{dr} - i_{ds} i'_{qr}\right) - \frac{B_m}{J}\omega_r - \frac{P}{2J}T_L,$$

$$\frac{d\theta_r}{dt} = \omega_r. \tag{5.3.30}$$

The last differential equation in the set (5.3.30) can be omitted in the analysis and simulations if induction micromotors are used in electric drive applications. That is, for electric drives one finds

$$\frac{di_{qs}}{dt}=\frac{1}{L_{SM}L_{RM}-M^2}\Big[-L_{RM}r_s i_{qs}-\left(L_{SM}L_{RM}-M^2\right)\omega i_{ds}+Mr_r' i_{qr}'$$
$$-M\left(Mi_{ds}+L_{RM}i_{dr}'\right)\omega_r+L_{RM}u_{qs}-Mu_{qr}'\Big],$$

$$\frac{di_{ds}}{dt}=\frac{1}{L_{SM}L_{RM}-M^2}\Big[\left(L_{SM}L_{RM}-M^2\right)\omega i_{qs}-L_{RM}r_s i_{ds}+Mr_r' i_{dr}'$$
$$+M\left(Mi_{qs}+L_{RM}i_{qr}'\right)\omega_r+L_{RM}u_{ds}-Mu_{dr}'\Big],$$

$$\frac{di_{os}}{dt}=\frac{1}{L_{ls}}\left(-r_s i_{os}+u_{os}\right),$$

$$\frac{di_{qr}'}{dt}=\frac{1}{L_{SM}L_{RM}-M^2}\Big[Mr_s i_{qs}-L_{SM}r_r' i_{qr}'-\left(L_{SM}L_{RM}-M^2\right)\omega i_{dr}'$$
$$+L_{SM}\left(Mi_{ds}+L_{RM}i_{dr}'\right)\omega_r-Mu_{qs}+L_{SM}u_{qr}'\Big],$$

$$\frac{di_{dr}'}{dt}=\frac{1}{L_{SM}L_{RM}-M^2}\Big[Mr_s i_{ds}+\left(L_{SM}L_{RM}-M^2\right)\omega i_{qr}'-L_{SM}r_r' i_{dr}'$$
$$-L_{SM}\left(Mi_{qs}+L_{RM}i_{qr}'\right)\omega_r-Mu_{ds}+L_{SM}u_{dr}'\Big],$$

$$\frac{di_{or}'}{dt}=\frac{1}{L_{lr}'}\left(-r_r' i_{or}'+u_{or}'\right),$$

$$\frac{d\omega_r}{dt}=\frac{3P^2}{8J}M\left(i_{qs}i_{dr}'-i_{ds}i_{qr}'\right)-\frac{B_m}{J}\omega_r-\frac{P}{2J}T_L. \quad (5.3.31)$$

In the state-space form, nonlinear differential equations (5.3.31) are given as

$$\begin{bmatrix} \frac{di_{qs}}{dt} \\ \frac{di_{ds}}{dt} \\ \frac{di_{os}}{dt} \\ \frac{di'_{qr}}{dt} \\ \frac{di'_{dr}}{dt} \\ \frac{di'_{or}}{dt} \\ \frac{d\omega_r}{dt} \end{bmatrix} = \begin{bmatrix} -\frac{L_{RM}r_s}{L_{SM}L_{RM}-M^2} & -\omega & 0 & \frac{Mr'_r}{L_{SM}L_{RM}-M^2} & 0 & 0 & 0 \\ \omega & -\frac{L_{RM}r_s}{L_{SM}L_{RM}-M^2} & 0 & 0 & \frac{Mr'_r}{L_{SM}L_{RM}-M^2} & 0 & 0 \\ 0 & 0 & -\frac{r_s}{L_{ls}} & 0 & 0 & 0 & 0 \\ \frac{Mr_s}{L_{SM}L_{RM}-M^2} & 0 & 0 & -\frac{L_{SM}r'_r}{L_{SM}L_{RM}-M^2} & -\omega & 0 & 0 \\ 0 & \frac{Mr_s}{L_{SM}L_{RM}-M^2} & 0 & \omega & -\frac{L_{SM}r'_r}{L_{SM}L_{RM}-M^2} & 0 & 0 \\ 0 & 0 & 0 & 0 & 0 & -\frac{r'_r}{L'_{lr}} & 0 \\ 0 & 0 & 0 & 0 & 0 & 0 & -\frac{B_m}{J} \end{bmatrix} \begin{bmatrix} i_{qs} \\ i_{ds} \\ i_{os} \\ i'_{qr} \\ i'_{dr} \\ i'_{or} \\ \omega_r \end{bmatrix}$$

$$+ \begin{bmatrix} -\frac{M\left(Mi_{ds}+L_{RM}i'_{dr}\right)\omega_r}{L_{SM}L_{RM}-M^2} \\ \frac{M\left(Mi_{qs}+L_{RM}i'_{qr}\right)\omega_r}{L_{SM}L_{RM}-M^2} \\ 0 \\ \frac{L_{SM}\left(Mi_{ds}+L_{RM}i'_{dr}\right)\omega_r}{L_{SM}L_{RM}-M^2} \\ -\frac{L_{SM}\left(Mi_{qs}+L_{RM}i'_{qr}\right)\omega_r}{L_{SM}L_{RM}-M^2} \\ 0 \\ \frac{3P^2}{8J}M\left(i_{qs}i'_{dr}-i_{ds}i'_{qr}\right) \end{bmatrix}$$

$$+ \begin{bmatrix} \frac{L_{RM}}{L_{SM}L_{RM}-M^2} & 0 & 0 & -\frac{M}{L_{SM}L_{RM}-M^2} & 0 & 0 \\ 0 & \frac{L_{RM}}{L_{SM}L_{RM}-M^2} & 0 & 0 & -\frac{M}{L_{SM}L_{RM}-M^2} & 0 \\ 0 & 0 & \frac{1}{L_{ls}} & 0 & 0 & 0 \\ -\frac{M}{L_{SM}L_{RM}-M^2} & 0 & 0 & \frac{L_{SM}}{L_{SM}L_{RM}-M^2} & 0 & 0 \\ 0 & -\frac{M}{L_{SM}L_{RM}-M^2} & 0 & 0 & \frac{L_{SM}}{L_{SM}L_{RM}-M^2} & 0 \\ 0 & 0 & 0 & 0 & 0 & \frac{1}{L'_{lr}} \\ 0 & 0 & 0 & 0 & 0 & 0 \end{bmatrix} \begin{bmatrix} u_{qs} \\ u_{ds} \\ u_{os} \\ u'_{qr} \\ u'_{dr} \\ u'_{or} \end{bmatrix} - \begin{bmatrix} 0 \\ 0 \\ 0 \\ 0 \\ 0 \\ 0 \\ \frac{P}{2J} \end{bmatrix} T_L.$$

The block diagram for three-phase induction micromotors, modeled in the *arbitrary* reference frame is developed using (5.3.31). In particular, applying the Laplace operator, one finds the block diagram as shown in Figure 5.3.5.

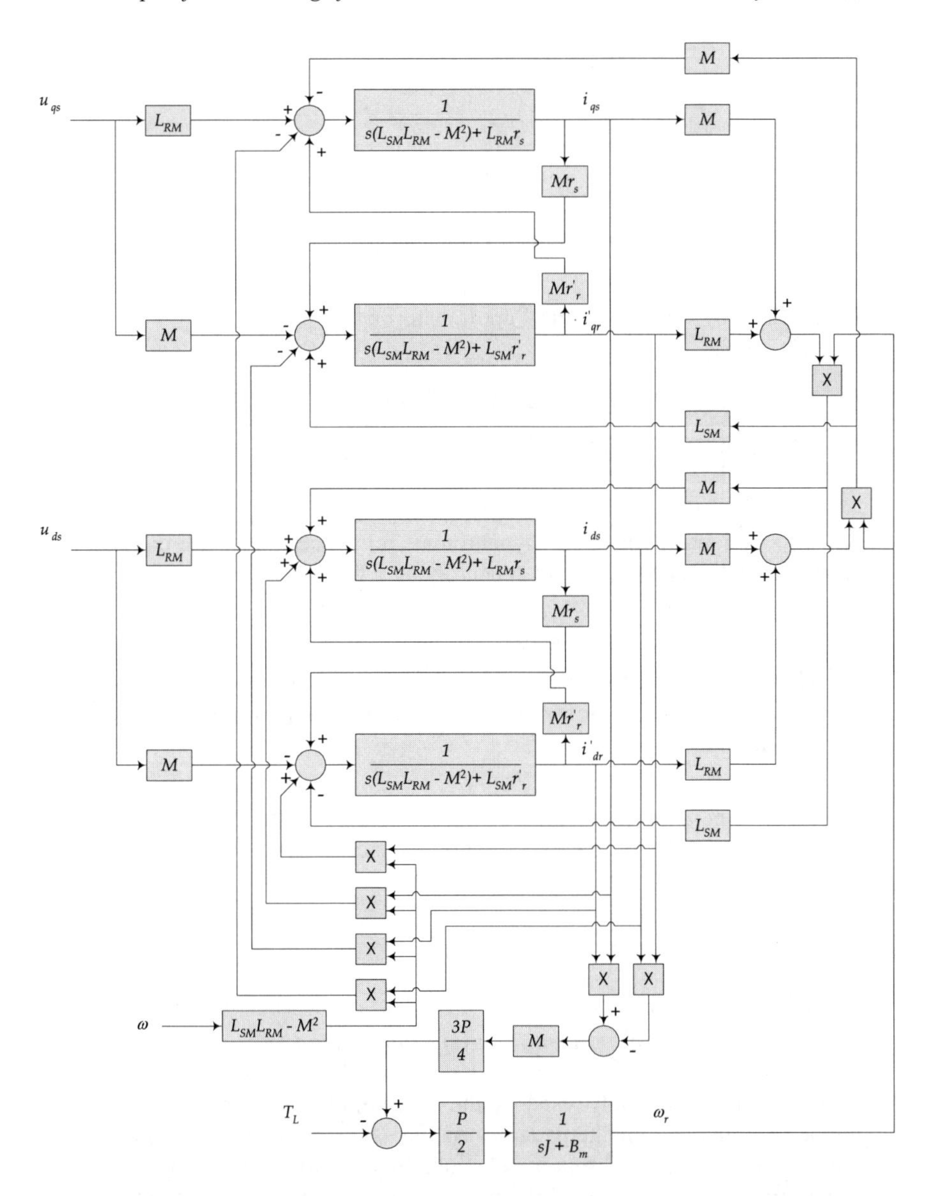


Figure 5.3.5 Block diagram of three-phase squirrel-cage induction micromotors in the *arbitrary* reference frame.

Micro- and miniscale induction motors are squirrel-cage motors, and the rotor windings are short-circuited. To guarantee the balanced operating conditions, one supplies the following balanced three-phase voltages

$$u_{as}(t)=\sqrt{2}u_M\cos\left(\omega_f t\right),$$
$$u_{bs}(t)=\sqrt{2}u_M\cos\left(\omega_f t-\tfrac{2}{3}\pi\right),$$
$$u_{cs}(t)=\sqrt{2}u_M\cos\left(\omega_f t+\tfrac{2}{3}\pi\right),$$

where the frequency of the applied voltage is $\omega_f = 2\pi f$ .

The *quadrature-*, *direct-*, and *zero-*axis components of stator voltages are obtained by using the stator Park transformation matrix as

$$\mathbf{u}_{qdos}=\mathbf{K}_s\mathbf{u}_{abcs},$$

$$\mathbf{K}_s=\frac{2}{3}\begin{bmatrix}\cos\theta & \cos\left(\theta-\frac{2}{3}\pi\right) & \cos\left(\theta+\frac{2}{3}\pi\right)\\ \sin\theta & \sin\left(\theta-\frac{2}{3}\pi\right) & \sin\left(\theta+\frac{2}{3}\pi\right)\\ \frac{1}{2} & \frac{1}{2} & \frac{1}{2}\end{bmatrix}.$$

The stationary, rotor, and synchronous reference frames are commonly used. For stationary, rotor, and synchronous reference frames, the reference frame angular velocities are

$$\omega=0,\ \omega=\omega_r \text{ and } \omega=\omega_e.$$

Hence, the corresponding angular displacement $\theta$ results. In particular, for zero initial conditions for stationary, rotor, and synchronous reference frames one finds

$$\theta=0\quad,\ \theta=\theta_r \text{ and } \theta=\theta_e.$$

Hence, the *quadrature-*, *direct-*, and *zero-*axis components of voltages can be obtained to guarantee the balance operation of induction micromotors.

### *Mathematical Model of Three-Phase Induction Micromotors in the Synchronous Reference Frame*

The most commonly used is the synchronous reference frame. The lumped-parameter mathematical model of three-phase induction micromotors in the synchronous reference frame is found by substituting the frame angular velocity in the differential equations obtained for the *arbitrary* reference frame (5.3.31).

Using

$$\omega=\omega_e$$

in equations (5.3.31), we have

$$\frac{di_{qs}^{e}}{dt}=\frac{1}{L_{SM}L_{RM}-M^{2}}\Big[-L_{RM}r_{s}i_{qs}^{e}-\left(L_{SM}L_{RM}-M^{2}\right)\omega_{e}i_{ds}^{e}+Mr_{r}^{'}i_{qr}^{'e}$$
$$-M\left(Mi_{ds}^{e}+L_{RM}i_{dr}^{'e}\right)\omega_{r}+L_{RM}u_{qs}^{e}-Mu_{qr}^{'e}\Big]$$
$$\frac{di_{ds}^{e}}{dt}=\frac{1}{L_{SM}L_{RM}-M^{2}}\Big[\left(L_{SM}L_{RM}-M^{2}\right)\omega_{e}i_{qs}^{e}-L_{RM}r_{s}i_{ds}^{e}+Mr_{r}^{'}i_{dr}^{'e}$$
$$+M\left(Mi_{qs}^{e}+L_{RM}i_{qr}^{'e}\right)\omega_{r}+L_{RM}u_{ds}^{e}-Mu_{dr}^{'e}\Big],$$
$$\frac{di_{os}^{e}}{dt}=\frac{1}{L_{ls}}\left(-r_{s}i_{os}^{e}+u_{os}^{e}\right),$$
$$\frac{di_{qr}^{'e}}{dt}=\frac{1}{L_{SM}L_{RM}-M^{2}}\Big[Mr_{s}i_{qs}^{e}-L_{SM}r_{r}^{'}i_{qr}^{'e}-\left(L_{SM}L_{RM}-M^{2}\right)\omega_{e}i_{dr}^{'e}$$
$$+L_{SM}\left(Mi_{ds}^{e}+L_{RM}i_{dr}^{'e}\right)\omega_{r}-Mu_{qs}^{e}+L_{SM}u_{qr}^{'e}\Big],$$
$$\frac{di_{dr}^{'e}}{dt}=\frac{1}{L_{SM}L_{RM}-M^{2}}\Big[Mr_{s}i_{ds}^{e}+\left(L_{SM}L_{RM}-M^{2}\right)\omega_{e}i_{qr}^{'e}-L_{SM}r_{r}^{'}i_{dr}^{'e}$$
$$-L_{SM}\left(Mi_{qs}^{e}+L_{RM}i_{qr}^{'e}\right)\omega_{r}-Mu_{ds}^{e}+L_{SM}u_{dr}^{'e}\Big],$$
$$\frac{di_{or}^{'e}}{dt}=\frac{1}{L_{lr}^{'}}\left(-r_{r}^{'}i_{or}^{'e}+u_{or}^{'e}\right),$$
$$\frac{d\omega_{r}}{dt}=\frac{3P^{2}}{8J}M\left(i_{qs}^{e}i_{dr}^{'e}-i_{ds}^{e}i_{qr}^{'e}\right)-\frac{B_{m}}{J}\omega_{r}-\frac{P}{2J}T_{L},$$
$$\frac{d\theta_{r}}{dt}=\omega_{r}. \qquad (5.3.32)$$

The superscript $e$ denotes the synchronous frame of reference. In the state-space form, using (5.3.32), we have the following differential equation for electric drives

$$
\begin{bmatrix}
\frac{di_{qs}^{e}}{dt}\\
\frac{di_{ds}^{e}}{dt}\\
\frac{di_{os}^{e}}{dt}\\
\frac{di_{qr}'^{e}}{dt}\\
\frac{di_{dr}'^{e}}{dt}\\
\frac{di_{or}'^{e}}{dt}\\
\frac{d\omega_r}{dt}
\end{bmatrix}
=
\begin{bmatrix}
-\frac{L_{RM}r_s}{L_{SM}L_{RM}-M^2} & -\omega_e & 0 & \frac{Mr_r'}{L_{SM}L_{RM}-M^2} & 0 & 0 & 0\\
\omega_e & -\frac{L_{RM}r_s}{L_{SM}L_{RM}-M^2} & 0 & 0 & \frac{Mr_r'}{L_{SM}L_{RM}-M^2} & 0 & 0\\
0 & 0 & -\frac{r_s}{L_{ls}} & 0 & 0 & 0 & 0\\
\frac{Mr_s}{L_{SM}L_{RM}-M^2} & 0 & 0 & -\frac{L_{SM}r_r'}{L_{SM}L_{RM}-M^2} & -\omega_e & 0 & 0\\
0 & \frac{Mr_s}{L_{SM}L_{RM}-M^2} & 0 & \omega_e & -\frac{L_{SM}r_r'}{L_{SM}L_{RM}-M^2} & 0 & 0\\
0 & 0 & 0 & 0 & 0 & -\frac{r_r'}{L_{lr}'} & 0\\
0 & 0 & 0 & 0 & 0 & 0 & -\frac{B_m}{J}
\end{bmatrix}
\begin{bmatrix}
i_{qs}^{e}\\ i_{ds}^{e}\\ i_{os}^{e}\\ i_{qr}'^{e}\\ i_{dr}'^{e}\\ i_{or}'^{e}\\ \omega_r
\end{bmatrix}
$$

$$
+\begin{bmatrix}
-\frac{M\left(Mi_{ds}^{e}+L_{RM}i_{dr}'^{e}\right)\omega_r}{L_{SM}L_{RM}-M^2}\\
\frac{M\left(Mi_{qs}^{e}+L_{RM}i_{qr}'^{e}\right)\omega_r}{L_{SM}L_{RM}-M^2}\\
0\\
\frac{L_{SM}\left(Mi_{ds}^{e}+L_{RM}i_{dr}'^{e}\right)\omega_r}{L_{SM}L_{RM}-M^2}\\
-\frac{L_{SM}\left(Mi_{qs}^{e}+L_{RM}i_{qr}'^{e}\right)\omega_r}{L_{SM}L_{RM}-M^2}\\
0\\
\frac{3P^2}{8J}M\left(i_{qs}^{e}i_{dr}'^{e}-i_{ds}^{e}i_{qr}'^{e}\right)
\end{bmatrix}
$$

$$
+\begin{bmatrix}
\frac{L_{RM}}{L_{SM}L_{RM}-M^2} & 0 & 0 & -\frac{M}{L_{SM}L_{RM}-M^2} & 0 & 0\\
0 & \frac{L_{RM}}{L_{SM}L_{RM}-M^2} & 0 & 0 & -\frac{M}{L_{SM}L_{RM}-M^2} & 0\\
0 & 0 & \frac{1}{L_{ls}} & 0 & 0 & 0\\
-\frac{M}{L_{SM}L_{RM}-M^2} & 0 & 0 & \frac{L_{SM}}{L_{SM}L_{RM}-M^2} & 0 & 0\\
0 & -\frac{M}{L_{SM}L_{RM}-M^2} & 0 & 0 & \frac{L_{SM}}{L_{SM}L_{RM}-M^2} & 0\\
0 & 0 & 0 & 0 & 0 & \frac{1}{L_{lr}'}\\
0 & 0 & 0 & 0 & 0 & 0
\end{bmatrix}
\begin{bmatrix}
u_{qs}^{e}\\ u_{ds}^{e}\\ u_{os}^{e}\\ u_{qr}'^{e}\\ u_{dr}'^{e}\\ u_{or}'^{e}
\end{bmatrix}
-\begin{bmatrix}
0\\0\\0\\0\\0\\0\\ \frac{P}{2J}
\end{bmatrix}T_L .
$$

The *quadrature*, *direct* and *zero* voltages $u_{qs}^{e}$, $u_{ds}^{e}$ and $u_{os}^{e}$ to guarantee the balanced operation of induction micromotors are found making use of the following relationship

$$\mathbf{u}_{qdos}^{e}=\mathbf{K}_{s}^{e}\mathbf{u}_{abcs}\,.$$

Taking note that $\theta=\theta_e$, we have

$$\mathbf{K}_s = \tfrac{2}{3}\begin{bmatrix} \cos\theta & \cos\left(\theta - \tfrac{2}{3}\pi\right) & \cos\left(\theta + \tfrac{2}{3}\pi\right) \\ \sin\theta & \sin\left(\theta - \tfrac{2}{3}\pi\right) & \sin\left(\theta + \tfrac{2}{3}\pi\right) \\ \tfrac{1}{2} & \tfrac{1}{2} & \tfrac{1}{2} \end{bmatrix}.$$

Therefore, one finds

$$\mathbf{K}_s^e = \tfrac{2}{3}\begin{bmatrix} \cos\theta_e & \cos\left(\theta_e - \tfrac{2}{3}\pi\right) & \cos\left(\theta_e + \tfrac{2}{3}\pi\right) \\ \sin\theta_e & \sin\left(\theta_e - \tfrac{2}{3}\pi\right) & \sin\left(\theta_e + \tfrac{2}{3}\pi\right) \\ \tfrac{1}{2} & \tfrac{1}{2} & \tfrac{1}{2} \end{bmatrix}.$$

Hence,

$$\begin{bmatrix} u_{qs}^e \\ u_{ds}^e \\ u_{os}^e \end{bmatrix} = \tfrac{2}{3}\begin{bmatrix} \cos\theta_e & \cos\left(\theta_e - \tfrac{2}{3}\pi\right) & \cos\left(\theta_e + \tfrac{2}{3}\pi\right) \\ \sin\theta_e & \sin\left(\theta_e - \tfrac{2}{3}\pi\right) & \sin\left(\theta_e + \tfrac{2}{3}\pi\right) \\ \tfrac{1}{2} & \tfrac{1}{2} & \tfrac{1}{2} \end{bmatrix}\begin{bmatrix} u_{as} \\ u_{bs} \\ u_{cs} \end{bmatrix},$$

and finally we obtain

$$u_{qs}^e(t) = \tfrac{2}{3}\left[u_{as}\cos\theta_e + u_{bs}\cos\left(\theta_e - \tfrac{2}{3}\pi\right) + u_{cs}\cos\left(\theta_e + \tfrac{2}{3}\pi\right)\right],$$

$$u_{ds}^e(t) = \tfrac{2}{3}\left[u_{as}\sin\theta_e + u_{bs}\sin\left(\theta_e - \tfrac{2}{3}\pi\right) + u_{cs}\sin\left(\theta_e + \tfrac{2}{3}\pi\right)\right],$$

$$u_{os}^e(t) = \tfrac{1}{3}\left(u_{as} + u_{bs} + u_{cs}\right).$$

Taking note of a balanced three-phase voltage set

$$u_{as}(t) = \sqrt{2}u_M\cos\left(\omega_f t\right),$$

$$u_{bs}(t) = \sqrt{2}u_M\cos\left(\omega_f t - \tfrac{2}{3}\pi\right),$$

$$u_{cs}(t) = \sqrt{2}u_M\cos\left(\omega_f t + \tfrac{2}{3}\pi\right),$$

and assuming that the initial displacement of the *quadrature* magnetic axis is zero, from $\theta_e = \omega_f t$, we have that the following *quadrature*, *direct*, and *zero* stator voltages must be supplied to guarantee the balance operation

$$u_{qs}^e(t) = \sqrt{2}u_M,\ u_{ds}^e(t) = 0,\ u_{os}^e(t) = 0. \tag{5.3.33}$$

It should be emphasized that the *quadrature*-, *direct*-, and *zero*-axis components of stator and rotor voltages, currents, and flux linkages have dc form. Furthermore, to control induction micromotors, only the dc *quadrature* voltage $u_{qs}^e(t)$ is regulated because $u_{ds}^e(t) = 0$ and $u_{os}^e(t) = 0$.

Using (5.3.32), the block diagram is developed, see Figure 5.3.6.

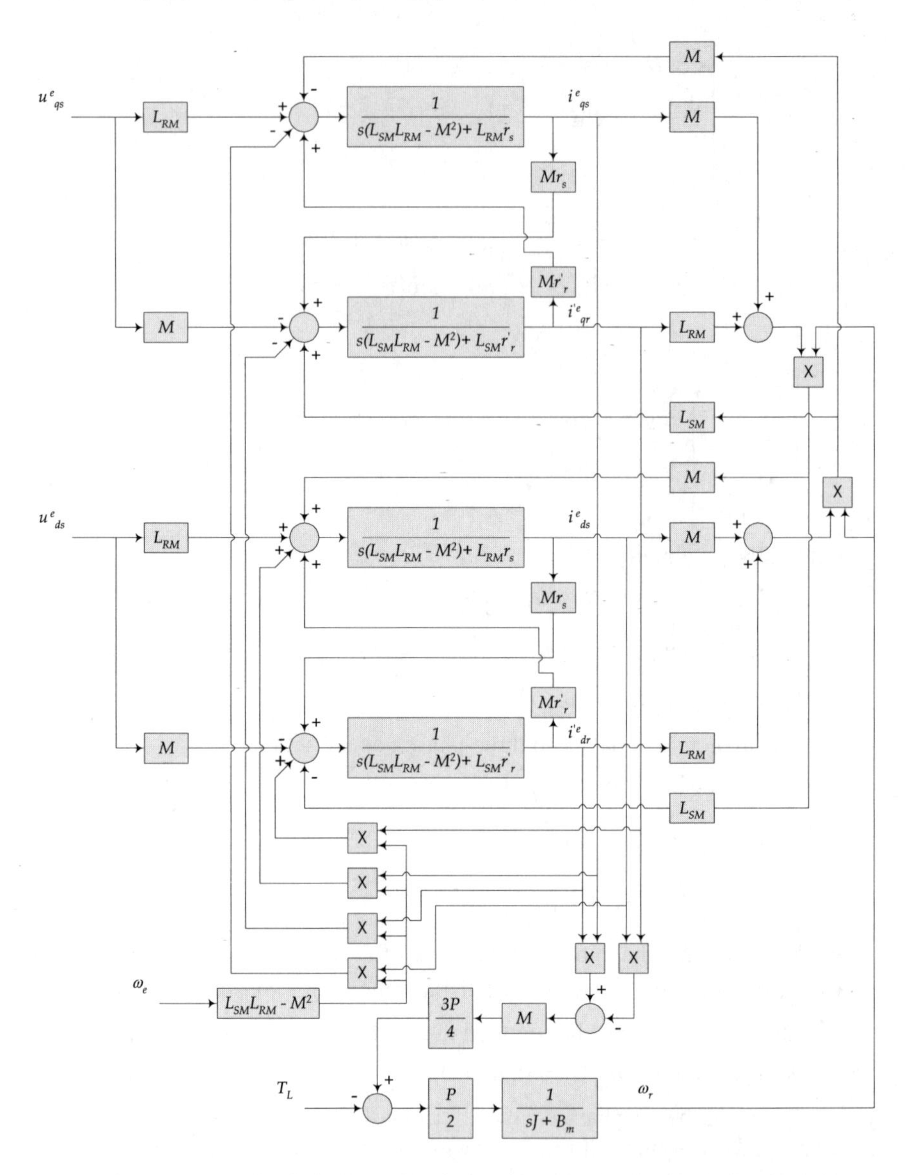


Figure 5.3.6 Block diagram for three-phase squirrel-cage induction micromotors modeled in the synchronous reference frame.

## 5.4 SYNCHRONOUS MICROTRANSDUCERS

Microscale synchronous transducers can be used as motors and generators. Microgenerators (velocity and position sensors) convert mechanical energy into electrical energy, while micromotors-microactuators convert electrical energy into mechanical energy. A broad spectrum of synchronous microtransducers can be used in MEMS as actuators, e.g., electric drives, *servos*, or microscale power systems. We will develop lumped-parameter nonlinear mathematical models, and perform nonlinear analysis of synchronous microtransducers.

In this section, the following variables and symbols are used:

$u_{as}, u_{bs}$ and $u_{cs}$ are the phase voltages in the stator windings *as, bs* and *cs*;

$u_{qs}, u_{ds}$ and $u_{os}$ are the *quadrature*-, *direct*-, and *zero*-axis stator voltages;

$i_{as}, i_{bs}$ and $i_{cs}$ are the phase currents in the stator windings *as, bs* and *cs*;

$i_{qs}, i_{ds}$ and $i_{os}$ are the *quadrature*-, *direct*-, and *zero*-axis stator currents;

$\psi_{as}, \psi_{bs}$ and $\psi_{cs}$ are the stator flux linkages;

$\psi_{qs}, \psi_{ds}$ and $\psi_{0s}$ are the *quadrature*-, *direct*-, and *zero*-axis stator flux linkages;

$\psi_m$ is the magnitude of the flux linkages established by the permanent-magnets;

$\omega_r$ and $\omega_{rm}$ are the electrical and rotor angular velocities;

$\theta_r$ and $\theta_{rm}$ are the electrical and rotor angular displacements;

$T_e$ is the electromagnetic torque developed;

$T_L$ is the load torque applied;

$B_m$ is the viscous friction coefficient;

$J$ is the equivalent moment of inertia;

$r_s$ is the resistances of the stator windings;

$L_{ss}$ is the self-inductances of the stator windings;

$L_{ms}$ and $L_{ls}$ are the stator magnetizing and leakage inductances;

$L_{mq}$ and $L_{md}$ are the magnetizing inductances in the *quadrature* and *direct* axes;

$\Re_{md}$ and $\Re_{mq}$ are the magnetizing reluctances in the *direct* and *quadrature* axes;

$N_s$ is the number of turns of the stator windings;

$P$ is the number of poles;

$\omega$ and $\theta$ are the angular velocity and displacement of the reference frame.

### 5.4.1 Single-Phase Reluctance Micromotors

We consider single-phase reluctance micromotors to study the operation of synchronous microtransducers, analyze important features, as well as to visualize mathematical model developments. It should be emphasized that microscale synchronous reluctance motors can be easily manufactured. A single-phase reluctance microtransducer is documented in Figure 5.4.1.

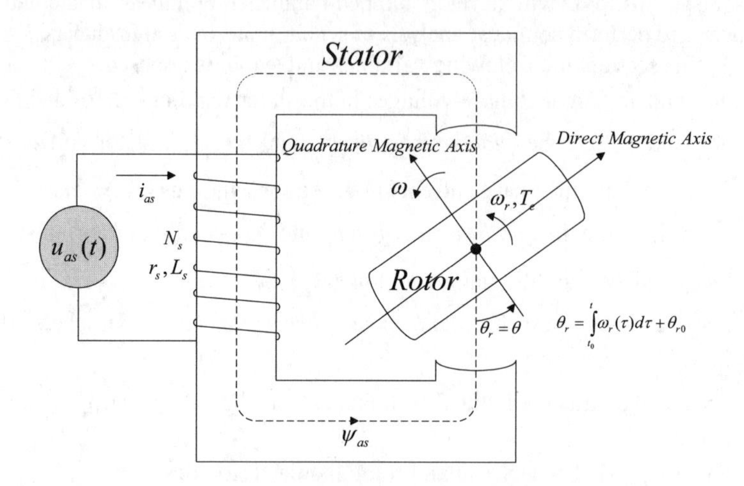


Figure 5.4.1 Microscale single-phase reluctance transducer.

The *quadrature* and *direct* magnetic axes are fixed with the rotor, which rotates with angular velocity $\omega_r$. These magnetic axes rotate with the angular velocity $\omega$. It should be emphasized that under normal operation the angular velocity of synchronous micromachines is equal to the synchronous angular velocity $\omega_e$. Hence, we have $\omega_r = \omega_e$ and $\omega = \omega_r = \omega_e$. Hence, the angular displacements of the rotor $\theta_r$ and the angular displacement of the *quadrature* magnetic axis $\theta$ are equal, and assuming that the initial conditions are zero

$$\theta_r = \theta = \int_{t_0}^{t} \omega_r(\tau)d\tau = \int_{t_0}^{t} \omega(\tau)d\tau.$$

The magnetizing reluctance $\Re_m$ is a function of the rotor angular displacement $\theta_r$. Using the number of turns $N_s$, the magnetizing inductance is $L_m(\theta_r) = \dfrac{N_s^2}{\Re_m(\theta_r)}$. This magnetizing inductance varies twice per one revolution of the rotor and has minimum and maximum values. Thus, we have

$$L_{m\min} = \left.\frac{N_s^2}{\Re_{m\max}(\theta_r)}\right|_{\theta_r = 0, \pi, 2\pi, \ldots} \text{ and } L_{m\max} = \left.\frac{N_s^2}{\Re_{m\min}(\theta_r)}\right|_{\theta_r = \frac{1}{2}\pi, \frac{3}{2}\pi, \frac{5}{2}\pi, \ldots}.$$

Assume that this variation is a sinusoidal function of the rotor angular displacement. Then,

$$L_m(\theta_r) = \bar{L}_m - L_{\Delta m} \cos 2\theta_r,$$

where $\bar{L}_m$ is the average value of the magnetizing inductance; $L_{\Delta m}$ is the half of amplitude of the sinusoidal variation of the magnetizing inductance.

The plot for $L_m(\theta_r)$ is documented in Figure 5.4.2.

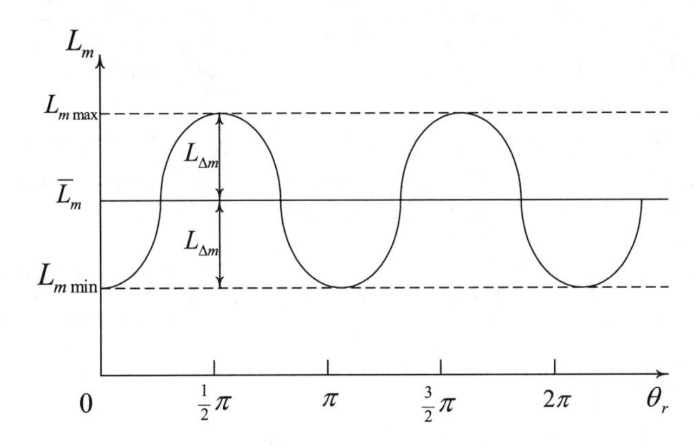


Figure 5.4.2 Magnetizing inductance $L_m(\theta_r)$.

The electromagnetic torque, developed by single-phase reluctance micromotors is found using the expression for the coenergy $W_c(i_{as}, \theta_r)$.

From

$$W_c(i_{as}, \theta_r) = \tfrac{1}{2}\left(L_{ls} + \bar{L}_m - L_{\Delta m}\cos 2\theta_r\right) i_{as}^2,$$

one finds

$$T_e = \frac{\partial W_c(i_{as}, \theta_r)}{\partial \theta_r} = \frac{\partial\left(\frac{1}{2} i_{as}^2 \left(L_{ls} + \bar{L}_m - L_{\Delta m}\cos 2\theta_r\right)\right)}{\partial \theta_r} = L_{\Delta m} i_{as}^2 \sin 2\theta_r.$$

It is clear that the electromagnetic torque is not developed by synchronous reluctance micromotors (microactuator) if one feeds the dc current or voltage to the winding. Hence, conventional control algorithms cannot be applied, and new methods, which are based upon electromagnetic–electromechanical features must be researched. The average value of $T_e$ is not equal to zero if the current is a function of $\theta_r$. As an illustration, we fed the following current to the motor winding

$$i_{as} = i_M \operatorname{Re}\left(\sqrt{\sin 2\theta_r}\right).$$

Then, the electromagnetic torque is

$$T_e = L_{\Delta m} i_{as}^2 \sin 2\theta_r = L_{\Delta m} i_M^2 \left(\mathrm{Re}\sqrt{\sin 2\theta_r}\right)^2 \sin 2\theta_r \neq 0,$$

and

$$T_{e\,av} = \frac{1}{\pi}\int_0^{\pi} L_{\Delta m} i_{as}^2 \sin 2\theta_r d\theta_r = \tfrac{1}{4} L_{\Delta m} i_M^2 .$$

The mathematical model of the single-phase reluctance micromotor is found by using Kirchhoff's and Newton's second laws

$$u_{as} = r_s i_{as} + \frac{d\psi_{as}}{dt}, \qquad \text{(circuitry equation)}$$

$$T_e - B_m \omega_r - T_L = J\frac{d^2\theta_r}{dt^2}. \qquad (\textit{torsional-mechanical}\text{ equation})$$

From $\psi_{as} = \left(L_{ls} + \bar{L}_m - L_{\Delta m}\cos 2\theta_r\right) i_{as}$, one obtains a set of three first-order nonlinear differential equations which models single-phase reluctance micromotors. In particular, we have

$$\frac{di_{as}}{dt} = -\frac{r_s}{L_{ls} + \bar{L}_m - L_{\Delta m}\cos 2\theta_r} i_{as} - \frac{2L_{\Delta m}}{L_{ls} + \bar{L}_m - L_{\Delta m}\cos 2\theta_r} i_{as}\omega_r \sin 2\theta_r$$

$$+ \frac{1}{L_{ls} + \bar{L}_m - L_{\Delta m}\cos 2\theta_r} u_{as},$$

$$\frac{d\omega_r}{dt} = \frac{1}{J}\left(L_{\Delta m} i_{as}^2 \sin 2\theta_r - B_m\omega_r - T_L\right),$$

$$\frac{d\theta_r}{dt} = \omega_r .$$

## 5.4.2 Permanent-Magnet Synchronous Microtransducers

Permanent-magnet synchronous microtransducers are brushless micromachines because the excitation flux is produced by permanent magnets deposited on the microrotor.

*Mathematical Model of Two-phase Permanent-Magnet Synchronous Micromotors*

Consider two-phase permanent-magnet synchronous micromotors. Using Kirchhoff's voltage law, we have the following two equations

$$u_{as} = r_s i_{as} + \frac{d\psi_{as}}{dt},$$

$$u_{bs} = r_s i_{bs} + \frac{d\psi_{bs}}{dt},$$

where the flux linkages are expressed as

$$\psi_{as} = L_{asas} i_{as} + L_{asbs} i_{bs} + \psi_{asm},$$

$$\psi_{bs} = L_{bsas} i_{as} + L_{bsbs} i_{bs} + \psi_{bsm}. .$$

Here, $u_{as}$ and $u_{bs}$ are the phase voltages in the stator microwindings *as* and *bs*; $i_{as}$ and $i_{bs}$ are the phase currents in the stator microwindings; $\psi_{as}$ and $\psi_{bs}$ are the stator flux linkages; $r_s$ is the resistances of the stator microwindings; $L_{asas}$, $L_{asbs}$, $L_{bsas}$ and $L_{bsbs}$ are the mutual inductances.

The flux linkages are periodic functions of the angular displacement (rotor position), and let

$$\psi_{asm} = \psi_m \sin\theta_{rm} \text{ and } \psi_{bsm} = -\psi_m \cos\theta_{rm}.$$

The self-inductances of the stator windings is found to be

$$L_{ss} = L_{asas} = L_{bsbs} = L_{ls} + \overline{L}_m.$$

The stator windings are displaced by 90 electrical degrees.

Hence, the mutual inductances between the stator windings are

$$L_{asbs} = L_{bsas} = 0.$$

Thus, we have

$$\psi_{as} = L_{ss} i_{as} + \psi_m \sin\theta_{rm},$$

$$\psi_{bs} = L_{ss} i_{bs} - \psi_m \cos\theta_{rm}.$$

Therefore, one finds

$$u_{as} = r_s i_{as} + \frac{d(L_{ss} i_{as} + \psi_m \sin\theta_{rm})}{dt} = r_s i_{as} + L_{ss}\frac{di_{as}}{dt} + \psi_m \omega_{rm} \cos\theta_{rm}$$

$$u_{bs} = r_s i_{bs} + \frac{d(L_{ss} i_{bs} - \psi_m \cos\theta_{rm})}{dt} = r_s i_{bs} + L_{ss}\frac{di_{bs}}{dt} - \psi_m \omega_{rm} \sin\theta_{rm}$$

Using Newton's second law $T_e - B_m\omega_{rm} - T_L = J\frac{d^2\theta_{rm}}{dt^2}$, we have

$$\frac{d\omega_{rm}}{dt} = \frac{1}{J}\left(T_e - B_m\omega_{rm} - T_L\right),$$

$$\frac{d\theta_{rm}}{dt} = \omega_{rm}.$$

The expression for the electromagnetic torque developed by permanent-magnet micromotors can be obtain by using the coenergy

$$W_c = \tfrac{1}{2}\left(L_{ss} i_{as}^2 + L_{ss} i_{bs}^2\right) + \psi_m i_{as} \sin\theta_{rm} - \psi_m i_{bs} \cos\theta_{rm} + W_{PM}.$$

Then, one has

$$T_e = \frac{\partial W_c}{\partial \theta_{rm}} = \frac{P\psi_m}{2}\left[i_{as}\cos\theta_{rm} + i_{bs}\sin\theta_{rm}\right].$$

Augmenting the circuitry transients with the *torsional-mechanical* dynamics, one finds the mathematical model of two-phase permanent-magnet micromotors in the following form

$$\frac{di_{as}}{dt} = -\frac{r_s}{L_{ss}} i_{as} - \frac{\psi_m}{L_{ss}} \omega_{rm} \cos\theta_{rm} + \frac{1}{L_{ss}} u_{as},$$

$$\frac{di_{bs}}{dt} = -\frac{r_s}{L_{ss}} i_{bs} + \frac{\psi_m}{L_{ss}} \omega_{rm} \sin\theta_{rm} + \frac{1}{L_{ss}} u_{bs},$$

$$\frac{d\omega_{rm}}{dt} = \frac{P\psi_m}{2J}\left(i_{as}\cos\theta_{rm} + i_{bs}\sin\theta_{rm}\right) - \frac{B_m}{J}\omega_{rm} - \frac{1}{J}T_L,$$

$$\frac{d\theta_{rm}}{dt} = \omega_{rm}.$$

For two-phase micromotors (assuming the sinusoidal winding distributions and the sinusoidal *mmf* waveforms), the electromagnetic torque is expressed as

$$T_e = \frac{P\psi_m}{2}\left(i_{as}\cos\theta_r + i_{bs}\sin\theta_r\right).$$

Hence, to guarantee the balanced operation, one feeds

$$i_{as} = \sqrt{2} i_M \cos\theta_r \text{ and } i_{bs} = \sqrt{2} i_M \sin\theta_r$$

to maximize the electromagnetic torque. In fact, one obtains

$$T_e = \frac{P\psi_m}{2}\left(i_{as}\cos\theta_r + i_{bs}\sin\theta_r\right) = \frac{P\psi_m}{2}\sqrt{2} i_M\left(\cos^2\theta_r + \sin^2\theta_r\right) = \frac{P\psi_m}{\sqrt{2}} i_M$$

### *Permanent-Magnet Synchronous Microtransducers in the Machine Variables*

Three-phase two-pole permanent-magnet synchronous microtransducers (microscale motors and generators) are illustrated in Figures 5.4.3 and 5.4.4.

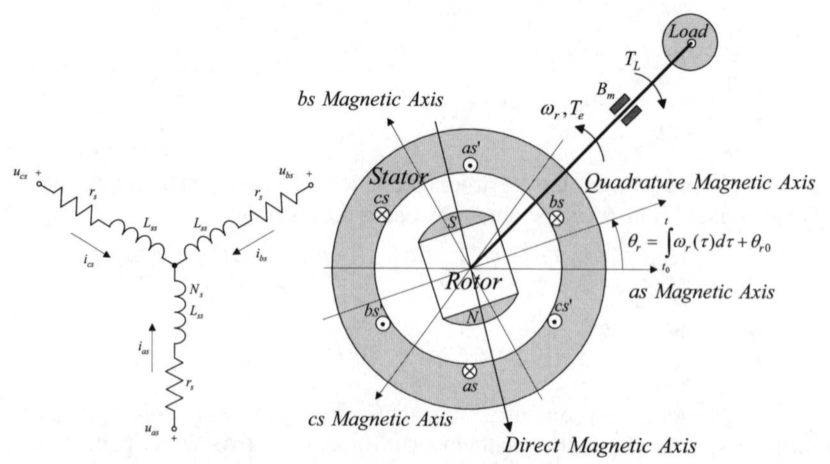


Figure 5.4.3 Two-pole permanent-magnet synchronous micromotor.

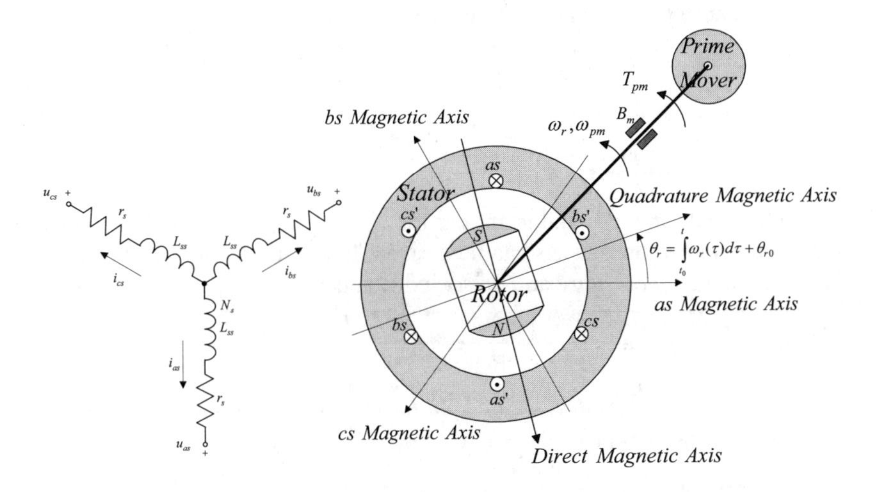


Figure 5.4.4 Three-phase wye-connected synchronous microgenerator.

From Kirchhoff's second law, one obtains three differential equations for the *as*, *bs* and *cs* stator windings. In particular,

$$u_{as} = r_s i_{as} + \frac{d\psi_{as}}{dt},$$

$$u_{bs} = r_s i_{bs} + \frac{d\psi_{bs}}{dt},$$

$$u_{cs} = r_s i_{cs} + \frac{d\psi_{cs}}{dt}, \qquad (5.4.1)$$

where the flux linkages $\psi_{as}, \psi_{bs}$ and $\psi_{cs}$ are

$$\psi_{as} = L_{asas} i_{as} + L_{asbs} i_{bs} + L_{ascs} i_{cs} + \psi_{asm},$$

$$\psi_{bs} = L_{bsas} i_{as} + L_{bsbs} i_{bs} + L_{bscs} i_{cs} + \psi_{bsm},$$

$$\psi_{cs} = L_{csas} i_{as} + L_{csbs} i_{bs} + L_{cscs} i_{cs} + \psi_{csm}.$$

From (5.4.1), one finds

$$\mathbf{u}_{abcs} = \mathbf{r}_s \mathbf{i}_{abcs} + \frac{d\boldsymbol{\psi}_{abcs}}{dt}, \quad \begin{bmatrix} u_{as} \\ u_{bs} \\ u_{cs} \end{bmatrix} = \begin{bmatrix} r_s & 0 & 0 \\ 0 & r_s & 0 \\ 0 & 0 & r_s \end{bmatrix} \begin{bmatrix} i_{as} \\ i_{bs} \\ i_{cs} \end{bmatrix} + \begin{bmatrix} \frac{d\psi_{as}}{dt} \\ \frac{d\psi_{bs}}{dt} \\ \frac{d\psi_{cs}}{dt} \end{bmatrix}.$$

The flux linkages $\psi_{asm}, \psi_{bsm}$, and $\psi_{csm}$, established by the permanent magnet, are periodic functions of $\theta_r$. We assume that $\psi_{asm}, \psi_{bsm}$, and $\psi_{csm}$ vary obeying the sine law. The stator windings are displaced by 120 electrical degrees. Denoting the magnitude of the flux linkages established by the permanent magnet as $\psi_m$, one has

$$\psi_{asm} = \psi_m \sin\theta_r,$$

$$\psi_{bsm} = \psi_m \sin\left(\theta_r - \tfrac{2}{3}\pi\right),$$

$$\psi_{csm} = \psi_m \sin\left(\theta_r + \tfrac{2}{3}\pi\right).$$

Self- and mutual inductances for three-phase permanent-magnet synchronous microtransducers can be derived. In particular, the equations for the magnetizing *quadrature* and *direct* inductances are

$$L_{mq} = \frac{N_s^2}{\Re_{mq}} \text{ and } L_{md} = \frac{N_s^2}{\Re_{md}}.$$

In general, the *quadrature* and *direct* magnetizing reluctances can be different, and $\Re_{mq} > \Re_{md}$. Hence, we have $L_{mq} < L_{md}$.

The minimum value of $L_{asas}$ occurs periodically at $\theta_r = 0, \pi, 2\pi, \ldots$, while the maximum value of $L_{asas}$ occurs at $\theta_r = \frac{1}{2}\pi, \frac{3}{2}\pi, \frac{5}{2}\pi, \ldots$.

One concludes that the self-inductance $L_{asas}(\theta_r)$, which is bounded as $L_{ls} + L_{mq} \le L_{asas} \le L_{ls} + L_{md}$, is a periodic function of $\theta_r$. Assuming that $L_{asas}(\theta_r)$ varies as a sine function with a dc component, we have

$$L_{asas} = L_{ls} + \bar{L}_m - L_{\Delta m} \cos 2\theta_r.$$

Here, $\bar{L}_m$ is the average value of the magnetizing inductance; $L_{\Delta m}$ is half the amplitude of the sinusoidal variation of the magnetizing inductance.

The relationships between $L_{mq}$, $L_{md}$, and $\bar{L}_m$, $L_{\Delta m}$ must be found. For three-phase synchronous microtransducers, one obtains

$$L_{mq} = \tfrac{3}{2}\left(\bar{L}_m - L_{\Delta m}\right),$$

$$L_{md} = \tfrac{3}{2}\left(\bar{L}_m + L_{\Delta m}\right).$$

Therefore,

$$\bar{L}_m = \tfrac{1}{3}\left(L_{mq} + L_{md}\right)$$

and $L_{\Delta m} = \tfrac{1}{3}\left(L_{md} - L_{mq}\right)$.

Using the expressions for $L_{mq}$ and $L_{md}$, we have

$$\bar{L}_m = \frac{1}{3}\left(\frac{N_s^2}{\Re_{mq}} + \frac{N_s^2}{\Re_{md}}\right) \text{ and } L_{\Delta m} = \frac{1}{3}\left(\frac{N_s^2}{\Re_{md}} - \frac{N_s^2}{\Re_{mq}}\right).$$

Therefore, the following equations for $\psi_{as}$, $\psi_{bs}$ and $\psi_{cs}$ result

$$\psi_{as}=\left(L_{ls}+\bar{L}_m-L_{\Delta m}\cos 2\theta_r\right)i_{as}+\left(-\tfrac{1}{2}\bar{L}_m-L_{\Delta m}\cos 2\left(\theta_r-\tfrac{1}{3}\pi\right)\right)i_{bs}$$
$$+\left(-\tfrac{1}{2}\bar{L}_m-L_{\Delta m}\cos 2\left(\theta_r+\tfrac{1}{3}\pi\right)\right)i_{cs}+\psi_m\sin\theta_r,$$
$$\psi_{bs}=\left(-\tfrac{1}{2}\bar{L}_m-L_{\Delta m}\cos 2\left(\theta_r-\tfrac{1}{3}\pi\right)\right)i_{as}+\left(L_{ls}+\bar{L}_m-L_{\Delta m}\cos 2\left(\theta_r-\tfrac{2}{3}\pi\right)\right)i_{bs}$$
$$+\left(-\tfrac{1}{2}\bar{L}_m-L_{\Delta m}\cos 2\theta_r\right)i_{cs}+\psi_m\sin\left(\theta_r-\tfrac{2}{3}\pi\right),$$
$$\psi_{cs}=\left(-\tfrac{1}{2}\bar{L}_m-L_{\Delta m}\cos 2\left(\theta_r+\tfrac{1}{3}\pi\right)\right)i_{as}+\left(-\tfrac{1}{2}\bar{L}_m-L_{\Delta m}\cos 2\theta_r\right)i_{bs}$$
$$+\left(L_{ls}+\bar{L}_m-L_{\Delta m}\cos 2\left(\theta_r+\tfrac{2}{3}\pi\right)\right)i_{cs}+\psi_m\sin\left(\theta_r+\tfrac{2}{3}\pi\right)$$
$$(5.4.2)$$

From (5.4.2), one has

$$\boldsymbol{\psi}_{abcs}=\mathbf{L}_s\mathbf{i}_{abcs}+\boldsymbol{\psi}_m$$
$$=\begin{bmatrix} L_{ls}+\bar{L}_m-L_{\Delta m}\cos 2\theta_r & -\frac{1}{2}\bar{L}_m-L_{\Delta m}\cos 2\left(\theta_r-\frac{1}{3}\pi\right) & -\frac{1}{2}\bar{L}_m-L_{\Delta m}\cos 2\left(\theta_r+\frac{1}{3}\pi\right)\\ -\frac{1}{2}\bar{L}_m-L_{\Delta m}\cos 2\left(\theta_r-\frac{1}{3}\pi\right) & L_{ls}+\bar{L}_m-L_{\Delta m}\cos 2\left(\theta_r-\frac{2}{3}\pi\right) & -\frac{1}{2}\bar{L}_m-L_{\Delta m}\cos 2\theta_r\\ -\frac{1}{2}\bar{L}_m-L_{\Delta m}\cos 2\left(\theta_r+\frac{1}{3}\pi\right) & -\frac{1}{2}\bar{L}_m-L_{\Delta m}\cos 2\theta_r & L_{ls}+\bar{L}_m-L_{\Delta m}\cos 2\left(\theta_r+\frac{2}{3}\pi\right)\end{bmatrix}\begin{bmatrix} i_{as}\\ i_{bs}\\ i_{cs}\end{bmatrix}+\psi_m\begin{bmatrix}\sin\theta_r\\ \sin\left(\theta_r-\frac{2}{3}\pi\right)\\ \sin\left(\theta_r+\frac{2}{3}\pi\right)\end{bmatrix}.$$

The inductance matrix $\mathbf{L}_s$ is given by

$$\mathbf{L}_s=\begin{bmatrix} L_{ls}+\bar{L}_m-L_{\Delta m}\cos 2\theta_r & -\frac{1}{2}\bar{L}_m-L_{\Delta m}\cos 2\left(\theta_r-\frac{1}{3}\pi\right) & -\frac{1}{2}\bar{L}_m-L_{\Delta m}\cos 2\left(\theta_r+\frac{1}{3}\pi\right)\\ -\frac{1}{2}\bar{L}_m-L_{\Delta m}\cos 2\left(\theta_r-\frac{1}{3}\pi\right) & L_{ls}+\bar{L}_m-L_{\Delta m}\cos 2\left(\theta_r-\frac{2}{3}\pi\right) & -\frac{1}{2}\bar{L}_m-L_{\Delta m}\cos 2\theta_r\\ -\frac{1}{2}\bar{L}_m-L_{\Delta m}\cos 2\left(\theta_r+\frac{1}{3}\pi\right) & -\frac{1}{2}\bar{L}_m-L_{\Delta m}\cos 2\theta_r & L_{ls}+\bar{L}_m-L_{\Delta m}\cos 2\left(\theta_r+\frac{2}{3}\pi\right)\end{bmatrix}.$$

It was shown that $\bar{L}_m$ and $L_{\Delta m}$ are expressed as

$$\bar{L}_m=\tfrac{1}{3}\left(\frac{N_s^2}{\Re_{mq}}+\frac{N_s^2}{\Re_{md}}\right)\text{ and }L_{\Delta m}=\tfrac{1}{3}\left(\frac{N_s^2}{\Re_{md}}-\frac{N_s^2}{\Re_{mq}}\right).$$

Permanent-magnet synchronous microtransducers are round-rotor electrical machines (the magnetic paths in the *quadrature* and *direct* magnetic axes are identical, and $\Re_{mq}=\Re_{md}$) [1]. Thus,

$$\bar{L}_m=\frac{2N_s^2}{3\Re_{mq}}=\frac{2N_s^2}{3\Re_{md}}\text{ and }L_{\Delta m}=0.$$

Therefore, the inductance matrix is

$$\mathbf{L}_s=\begin{bmatrix} L_{ls}+\bar{L}_m & -\frac{1}{2}\bar{L}_m & -\frac{1}{2}\bar{L}_m\\ -\frac{1}{2}\bar{L}_m & L_{ls}+\bar{L}_m & -\frac{1}{2}\bar{L}_m\\ -\frac{1}{2}\bar{L}_m & -\frac{1}{2}\bar{L}_m & L_{ls}+\bar{L}_m\end{bmatrix}.$$

From (5.4.2) the expressions for the flux linkages are

$$\psi_{as}=\left(L_{ls}+\bar{L}_m\right)i_{as}-\tfrac{1}{2}\bar{L}_m i_{bs}-\tfrac{1}{2}\bar{L}_m i_{cs}+\psi_m\sin\theta_r,$$
$$\psi_{bs}=-\tfrac{1}{2}\bar{L}_m i_{as}+\left(L_{ls}+\bar{L}_m\right)i_{bs}-\tfrac{1}{2}\bar{L}_m i_{cs}+\psi_m\sin\left(\theta_r-\tfrac{2}{3}\pi\right),$$
$$\psi_{cs}=-\tfrac{1}{2}\bar{L}_m i_{as}-\tfrac{1}{2}\bar{L}_m i_{bs}+\left(L_{ls}+\bar{L}_m\right)i_{cs}+\psi_m\sin\left(\theta_r+\tfrac{2}{3}\pi\right),\quad(5.4.3)$$

or in matrix form

$$\boldsymbol{\psi}_{abcs} = \mathbf{L}_s \mathbf{i}_{abcs} + \boldsymbol{\psi}_m = \begin{bmatrix} L_{ls} + \bar{L}_m & -\frac{1}{2}\bar{L}_m & -\frac{1}{2}\bar{L}_m \\ -\frac{1}{2}\bar{L}_m & L_{ls} + \bar{L}_m & -\frac{1}{2}\bar{L}_m \\ -\frac{1}{2}\bar{L}_m & -\frac{1}{2}\bar{L}_m & L_{ls} + \bar{L}_m \end{bmatrix} \begin{bmatrix} i_{as} \\ i_{bs} \\ i_{cs} \end{bmatrix} + \psi_m \begin{bmatrix} \sin\theta_r \\ \sin\left(\theta_r - \frac{2}{3}\pi\right) \\ \sin\left(\theta_r + \frac{2}{3}\pi\right) \end{bmatrix}.$$

Using (5.4.1) and (5.4.3), we have

$$\mathbf{u}_{abcs} = \mathbf{r}_s \mathbf{i}_{abcs} + \frac{d\boldsymbol{\psi}_{abcs}}{dt} = \mathbf{r}_s \mathbf{i}_{abcs} + \mathbf{L}_s \frac{d\mathbf{i}_{abcs}}{dt} + \frac{d\boldsymbol{\psi}_m}{dt},$$

where $\frac{d\boldsymbol{\psi}_m}{dt} = \psi_m \begin{bmatrix} \omega_r \cos\theta_r \\ \omega_r \cos\left(\theta_r - \frac{2}{3}\pi\right) \\ \omega_r \cos\left(\theta_r + \frac{2}{3}\pi\right) \end{bmatrix}$.

Cauchy's form can be found by making use of $\mathbf{L}_s^{-1}$. In particular,

$$\frac{d\mathbf{i}_{abcs}}{dt} = -\mathbf{L}_s^{-1}\mathbf{r}_s \mathbf{i}_{abcs} - \mathbf{L}_s^{-1}\frac{d\boldsymbol{\psi}_m}{dt} + \mathbf{L}_s^{-1}\mathbf{u}_{abcs}.$$

The state-space stator circuitry dynamics in Cauchy's form is given as

$$\begin{bmatrix} \frac{di_{as}}{dt} \\ \frac{di_{bs}}{dt} \\ \frac{di_{cs}}{dt} \end{bmatrix} = \begin{bmatrix} -\frac{r_s\left(2L_{ss} - \bar{L}_m\right)}{2L_{ss}^2 - L_{ss}\bar{L}_m - \bar{L}_m^2} & -\frac{r_s\bar{L}_m}{2L_{ss}^2 - L_{ss}\bar{L}_m - \bar{L}_m^2} & -\frac{r_s\bar{L}_m}{2L_{ss}^2 - L_{ss}\bar{L}_m - \bar{L}_m^2} \\ -\frac{r_s\bar{L}_m}{2L_{ss}^2 - L_{ss}\bar{L}_m - \bar{L}_m^2} & -\frac{r_s\left(2L_{ss} - \bar{L}_m\right)}{2L_{ss}^2 - L_{ss}\bar{L}_m - \bar{L}_m^2} & -\frac{r_s\bar{L}_m}{2L_{ss}^2 - L_{ss}\bar{L}_m - \bar{L}_m^2} \\ -\frac{r_s\bar{L}_m}{2L_{ss}^2 - L_{ss}\bar{L}_m - \bar{L}_m^2} & -\frac{r_s\bar{L}_m}{2L_{ss}^2 - L_{ss}\bar{L}_m - \bar{L}_m^2} & -\frac{r_s\left(2L_{ss} - \bar{L}_m\right)}{2L_{ss}^2 - L_{ss}\bar{L}_m - \bar{L}_m^2} \end{bmatrix} \begin{bmatrix} i_{as} \\ i_{bs} \\ i_{cs} \end{bmatrix}$$

$$+ \begin{bmatrix} -\frac{\psi_m\left(2L_{ss} - \bar{L}_m\right)}{2L_{ss}^2 - L_{ss}\bar{L}_m - \bar{L}_m^2} & -\frac{\psi_m\bar{L}_m}{2L_{ss}^2 - L_{ss}\bar{L}_m - \bar{L}_m^2} & -\frac{\psi_m\bar{L}_m}{2L_{ss}^2 - L_{ss}\bar{L}_m - \bar{L}_m^2} \\ -\frac{\psi_m\bar{L}_m}{2L_{ss}^2 - L_{ss}\bar{L}_m - \bar{L}_m^2} & -\frac{\psi_m\left(2L_{ss} - \bar{L}_m\right)}{2L_{ss}^2 - L_{ss}\bar{L}_m - \bar{L}_m^2} & -\frac{\psi_m\bar{L}_m}{2L_{ss}^2 - L_{ss}\bar{L}_m - \bar{L}_m^2} \\ -\frac{\psi_m\bar{L}_m}{2L_{ss}^2 - L_{ss}\bar{L}_m - \bar{L}_m^2} & -\frac{\psi_m\bar{L}_m}{2L_{ss}^2 - L_{ss}\bar{L}_m - \bar{L}_m^2} & -\frac{\psi_m\left(2L_{ss} - \bar{L}_m\right)}{2L_{ss}^2 - L_{ss}\bar{L}_m - \bar{L}_m^2} \end{bmatrix} \begin{bmatrix} \omega_r \cos\theta_r \\ \omega_r \cos\left(\theta_r - \frac{2}{3}\pi\right) \\ \omega_r \cos\left(\theta_r + \frac{2}{3}\pi\right) \end{bmatrix}$$

$$+ \begin{bmatrix} \frac{2L_{ss} - \bar{L}_m}{2L_{ss}^2 - L_{ss}\bar{L}_m - \bar{L}_m^2} & \frac{\bar{L}_m}{2L_{ss}^2 - L_{ss}\bar{L}_m - \bar{L}_m^2} & \frac{\bar{L}_m}{2L_{ss}^2 - L_{ss}\bar{L}_m - \bar{L}_m^2} \\ \frac{\bar{L}_m}{2L_{ss}^2 - L_{ss}\bar{L}_m - \bar{L}_m^2} & \frac{2L_{ss} - \bar{L}_m}{2L_{ss}^2 - L_{ss}\bar{L}_m - \bar{L}_m^2} & \frac{\bar{L}_m}{2L_{ss}^2 - L_{ss}\bar{L}_m - \bar{L}_m^2} \\ \frac{\bar{L}_m}{2L_{ss}^2 - L_{ss}\bar{L}_m - \bar{L}_m^2} & \frac{\bar{L}_m}{2L_{ss}^2 - L_{ss}\bar{L}_m - \bar{L}_m^2} & \frac{2L_{ss} - \bar{L}_m}{2L_{ss}^2 - L_{ss}\bar{L}_m - \bar{L}_m^2} \end{bmatrix} \begin{bmatrix} u_{as} \\ u_{bs} \\ u_{cs} \end{bmatrix}.$$

Here, $L_{ss} = L_{ls} + \bar{L}_m$.

In expanded form, we have the following nonlinear differential equations which allow the designer to model the circuitry transient behavior

$$\frac{di_{as}}{dt}=-\frac{r_s\left(2L_{ss}-\overline{L}_m\right)}{2L_{ss}^2-L_{ss}\overline{L}_m-\overline{L}_m^2}i_{as}-\frac{r_s\overline{L}_m}{2L_{ss}^2-L_{ss}\overline{L}_m-\overline{L}_m^2}i_{bs}-\frac{r_s\overline{L}_m}{2L_{ss}^2-L_{ss}\overline{L}_m-\overline{L}_m^2}i_{cs}$$

$$-\frac{\psi_m\left(2L_{ss}-\overline{L}_m\right)}{2L_{ss}^2-L_{ss}\overline{L}_m-\overline{L}_m^2}\omega_r\cos\theta_r-\frac{\psi_m\overline{L}_m}{2L_{ss}^2-L_{ss}\overline{L}_m-\overline{L}_m^2}\omega_r\cos\left(\theta_r-\tfrac{2}{3}\pi\right)$$

$$-\frac{\psi_m\overline{L}_m}{2L_{ss}^2-L_{ss}\overline{L}_m-\overline{L}_m^2}\omega_r\cos\left(\theta_r+\tfrac{2}{3}\pi\right)$$

$$+\frac{2L_{ss}-\overline{L}_m}{2L_{ss}^2-L_{ss}\overline{L}_m-\overline{L}_m^2}u_{as}+\frac{\overline{L}_m}{2L_{ss}^2-L_{ss}\overline{L}_m-\overline{L}_m^2}u_{bs}+\frac{\overline{L}_m}{2L_{ss}^2-L_{ss}\overline{L}_m-\overline{L}_m^2}u_{cs},$$

$$\frac{di_{bs}}{dt}=-\frac{r_s\overline{L}_m}{2L_{ss}^2-L_{ss}\overline{L}_m-\overline{L}_m^2}i_{as}-\frac{r_s\left(2L_{ss}-\overline{L}_m\right)}{2L_{ss}^2-L_{ss}\overline{L}_m-\overline{L}_m^2}i_{bs}-\frac{r_s\overline{L}_m}{2L_{ss}^2-L_{ss}\overline{L}_m-\overline{L}_m^2}i_{cs}$$

$$-\frac{\psi_m\overline{L}_m}{2L_{ss}^2-L_{ss}\overline{L}_m-\overline{L}_m^2}\omega_r\cos\theta_r-\frac{\psi_m\left(2L_{ss}-\overline{L}_m\right)}{2L_{ss}^2-L_{ss}\overline{L}_m-\overline{L}_m^2}\omega_r\cos\left(\theta_r-\tfrac{2}{3}\pi\right)$$

$$-\frac{\psi_m\overline{L}_m}{2L_{ss}^2-L_{ss}\overline{L}_m-\overline{L}_m^2}\omega_r\cos\left(\theta_r+\tfrac{2}{3}\pi\right)$$

$$+\frac{\overline{L}_m}{2L_{ss}^2-L_{ss}\overline{L}_m-\overline{L}_m^2}u_{as}+\frac{2L_{ss}-\overline{L}_m}{2L_{ss}^2-L_{ss}\overline{L}_m-\overline{L}_m^2}u_{bs}+\frac{\overline{L}_m}{2L_{ss}^2-L_{ss}\overline{L}_m-\overline{L}_m^2}u_{cs},$$

$$\frac{di_{cs}}{dt}=-\frac{r_s\overline{L}_m}{2L_{ss}^2-L_{ss}\overline{L}_m-\overline{L}_m^2}i_{as}-\frac{r_s\overline{L}_m}{2L_{ss}^2-L_{ss}\overline{L}_m-\overline{L}_m^2}i_{bs}-\frac{r_s\left(2L_{ss}-\overline{L}_m\right)}{2L_{ss}^2-L_{ss}\overline{L}_m-\overline{L}_m^2}i_{cs}$$

$$-\frac{\psi_m\overline{L}_m}{2L_{ss}^2-L_{ss}\overline{L}_m-\overline{L}_m^2}\omega_r\cos\theta_r-\frac{\psi_m\overline{L}_m}{2L_{ss}^2-L_{ss}\overline{L}_m-\overline{L}_m^2}\omega_r\cos\left(\theta_r-\tfrac{2}{3}\pi\right)$$

$$-\frac{\psi_m\left(2L_{ss}-\overline{L}_m\right)}{2L_{ss}^2-L_{ss}\overline{L}_m-\overline{L}_m^2}\omega_r\cos\left(\theta_r+\tfrac{2}{3}\pi\right)$$

$$+\frac{\overline{L}_m}{2L_{ss}^2-L_{ss}\overline{L}_m-\overline{L}_m^2}u_{as}+\frac{\overline{L}_m}{2L_{ss}^2-L_{ss}\overline{L}_m-\overline{L}_m^2}u_{bs}+\frac{2L_{ss}-\overline{L}_m}{2L_{ss}^2-L_{ss}\overline{L}_m-\overline{L}_m^2}u_{cs}. \tag{5.4.4}$$

Having derived the differential equations to model the circuitry dynamics, the transient behavior of the rotor (mechanical system) must be incorporated. One cannot solve (5.4.4) where the electrical angular velocity $\omega_r$ and angular displacement $\theta_r$ are used as the state variables.

Making use of Newton's second law

$$T_e-B_m\omega_{rm}-T_L=J\frac{d^2\theta_{rm}}{dt^2},$$

we have a set of two differential equations. In particular,

$$\frac{d\omega_{rm}}{dt} = \frac{1}{J}\left(T_e - B_m\omega_{rm} - T_L\right),$$
$$\frac{d\theta_{rm}}{dt} = \omega_{rm}.$$

The expression for the electromagnetic torque developed must be found using the coenergy

$$W_c = \tfrac{1}{2}\begin{bmatrix} i_{as} & i_{bs} & i_{cs} \end{bmatrix}\mathbf{L}_s\begin{bmatrix} i_{as} \\ i_{bs} \\ i_{cs} \end{bmatrix} + \begin{bmatrix} i_{as} & i_{bs} & i_{cs} \end{bmatrix}\begin{bmatrix} \psi_m \sin\theta_r \\ \psi_m \sin\left(\theta_r - \tfrac{2}{3}\pi\right) \\ \psi_m \sin\left(\theta_r + \tfrac{2}{3}\pi\right) \end{bmatrix} + W_{PM}.$$

Here, $W_{PM}$ is the energy stored in the permanent magnet.

For round-rotor synchronous microtransducers, we have

$$\mathbf{L}_s = \begin{bmatrix} L_{ls} + \bar{L}_m & -\tfrac{1}{2}\bar{L}_m & -\tfrac{1}{2}\bar{L}_m \\ -\tfrac{1}{2}\bar{L}_m & L_{ls} + \bar{L}_m & -\tfrac{1}{2}\bar{L}_m \\ -\tfrac{1}{2}\bar{L}_m & -\tfrac{1}{2}\bar{L}_m & L_{ls} + \bar{L}_m \end{bmatrix}.$$

The inductance matrix $\mathbf{L}_s$ and $W_{PM}$ are not functions of $\theta_r$. One obtains the following formula to calculate the electromagnetic torque for three-phase *P*-pole permanent-magnet synchronous micromotors

$$T_e = \frac{P}{2}\frac{\partial W_c}{\partial \theta_r} = \frac{P\psi_m}{2}\left(i_{as}\cos\theta_r + i_{bs}\cos\left(\theta_r - \tfrac{2}{3}\pi\right) + i_{cs}\cos\left(\theta_r + \tfrac{2}{3}\pi\right)\right).$$

Therefore, we have

$$\frac{d\omega_{rm}}{dt} = \frac{P\psi_m}{2J}\left(i_{as}\cos\theta_r + i_{bs}\cos\left(\theta_r - \tfrac{2}{3}\pi\right) + i_{cs}\cos\left(\theta_r + \tfrac{2}{3}\pi\right)\right) - \frac{B_m}{J}\omega_{rm} - \frac{1}{J}T_L,$$
$$\frac{d\theta_{rm}}{dt} = \omega_{rm}.$$

Using the electrical angular velocity $\omega_r$ and displacement $\theta_r$, related to the mechanical angular velocity and displacement as $\omega_{rm} = \frac{2}{P}\omega_r$ and $\theta_{rm} = \frac{2}{P}\theta_r$, the following differential equations to model the *torsional-mechanical* transient dynamics finally result

$$\frac{d\omega_r}{dt} = \frac{P^2\psi_m}{4J}\left(i_{as}\cos\theta_r + i_{bs}\cos\left(\theta_r - \tfrac{2}{3}\pi\right) + i_{cs}\cos\left(\theta_r + \tfrac{2}{3}\pi\right)\right) - \frac{B_m}{J}\omega_r - \frac{P}{2J}T_L,$$
$$\frac{d\theta_r}{dt} = \omega_r. \qquad (5.4.5)$$

From (5.4.4) and (5.4.5), one obtains a nonlinear mathematical model of permanent-magnet synchronous micromotors in Cauchy's form as given by a system of five highly nonlinear differential equations

$$\frac{di_{as}}{dt}=-\frac{r_s\left(2L_{ss}-\bar{L}_m\right)}{2L_{ss}^2-L_{ss}\bar{L}_m-\bar{L}_m^2}i_{as}-\frac{r_s\bar{L}_m}{2L_{ss}^2-L_{ss}\bar{L}_m-\bar{L}_m^2}i_{bs}-\frac{r_s\bar{L}_m}{2L_{ss}^2-L_{ss}\bar{L}_m-\bar{L}_m^2}i_{cs}$$
$$-\frac{\psi_m\left(2L_{ss}-\bar{L}_m\right)}{2L_{ss}^2-L_{ss}\bar{L}_m-\bar{L}_m^2}\omega_r\cos\theta_r-\frac{\psi_m\bar{L}_m}{2L_{ss}^2-L_{ss}\bar{L}_m-\bar{L}_m^2}\omega_r\cos\left(\theta_r-\tfrac{2}{3}\pi\right)$$
$$-\frac{\psi_m\bar{L}_m}{2L_{ss}^2-L_{ss}\bar{L}_m-\bar{L}_m^2}\omega_r\cos\left(\theta_r+\tfrac{2}{3}\pi\right)$$
$$+\frac{2L_{ss}-\bar{L}_m}{2L_{ss}^2-L_{ss}\bar{L}_m-\bar{L}_m^2}u_{as}+\frac{\bar{L}_m}{2L_{ss}^2-L_{ss}\bar{L}_m-\bar{L}_m^2}u_{bs}+\frac{\bar{L}_m}{2L_{ss}^2-L_{ss}\bar{L}_m-\bar{L}_m^2}u_{cs},$$
$$\frac{di_{bs}}{dt}=-\frac{r_s\bar{L}_m}{2L_{ss}^2-L_{ss}\bar{L}_m-\bar{L}_m^2}i_{as}-\frac{r_s\left(2L_{ss}-\bar{L}_m\right)}{2L_{ss}^2-L_{ss}\bar{L}_m-\bar{L}_m^2}i_{bs}-\frac{r_s\bar{L}_m}{2L_{ss}^2-L_{ss}\bar{L}_m-\bar{L}_m^2}i_{cs}$$
$$-\frac{\psi_m\bar{L}_m}{2L_{ss}^2-L_{ss}\bar{L}_m-\bar{L}_m^2}\omega_r\cos\theta_r-\frac{\psi_m\left(2L_{ss}-\bar{L}_m\right)}{2L_{ss}^2-L_{ss}\bar{L}_m-\bar{L}_m^2}\omega_r\cos\left(\theta_r-\tfrac{2}{3}\pi\right)$$
$$-\frac{\psi_m\bar{L}_m}{2L_{ss}^2-L_{ss}\bar{L}_m-\bar{L}_m^2}\omega_r\cos\left(\theta_r+\tfrac{2}{3}\pi\right)$$
$$+\frac{\bar{L}_m}{2L_{ss}^2-L_{ss}\bar{L}_m-\bar{L}_m^2}u_{as}+\frac{2L_{ss}-\bar{L}_m}{2L_{ss}^2-L_{ss}\bar{L}_m-\bar{L}_m^2}u_{bs}+\frac{\bar{L}_m}{2L_{ss}^2-L_{ss}\bar{L}_m-\bar{L}_m^2}u_{cs},$$
$$\frac{di_{cs}}{dt}=-\frac{r_s\bar{L}_m}{2L_{ss}^2-L_{ss}\bar{L}_m-\bar{L}_m^2}i_{as}-\frac{r_s\bar{L}_m}{2L_{ss}^2-L_{ss}\bar{L}_m-\bar{L}_m^2}i_{bs}-\frac{r_s\left(2L_{ss}-\bar{L}_m\right)}{2L_{ss}^2-L_{ss}\bar{L}_m-\bar{L}_m^2}i_{cs}$$
$$-\frac{\psi_m\bar{L}_m}{2L_{ss}^2-L_{ss}\bar{L}_m-\bar{L}_m^2}\omega_r\cos\theta_r-\frac{\psi_m\bar{L}_m}{2L_{ss}^2-L_{ss}\bar{L}_m-\bar{L}_m^2}\omega_r\cos\left(\theta_r-\tfrac{2}{3}\pi\right)$$
$$-\frac{\psi_m\left(2L_{ss}-\bar{L}_m\right)}{2L_{ss}^2-L_{ss}\bar{L}_m-\bar{L}_m^2}\omega_r\cos\left(\theta_r+\tfrac{2}{3}\pi\right)$$
$$+\frac{\bar{L}_m}{2L_{ss}^2-L_{ss}\bar{L}_m-\bar{L}_m^2}u_{as}+\frac{\bar{L}_m}{2L_{ss}^2-L_{ss}\bar{L}_m-\bar{L}_m^2}u_{bs}+\frac{2L_{ss}-\bar{L}_m}{2L_{ss}^2-L_{ss}\bar{L}_m-\bar{L}_m^2}u_{cs},$$
$$\frac{d\omega_r}{dt}=\frac{P^2\psi_m}{4J}\left(i_{as}\cos\theta_r+i_{bs}\cos\left(\theta_r-\tfrac{2}{3}\pi\right)+i_{cs}\cos\left(\theta_r+\tfrac{2}{3}\pi\right)\right)-\frac{B_m}{J}\omega_r-\frac{P}{2J}T_L,$$
$$\frac{d\theta_r}{dt}=\omega_r. \tag{5.4.6}$$

The state-space form of the lumped-parameter mathematical model can be derived. In particular, from (5.4.6), we have

$$\begin{bmatrix} \frac{di_{as}}{dt} \\ \frac{di_{bs}}{dt} \\ \frac{di_{cs}}{dt} \\ \frac{d\omega_r}{dt} \\ \frac{d\theta_r}{dt} \end{bmatrix} = \begin{bmatrix} -\frac{r_s(2L_{ss}-\bar{L}_m)}{2L_{ss}^2 - L_{ss}\bar{L}_m - \bar{L}_m^2} & -\frac{r_s\bar{L}_m}{2L_{ss}^2 - L_{ss}\bar{L}_m - \bar{L}_m^2} & -\frac{r_s\bar{L}_m}{2L_{ss}^2 - L_{ss}\bar{L}_m - \bar{L}_m^2} & 0 & 0 \\ -\frac{r_s\bar{L}_m}{2L_{ss}^2 - L_{ss}\bar{L}_m - \bar{L}_m^2} & -\frac{r_s(2L_{ss}-\bar{L}_m)}{2L_{ss}^2 - L_{ss}\bar{L}_m - \bar{L}_m^2} & -\frac{r_s\bar{L}_m}{2L_{ss}^2 - L_{ss}\bar{L}_m - \bar{L}_m^2} & 0 & 0 \\ -\frac{r_s\bar{L}_m}{2L_{ss}^2 - L_{ss}\bar{L}_m - \bar{L}_m^2} & -\frac{r_s\bar{L}_m}{2L_{ss}^2 - L_{ss}\bar{L}_m - \bar{L}_m^2} & -\frac{r_s(2L_{ss}-\bar{L}_m)}{2L_{ss}^2 - L_{ss}\bar{L}_m - \bar{L}_m^2} & 0 & 0 \\ 0 & 0 & 0 & -\frac{B_m}{J} & 0 \\ 0 & 0 & 0 & 1 & 0 \end{bmatrix} \begin{bmatrix} i_{as} \\ i_{bs} \\ i_{cs} \\ \omega_r \\ \theta_r \end{bmatrix}$$

$$+ \begin{bmatrix} -\frac{\psi_m(2L_{ss}-\bar{L}_m)}{2L_{ss}^2 - L_{ss}\bar{L}_m - \bar{L}_m^2}\omega_r & -\frac{\psi_m\bar{L}_m}{2L_{ss}^2 - L_{ss}\bar{L}_m - \bar{L}_m^2}\omega_r & -\frac{\psi_m\bar{L}_m}{2L_{ss}^2 - L_{ss}\bar{L}_m - \bar{L}_m^2}\omega_r \\ -\frac{\psi_m\bar{L}_m}{2L_{ss}^2 - L_{ss}\bar{L}_m - \bar{L}_m^2}\omega_r & -\frac{\psi_m(2L_{ss}-\bar{L}_m)}{2L_{ss}^2 - L_{ss}\bar{L}_m - \bar{L}_m^2}\omega_r & -\frac{\psi_m\bar{L}_m}{2L_{ss}^2 - L_{ss}\bar{L}_m - \bar{L}_m^2}\omega_r \\ -\frac{\psi_m\bar{L}_m}{2L_{ss}^2 - L_{ss}\bar{L}_m - \bar{L}_m^2}\omega_r & -\frac{\psi_m\bar{L}_m}{2L_{ss}^2 - L_{ss}\bar{L}_m - \bar{L}_m^2}\omega_r & -\frac{\psi_m(2L_{ss}-\bar{L}_m)}{2L_{ss}^2 - L_{ss}\bar{L}_m - \bar{L}_m^2}\omega_r \\ \frac{P^2\psi_m}{4J}i_{as} & \frac{P^2\psi_m}{4J}i_{bs} & \frac{P^2\psi_m}{4J}i_{cs} \\ 0 & 0 & 0 \end{bmatrix} \begin{bmatrix} \cos\theta_r \\ \cos\left(\theta_r - \frac{2}{3}\pi\right) \\ \cos\left(\theta_r + \frac{2}{3}\pi\right) \end{bmatrix}$$

$$+ \begin{bmatrix} \frac{2L_{ss}-\bar{L}_m}{2L_{ss}^2 - L_{ss}\bar{L}_m - \bar{L}_m^2} & \frac{\bar{L}_m}{2L_{ss}^2 - L_{ss}\bar{L}_m - \bar{L}_m^2} & \frac{\bar{L}_m}{2L_{ss}^2 - L_{ss}\bar{L}_m - \bar{L}_m^2} \\ \frac{\bar{L}_m}{2L_{ss}^2 - L_{ss}\bar{L}_m - \bar{L}_m^2} & \frac{2L_{ss}-\bar{L}_m}{2L_{ss}^2 - L_{ss}\bar{L}_m - \bar{L}_m^2} & \frac{\bar{L}_m}{2L_{ss}^2 - L_{ss}\bar{L}_m - \bar{L}_m^2} \\ \frac{\bar{L}_m}{2L_{ss}^2 - L_{ss}\bar{L}_m - \bar{L}_m^2} & \frac{\bar{L}_m}{2L_{ss}^2 - L_{ss}\bar{L}_m - \bar{L}_m^2} & \frac{2L_{ss}-\bar{L}_m}{2L_{ss}^2 - L_{ss}\bar{L}_m - \bar{L}_m^2} \\ 0 & 0 & 0 \\ 0 & 0 & 0 \end{bmatrix} \begin{bmatrix} u_{as} \\ u_{bs} \\ u_{cs} \end{bmatrix} - \begin{bmatrix} 0 \\ 0 \\ 0 \\ \frac{P}{2J} \\ 0 \end{bmatrix} T_L .$$

To control the angular velocity, one regulates the currents fed or voltages applied to the stator *abc* windings. Neglecting the viscous friction coefficient, the analysis of Newton's second law

$$T_e - T_L = J\frac{d\omega_{rm}}{dt}$$

indicates that

- The angular velocity $\omega_{rm}$ increases if $T_e > T_L$,
- The angular velocity $\omega_{rm}$ decreases if $T_e < T_L$,
- The angular velocity $\omega_{rm}$ is constant ($\omega_{rm} = const$) if $T_e = T_L$.

That is, to regulate the electromagnetic torque, which was found as

$$T_e = \frac{P\psi_m}{2}\left(i_{as}\cos\theta_r + i_{bs}\cos\left(\theta_r - \tfrac{2}{3}\pi\right) + i_{cs}\cos\left(\theta_r + \tfrac{2}{3}\pi\right)\right),$$

must be changed.

If, using ICs, the *abc* windings are fed by a balanced three-phase current set

$$i_{as}(t)=\sqrt{2}i_M\cos(\omega_r t)=\sqrt{2}i_M\cos(\omega_e t)=\sqrt{2}i_M\cos\theta_r,$$
$$i_{bs}(t)=\sqrt{2}i_M\cos\left(\omega_r t-\tfrac{2}{3}\pi\right)=\sqrt{2}i_M\cos\left(\omega_e t-\tfrac{2}{3}\pi\right)=\sqrt{2}i_M\cos\left(\theta_r-\tfrac{2}{3}\pi\right),$$
$$i_{cs}(t)=\sqrt{2}i_M\cos\left(\omega_r t+\tfrac{2}{3}\pi\right)=\sqrt{2}i_M\cos\left(\omega_e t+\tfrac{2}{3}\pi\right)=\sqrt{2}i_M\cos\left(\theta_r+\tfrac{2}{3}\pi\right),$$

taking note of the trigonometric identity

$$\cos^2\theta_r+\cos^2\left(\theta_r-\tfrac{2}{3}\pi\right)+\cos^2\left(\theta_r+\tfrac{2}{3}\pi\right)=\tfrac{3}{2},$$

one obtains

$$T_e=\frac{P\psi_m}{2}\sqrt{2}i_M\left(\cos^2\theta_r+\cos^2\left(\theta_r-\tfrac{2}{3}\pi\right)+\cos^2\left(\theta_r+\tfrac{2}{3}\pi\right)\right)=\frac{3P\psi_m}{2\sqrt{2}}i_M.$$

One concludes that to regulate the angular velocity, $i_M$ must be changed. Furthermore, the phase currents $i_{as}(t)$, $i_{bs}(t)$ and $i_{cs}(t)$, which are shifted by $\frac{2}{3}\pi$, are the functions of the electrical angular displacement $\theta_r$ (measured using the Hall-effect microsensors).

If the voltage-fed ICs are used, one changes the magnitude of voltages $u_{as}(t)$, $u_{bs}(t)$ and $u_{cs}(t)$. The angular displacement $\theta_r$ is needed to be measured (or estimated) in order to generate phase voltages.

In particular, the *abc* voltages needed to be supplied are

$$u_{as}(t)=\sqrt{2}u_M\cos\left(\theta_r+\varphi_u\right),$$
$$u_{bs}(t)=\sqrt{2}u_M\cos\left(\theta_r-\tfrac{2}{3}\pi+\varphi_u\right),$$
$$u_{cs}(t)=\sqrt{2}u_M\cos\left(\theta_r+\tfrac{2}{3}\pi+\varphi_u\right).$$

Neglecting the circuitry transients (assuming that inductances are negligible small), we have

$$u_{as}(t)=\sqrt{2}u_M\cos\theta_r,$$
$$u_{bs}(t)=\sqrt{2}u_M\cos\left(\theta_r-\tfrac{2}{3}\pi\right),$$
$$u_{cs}(t)=\sqrt{2}u_M\cos\left(\theta_r+\tfrac{2}{3}\pi\right).$$

Using a set of nonlinear differential equations (5.4.6), the block diagram is developed and documented in Figure 5.4.5 (here, $T_s=\dfrac{r_s\left(2L_{ss}-\overline{L}_m\right)}{2L_{ss}^2-L_{ss}\overline{L}_m-\overline{L}_m^2}$).

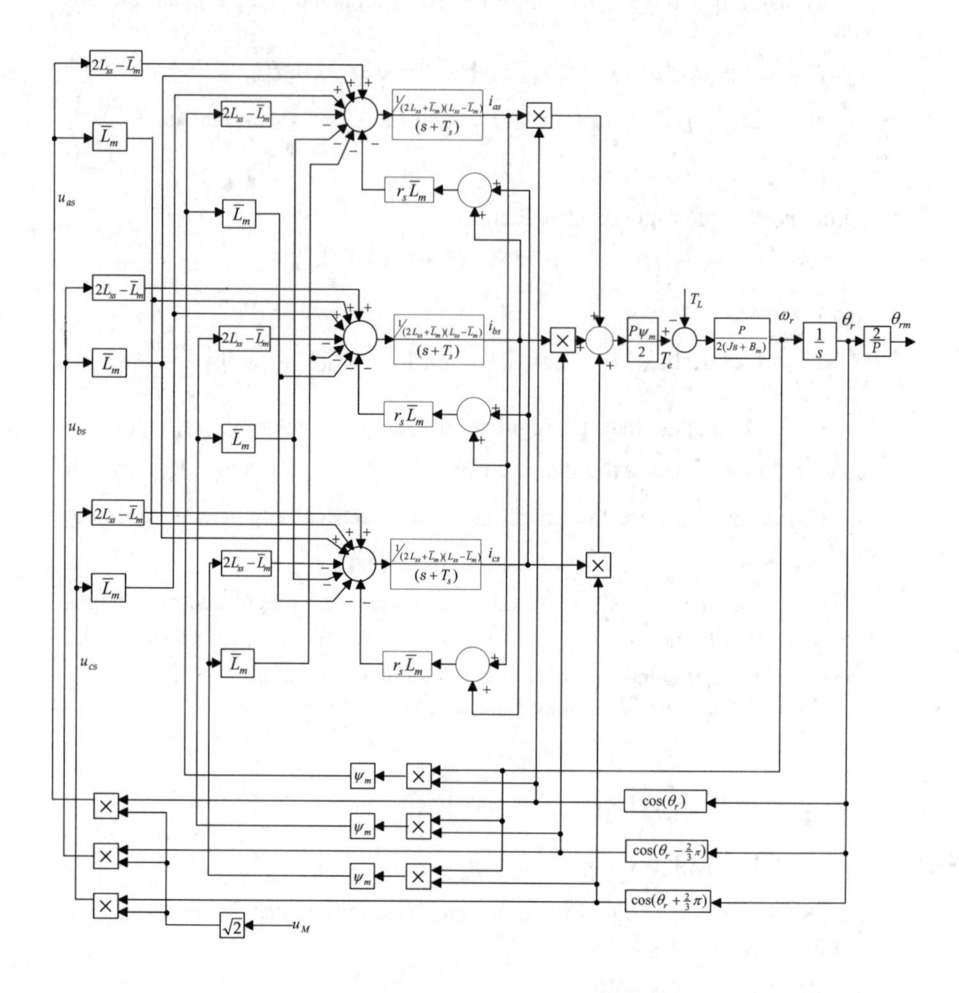


Figure 5.4.5 Block diagram of three-phase permanent-magnet synchronous micromotors controlled by applying three-phase balanced voltage set $u_{as}(t)=\sqrt{2}u_M\cos\theta_r$, $u_{bs}(t)=\sqrt{2}u_M\cos\left(\theta_r-\frac{2}{3}\pi\right)$ and $u_{cs}(t)=\sqrt{2}u_M\cos\left(\theta_r+\frac{2}{3}\pi\right)$.

## *The Lagrange Equations of Motion and Dynamics of Permanent-Magnet Synchronous Micromotors*

Having derived the mathematical model for three-phase permanent-magnet synchronous micromotors using Kirchhoff's voltage law (to model the circuitry dynamics), Newtonian mechanics (to model the *torsional-mechanical* dynamics), and the coenergy concept (to find the electromagnetic torque), let us attack the problem of model development using Lagrange's concept.

The generalized coordinates are the electric charges in the *abc* stator windings

$q_1 = \frac{i_{as}}{s}$, $\dot{q}_1 = i_{as}$, $q_2 = \frac{i_{bs}}{s}$, $\dot{q}_2 = i_{bs}$, $q_3 = \frac{i_{cs}}{s}$, $\dot{q}_3 = i_{cs}$, and the angular displacement $q_4 = \theta_r$, $\dot{q}_4 = \omega_r$.

The generalized forces are the applied voltages to the *abc* windings $Q_1 = u_{as}$, $Q_2 = u_{bs}$, $Q_3 = u_{cs}$ and the load torque $Q_4 = -T_L$.

The resulting Lagrange equations are

$$\frac{d}{dt}\left(\frac{\partial \Gamma}{\partial \dot{q}_1}\right) - \frac{\partial \Gamma}{\partial q_1} + \frac{\partial D}{\partial \dot{q}_1} + \frac{\partial \Pi}{\partial q_1} = Q_1,$$

$$\frac{d}{dt}\left(\frac{\partial \Gamma}{\partial \dot{q}_2}\right) - \frac{\partial \Gamma}{\partial q_2} + \frac{\partial D}{\partial \dot{q}_2} + \frac{\partial \Pi}{\partial q_2} = Q_2,$$

$$\frac{d}{dt}\left(\frac{\partial \Gamma}{\partial \dot{q}_3}\right) - \frac{\partial \Gamma}{\partial q_3} + \frac{\partial D}{\partial \dot{q}_3} + \frac{\partial \Pi}{\partial q_3} = Q_3,$$

$$\frac{d}{dt}\left(\frac{\partial \Gamma}{\partial \dot{q}_4}\right) - \frac{\partial \Gamma}{\partial q_4} + \frac{\partial D}{\partial \dot{q}_4} + \frac{\partial \Pi}{\partial q_4} = Q_4.$$

The total kinetic energy includes kinetic energies of electrical and mechanical systems. In particular,

$$\Gamma = \Gamma_E + \Gamma_M = \tfrac{1}{2}L_{asas}\dot{q}_1^2 + \tfrac{1}{2}\left(L_{asbs} + L_{bsas}\right)\dot{q}_1\dot{q}_2 + \tfrac{1}{2}\left(L_{ascs} + L_{csas}\right)\dot{q}_1\dot{q}_3$$
$$+ \tfrac{1}{2}L_{bsbs}\dot{q}_2^2 + \tfrac{1}{2}\left(L_{bscs} + L_{csbs}\right)\dot{q}_2\dot{q}_3 + \tfrac{1}{2}L_{cscs}\dot{q}_3^2$$
$$+ \psi_m\dot{q}_1 \sin q_4 + \psi_m\dot{q}_2 \sin\left(q_4 - \tfrac{2}{3}\pi\right) + \psi_m\dot{q}_3 \sin\left(q_4 + \tfrac{2}{3}\pi\right) + \tfrac{1}{2}J\dot{q}_4^2.$$

Then, we have

$$\frac{\partial \Gamma}{\partial q_1} = 0,$$

$$\frac{\partial \Gamma}{\partial \dot{q}_1} = L_{asas}\dot{q}_1 + \tfrac{1}{2}\left(L_{asbs} + L_{bsas}\right)\dot{q}_2 + \tfrac{1}{2}\left(L_{ascs} + L_{csas}\right)\dot{q}_3 + \psi_m \sin q_4,$$

$$\frac{\partial\Gamma}{\partial q_2}=0,$$

$$\frac{\partial\Gamma}{\partial\dot{q}_2}=\tfrac{1}{2}\left(L_{asbs}+L_{bsas}\right)\dot{q}_1+L_{bsbs}\dot{q}_2+\tfrac{1}{2}\left(L_{bscs}+L_{csbs}\right)\dot{q}_3+\psi_m\sin\left(q_4-\tfrac{2}{3}\pi\right),$$

$$\frac{\partial\Gamma}{\partial q_3}=0,$$

$$\frac{\partial\Gamma}{\partial\dot{q}_3}=\tfrac{1}{2}\left(L_{ascs}+L_{csas}\right)\dot{q}_1+\tfrac{1}{2}\left(L_{bscs}+L_{csbs}\right)\dot{q}_2+L_{cscs}\dot{q}_3+\psi_m\sin\left(q_4+\tfrac{2}{3}\pi\right),$$

$$\frac{\partial\Gamma}{\partial q_4}=\psi_m\dot{q}_1\cos q_4+\psi_m\dot{q}_2\cos\left(q_4-\tfrac{2}{3}\pi\right)+\psi_m\dot{q}_3\cos\left(q_4+\tfrac{2}{3}\pi\right),$$

$$\frac{\partial\Gamma}{\partial\dot{q}_4}=J\dot{q}_4.$$

The total potential energy is $\Pi=0$.

The total dissipated energy is found as a sum of the heat energy dissipated by the electrical system and the heat energy dissipated by the mechanical system. That is,

$$D=\tfrac{1}{2}\left(r_s\dot{q}_1^2+r_s\dot{q}_2^2+r_s\dot{q}_3^2+B_m\dot{q}_4^2\right).$$

One obtains

$$\frac{\partial D}{\partial\dot{q}_1}=r_s\dot{q}_1,\quad \frac{\partial D}{\partial\dot{q}_2}=r_s\dot{q}_2,\quad \frac{\partial D}{\partial\dot{q}_3}=r_s\dot{q}_3 \text{ and } \frac{\partial D}{\partial\dot{q}_4}=B_m\dot{q}_4.$$

Taking note of $\dot{q}_1=i_{as}$, $\dot{q}_2=i_{bs}$, $\dot{q}_3=i_{cs}$ and $\dot{q}_4=\omega_r$, the Lagrange equations lead us to four differential equations

$$L_{asas}\frac{di_{as}}{dt}+\tfrac{1}{2}\left(L_{asbs}+L_{bsas}\right)\frac{di_{bs}}{dt}+\tfrac{1}{2}\left(L_{ascs}+L_{csas}\right)\frac{di_{cs}}{dt}$$
$$+\psi_m\omega_r\cos\theta_r+r_si_{as}=u_{as},$$

$$\tfrac{1}{2}\left(L_{asbs}+L_{bsas}\right)\frac{di_{as}}{dt}+L_{bsbs}\frac{di_{bs}}{dt}+\tfrac{1}{2}\left(L_{bscs}+L_{csbs}\right)\frac{di_{cs}}{dt}$$
$$+\psi_m\omega_r\cos\left(\theta_r-\tfrac{2}{3}\pi\right)+r_si_{bs}=u_{bs},$$

$$\tfrac{1}{2}\left(L_{ascs}+L_{csas}\right)\frac{di_{as}}{dt}+\tfrac{1}{2}\left(L_{bscs}+L_{csbs}\right)\frac{di_{bs}}{dt}+L_{cscs}\frac{di_{cs}}{dt}$$
$$+\psi_m\omega_r\cos\left(\theta_r+\tfrac{2}{3}\pi\right)+r_si_{cs}=u_{cs},$$

$$J\frac{d^2\theta_r}{dt^2}-\psi_mi_{as}\cos\theta_r-\psi_mi_{bs}\cos\left(\theta_r-\tfrac{2}{3}\pi\right)-\psi_mi_{cs}\cos\left(\theta_r+\tfrac{2}{3}\pi\right)+B_m\frac{d\theta_r}{dt}=-T_L$$

For round-rotor permanent-magnet synchronous microtransducers, one obtains

$$\left(L_{ls}+\bar{L}_m\right)\frac{di_{as}}{dt}-\tfrac{1}{2}\bar{L}_m\frac{di_{bs}}{dt}-\tfrac{1}{2}\bar{L}_m\frac{di_{cs}}{dt}+\psi_m\omega_r\cos\theta_r+r_s i_{as}=u_{as},$$

$$-\tfrac{1}{2}\bar{L}_m\frac{di_{as}}{dt}+\left(L_{ls}+\bar{L}_m\right)\frac{di_{bs}}{dt}-\tfrac{1}{2}\bar{L}_m\frac{di_{cs}}{dt}+\psi_m\omega_r\cos\left(\theta_r-\tfrac{2}{3}\pi\right)+r_s i_{bs}=u_{bs},$$

$$-\tfrac{1}{2}\bar{L}_m\frac{di_{as}}{dt}-\tfrac{1}{2}\bar{L}_m\frac{di_{bs}}{dt}+\left(L_{ls}+\bar{L}_m\right)\frac{di_{cs}}{dt}+\psi_m\omega_r\cos\left(\theta_r+\tfrac{2}{3}\pi\right)+r_s i_{cs}=u_{cs},$$

$$J\frac{d\omega_r}{dt}+B_m\omega_r-\psi_m\left[i_{as}\cos\theta_r+i_{bs}\cos\left(\theta_r-\tfrac{2}{3}\pi\right)+i_{cs}\cos\left(\theta_r+\tfrac{2}{3}\pi\right)\right]=-T_L,$$

$$\frac{d\theta_r}{dt}=\omega_r.$$

From the fourth differential equation one finds that the electromagnetic torque as

$$T_e=\psi_m\left[i_{as}\cos\theta_r+i_{bs}\cos\left(\theta_r-\tfrac{2}{3}\pi\right)+i_{cs}\cos\left(\theta_r+\tfrac{2}{3}\pi\right)\right].$$

Differential equations in Cauchy's form result, as given in the state-space form by (5.4.6) for $P$-pole permanent-magnet synchronous micromotors. It was demonstrated that applying Lagrange's concept, a complete mathematical model for permanent-magnet synchronous microtransducers can be straightforwardly developed.

### *Three-Phase Permanent-Magnet Synchronous Microgenerators*

For permanent-magnet synchronous microgenerators, as shown in Figure 5.4.4, the mathematical model can be developed using Kirchhoff's second law

$$\mathbf{u}_{abcs}=-\mathbf{r}_s\mathbf{i}_{abcs}+\frac{d\boldsymbol{\psi}_{abcs}}{dt},\quad \begin{bmatrix}u_{as}\\u_{bs}\\u_{cs}\end{bmatrix}=-\begin{bmatrix}r_s&0&0\\0&r_s&0\\0&0&r_s\end{bmatrix}\begin{bmatrix}i_{as}\\i_{bs}\\i_{cs}\end{bmatrix}+\begin{bmatrix}\frac{d\psi_{as}}{dt}\\\frac{d\psi_{bs}}{dt}\\\frac{d\psi_{cs}}{dt}\end{bmatrix},$$

$$\boldsymbol{\psi}_{abcs}=-\mathbf{L}_s\mathbf{i}_{abcs}+\boldsymbol{\psi}_m$$

$$=-\begin{bmatrix}L_{ls}+\bar{L}_m-L_{\Delta m}\cos 2\theta_r & -\tfrac{1}{2}\bar{L}_m-L_{\Delta m}\cos 2\left(\theta_r-\tfrac{1}{3}\pi\right) & -\tfrac{1}{2}\bar{L}_m-L_{\Delta m}\cos 2\left(\theta_r+\tfrac{1}{3}\pi\right)\\ -\tfrac{1}{2}\bar{L}_m-L_{\Delta m}\cos 2\left(\theta_r-\tfrac{1}{3}\pi\right) & L_{ls}+\bar{L}_m-L_{\Delta m}\cos 2\left(\theta_r-\tfrac{2}{3}\pi\right) & -\tfrac{1}{2}\bar{L}_m-L_{\Delta m}\cos 2\theta_r\\ -\tfrac{1}{2}\bar{L}_m-L_{\Delta m}\cos 2\left(\theta_r+\tfrac{1}{3}\pi\right) & -\tfrac{1}{2}\bar{L}_m-L_{\Delta m}\cos 2\theta_r & L_{ls}+\bar{L}_m-L_{\Delta m}\cos 2\left(\theta_r+\tfrac{2}{3}\pi\right)\end{bmatrix}\begin{bmatrix}i_{as}\\i_{bs}\\i_{cs}\end{bmatrix}+\psi_m\begin{bmatrix}\sin\theta_r\\\sin\left(\theta_r-\tfrac{2}{3}\pi\right)\\\sin\left(\theta_r+\tfrac{2}{3}\pi\right)\end{bmatrix},$$

and Newton's second law of motion $-T_e-B_m\omega_{rm}+T_{pm}=J\dfrac{d^2\theta_{rm}}{dt^2}$, which gives

$$\frac{d\omega_{rm}}{dt}=\frac{1}{J}\left(-T_e-B_m\omega_{rm}+T_{pm}\right),\quad \frac{d\theta_{rm}}{dt}=\omega_{rm}.$$

The application of the results presented for the permanent-magnet synchronous micromotors results in the following set of differential equations

$$\frac{di_{as}}{dt}=-\frac{r_s\left(2L_{ss}-\bar{L}_m\right)}{2L_{ss}^2-L_{ss}\bar{L}_m-\bar{L}_m^2}i_{as}-\frac{r_s\bar{L}_m}{2L_{ss}^2-L_{ss}\bar{L}_m-\bar{L}_m^2}i_{bs}-\frac{r_s\bar{L}_m}{2L_{ss}^2-L_{ss}\bar{L}_m-\bar{L}_m^2}i_{cs}$$

$$+\frac{\psi_m\left(2L_{ss}-\bar{L}_m\right)}{2L_{ss}^2-L_{ss}\bar{L}_m-\bar{L}_m^2}\omega_r\cos\theta_r+\frac{\psi_m\bar{L}_m}{2L_{ss}^2-L_{ss}\bar{L}_m-\bar{L}_m^2}\omega_r\cos\left(\theta_r-\tfrac{2}{3}\pi\right)$$

$$+\frac{\psi_m\bar{L}_m}{2L_{ss}^2-L_{ss}\bar{L}_m-\bar{L}_m^2}\omega_r\cos\left(\theta_r+\tfrac{2}{3}\pi\right)$$

$$-\frac{2L_{ss}-\bar{L}_m}{2L_{ss}^2-L_{ss}\bar{L}_m-\bar{L}_m^2}u_{as}-\frac{\bar{L}_m}{2L_{ss}^2-L_{ss}\bar{L}_m-\bar{L}_m^2}u_{bs}-\frac{\bar{L}_m}{2L_{ss}^2-L_{ss}\bar{L}_m-\bar{L}_m^2}u_{cs},$$

$$\frac{di_{bs}}{dt}=-\frac{r_s\bar{L}_m}{2L_{ss}^2-L_{ss}\bar{L}_m-\bar{L}_m^2}i_{as}-\frac{r_s\left(2L_{ss}-\bar{L}_m\right)}{2L_{ss}^2-L_{ss}\bar{L}_m-\bar{L}_m^2}i_{bs}-\frac{r_s\bar{L}_m}{2L_{ss}^2-L_{ss}\bar{L}_m-\bar{L}_m^2}i_{cs}$$

$$+\frac{\psi_m\bar{L}_m}{2L_{ss}^2-L_{ss}\bar{L}_m-\bar{L}_m^2}\omega_r\cos\theta_r+\frac{\psi_m\left(2L_{ss}-\bar{L}_m\right)}{2L_{ss}^2-L_{ss}\bar{L}_m-\bar{L}_m^2}\omega_r\cos\left(\theta_r-\tfrac{2}{3}\pi\right)$$

$$+\frac{\psi_m\bar{L}_m}{2L_{ss}^2-L_{ss}\bar{L}_m-\bar{L}_m^2}\omega_r\cos\left(\theta_r+\tfrac{2}{3}\pi\right)$$

$$-\frac{\bar{L}_m}{2L_{ss}^2-L_{ss}\bar{L}_m-\bar{L}_m^2}u_{as}-\frac{2L_{ss}-\bar{L}_m}{2L_{ss}^2-L_{ss}\bar{L}_m-\bar{L}_m^2}u_{bs}-\frac{\bar{L}_m}{2L_{ss}^2-L_{ss}\bar{L}_m-\bar{L}_m^2}u_{cs},$$

$$\frac{di_{cs}}{dt}=-\frac{r_s\bar{L}_m}{2L_{ss}^2-L_{ss}\bar{L}_m-\bar{L}_m^2}i_{as}-\frac{r_s\bar{L}_m}{2L_{ss}^2-L_{ss}\bar{L}_m-\bar{L}_m^2}i_{bs}-\frac{r_s\left(2L_{ss}-\bar{L}_m\right)}{2L_{ss}^2-L_{ss}\bar{L}_m-\bar{L}_m^2}i_{cs}$$

$$+\frac{\psi_m\bar{L}_m}{2L_{ss}^2-L_{ss}\bar{L}_m-\bar{L}_m^2}\omega_r\cos\theta_r+\frac{\psi_m\bar{L}_m}{2L_{ss}^2-L_{ss}\bar{L}_m-\bar{L}_m^2}\omega_r\cos\left(\theta_r-\tfrac{2}{3}\pi\right)$$

$$+\frac{\psi_m\left(2L_{ss}-\bar{L}_m\right)}{2L_{ss}^2-L_{ss}\bar{L}_m-\bar{L}_m^2}\omega_r\cos\left(\theta_r+\tfrac{2}{3}\pi\right)$$

$$-\frac{\bar{L}_m}{2L_{ss}^2-L_{ss}\bar{L}_m-\bar{L}_m^2}u_{as}-\frac{\bar{L}_m}{2L_{ss}^2-L_{ss}\bar{L}_m-\bar{L}_m^2}u_{bs}-\frac{2L_{ss}-\bar{L}_m}{2L_{ss}^2-L_{ss}\bar{L}_m-\bar{L}_m^2}u_{cs},$$

$$\frac{d\omega_r}{dt}=-\frac{P^2\psi_m}{4J}\left(i_{as}\cos\theta_r+i_{bs}\cos\left(\theta_r-\tfrac{2}{3}\pi\right)+i_{cs}\cos\left(\theta_r+\tfrac{2}{3}\pi\right)\right)-\frac{B_m}{J}\omega_r+\frac{P}{2J}T_{pm},$$

$$\frac{d\theta_r}{dt}=\omega_r\,. \tag{5.4.7}$$

In the state-space form, from (5.4.7), we have the following mathematical model of three-phase permanent-magnet synchronous microgenerators

$$
\begin{bmatrix} \frac{di_{as}}{dt} \\ \frac{di_{bs}}{dt} \\ \frac{di_{cs}}{dt} \\ \frac{d\omega_r}{dt} \\ \frac{d\theta_r}{dt} \end{bmatrix}
=
\begin{bmatrix}
-\frac{r_s\left(2L_{ss}-\bar{L}_m\right)}{2L_{ss}^2-L_{ss}\bar{L}_m-\bar{L}_m^2} & -\frac{r_s\bar{L}_m}{2L_{ss}^2-L_{ss}\bar{L}_m-\bar{L}_m^2} & -\frac{r_s\bar{L}_m}{2L_{ss}^2-L_{ss}\bar{L}_m-\bar{L}_m^2} & 0 & 0 \\
-\frac{r_s\bar{L}_m}{2L_{ss}^2-L_{ss}\bar{L}_m-\bar{L}_m^2} & -\frac{r_s\left(2L_{ss}-\bar{L}_m\right)}{2L_{ss}^2-L_{ss}\bar{L}_m-\bar{L}_m^2} & -\frac{r_s\bar{L}_m}{2L_{ss}^2-L_{ss}\bar{L}_m-\bar{L}_m^2} & 0 & 0 \\
-\frac{r_s\bar{L}_m}{2L_{ss}^2-L_{ss}\bar{L}_m-\bar{L}_m^2} & -\frac{r_s\bar{L}_m}{2L_{ss}^2-L_{ss}\bar{L}_m-\bar{L}_m^2} & -\frac{r_s\left(2L_{ss}-\bar{L}_m\right)}{2L_{ss}^2-L_{ss}\bar{L}_m-\bar{L}_m^2} & 0 & 0 \\
0 & 0 & 0 & -\frac{B_m}{J} & 0 \\
0 & 0 & 0 & 1 & 0
\end{bmatrix}
\begin{bmatrix} i_{as} \\ i_{bs} \\ i_{cs} \\ \omega_r \\ \theta_r \end{bmatrix}
$$

$$
+\begin{bmatrix}
\frac{\psi_m(2L_{ss}-\bar{L}_m)}{2L_{ss}^2-L_{ss}\bar{L}_m-\bar{L}_m^2}\omega_r & \frac{\psi_m\bar{L}_m}{2L_{ss}^2-L_{ss}\bar{L}_m-\bar{L}_m^2}\omega_r & \frac{\psi_m\bar{L}_m}{2L_{ss}^2-L_{ss}\bar{L}_m-\bar{L}_m^2}\omega_r \\
\frac{\psi_m\bar{L}_m}{2L_{ss}^2-L_{ss}\bar{L}_m-\bar{L}_m^2}\omega_r & \frac{\psi_m(2L_{ss}-\bar{L}_m)}{2L_{ss}^2-L_{ss}\bar{L}_m-\bar{L}_m^2}\omega_r & \frac{\psi_m\bar{L}_m}{2L_{ss}^2-L_{ss}\bar{L}_m-\bar{L}_m^2}\omega_r \\
\frac{\psi_m\bar{L}_m}{2L_{ss}^2-L_{ss}\bar{L}_m-\bar{L}_m^2}\omega_r & \frac{\psi_m\bar{L}_m}{2L_{ss}^2-L_{ss}\bar{L}_m-\bar{L}_m^2}\omega_r & \frac{\psi_m(2L_{ss}-\bar{L}_m)}{2L_{ss}^2-L_{ss}\bar{L}_m-\bar{L}_m^2}\omega_r \\
-\frac{P^2\psi_m}{4J}i_{as} & -\frac{P^2\psi_m}{4J}i_{bs} & -\frac{P^2\psi_m}{4J}i_{cs} \\
0 & 0 & 0
\end{bmatrix}
\begin{bmatrix} \cos\theta_r \\ \cos\left(\theta_r-\frac{2}{3}\pi\right) \\ \cos\left(\theta_r+\frac{2}{3}\pi\right) \end{bmatrix}
$$

$$
-\begin{bmatrix}
\frac{2L_{ss}-\bar{L}_m}{2L_{ss}^2-L_{ss}\bar{L}_m-\bar{L}_m^2} & \frac{\bar{L}_m}{2L_{ss}^2-L_{ss}\bar{L}_m-\bar{L}_m^2} & \frac{\bar{L}_m}{2L_{ss}^2-L_{ss}\bar{L}_m-\bar{L}_m^2} \\
\frac{\bar{L}_m}{2L_{ss}^2-L_{ss}\bar{L}_m-\bar{L}_m^2} & \frac{2L_{ss}-\bar{L}_m}{2L_{ss}^2-L_{ss}\bar{L}_m-\bar{L}_m^2} & \frac{\bar{L}_m}{2L_{ss}^2-L_{ss}\bar{L}_m-\bar{L}_m^2} \\
\frac{\bar{L}_m}{2L_{ss}^2-L_{ss}\bar{L}_m-\bar{L}_m^2} & \frac{\bar{L}_m}{2L_{ss}^2-L_{ss}\bar{L}_m-\bar{L}_m^2} & \frac{2L_{ss}-\bar{L}_m}{2L_{ss}^2-L_{ss}\bar{L}_m-\bar{L}_m^2} \\
0 & 0 & 0 \\
0 & 0 & 0
\end{bmatrix}
\begin{bmatrix} u_{as} \\ u_{bs} \\ u_{cs} \end{bmatrix}
+\begin{bmatrix} 0 \\ 0 \\ 0 \\ \frac{P}{2J} \\ 0 \end{bmatrix} T_{pm}.
$$

One concludes that the lumped-parameter nonlinear mathematical model of permanent-magnet synchronous microgenerators is derived to be used in analysis, modeling, and control.

## *Mathematical Models of Permanent-Magnet Synchronous Micromachines in the Arbitrary, Rotor, and Synchronous Reference Frames*

### *Arbitrary Reference Frame*

By fixing the reference frame with the rotor and making use of the *direct* Park transformations

$$\mathbf{u}_{qd0s}=\mathbf{K}_s\mathbf{u}_{abcs},\ \mathbf{i}_{qd0s}=\mathbf{K}_s\mathbf{i}_{abcs},\ \boldsymbol{\psi}_{qd0s}=\mathbf{K}_s\boldsymbol{\psi}_{abcs},$$

$$\mathbf{K}_s = \frac{2}{3}\begin{bmatrix} \cos\theta & \cos\left(\theta - \frac{2}{3}\pi\right) & \cos\left(\theta + \frac{2}{3}\pi\right) \\ \sin\theta & \sin\left(\theta - \frac{2}{3}\pi\right) & \sin\left(\theta + \frac{2}{3}\pi\right) \\ \frac{1}{2} & \frac{1}{2} & \frac{1}{2} \end{bmatrix},$$

circuitry differential equation (5.4.1) $\mathbf{u}_{abcs} = \mathbf{r}_s \mathbf{i}_{abcs} + \frac{d\boldsymbol{\psi}_{abcs}}{dt}$ is rewritten in the $qd0$ variables as

$$\mathbf{K}_s^{-1}\mathbf{u}_{qd0s} = \mathbf{r}_s \mathbf{K}_s^{-1}\mathbf{i}_{qd0s} + \frac{d\left(\mathbf{K}_s^{-1}\boldsymbol{\psi}_{qd0s}\right)}{dt},$$

$$\mathbf{K}_s^{-1} = \begin{bmatrix} \cos\theta & \sin\theta & 1 \\ \cos\left(\theta - \frac{2}{3}\pi\right) & \sin\left(\theta - \frac{2}{3}\pi\right) & 1 \\ \cos\left(\theta + \frac{2}{3}\pi\right) & \sin\left(\theta + \frac{2}{3}\pi\right) & 1 \end{bmatrix}.$$

Multiplying left and right sides by $\mathbf{K}_s$, one obtains

$$\mathbf{K}_s \mathbf{K}_s^{-1}\mathbf{u}_{qdos} = \mathbf{K}_s \mathbf{r}_s \mathbf{K}_s^{-1}\mathbf{i}_{qd0s} + \mathbf{K}_s \frac{d\mathbf{K}_s^{-1}}{dt}\boldsymbol{\psi}_{qd0s} + \mathbf{K}_s \mathbf{K}_s^{-1}\frac{d\boldsymbol{\psi}_{qd0s}}{dt}.$$

The matrix $\mathbf{r}_s$ is diagonal, and thus $\mathbf{K}_s \mathbf{r}_s \mathbf{K}_s^{-1} = \mathbf{r}_s$.

From $\frac{d\mathbf{K}_s^{-1}}{dt} = \omega \begin{bmatrix} -\sin\theta & \cos\theta & 0 \\ -\sin\left(\theta - \frac{2}{3}\pi\right) & \cos\left(\theta - \frac{2}{3}\pi\right) & 0 \\ -\sin\left(\theta + \frac{2}{3}\pi\right) & \cos\left(\theta + \frac{2}{3}\pi\right) & 0 \end{bmatrix}$, we have

$$\mathbf{K}_s \frac{d\mathbf{K}_s^{-1}}{dt} = \omega \begin{bmatrix} 0 & 1 & 0 \\ -1 & 0 & 0 \\ 0 & 0 & 0 \end{bmatrix}.$$

Hence, (5.4.1) is rewritten in the $qd0$ variables as

$$\mathbf{u}_{qd0s} = \mathbf{r}_s \mathbf{i}_{qd0s} + \omega \begin{bmatrix} \psi_{ds} \\ -\psi_{qs} \\ 0 \end{bmatrix} + \frac{d\boldsymbol{\psi}_{qd0s}}{dt}. \qquad (5.4.8)$$

Using the Park transformation, the *quadrature*-, *direct*-, and *zero*-axis components of stator flux linkages are found as

$$\boldsymbol{\psi}_{qd0s} = \mathbf{K}_s \boldsymbol{\psi}_{abcs},$$

where

$$\boldsymbol{\psi}_{abcs} = \mathbf{L}_s \mathbf{i}_{abcs} + \boldsymbol{\psi}_m = \begin{bmatrix} L_{ls} + \bar{L}_m & -\frac{1}{2}\bar{L}_m & -\frac{1}{2}\bar{L}_m \\ -\frac{1}{2}\bar{L}_m & L_{ls} + \bar{L}_m & -\frac{1}{2}\bar{L}_m \\ -\frac{1}{2}\bar{L}_m & -\frac{1}{2}\bar{L}_m & L_{ls} + \bar{L}_m \end{bmatrix} \begin{bmatrix} i_{as} \\ i_{bs} \\ i_{cs} \end{bmatrix} + \psi_m \begin{bmatrix} \sin\theta_r \\ \sin\left(\theta_r - \frac{2}{3}\pi\right) \\ \sin\left(\theta_r + \frac{2}{3}\pi\right) \end{bmatrix}.$$

Hence,

$$\boldsymbol{\Psi}_{qd0s} = \mathbf{K}_s \mathbf{L}_s \mathbf{K}_s^{-1} \mathbf{i}_{qd0s} + \mathbf{K}_s \boldsymbol{\Psi}_m, \tag{5.4.9}$$

where $\mathbf{K}_s \mathbf{L}_s \mathbf{K}_s^{-1} = \begin{bmatrix} L_{ls} + \frac{3}{2}\bar{L}_m & 0 & 0 \\ 0 & L_{ls} + \frac{3}{2}\bar{L}_m & 0 \\ 0 & 0 & L_{ls} \end{bmatrix}$;

$$\mathbf{K}_s \boldsymbol{\Psi}_m = \frac{2}{3}\begin{bmatrix} \cos\theta & \cos\left(\theta - \frac{2}{3}\pi\right) & \cos\left(\theta + \frac{2}{3}\pi\right) \\ \sin\theta & \sin\left(\theta - \frac{2}{3}\pi\right) & \sin\left(\theta + \frac{2}{3}\pi\right) \\ \frac{1}{2} & \frac{1}{2} & \frac{1}{2} \end{bmatrix} \psi_m \begin{bmatrix} \sin\theta_r \\ \sin\left(\theta_r - \frac{2}{3}\pi\right) \\ \sin\left(\theta_r + \frac{2}{3}\pi\right) \end{bmatrix} = \psi_m \begin{bmatrix} -\sin(\theta - \theta_r) \\ \cos(\theta - \theta_r) \\ 0 \end{bmatrix}.$$

From (5.4.9) we obtain

$$\boldsymbol{\Psi}_{qd0s} = \begin{bmatrix} L_{ls} + \frac{3}{2}\bar{L}_m & 0 & 0 \\ 0 & L_{ls} + \frac{3}{2}\bar{L}_m & 0 \\ 0 & 0 & L_{ls} \end{bmatrix} \mathbf{i}_{qd0s} + \psi_m \begin{bmatrix} -\sin(\theta - \theta_r) \\ \cos(\theta - \theta_r) \\ 0 \end{bmatrix}.$$

Using (5.4.8) one finds

$$\mathbf{u}_{qd0s} = \mathbf{r}_s \mathbf{i}_{qd0s} + \omega \begin{bmatrix} \psi_{ds} \\ -\psi_{qs} \\ 0 \end{bmatrix} + \begin{bmatrix} L_{ls} + \frac{3}{2}\bar{L}_m & 0 & 0 \\ 0 & L_{ls} + \frac{3}{2}\bar{L}_m & 0 \\ 0 & 0 & L_{ls} \end{bmatrix} \frac{d\mathbf{i}_{qd0s}}{dt} + \psi_m \frac{d\begin{bmatrix} -\sin(\theta - \theta_r) \\ \cos(\theta - \theta_r) \\ 0 \end{bmatrix}}{dt}$$

Three differential equations which model the permanent-magnet circuitry dynamics in the *arbitrary* reference frame are found as

$$\mathbf{u}_{qd0s} = \mathbf{r}_s \mathbf{i}_{qd0s} + \omega \begin{bmatrix} \psi_{ds} \\ -\psi_{qs} \\ 0 \end{bmatrix} + \begin{bmatrix} L_{ls} + \frac{3}{2}\bar{L}_m & 0 & 0 \\ 0 & L_{ls} + \frac{3}{2}\bar{L}_m & 0 \\ 0 & 0 & L_{ls} \end{bmatrix} \frac{d\mathbf{i}_{qd0s}}{dt}$$

$$+ \psi_m \frac{d\begin{bmatrix} -\sin(\theta - \theta_r) \\ \cos(\theta - \theta_r) \\ 0 \end{bmatrix}}{dt}.$$

### *Rotor Reference Frame*

The electrical angular velocity is equal to the synchronous angular velocity. We assign the angular velocity of the reference frame to be $\omega = \omega_r = \omega_e$. Then, taking note of $\theta = \theta_r$, we have the Park transformation matrix

$$\mathbf{K}_s^r = \frac{2}{3}\begin{bmatrix} \cos\theta_r & \cos\left(\theta_r - \frac{2}{3}\pi\right) & \cos\left(\theta_r + \frac{2}{3}\pi\right) \\ \sin\theta_r & \sin\left(\theta_r - \frac{2}{3}\pi\right) & \sin\left(\theta_r + \frac{2}{3}\pi\right) \\ \frac{1}{2} & \frac{1}{2} & \frac{1}{2} \end{bmatrix}.$$

One finds

$$\mathbf{K}_s^r \boldsymbol{\psi}_m = \frac{2}{3}\begin{bmatrix} \cos\theta_r & \cos\left(\theta_r - \frac{2}{3}\pi\right) & \cos\left(\theta_r + \frac{2}{3}\pi\right) \\ \sin\theta_r & \sin\left(\theta_r - \frac{2}{3}\pi\right) & \sin\left(\theta_r + \frac{2}{3}\pi\right) \\ \frac{1}{2} & \frac{1}{2} & \frac{1}{2} \end{bmatrix} \psi_m \begin{bmatrix} \sin\theta_r \\ \sin\left(\theta_r - \frac{2}{3}\pi\right) \\ \sin\left(\theta_r + \frac{2}{3}\pi\right) \end{bmatrix} = \begin{bmatrix} 0 \\ \psi_m \\ 0 \end{bmatrix}.$$

From (5.4.9) we have

$$\boldsymbol{\psi}_{qd0s}^r = \begin{bmatrix} L_{ls} + \frac{3}{2}\bar{L}_m & 0 & 0 \\ 0 & L_{ls} + \frac{3}{2}\bar{L}_m & 0 \\ 0 & 0 & L_{ls} \end{bmatrix} \mathbf{i}_{qd0s}^r + \begin{bmatrix} 0 \\ \psi_m \\ 0 \end{bmatrix}.$$

In expanded form, the *quadrature*, *direct*, and *zero* flux linkages are found to be

$$\psi_{qs}^r = \left(L_{ls} + \tfrac{3}{2}\bar{L}_m\right) i_{qs}^r,$$

$$\psi_{ds}^r = \left(L_{ls} + \tfrac{3}{2}\bar{L}_m\right) i_{ds}^r + \psi_m,$$

$$\psi_{0s}^r = L_{ls} i_{0s}^r.$$

In the rotor reference frame using (5.4.8), one finds

$$\frac{di_{qs}^r}{dt} = -\frac{r_s}{L_{ls} + \frac{3}{2}\bar{L}_m} i_{qs}^r - \frac{\psi_m}{L_{ls} + \frac{3}{2}\bar{L}_m}\omega_r - i_{ds}^r \omega_r + \frac{1}{L_{ls} + \frac{3}{2}\bar{L}_m} u_{qs}^r,$$

$$\frac{di_{ds}^r}{dt} = -\frac{r_s}{L_{ls} + \frac{3}{2}\bar{L}_m} i_{ds}^r + i_{qs}^r \omega_r + \frac{1}{L_{ls} + \frac{3}{2}\bar{L}_m} u_{ds}^r,$$

$$\frac{di_{0s}^r}{dt} = -\frac{r_s}{L_{ls}} i_{0s}^r + \frac{1}{L_{ls}} u_{0s}^r. \qquad (5.4.10)$$

The electromagnetic torque

$$T_e = \frac{P\psi_m}{2}\left[i_{as}\cos\theta_r + i_{bs}\cos\left(\theta_r - \tfrac{2}{3}\pi\right) + i_{cs}\cos\left(\theta_r + \tfrac{2}{3}\pi\right)\right]$$

should be found in terms of the *quadrature*, *direct* and *zero* currents.

Using the Park transformation

$$\begin{bmatrix} i_{as} \\ i_{bs} \\ i_{cs} \end{bmatrix} = \begin{bmatrix} \cos\theta_r & \sin\theta_r & 1 \\ \cos\left(\theta_r - \frac{2}{3}\pi\right) & \sin\left(\theta_r - \frac{2}{3}\pi\right) & 1 \\ \cos\left(\theta_r + \frac{2}{3}\pi\right) & \sin\left(\theta_r + \frac{2}{3}\pi\right) & 1 \end{bmatrix} \begin{bmatrix} i_{qs}^r \\ i_{ds}^r \\ i_{0s}^r \end{bmatrix},$$

and substituting

$$i_{as} = \cos\theta_r i_{qs}^r + \sin\theta_r i_{ds}^r + i_{0s}^r,$$
$$i_{bs} = \cos\left(\theta_r - \tfrac{2}{3}\pi\right) i_{qs}^r + \sin\left(\theta_r - \tfrac{2}{3}\pi\right) i_{ds}^r + i_{0s}^r,$$
$$i_{cs} = \cos\left(\theta_r + \tfrac{2}{3}\pi\right) i_{qs}^r + \sin\left(\theta_r + \tfrac{2}{3}\pi\right) i_{ds}^r + i_{0s}^r$$

in the expression for $T_e$, one finds

$$T_e = \frac{3P\psi_m}{4} i_{qs}^r.$$

For *P*-pole permanent-magnet synchronous micromotors, the *torsional-mechanical* dynamics is

$$\frac{d\omega_r}{dt} = \frac{3P^2\psi_m}{8J} i_{qs}^r - \frac{B_m}{J}\omega_r - \frac{P}{2J}T_L,$$
$$\frac{d\theta_r}{dt} = \omega_r. \tag{5.4.11}$$

Augmenting differential equations (5.4.10) and (5.4.11), we have the lumped-parameter mathematical model of three-phase permanent-magnet synchronous micromotors in the rotor reference frame

$$\frac{di_{qs}^r}{dt} = -\frac{r_s}{L_{ls} + \frac{3}{2}\bar{L}_m} i_{qs}^r - \frac{\psi_m}{L_{ls} + \frac{3}{2}\bar{L}_m}\omega_r - i_{ds}^r\omega_r + \frac{1}{L_{ls} + \frac{3}{2}\bar{L}_m} u_{qs}^r,$$

$$\frac{di_{ds}^r}{dt} = -\frac{r_s}{L_{ls} + \frac{3}{2}\bar{L}_m} i_{ds}^r + i_{qs}^r\omega_r + \frac{1}{L_{ls} + \frac{3}{2}\bar{L}_m} u_{ds}^r,$$

$$\frac{di_{0s}^r}{dt} = -\frac{r_s}{L_{ls}} i_{0s}^r + \frac{1}{L_{ls}} u_{0s}^r,$$

$$\frac{d\omega_r}{dt} = \frac{3P^2\psi_m}{8J} i_{qs}^r - \frac{B_m}{J}\omega_r - \frac{P}{2J}T_L,$$
$$\frac{d\theta_r}{dt} = \omega_r. \tag{5.4.12}$$

In the state-space form, the mathematical model of permanent-magnet synchronous micromotors in the rotor reference frame is given by

$$\begin{bmatrix} \frac{di_{qs}^r}{dt} \\ \frac{di_{ds}^r}{dt} \\ \frac{di_{0s}^r}{dt} \\ \frac{d\omega_r}{dt} \\ \frac{d\theta_r}{dt} \end{bmatrix} = \begin{bmatrix} -\frac{r_s}{L_{ls}+\frac{3}{2}\bar{L}_m} & 0 & 0 & -\frac{\psi_m}{L_{ls}+\frac{3}{2}\bar{L}_m} & 0 \\ 0 & -\frac{r_s}{L_{ls}+\frac{3}{2}\bar{L}_m} & 0 & 0 & 0 \\ 0 & 0 & -\frac{r_s}{L_{ls}} & 0 & 0 \\ \frac{3P^2\psi_m}{8J} & 0 & 0 & -\frac{B_m}{J} & 0 \\ 0 & 0 & 0 & 1 & 0 \end{bmatrix} \begin{bmatrix} i_{qs}^r \\ i_{ds}^r \\ i_{0s}^r \\ \omega_r \\ \theta_r \end{bmatrix}$$

$$+ \begin{bmatrix} -i_{ds}^r\omega_r \\ i_{qs}^r\omega_r \\ 0 \\ 0 \\ 0 \end{bmatrix} + \begin{bmatrix} \frac{1}{L_{ls}+\frac{3}{2}\bar{L}_m} & 0 & 0 \\ 0 & \frac{1}{L_{ls}+\frac{3}{2}\bar{L}_m} & 0 \\ 0 & 0 & \frac{1}{L_{ls}} \\ 0 & 0 & 0 \\ 0 & 0 & 0 \end{bmatrix} \begin{bmatrix} u_{qs}^r \\ u_{ds}^r \\ u_{0s}^r \end{bmatrix} - \begin{bmatrix} 0 \\ 0 \\ 0 \\ \frac{P}{2J} \\ 0 \end{bmatrix} T_L .$$

A balanced three-phase current set, to be fed to the stator windings, is

$i_{as}(t) = \sqrt{2} i_M \cos\theta_r$,

$i_{bs}(t) = \sqrt{2} i_M \cos\left(\theta_r - \frac{2}{3}\pi\right)$,

$i_{cs}(t) = \sqrt{2} i_M \cos\left(\theta_r + \frac{2}{3}\pi\right)$.

Using the *direct* Park transformation

$$\begin{bmatrix} i_{qs}^r \\ i_{ds}^r \\ i_{0s}^r \end{bmatrix} = \frac{2}{3} \begin{bmatrix} \cos\theta_r & \cos\left(\theta_r - \frac{2}{3}\pi\right) & \cos\left(\theta_r + \frac{2}{3}\pi\right) \\ \sin\theta_r & \sin\left(\theta_r - \frac{2}{3}\pi\right) & \sin\left(\theta_r + \frac{2}{3}\pi\right) \\ \frac{1}{2} & \frac{1}{2} & \frac{1}{2} \end{bmatrix} \begin{bmatrix} i_{as} \\ i_{bs} \\ i_{cs} \end{bmatrix},$$

one obtains the *quadrature*, *direct* and *zero* currents to regulate the angular velocity of permanent-magnet synchronous micromotors and guarantee the balanced operating conditions. We have

$$\begin{bmatrix} i_{qs}^r \\ i_{ds}^r \\ i_{0s}^r \end{bmatrix} = \frac{2}{3} \begin{bmatrix} \cos\theta_r & \cos\left(\theta_r - \frac{2}{3}\pi\right) & \cos\left(\theta_r + \frac{2}{3}\pi\right) \\ \sin\theta_r & \sin\left(\theta_r - \frac{2}{3}\pi\right) & \sin\left(\theta_r + \frac{2}{3}\pi\right) \\ \frac{1}{2} & \frac{1}{2} & \frac{1}{2} \end{bmatrix} \begin{bmatrix} \sqrt{2} i_M \cos\theta_r \\ \sqrt{2} i_M \cos\left(\theta_r - \frac{2}{3}\pi\right) \\ \sqrt{2} i_M \cos\left(\theta_r + \frac{2}{3}\pi\right) \end{bmatrix}.$$

Hence, one obtains

$i_{qs}^r(t) = \sqrt{2} i_M$, $i_{ds}^r(t) = 0$, $i_{0s}^r(t) = 0$.

Due to the self-inductances, the *abc* voltages should be supplied with advanced phase shifting. One supplies the following phase voltages

$$u_{as}(t)=\sqrt{2}u_M\cos\left(\theta_r+\varphi_u\right),\ u_{bs}(t)=\sqrt{2}u_M\cos\left(\theta_r-\tfrac{2}{3}\pi+\varphi_u\right),$$

$$u_{cs}(t)=\sqrt{2}u_M\cos\left(\theta_r+\tfrac{2}{3}\pi+\varphi_u\right).$$

Taking note of the *direct* Park transformation

$$\begin{bmatrix}u_{qs}^r\\u_{ds}^r\\u_{0s}^r\end{bmatrix}=\frac{2}{3}\begin{bmatrix}\cos\theta_r & \cos\left(\theta_r-\frac{2}{3}\pi\right) & \cos\left(\theta_r+\frac{2}{3}\pi\right)\\ \sin\theta_r & \sin\left(\theta_r-\frac{2}{3}\pi\right) & \sin\left(\theta_r+\frac{2}{3}\pi\right)\\ \frac{1}{2} & \frac{1}{2} & \frac{1}{2}\end{bmatrix}\begin{bmatrix}u_{as}\\u_{bs}\\u_{cs}\end{bmatrix},$$

one finds

$$\begin{bmatrix}u_{qs}^r\\u_{ds}^r\\u_{0s}^r\end{bmatrix}=\frac{2}{3}\begin{bmatrix}\cos\theta_r & \cos\left(\theta_r-\frac{2}{3}\pi\right) & \cos\left(\theta_r+\frac{2}{3}\pi\right)\\ \sin\theta_r & \sin\left(\theta_r-\frac{2}{3}\pi\right) & \sin\left(\theta_r+\frac{2}{3}\pi\right)\\ \frac{1}{2} & \frac{1}{2} & \frac{1}{2}\end{bmatrix}\begin{bmatrix}\sqrt{2}u_M\cos\left(\theta_r+\varphi_u\right)\\ \sqrt{2}u_M\cos\left(\theta_r-\frac{2}{3}\pi+\varphi_u\right)\\ \sqrt{2}u_M\cos\left(\theta_r+\frac{2}{3}\pi+\varphi_u\right)\end{bmatrix}.$$

Using the trigonometric identities, we have

$$u_{qs}^r(t)=\sqrt{2}u_M\cos\varphi_u,\ u_{ds}^r(t)=-\sqrt{2}u_M\sin\varphi_u,\ u_{0s}^r(t)=0.$$

Due to small inductances, $\varphi_u\approx 0$, and the following voltages must be applied

$$u_{qs}^r(t)=\sqrt{2}u_M,\ u_{ds}^r(t)=0,\ u_{0s}^r(t)=0.$$

To visualize the results, an *s*-domain block diagram in the *qd*0 variables is developed using (5.4.12), see Figure 5.4.6.

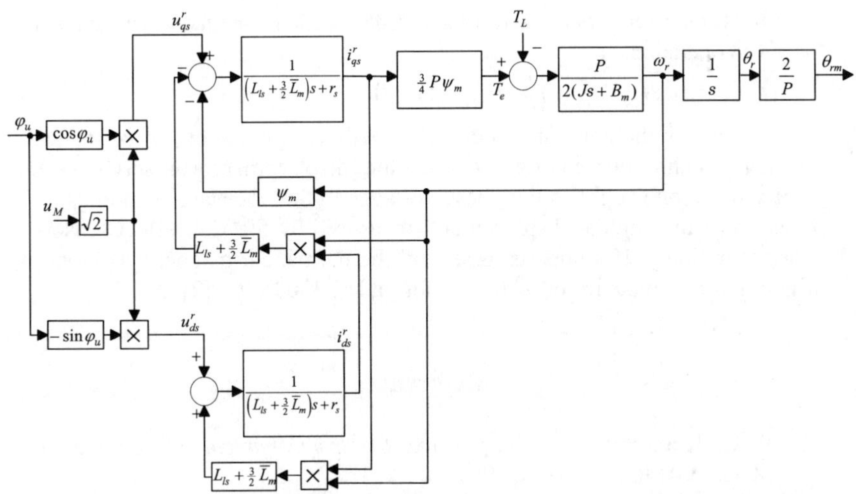


Figure 5.4.6 *s*-domain block diagram of permanent-magnet synchronous micromotors in the rotor reference frame.

*Synchronous Reference Frame*

Analyzing permanent-magnet synchronous microtransducers in the synchronous reference frame, one specifies the angular velocity of the reference frame to be $\omega = \omega_e$. Hence, $\theta = \theta_e$, and the Park transformation matrix is given as $\mathbf{K}_s^e = \frac{2}{3}\begin{bmatrix} \cos\theta_e & \cos\left(\theta_e - \frac{2}{3}\pi\right) & \cos\left(\theta_e + \frac{2}{3}\pi\right) \\ \sin\theta_e & \sin\left(\theta_e - \frac{2}{3}\pi\right) & \sin\left(\theta_e + \frac{2}{3}\pi\right) \\ \frac{1}{2} & \frac{1}{2} & \frac{1}{2} \end{bmatrix}$.

Substituting $\omega_r = \omega_e$ in (5.4.12) we have the following system of differential equations which model the permanent-magnet micromotor dynamics in the synchronous reference frame

$$\frac{di_{qs}^e}{dt} = -\frac{r_s}{L_{ls} + \frac{3}{2}\bar{L}_m} i_{qs}^e - \frac{\psi_m}{L_{ls} + \frac{3}{2}\bar{L}_m}\omega_r - i_{ds}^e\omega_r + \frac{1}{L_{ls} + \frac{3}{2}\bar{L}_m} u_{qs}^e,$$

$$\frac{di_{ds}^e}{dt} = -\frac{r_s}{L_{ls} + \frac{3}{2}\bar{L}_m} i_{ds}^e + i_{qs}^e\omega_r + \frac{1}{L_{ls} + \frac{3}{2}\bar{L}_m} u_{ds}^e,$$

$$\frac{di_{0s}^e}{dt} = -\frac{r_s}{L_{ls}} i_{0s}^e + \frac{1}{L_{ls}} u_{0s}^e,$$

$$\frac{d\omega_r}{dt} = \frac{3P^2\psi_m}{8J} i_{qs}^e - \frac{B_m}{J}\omega_r - \frac{P}{2J}T_L,$$

$$\frac{d\theta_r}{dt} = \omega_r .$$

The *quadrature, direct,* and *zero* currents, needed to be fed to guarantee the balanced operation, are

$$i_{qs}^e(t) = \sqrt{2} i_M,\ i_{ds}^e(t) = 0,\ i_{0s}^e(t) = 0 .$$

To control the angular velocity (in the drive application) of permanent-magnet synchronous micromotors or the displacement (in servo-system application), one supplies the phase voltages to the *abc* stator windings as a function of the angular displacement (measured by the Hall-effect sensors). Correspondingly, ICs must is used, and the permanent-magnet synchronous micromotors can be driver by the Motorola MC33035 ICs [2].

## References

1. P. C. Krause and O. Wasynczuk, *Electromechanical Motion Devices*, McGraw-Hill, New York, 1989.
2. S. E. Lyshevski, *Nano- and Microelectromechanical Systems, Fundamentals of Nano- and Microengineering*, CRC Press, Boca Raton, FL, 2000.

## 5.5 MICROSCALE PERMANENT-MAGNET STEPPER MICROMOTORS

In MEMS and microscale devices, permanent-magnet stepper micromotors can be used. Translational and rotational stepper micromotors (which are synchronous microtransducers-micromachines) have been designed, fabricated, and tested. These micromotors (microactuators) develop high electromagnetic torque, while the mechanical angular velocity is relatively low. Therefore, permanent-magnet stepper micromotors can be easily integrated into *servos* as direct-drive microservos. This direct connection of micromotors without matching mechanical coupling allows one to achieve a remarkable level of efficiency, reliability, and performance.

Stepper micromotors must be controlled to ensure stability, precision tracking, desired steady-state and dynamic performance, disturbance rejection, and zero steady-state error. To approach the analysis, simulation and control, complete nonlinear mathematical models of stepper micromotors must be found. The operating principles in control of stepper micromotors are wel-known. In particular, by energizing the stator windings in the proper sequence, rotor rotates in the *counterclockwise* or *clockwise* direction due to the electromagnetic torque developed. In particular, the rotor displaces by full or half step. Hence, energizing windings one achieves the angular increment equal to full or half step. The angular velocity is regulated by changing the frequency of the phase currents fed or voltages supplied to the phase windings as was shown for permanent-magnet synchronous micromotors. Due to the possibilities to operate stepper motors in the open-loop modes properly energizing the windings, the stepper motors were among the first electric machines to be fabricated and tested in the early 90s (Technical University of Berlin, Kiev Polytechnic Institute, and University of Madison), see Figure 5.5.1.

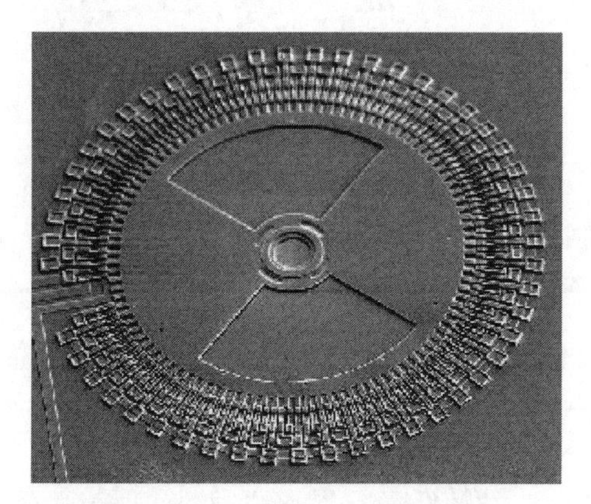

Figure 5.5.1 Micromachined stepper motor.

### 5.5.1 Mathematical Model in the *Machine* Variables

For two-phase permanent-magnet stepper micromotors, we have

$$u_{as} = r_s i_{as} + \frac{d\psi_{as}}{dt},$$
$$u_{bs} = r_s i_{bs} + \frac{d\psi_{bs}}{dt}, \quad (5.5.1)$$

where the flux linkages are

$$\psi_{as} = L_{asas} i_{as} + L_{asbs} i_{bs} + \psi_{asm},$$
$$\psi_{bs} = L_{bsas} i_{as} + L_{bsbs} i_{bs} + \psi_{bsm}. \quad (5.5.2)$$

The electrical angular velocity and displacement are found using the number of rotor tooth $RT$, $\omega_r = RT\omega_{rm}$ and $\theta_r = RT\theta_{rm}$. The flux linkages are the functions of the number of the rotor tooth and displacement,

$$\psi_{asm} = \psi_m \cos(RT\theta_{rm}),$$
$$\psi_{bsm} = \psi_m \sin(RT\theta_{rm}). \quad (5.5.3)$$

The self-inductance of the stator windings is

$$L_{ss} = L_{asas} = L_{bsbs} = L_{ls} + \bar{L}_m. \quad (5.5.4)$$

The stator windings are displaced by 90 electrical degrees. Hence, the mutual inductances between the stator windings are zero, $L_{asbs} = L_{bsas} = 0$.

From (5.5.2), (5.5.3) and (5.5.4), we have

$$\psi_{as} = L_{ss} i_{as} + \psi_m \cos(RT\theta_{rm}),$$
$$\psi_{bs} = L_{ss} i_{bs} + \psi_m \sin(RT\theta_{rm}). \quad (5.5.5)$$

Taking note of (5.5.1) and (5.5.5), one has

$$u_{as} = r_s i_{as} + \frac{d(L_{ss} i_{as} + \psi_m \cos(RT\theta_{rm}))}{dt}$$
$$= r_s i_{as} + L_{ss}\frac{di_{as}}{dt} - RT\psi_m \omega_{rm} \sin(RT\theta_{rm}),$$
$$u_{bs} = r_s i_{bs} + \frac{d(L_{ss} i_{bs} + \psi_m \sin(RT\theta_{rm}))}{dt}$$
$$= r_s i_{bs} + L_{ss}\frac{di_{bs}}{dt} + RT\psi_m \omega_{rm} \cos(RT\theta_{rm}).$$

Therefore,

$$\frac{di_{as}}{dt} = -\frac{r_s}{L_{ss}} i_{as} + \frac{RT\psi_m}{L_{ss}} \omega_{rm} \sin(RT\theta_{rm}) + \frac{1}{L_{ss}} u_{as},$$
$$\frac{di_{bs}}{dt} = -\frac{r_s}{L_{ss}} i_{bs} - \frac{RT\psi_m}{L_{ss}} \omega_{rm} \cos(RT\theta_{rm}) + \frac{1}{L_{ss}} u_{bs}. \quad (5.5.6)$$

Using Newton's second law we have

$$\frac{d\omega_{rm}}{dt} = \frac{1}{J}\left(T_e - B_m\omega_{rm} - T_L\right),$$

$$\frac{d\theta_{rm}}{dt} = \omega_{rm}.$$

The expression for the electromagnetic torque developed by permanent-magnet stepper micromotors must be found. Taking note of

$$W_c = \tfrac{1}{2}\left(L_{ss}i_{as}^2 + L_{ss}i_{bs}^2\right) + \psi_m i_{as}\cos\left(RT\theta_{rm}\right) + \psi_m i_{bs}\sin\left(RT\theta_{rm}\right) + W_{PM},$$

one finds the electromagnetic torque

$$T_e = \frac{\partial W_c}{\partial \theta_{rm}} = -RT\psi_m\left[i_{as}\sin\left(RT\theta_{rm}\right) - i_{bs}\cos\left(RT\theta_{rm}\right)\right].$$

Hence, the transient evolution of the rotor angular velocity $\omega_{rm}$ and displacement $\theta_{rm}$ is modeled by the following differential equations

$$\frac{d\omega_{rm}}{dt} = -\frac{RT\psi_m}{J}\left[i_{as}\sin\left(RT\theta_{rm}\right) - i_{bs}\cos\left(RT\theta_{rm}\right)\right] - \frac{B_m}{J}\omega_{rm} - \frac{1}{J}T_L,$$

$$\frac{d\theta_{rm}}{dt} = \omega_{rm}. \qquad (5.5.7)$$

Augmenting (5.5.6) and (5.5.7), one has

$$\frac{di_{as}}{dt} = -\frac{r_s}{L_{ss}}i_{as} + \frac{RT\psi_m}{L_{ss}}\omega_{rm}\sin\left(RT\theta_{rm}\right) + \frac{1}{L_{ss}}u_{as},$$

$$\frac{di_{bs}}{dt} = -\frac{r_s}{L_{ss}}i_{bs} - \frac{RT\psi_m}{L_{ss}}\omega_{rm}\cos\left(RT\theta_{rm}\right) + \frac{1}{L_{ss}}u_{bs},$$

$$\frac{d\omega_{rm}}{dt} = -\frac{RT\psi_m}{J}\left[i_{as}\sin\left(RT\theta_{rm}\right) - i_{bs}\cos\left(RT\theta_{rm}\right)\right] - \frac{B_m}{J}\omega_{rm} - \frac{1}{J}T_L,$$

$$\frac{d\theta_{rm}}{dt} = \omega_{rm}. \qquad (5.5.8)$$

These four nonlinear differential equations are rewritten in the state-space form as

$$\begin{bmatrix} \frac{di_{as}}{dt} \\ \frac{di_{bs}}{dt} \\ \frac{d\omega_{rm}}{dt} \\ \frac{d\theta_{rm}}{dt} \end{bmatrix} = \begin{bmatrix} -\frac{r_s}{L_{ss}} & 0 & 0 & 0 \\ 0 & -\frac{r_s}{L_{ss}} & 0 & 0 \\ 0 & 0 & -\frac{B_m}{J} & 0 \\ 0 & 0 & 1 & 0 \end{bmatrix} \begin{bmatrix} i_{as} \\ i_{bs} \\ \omega_{rm} \\ \theta_{rm} \end{bmatrix}$$

$$+ \begin{bmatrix} \frac{RT\psi_m}{L_{ss}} \omega_{rm} \sin(RT\theta_{rm}) \\ -\frac{RT\psi_m}{L_{ss}} \omega_{rm} \cos(RT\theta_{rm}) \\ -\frac{RT\psi_m}{J} [i_{as} \sin(RT\theta_{rm}) - i_{bs} \cos(RT\theta_{rm})] \\ 0 \end{bmatrix} + \begin{bmatrix} \frac{1}{L_{ss}} & 0 \\ 0 & \frac{1}{L_{ss}} \\ 0 & 0 \\ 0 & 0 \end{bmatrix} \begin{bmatrix} u_{as} \\ u_{bs} \end{bmatrix} - \begin{bmatrix} 0 \\ 0 \\ \frac{1}{J} \\ 0 \end{bmatrix} T_L.$$

From (5.5.8), an $s$-domain block diagram is developed and illustrated in Figure 5.5.2.

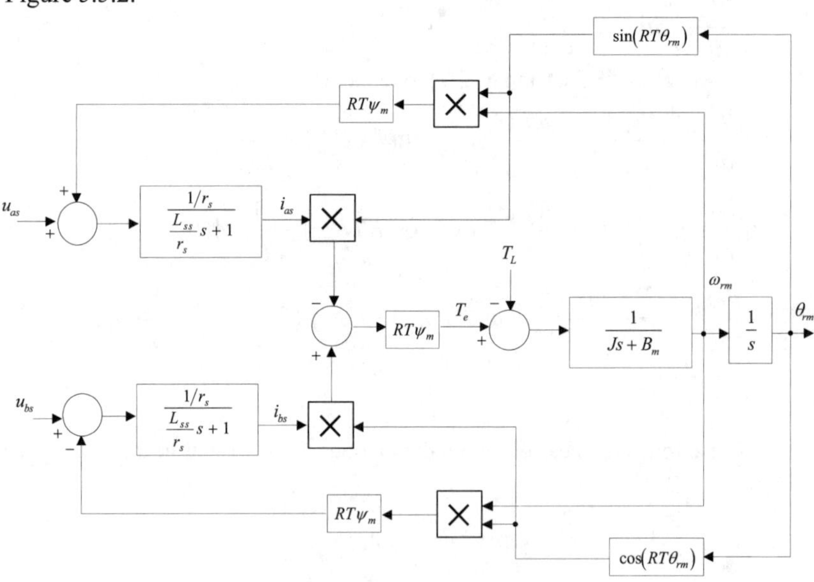


Figure 5.5.2 Block diagram of permanent-magnet stepper micromotors.

The analysis of the torque equation

$$T_e = -RT\psi_m \left[ i_{as} \sin(RT\theta_{rm}) - i_{bs} \cos(RT\theta_{rm}) \right]$$

guides one to the conclusion that the expressions for a balanced two-phase current sinusoidal set is

$$i_{as} = -\sqrt{2}i_M \sin(RT\theta_{rm}),$$
$$i_{bs} = \sqrt{2}i_M \cos(RT\theta_{rm}), \qquad (5.5.9)$$

because the electromagnetic torque is a function of the current magnitude $i_M$, and

$$T_e = \sqrt{2}RT\psi_m i_M .$$

Using ICs, the phase currents (5.5.9) needed to be fed are the functions of the rotor angular displacement. Assuming that the inductances are negligibly small, we have the following phase voltages needed to be supplied

$$u_{as} = -\sqrt{2}u_M \sin(RT\theta_{rm}),$$
$$u_{bs} = \sqrt{2}u_M \cos(RT\theta_{rm}). \qquad (5.5.10)$$

An *s*-domain block diagram of permanent-magnet stepper micromotors which is controlled by changing the phase voltages using ICs, as given by (5.5.10), is shown in Figure 5.5.3.

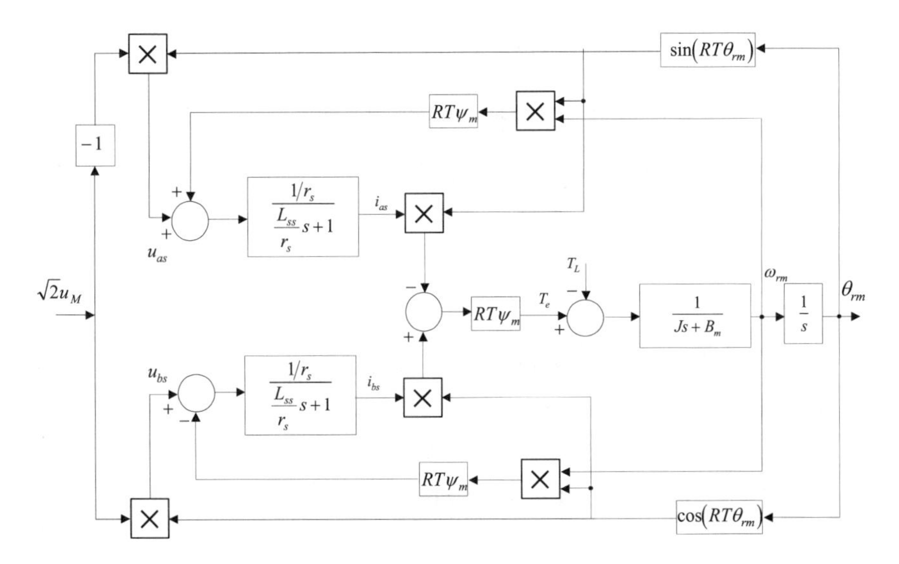


Figure 5.5.3 *s*-domain diagram of permanent-magnet stepper micromotors, $u_{as} = -\sqrt{2}u_M \sin(RT\theta_{rm})$ and $u_{bs} = \sqrt{2}u_M \cos(RT\theta_{rm})$.

## 5.5.2 Mathematical Models of Permanent-Magnet Stepper Micromotors in the Rotor and Synchronous Reference Frames

It was shown that using the *machine* variables, Kirchhoff's voltage law results in two nonlinear differential equations

$$u_{as} = r_s i_{as} + L_{ss}\frac{di_{as}}{dt} - RT\psi_m\omega_{rm}\sin\left(RT\theta_{rm}\right),$$

$$u_{bs} = r_s i_{bs} + L_{ss}\frac{di_{bs}}{dt} + RT\psi_m\omega_{rm}\cos\left(RT\theta_{rm}\right).$$

Applying the *direct* Park formation, which in the rotor reference frame is given as

$$\begin{bmatrix} u_{qs}^r \\ u_{ds}^r \end{bmatrix} = \begin{bmatrix} -\sin(RT\theta_{rm}) & \cos(RT\theta_{rm}) \\ \cos(RT\theta_{rm}) & \sin(RT\theta_{rm}) \end{bmatrix}\begin{bmatrix} u_{as} \\ u_{bs} \end{bmatrix},$$

$$\begin{bmatrix} i_{qs}^r \\ i_{ds}^r \end{bmatrix} = \begin{bmatrix} -\sin(RT\theta_{rm}) & \cos(RT\theta_{rm}) \\ \cos(RT\theta_{rm}) & \sin(RT\theta_{rm}) \end{bmatrix}\begin{bmatrix} i_{as} \\ i_{bs} \end{bmatrix},$$

the following differential equations in the *qd* quantities are found

$$u_{qs}^r = r_s i_{qs}^r + L_{ss}\frac{di_{qs}^r}{dt} + RT\psi_m\omega_{rm} + RTL_{ss}i_{ds}^r\omega_{rm},$$

$$u_{ds}^r = r_s i_{ds}^r + L_{ss}\frac{di_{ds}^r}{dt} - RTL_{ss}i_{qs}^r\omega_{rm}.$$

Hence, the resulting nonlinear circuitry dynamics is

$$\frac{di_{qs}^r}{dt} = -\frac{r_s}{L_{ss}}i_{qs}^r - \frac{RT\psi_m}{L_{ss}}\omega_{rm} - RTi_{ds}^r\omega_{rm} + \frac{1}{L_{ss}}u_{qs}^r,$$

$$\frac{di_{ds}^r}{dt} = -\frac{r_s}{L_{ss}}i_{ds}^r + RTi_{qs}^r\omega_{rm} + \frac{1}{L_{ss}}u_{ds}^r. \tag{5.5.11}$$

From

$$T_e = -RT\psi_m\left[i_{as}\sin\left(RT\theta_{rm}\right) - i_{bs}\cos\left(RT\theta_{rm}\right)\right],$$

using the *inverse* Park transformation

$$\begin{bmatrix} i_{as} \\ i_{bs} \end{bmatrix} = \begin{bmatrix} -\sin(RT\theta_{rm}) & \cos(RT\theta_{rm}) \\ \cos(RT\theta_{rm}) & \sin(RT\theta_{rm}) \end{bmatrix}\begin{bmatrix} i_{qs}^r \\ i_{ds}^r \end{bmatrix},$$

we have

$$T_e = RT\psi_m i_{qs}^r.$$

From Newton's second law of motions, one has

$$\frac{d\omega_{rm}}{dt} = \frac{RT\psi_m}{J}i_{qs}^r - \frac{B_m}{J}\omega_{rm} - \frac{1}{J}T_L,$$

$$\frac{d\theta_{rm}}{dt} = \omega_{rm}. \tag{5.5.12}$$

Augmenting differential equations (5.5.11) and (5.5.12), the following lumped-parameter mathematical model of permanent-magnet synchronous micromotors in the rotor reference frame results

$$\frac{di_{qs}^r}{dt} = -\frac{r_s}{L_{ss}} i_{qs}^r - \frac{RT\psi_m}{L_{ss}} \omega_{rm} - RT i_{ds}^r \omega_{rm} + \frac{1}{L_{ss}} u_{qs}^r,$$

$$\frac{di_{ds}^r}{dt} = -\frac{r_s}{L_{ss}} i_{ds}^r + RT i_{qs}^r \omega_{rm} + \frac{1}{L_{ss}} u_{ds}^r,$$

$$\frac{d\omega_{rm}}{dt} = \frac{RT\psi_m}{J} i_{qs}^r - \frac{B_m}{J} \omega_{rm} - \frac{1}{J} T_L,$$

$$\frac{d\theta_{rm}}{dt} = \omega_{rm}. \tag{5.5.13}$$

Four nonlinear differential equations, which describe the circuitry and *torsional-mechanical* dynamics, are derived. These nonlinear differential equations can be used for analysis, simulation, control, and performance analysis of permanent-magnet stepper micromotors.

In the state-space form, we have

$$\begin{bmatrix} \frac{di_{qs}^r}{dt} \\ \frac{di_{ds}^r}{dt} \\ \frac{d\omega_{rm}}{dt} \\ \frac{d\theta_{rm}}{dt} \end{bmatrix} = \begin{bmatrix} -\frac{r_s}{L_{ss}} & 0 & -\frac{RT\psi_m}{L_{ss}} & 0 \\ 0 & -\frac{r_s}{L_{ss}} & 0 & 0 \\ \frac{RT\psi_m}{J} & 0 & -\frac{B_m}{J} & 0 \\ 0 & 0 & 1 & 0 \end{bmatrix} \begin{bmatrix} i_{qs}^r \\ i_{ds}^r \\ \omega_{rm} \\ \theta_{rm} \end{bmatrix}$$

$$+ \begin{bmatrix} -RT i_{ds}^r \omega_{rm} \\ RT i_{qs}^r \omega_{rm} \\ 0 \\ 0 \end{bmatrix} + \begin{bmatrix} \frac{1}{L_{ss}} & 0 \\ 0 & \frac{1}{L_{ss}} \\ 0 & 0 \\ 0 & 0 \end{bmatrix} \begin{bmatrix} u_{qs}^r \\ u_{ds}^r \end{bmatrix} - \begin{bmatrix} 0 \\ 0 \\ \frac{1}{J} \\ 0 \end{bmatrix} T_L.$$

It is evident that these nonlinear differential equations cannot be linearized. Straightforward analytical and numerical analysis can be performed using the developed lumped-parameter mathematical models.

The phase currents and voltages supplied to the *ab* windings must be fed using the rotor angular displacement. We have the two-phase current

$$i_{as} = -\sqrt{2} i_M \sin(RT\theta_{rm}),\ i_{bs} = \sqrt{2} i_M \cos(RT\theta_{rm}),$$

and two-phase voltage

$$u_{as} = -\sqrt{2} u_M \sin(RT\theta_{rm}),\ \ u_{bs} = \sqrt{2} u_M \cos(RT\theta_{rm})$$

balanced sets.

From the Park transformation, as given by,

$$\begin{bmatrix} i_{qs}^r \\ i_{ds}^r \end{bmatrix} = \begin{bmatrix} -\sin(RT\theta_{rm}) & \cos(RT\theta_{rm}) \\ \cos(RT\theta_{rm}) & \sin(RT\theta_{rm}) \end{bmatrix} \begin{bmatrix} i_{as} \\ i_{bs} \end{bmatrix},$$

one obtains

$$i_{qs}^r = -i_{as}\sin(RT\theta_{rm}) + i_{bs}\cos(RT\theta_{rm}),$$

$$i_{ds}^r = i_{as}\cos(RT\theta_{rm}) + i_{bs}\sin(RT\theta_{rm}).$$

Therefore,

$$i_{qs}^r = \sqrt{2}i_M \sin^2\left(RT\theta_{rm}\right) + \sqrt{2}i_M \cos^2\left(RT\theta_{rm}\right) = \sqrt{2}i_M,$$

and

$$i_{ds}^r = -\sqrt{2}i_M \sin\left(RT\theta_{rm}\right)\cos\left(RT\theta_{rm}\right) + \sqrt{2}i_M \sin\left(RT\theta_{rm}\right)\cos\left(RT\theta_{rm}\right) = 0.$$

Thus,

$$i_{qs}^r = \sqrt{2}i_M$$

and $i_{ds}^r = 0$.

Similarly, for the *quadrature* and *direct* voltages, from the following relationship derived using the Park transformation matrix

$$\begin{bmatrix} u_{qs}^r \\ u_{ds}^r \end{bmatrix} = \begin{bmatrix} -\sin(RT\theta_{rm}) & \cos(RT\theta_{rm}) \\ \cos(RT\theta_{rm}) & \sin(RT\theta_{rm}) \end{bmatrix} \begin{bmatrix} u_{as} \\ u_{bs} \end{bmatrix},$$

one has the expressions for the *quadrature* and *direct* voltages to guarantee the balance operation. The multiplication and simplification gives

$$u_{qs}^r = \sqrt{2}u_M$$

and $u_{ds}^r = 0$.

If the advanced shifting is used, we obtain

$$u_{qs}^r = \sqrt{2}u_M \cos\varphi_u,$$

$$u_{ds}^r = -\sqrt{2}u_M \sin\varphi_u. \tag{5.5.14}$$

Using the nonlinear differential equations (5.5.13), the block diagram of permanent-magnet stepper micromotors, modeled in the rotor reference frame and controlled by changing the *quadrature* and *direct* voltages, is developed. The resulting block diagram is illustrated in Figure 5.5.4.

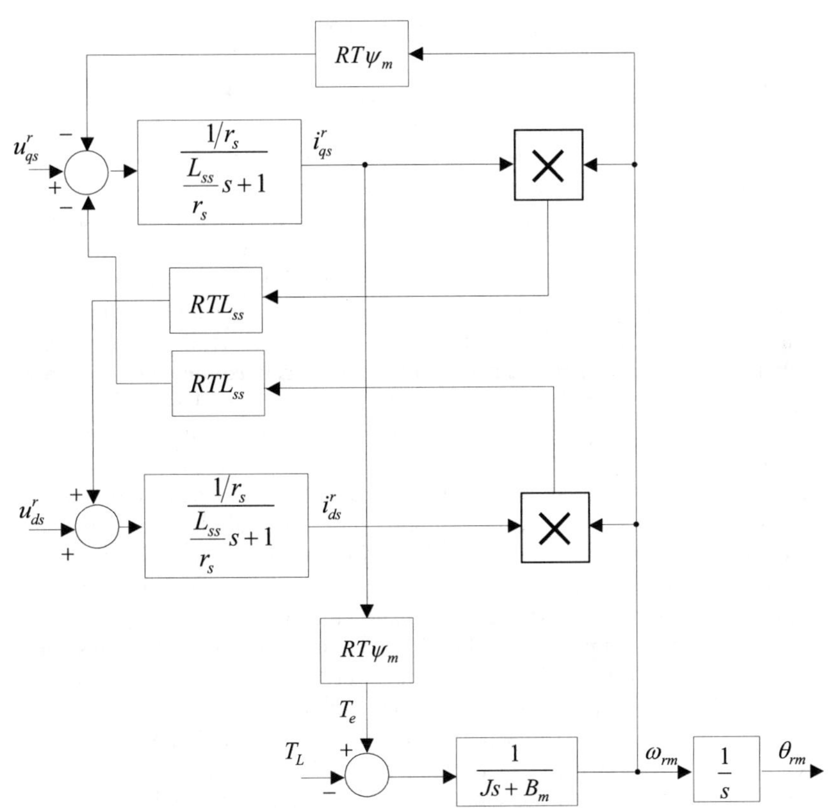


Figure 5.5.4 Block diagram of permanent-magnet stepper micromotors modeled in the rotor reference frame.

Synchronous micromotors rotate with the synchronous angular velocity, $\omega_r = \omega_e$. From (5.5.13), the resulting model of permanent-magnet stepper micromotors in the synchronous reference frame is

$$\frac{di_{qs}^e}{dt} = -\frac{r_s}{L_{ss}} i_{qs}^e - \frac{RT\psi_m}{L_{ss}} \omega_{rm} - RTi_{ds}^e \omega_{rm} + \frac{1}{L_{ss}} u_{qs}^e,$$

$$\frac{di_{ds}^e}{dt} = -\frac{r_s}{L_{ss}} i_{ds}^e + RTi_{qs}^e \omega_{rm} + \frac{1}{L_{ss}} u_{ds}^e,$$

$$\frac{d\omega_{rm}}{dt} = \frac{RT\psi_m}{J} i_{qs}^e - \frac{B_m}{J}\omega_{rm} - \frac{1}{J}T_L,$$

$$\frac{d\theta_{rm}}{dt} = \omega_{rm}.$$

To control the angular velocity, the ICs energizes the *as* and *bs* windings (the so-called step-by-step open-loop operation). As an example, the Motorola monolithic MC3479 ICs driver-controller can be used.

## 5.6 PIEZOTRANSDUCERS

The generation of a potential difference across the opposite faces of certain nonconducting crystals as a result of the applied mechanical stress between these faces is called the piezoelectric effect. The piezoelectric effect, exhibited by quartz, tourmaline and Rochelle salt is quite modest, however, polycrystalline ferroelectric ceramic materials (for example, $BaTiO_3$ and lead zirconate titanate) exhibit strong piezoelectric effect.

Ferroelectric ceramics, which become piezoelectric when poled, are available in many variations and widely used in high-performance transducers for actuation and sensing. The ceramic materials are composed of many randomly oriented crystals or grains, each having one or a few domains. With the dipoles randomly oriented, the material is isotropic and does not exhibit the piezoelectric effect. By applying a strong dc electric field (using electrodes), the dipoles will tend to align themselves parallel to the field, and the material will have a permanent (or remanent) polarization becoming piezoelectric. After poling, the material has a remanent polarization $P_r$ and remanent stress $S_r$. For example, the lead zirconate titanate (PZT) crystallites are centrosymmetric cubic (isotropic) before poling, and after poling exhibit tetragonal symmetry (anisotropic structure) below the Curie temperature. Above the Curie temperature they lose the piezoelectric properties. The Curie temperature is the temperature at which the crystal structure changes from a non-symmetric to a symmetric form. The charge separation between the positive and negative ions results in electric dipole behavior (dipole groups with parallel orientation are called the Weiss domains). The Weiss domains are randomly oriented in the raw piezoelectric materials before poling is done (electric field usually greater than 2000 V/mm is applied to the heated PZT). With the field applied, the material expands along the axis of the electric field and contracts perpendicular to that axis. The electric dipoles align and nearly stay in alignment upon cooling. Thus, the material has a remanent polarization (which, in general, can be degraded by exceeding the mechanical, thermal and electrical limits of the material). When the voltage (electric field) is applied to a poled piezoelectric material, the Weiss domains increase their alignment as a nonlinear function of the voltage applied. The dimensional changes (expansion or contraction) in the piezoelectric materials result.

Piezoelectricity was discovered by Pierre and Jacques Curie in 1880 based upon their study of the symmetry of crystalline matter of quartz, topaz, tourmaline and Rochelle salt (sodium potassium tartrate). In particular, they measured the surface crystal charges subject to the mechanical stress. However, discovering the direct piezoelectric effect (electric field due to the applied stress), they did not study the converse piezoelectric effect (stress in response to the applied electric field). This phenomena was mathematically proven using thermodynamic principles by Lippmann in 1881, and the Curies experimentally confirmed the converse effect, thus illustrating a complete reversibility of electro-elasto-mechanical deformations in piezoelectric crystals. The reversible exchange of electrical and mechanical energy, and the application of thermodynamics in analysis of complex phenomena and relationships among mechanical, thermal, and electrical variables were made. In addition, the classification of piezoelectric crystals on the basis of asymmetric crystal structure was made. Twenty crystal classes (in which piezoelectric effects occur) and eighteen piezoelectric coefficients were published by Voigt in 1910. Based upon these fundamental results, the application of the piezoelectric phenomena started. In 1916, to detect the submarines, Paul Langevin designed an underwater ultrasonic detector using piezoelectric quartz sandwiched between two steel plates (the resonant frequency of the detector was 50 kHz). The success in the sonar developments stimulated intensive activities in resonating and nonresonating piezoelectric-based devices, however, the time for advanced actuators and sensors, microphones, accelerometers, ultrasonic transducers, filters, and other devices was not come yet.

During World War II, scientists from the US, USSR, Japan and Germany discovered ferroelectric ceramics (prepared by sintering metallic oxide powders) with improved piezoelectric characteristics and with dielectric constants hundreds of times higher than common crystals. In particular, the barium titanate piezoceramics were devised (this led to the PZT), the perovskite crystal structures were devised with the corresponding studies of their effect on the electromechanical characteristics, the doping with metallic impurities were performed in order to achieve desired properties (dielectric constant, stiffness, piezoelectric coupling coefficients, poling, robustness, etc.). These advances significantly contributed to establishing entirely new techniques and perspectives in piezoelectric device development through controlling (determining) the material properties and characteristics to specific applications such as sonars, smart piezotransducers (actuators-sensors), filter devices, etc.

The piezoelectric transducers are widely used in optic devices (image and beam stabilization, scanning microscopy, autofocusing, interferometry, alignment, switching, mirror scanners and petitioners, active optics, tuning, vibroacoustic attenuation and control), disk drives (actuation, testing, calibration, alignment, vibration attenuation), microelectronics (nanoactuation and measurements, wafer and mask positioning and alignment, lithography and etching), precision electromechanics (vibroacoustic attenuation and control, high-accuracy actuation and sensing, nano- and microrobotics and mechatronics, assembling, nanometrology), medicine (micro drives and actuators, sensors, micromanipulators and robots, mircostumulus devices), etc. Piezoelectric materials (actuators and sensors) are typically integrated in the controlled structures as the patched and embedded composite structures, and these piezotransducers are controlled using ICs.

Piezoelectric materials are used to convert electrical energy into mechanical energy and vice versa. For nano- and micropositioning, the precise (high-accuracy) and fast motion results due to the electric field (voltage) applied to the piezoelectric (the resulting strain in the order of 1/1000, e.g. 0.1%, leads that the 10 mm long actuator has 10 micrometers displacement). The piezoelectric actuator is illustrated in Figure 5.6.1. The advantages of the piezoelectric actuators are repeatable high-accuracy (nanometers) fast positioning achieved without moving parts (which lead to robustness, durability, efficiency), high force and torque (high force and torque density), high efficiency, integrity, affordability, controllability, etc. The piezoelectric actuators can generate the force and acceleration in the order of thousands of Newton and g's directly converting electrical energy into mechanical energy. The piezotransducers are widely used in active vibroacoustic and noise control devices, vibration and noise cancelation applications, geometry control, accurate pointing (including multi-degree of freedom systems such as the Stewart platforms illustrated in Figure 5.6.1), etc. The Stewart and inverted Stewart platforms are kinematically designed platforms to attain the control in six degrees of freedom ($x$, $y$, $z$, pitch, roll, and yaw). Therefore, Stewart platforms found applications in aerospace, manufacturing, medicine, robotics, etc. In general, Stewart platforms are based on the triangle-mounted frames actuated by high-performance actuators. Usually, two actuators are attached at each corner of the platform to guarantee the desired actuation with six degrees of freedom manipulation features.

Figure 5.6.1 Piezoelectric actuator and smart piezotransducers application in the Stewart platforms.

The relationships between the applied voltage and the resulting forces and displacement, and vice versa, depend upon the piezoelectromechanical properties of the ceramic, size and shape, direction of the electrical and mechanical excitation, temperature, etc. The piezoceramics are usually studied in the Cartesian coordinate system (three-dimensional orthogonal set of axes) using $x$, $y$ and $z$ axes. The polarization direction is established during manufacturing by applying a dc voltage (electric field) to a pair of electrodes. These poling electrodes are then removed and replaced by electrodes deposited on a second pair of faces. Piezoelectric materials are anisotropic, and their electromechanical properties differ for electrical and/or mechanical excitation along different directions. Therefore, for the systematic tabulation of electromechanical properties, the directions are standardized and specified. In particular, one defines the axes by numerals: 1 corresponds to $x$ axis, 2 corresponds to $y$ axis, and 3 corresponds to $z$ axis.

Piezoelectric ceramics are isotropic and become piezoelectric after poling (once polarized, piezoceramics are anisotropic). The direction of the poling electric field (applied dc voltage) is distinguished in three directions. The poling electric field can be applied in such ways that the ceramic will exhibit piezoelectric responses in various directions or a combination of directions. The poling process permanently changes the properties of the ceramics.

## 5.6.1 Piezoactuators: Steady-State Models and Characteristics

*Preliminaries* Simplifying the analysis, we consider the transverse effect in the rectangular piezoelectric film, see Figure 5.6.2. The first (1) and second (2) axes define a plane which is parallel to the film surface, while the third (3) axis is perpendicular to the surface (third axis points in the opposite direction to the electric field applied to pole the piezoelectric films). The film length, width, and thickness are within the first, second and third axes, respectively. A voltage is measured (or applied) across the upper and lower electrodes, which have free charges $+q$ and $-q$. A tensile force $f_1$ and an elongation displacement $y_1$ are measured along the first axis. It must be emphasized that if piezoelectric films are used as actuators, a voltage $V_3$ is applied across the film, and the induced polarization in the piezoelectric generates a tensile force $f_1$ and a displacement $y_1$. If piezoelectric films are used as sensors, a tensile force $f_1$ is applied stretching the film by $y_1$, and the induced polarization in the piezoelectric causes an increase of free charges generating a voltage $V_3$.

Thus, the steady-state piezoelectric behavior can be modeled as

$y_1 = s^E f_1 + dV_3$, $q = df_1 + c^T V_3$, (force and voltage are the left-side variables)

$f_1 = \frac{1}{s^E} y_1 - eV_3$, $q = ey_1 + c^S V_3$, (displacement and voltage are the left-side variables)

where $s^E$ is the piezoelectric film compliance at the constant electric field; $d$ is the piezoelectric film charge to force ratio; $c^T$ is the piezoelectric film capacitance at the constant stress; $e$ is the piezoelectric film charge to displacement ratio; $c^S$ is the piezoelectric film capacitance at the constant strain.

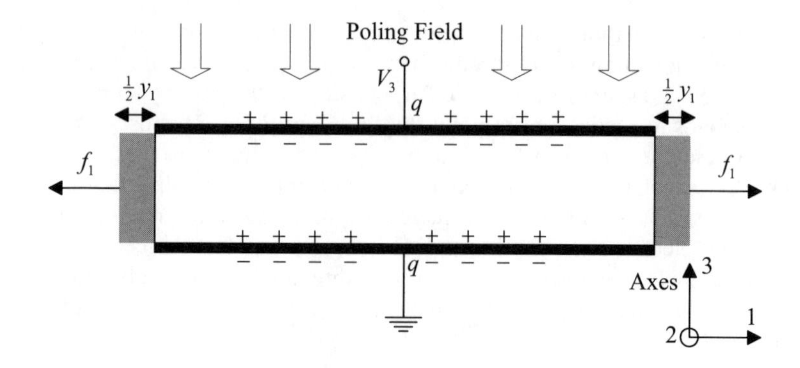


Figure 5.6.2 Piezoelectric thin film.

It is important to perform high-fidelity modeling and analysis of piezotransducers. In general, piezoelectricity and piezoelectromechanical properties are described by the constitutive equations which define how the vectors of the stress ($\mathbf{T}\in\mathbb{R}^{6\times 1}$, N/m$^2$), strain ($\mathbf{S}\in\mathbb{R}^{6\times 1}$), electric charge-density displacement ($\mathbf{D}\in\mathbb{R}^{3\times 1}$, C/m$^2$), and electric field ($\mathbf{E}\in\mathbb{R}^{3\times 1}$, N/C) are related.

In the conventional form, there are 21 independent elastic constant, 18 independent piezoelectric constants, and 6 independent dielectric constants. Four forms of the steady-state piezoelectric constitutive equations are given in Table 5.6.1.

Table 5.6.1 Steady-state piezoelectric constitutive equations.

| Strain-Charge Form | Stress-Charge Form | Strain-Voltage Form | Stress-Voltage Form |
|---|---|---|---|
| $\mathbf{S}=\mathbf{s}_E\cdot\mathbf{T}+\mathbf{d}^T\cdot\mathbf{E}$ | $\mathbf{T}=\mathbf{c}_E\cdot\mathbf{S}-\mathbf{e}^T\cdot\mathbf{E}$ | $\mathbf{S}=\mathbf{s}_D\cdot\mathbf{T}+\mathbf{g}^T\cdot\mathbf{D}$ | $\mathbf{T}=\mathbf{c}_D\cdot\mathbf{S}-\mathbf{d}^T\cdot\mathbf{D}$ |
| $\mathbf{D}=\mathbf{d}\cdot\mathbf{T}+\boldsymbol{\varepsilon}_T\cdot\mathbf{E}$ | $\mathbf{D}=\mathbf{e}\cdot\mathbf{S}-\boldsymbol{\varepsilon}_S\cdot\mathbf{E}$ | $\mathbf{E}=-\mathbf{g}\cdot\mathbf{T}+\varepsilon_T^{-1}\cdot\mathbf{D}$ | $\mathbf{E}=-\mathbf{q}\cdot\mathbf{S}+\varepsilon_S^{-1}\cdot\mathbf{D}$ |

Here, $\mathbf{s}\in\mathbb{R}^{6\times 6}$ is the matrix of compliance coefficients (m$^2$/N); $\mathbf{d}\in\mathbb{R}^{3\times 6}$ is the matrix of the piezoelectric coupling coefficients (C/N); $\mathbf{c}\in\mathbb{R}^{6\times 6}$ is the matrix of stiffness coefficients (N/m$^2$); $\mathbf{e}\in\mathbb{R}^{3\times 6}$ is the matrix of the piezoelectric coupling coefficients (C/m$^2$); $\mathbf{g}\in\mathbb{R}^{3\times 6}$ is the matrix of the piezoelectric coupling coefficients (m$^2$/C); $\boldsymbol{\varepsilon}\in\mathbb{R}^{3\times 6}$ is the matrix of electric permittivity (F/m); $\mathbf{q}\in\mathbb{R}^{3\times 6}$ is the matrix of the piezoelectric coupling coefficients (N/C).

The matrix transformations for converting piezoelectric constitutive relationships from one form into another form are given in Table 5.6.2.

Table 5.6.2 Matrix transformations for converting piezoelectric constitutive equations.

| Strain-Charge to Stress-Charge | Strain-Charge to Strain-Voltage | Stress-Charge to Stress-Voltage | Strain-Voltage to Stress-Voltage |
|---|---|---|---|
| $\mathbf{c}_E = \mathbf{s}_E^{-1}$, $\mathbf{e} = \mathbf{d}\cdot\mathbf{s}_E^{-1}$ $\boldsymbol{\varepsilon}_S = \boldsymbol{\varepsilon}_T - \mathbf{d}\cdot\mathbf{s}_E^{-1}\cdot\mathbf{d}^T$ | $\mathbf{s}_D = \mathbf{s}_E - \mathbf{d}^T\cdot\boldsymbol{\varepsilon}_T^{-1}\cdot\mathbf{d}$ $\mathbf{g} = \boldsymbol{\varepsilon}_T^{-1}\cdot\mathbf{d}$ | $\mathbf{c}_D = \mathbf{c}_E + \mathbf{e}^T\cdot\boldsymbol{\varepsilon}_S^{-1}\cdot\mathbf{e}$ $\mathbf{q} = \boldsymbol{\varepsilon}_S^{-1}\cdot\mathbf{e}$ | $\mathbf{c}_D = \mathbf{s}_D^{-1}$, $\mathbf{q} = \mathbf{g}\cdot\mathbf{s}_D^{-1}$ $\boldsymbol{\varepsilon}_S^{-1} = \boldsymbol{\varepsilon}_T^{-1} + \mathbf{g}\cdot\mathbf{s}_D^{-1}\cdot\mathbf{g}^T$ |

The subscripts describe the conditions under which the piezoelectric material property data was measured. For example, the subscript $E$ on the compliance matrix $\mathbf{s}_E$ means that the compliance data was measured under the constant (zero) electric field. The subscript $S$ on the permittivity matrix $\boldsymbol{\varepsilon}_S$ means that the permittivity data was measured under the constant (zero) strain.

Other commonly used pairs of the constitutive equations are given in the tensor form as (see the ANSI/IEEE Standard 176-1987 and references therein, [1])

$$S_{ij} = s_{ijkl}^{E} T_{kl} + d_{kij} E_k, \qquad \text{(strain-charge form)}$$
$$D_i = d_{ikl} T_{kl} + \varepsilon_{ik}^{T} E_k,$$

or

$$T_{ij} = c_{ijkl}^{E} S_{kl} - e_{kij} E_k, \qquad \text{(stress-charge form)}$$
$$D_i = e_{ikl} S_{kl} + \varepsilon_{ij}^{S} E_k,$$

or

$$S_{ij} = s_{ijkl}^{D} T_{kl} + g_{kij} D_k,$$
$$E_i = -g_{ikl} T_{kl} + \beta_{ik}^{T} D_k,$$

or

$$T_{ij} = c_{ijkl}^{D} S_{kl} - h_{kij} D_k,$$
$$E_i = -h_{ikl} S_{kl} + \beta_{ik}^{S} D_k,$$

where $S_{ij}$ is the mechanical strain; $T_{kl}$ is the mechanical stress; $E_k$ is the electric field; $D_j$ is the electrical displacement; $s_{ijkl}^{E}$ is the mechanical compliance of the material measured at the zero electric field; $d_{kij}$ is the piezoelectric coupling between the electrical and mechanical variables; $\varepsilon_{jk}^{T}$ is the dielectric permittivity measured at the zero mechanical stress; $c_{ijkl}^{E}$ is the stiffness of the material measured at the zero electric field; $e_{ikl}$ and $h_{ikl}$ are the piezoelectric constants; $i, j, k = 1, 2, 3$ and $p, q, r = 1, 2, 3, 4, 5, 6$.

In general, the linear piezoelectricity theory is based upon the first law of thermodynamics

$$\dot{U} = T_{ij}\dot{S}_{ij} + E_i\dot{D}_i,$$

where $U$ is the stored energy density for the piezoelectric continuum.

Defining the electric enthalpy density as

$$H = U - E_i D_i,$$

we have $\dot{H} = T_{ij}\dot{S}_{ij} - D_i\dot{E}_i$, and

$$T_{ij} = \frac{\partial H}{\partial S_{ij}} \text{ and } D_i = -\frac{\partial H}{\partial E_i}.$$

Thus, we have

$$U = \tfrac{1}{2} c_{ijkl}^{E} S_{ij} S_{kl} + \tfrac{1}{2} \varepsilon_{ij}^{S} E_i E_j$$

and $H = \tfrac{1}{2} c_{ijkl}^{E} S_{ij} S_{kl} - e_{kij} E_k S_{ij} - \tfrac{1}{2} \varepsilon_{ij}^{S} E_i E_j$.

The catalog data, provided by the piezoelectric ceramics manufacturers, is given using the reported notations. The double subscript of piezoelectric coefficients relate electrical and mechanical quantities. In particular, the first subscript specifies the direction of the electrical field associated with the voltage applied or the charge produced. The second subscript specifies the direction of the mechanical stress or strain. The piezoceramic material constants are given using superscripts which specify either a mechanical or electrical boundary condition. The superscripts are $T$ (constant or zero stress), $E$ (constant or zero electric field), $D$ (constant or zero charge-density displacement), and $S$ (constant or zero strain). The piezoelectric constants relating the mechanical strain produced by an applied electric field are called the strain constants or $d$ coefficients. We have,

$$d = \frac{\text{mechanical strain developed}}{\text{electric field applied}}.$$

The superscripts describe external factors (electrical and mechanical conditions) that effect the piezoelectric (piezoelectromechanical) characteristics. The subscripts describe the relationship of the piezoelectric (piezoelectromechanical) characteristics with respect to the poling axis. The subscripts define the axes of a component in terms of orthogonal axes (1 corresponds to the $x$ axis, 2 to the $y$ axis, and 3 to the $z$ axis). The first subscript gives the direction of the action, and the second subscript gives the direction of the response. For example, for the piezoelectric coupling constant $d$, the first subscript refers to the direction of the electric field and the second refers to the direction of the strain. For the converse piezoelectric constant $g$, the first subscript refers to the stress and the second subscript refers to the voltage (electric field).

The mechanical stiffness properties of the piezoelectric ceramics are described using Young's modulus which is the ratio of stress (force per unit area) to strain (change in length per unit length), and $Y=\frac{\text{stress}}{\text{strain}}$.

Mechanical stress produces the electric response (voltage) which opposes the resultant strain. Therefore, the effective Young's modulus with short-circuited electrodes is lower than with the open-circuited electrodes. It should be emphasized that the stiffness is different in the third direction from that in the first and second directions. These effects must be distinguished, described, and characterized. For example, $Y_{33}^E$ and $Y_{33}^D$ are the ratios of stress to strain (Young's modulus) in the third direction at constant electric field $E$ (electrodes short-circuited) and if the electrodes are open-circuited.

Due to the wide range of operating conditions, the piezoelectric materials must be properly chosen studying dielectric, ferroelectric, electromechanical, thermal, and other characteristics. The piezoelectric coefficients can be measured in accordance with the ANSI/IEEE 176-1987 Standard. In engineering applications, using the Impedance Analyzer, one finds the effective electromechanical coupling coefficient $k_e$ as

$$k_e=\sqrt{\frac{f_{\max}^2-f_{\min}^2}{f_{\max}^2}},$$

where $f_{\max}$ and $f_{\min}$ are the maximum and minimum impedance frequencies.

The piezoelectric coefficients for the transverse ($k_{31}$, $d_{31}$ and $g_{31}$) and longitudinal ($k_{33}$, $d_{33}$ and $g_{33}$) modes of operation are found by measuring the resonance properties of the thickness-poled and length-poled ceramics. In particular, these constants are found as

$$k_{31}=\sqrt{\frac{\frac{1}{2}\pi\frac{f_{\min}}{f_{\max}}\tan\left(\frac{1}{2}\pi\frac{f_{\max}-f_{\min}}{f_{\min}}\right)}{1+\frac{1}{2}\pi\frac{f_{\min}}{f_{\max}}\tan\left(\frac{1}{2}\pi\frac{f_{\max}-f_{\min}}{f_{\min}}\right)}},\quad s_{11}^E=\frac{1}{4\rho f_{\min}^2 l^2},$$

$$d_{31}=k_{31}\sqrt{\varepsilon_0 k_3 s_{11}^E},\quad g_{31}=\frac{d_{31}}{\varepsilon_0 k_3},$$

$$k_{33}=\sqrt{\frac{\pi}{2}\frac{f_m}{f_n}\tan\left(\frac{\pi}{2}\frac{f_n-f_m}{f_n}\right)},\quad s_{33}^D=\frac{1}{4\rho f_{\max}^2 l^2},\quad s_{33}^E=\frac{s_{33}^D}{1-k_{33}^2},$$

$$d_{33}=k_{33}\sqrt{\varepsilon_0 k_3 s_{33}^E},\quad g_{33}=\frac{d_{33}}{\varepsilon_0 k_3}.$$

Here, $k_{31}$ and $k_{33}$ are the transverse and extensional coupling coefficients; $s_{11}$ and $s_{33}$ are the elastic compliance constants [m$^2$/N] (superscripts $D$ and $E$ denote the measurements at the constant electric displacement and electric field); $d_{31}$ and $d_{33}$ are the transverse and extensional strain constants [m/V]; $g_{31}$ and $g_{33}$ are the transverse and extensional voltage constants [V-m/N]; $k_3$ is the dielectric constant; $l$ is the specimen length [m]; $\varepsilon_0$ is the permittivity of free space, $\varepsilon_0 = 8.85\times10^{-12}$ F/m; ρ is the density [kg/m$^3$].

The dielectric, ferroelectric, and piezoelectric properties of piezoelectric ceramics (PZT-4 and PZT-5 with the density ρ=7.5 kg/cm$^3$) for $T$=25$^0$C assuming the dimensional requirements (the diameter of the circular piezoelectric material is at least ten times greater than the thickness which is in the micrometer range, and the length of the rectangular piezoelectric material is at least three times greater than the width) are given below in Table 5.6.3.

Table 5.6.3 Dielectric (at 1000 Hz), ferroelectric, and piezoelectric properties of the piezoelectric ceramics.

| | $k_3$ | $E_c$ kV/cm | $P_R$ μC/cm$^2$ | $P_{sat}$ μC/cm$^2$ | $k_e$ | $d_{33}$ m/V | $g_{33}$ V-m/N | $k_{33}$ | $d_{31}$ m/V | $G_{31}$ V-m/N | $k_{31}$ |
|---|---|---|---|---|---|---|---|---|---|---|---|
| PZT4 | 1400 | 14 | 31 | 40 | 0.5 | $225\times10^{-12}$ | 0.009 | 0.35 | $-85\times10^{-12}$ | -0.008 | 0.22 |
| PZT5 | 3400 | 5.8 | 13 | 20 | 0.5 | $590\times10^{-12}$ | 0.0013 | 0.6 | $-270\times10^{-12}$ | -0.009 | 0.37 |

Almost all piezoelectric materials exhibit lower dielectric constant values at low temperature, and as the temperature increases, the dielectric constant increases. For example, for the PZT-4 and PZT-5A, the dielectric constant increases steadily as a function of temperature up to the Curie temperature, and the temperature range is from –150 to 250$^0$C. It must be emphasized that in general, the dielectric constants are nonlinear functions of the temperature. For example, for the PZT-5H ceramics, the maximum dielectric constant is observed at 185°C. The dissipation factors depend upon the temperature and the measurement frequency. In particular, for the PZT-4 tanδ=0.05 (measured from 1 to 100 kHz) over the temperature range from –150 to 180$^0$C, and for PZT-5A, the tanδ varies from 0.015 to 0.025 (measured at 100 Hz to 10 kHz) over the entire temperature range (from –150 to 250$^0$C). For PZT-5H, the tanδ is a nonlinear function of the temperature, and the tanδ varies from 0.01 to 0.05 (at 100 Hz to 10 kHz) with the peak value 0.05 at 165$^0$C (10 kHz). It must be emphasized that for PZT-5H, the tanδ varies from 0.035 to 0.17 (at 100 kHz) with the peak value 0.17 at 185$^0$C. The resistivity of the PZT-4 and PZT-5 also varies as a function of temperature. For example, though relatively small (up to 20%) variations observed for the temperature range from –150 to 100$^0$C for the PZT-4, the resistivity decreases in the high temperature region. The polarization is significantly influenced by the temperature. Usually, the maximum remanent polarization occurs in the temperature range from 0 to 50$^0$C. The polarization versus electric field characteristics, given as the $P-E$ curves, of the PZT-5A at temperatures 25$^0$C and 200$^0$C are illustrated in Figure 5.6.3.

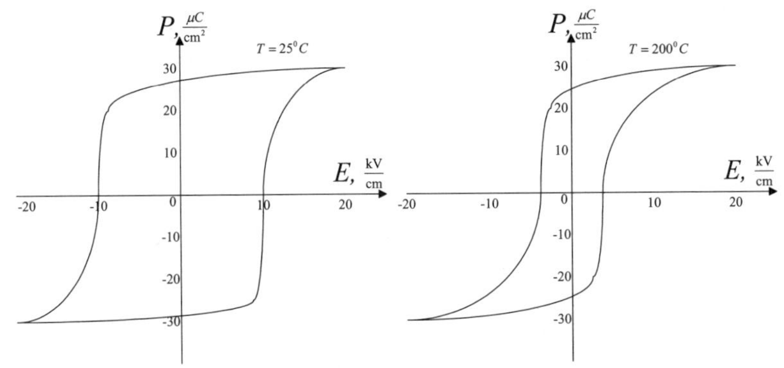


Figure 5.6.3 Ferroelectric polarization versus electric field characteristics ($P$–$E$ curves) of the PZT-5A piezoceramics and $25^0$C and $200^0$C.

The discussion and data provided indicate that the thorough consideration must be carried out selecting the piezoelectric actuators for given operating conditions. To illustrate an other trade-off, it should be emphasized that the power consumption also must be studied. In particular, the power ($P$), consumed by the piezoelectric actuators at the sinusoidal operations is proportional to the capacitance ($C$), operating frequency ($f$), and the squired peak-to-peak operating voltage ($V_{pp}$). In particular,

$$P = \pi C f V_{pp}^2 .$$

In micro- and nanoelectromechanical systems, devices, and structures, PZT thin films are widely used. The characteristics of sol-gel thin films depend upon the orientation of $PbZr_{1-x}Ti_x$ substrates. For example, sol-gel PZT films on Pt(111)/Ti/$SiO_2$/Si are 100-oriented. However, 111-oriented PZT thin films can be made on the same substrate when the films are pyrolyzed at $350^0$C. The PZT ($PbZr_{65}Ti_{35}$, $PbZr_{53}Ti_{47}$ and $PbZr_{35}Ti_{65}$) thin films, made on Pt(111)/$SiO_2$/Si and Pt(200)/$SiO_2$/Si substrates, exhibit the hysteresis (it must be emphasized that all piezoelectric ceramics exhibit the hysteresis effect). The ferroelectric polarization versus voltage (electric field) characteristics ($P$–$V$ or $P$–$E$ curves) are illustrated in Figure 5.6.4.

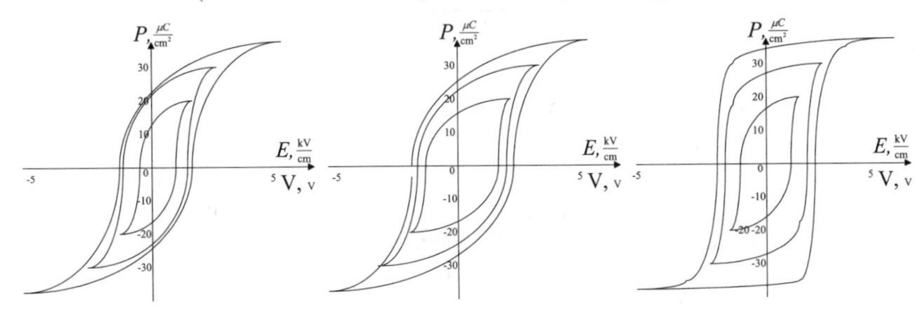


Figure 5.6.4 Ferroelectric polarization versus applied voltage (electric field) characteristics ($P$–$V$ or $P$–$E$ curves) for the sol-gel PZT ($PbZr_{65}Ti_{35}$, $PbZr_{53}Ti_{47}$ and $PbZr_{35}Ti_{65}$) thin films.

The saturation magnetization (*B–H* curves) and the force-displacement curves are similar to the *P–E* curves. Thus, the hysteresis is the important effect to be thoroughly studied. In addition to the nonlinear effects, parameter variations, and other phenomena and effects observed in piezotransducers, it is extremely important to integrate the piezotransducer dynamics. In fact, the steady-state analysis does not allow one to fully examine the system performance and make a conclusion based upon requirements and specifications imposed. This is particularly important in micro- and nanopositioning applications (vibration and noise attenuation and control, actuators and drives, etc.).

## 5.6.2 Mathematical Models of Piezoactuators: Dynamics and Nonlinear Equations of Motion

The mathematical models of piezotransducers are given in the form of differential equations which allow the designer to attain the steady-state and dynamic analysis.

Consider a circular plate piezoactuator (polarized ceramics of classes $C_3$, $C_{3v}$, $C_6$ or $C_{6v}$, circular surface normal to the three- or sixfold axis) assuming that the actuator surfaces are completely covered with electrodes, see Figure 5.6.5.

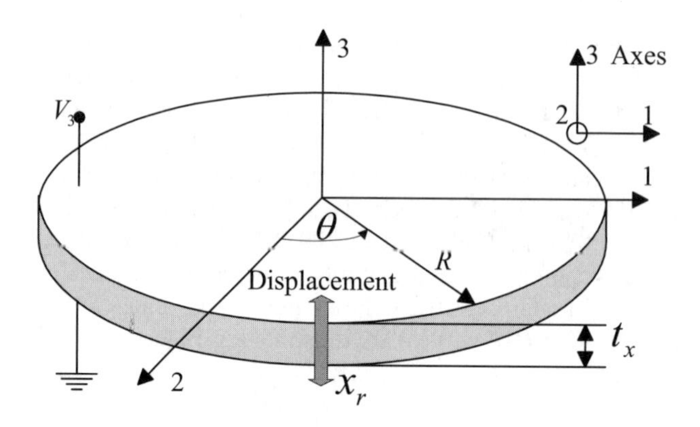


Figure 5.6.5 Circular plate piezoactuator.

The following partial differential equation models the dynamic motion (radial displacement $x_r$) of the circular piezoactuator [1]

$$\frac{\partial^2 x_r}{\partial r^2}+\frac{1}{r}\frac{\partial x_r}{\partial r}-\frac{x_r}{r^2}=\rho\frac{\left(s_{11}^E\right)^2-\left(s_{12}^E\right)^2}{s_{11}^E}\frac{\partial^2 x_r}{\partial t^2}$$

with the constitutive equations

$$T_{rr} = \frac{s_{11}^E}{\left(s_{11}^E\right)^2 - \left(s_{12}^E\right)^2} \frac{\partial x_r}{\partial r} - \frac{s_{12}^E}{\left(s_{11}^E\right)^2 - \left(s_{12}^E\right)^2} \frac{x_r}{r} + \frac{1}{t_x} \frac{d_{31}}{s_{11}^E + s_{12}^E} V_3,$$

$$T_{rr}\big|_{r=R} = 0,$$

$$D_3 = \frac{d_{31}}{s_{11}^E + s_{12}^E} \frac{1}{r} \frac{\partial (r x_r)}{\partial r} + \frac{1}{t_x} \left( \frac{2 d_{31}^2}{s_{11}^E + s_{12}^E} - \varepsilon_{33}^T \right) V_3 .$$

These equations can be simulated in the high-performance environments (e.g., MATLAB and MATEMATICA), and, making use of the piezoceramic parameters, the coefficients of the equations are straightforwardly found.

In addition to the high-fidelity modeling, which is based upon the solution of the partial differential equations, lumped-parameter models of piezotransducers provide manageable and practical results. It must be emphasized that the hysteresis phenomena cannot be neglected and thus, must be integrated in the mathematical modeling, simulation, analysis, and control.

The differential equations of motion of piezoelectric actuators are represented as

$$m \frac{d^2 x}{dt^2} + k_x \frac{dx}{dt} + k_1 x + k_2 x^2 + k_3 x^3 = k_v V - k_z z ,$$

and the hysteresis equation is

$$\frac{dz}{dt} = k_{z1} \frac{dV}{dt} - k_{z2} \left| \frac{dV}{dt} \right| z - k_{z3} \frac{dV}{dt} | z | .$$

Here, $x$ is the actuator displacement; $V$ is the applied voltage; $z$ is the hysteresis variable; $k_l$ are the time-varying coefficients.

Other contemporary equations of motion also can be derived and used. to attain accurate analysis and design. For example, the nonlinear dynamics with the hysteresis effect can be modeled as

$$m \frac{d^2 x}{dt^2} + k_x \frac{dx}{dt} + k_1 x + k_2 x^2 + k_3 x^3 = k_v V - k_f f ,$$

$$\frac{df}{dt} = \frac{dx}{dt} - \frac{1}{k_{fx}} \left| \frac{dx}{dt} \right| f ,$$

where $k_f$ and $k_{fx}$ are the coefficients which form and shape the hysteresis curve.

It must be emphasized that using the available high-performance software environments, the documented differential equations can be easily solved. The challenges result in the design of control algorithms. In particular, the hysteresis effect degrades the piezoactuators performance. Therefore, nonlinear phenomena must be integrated in the analysis, control, and optimization.

In the *xyz* plane, if three actuators are used, one has sets of nonlinear differential equations. For example, for the identical PZT-4 piezoactuators with diameter 2 mm and thickness 0.01 mm, we have the following set of nonlinear differential equations (using lumped-parameter model of piezoactuators)

$$\frac{d^2x}{dt^2}+k_x\frac{dx}{dt}+k_1x+k_2x^2+k_3x^3=k_vV_x,$$

$$\frac{d^2y}{dt^2}+k_y\frac{dy}{dt}+k_1y+k_2y^2+k_3y^3=k_vV_y,$$

$$\frac{d^2z}{dt^2}+k_y\frac{dz}{dt}+k_1z+k_2z^2+k_3z^3=k_vV_z$$

where $V_x$, $V_y$ and $V_z$ are the applied voltages to the *x*-, *y*- and *z*-axis piezoactuators to displace actuators.

The coefficients of the differential equations are

$k_x=k_y=1000$, $k_1=1\times10^7$, $k_2=-1\times10^6$, $k_3=5\times10^4$ and $k_v=1.5\times10^5$.

The simulation, using the MATLAB environment, can be straightforwardly performed to study the dynamic and steady-state responses of the piezotransducers using the piezoelectromechanical characteristics reported as well as differential equations given.

The giant magnetostrictive materials, fabricated using the rare-earth alloys, have enhanced electromechanical characteristics, compared with PZT. Table 5.6.4 documents some Terfenol-D ($Tb_{0.3}Dy_{0.7}Fe_2$) characteristics for $T$=25$^0$C. It must be emphasized that Terfenol-D (which has the Curie temperature 380$^0$C and density $\rho=9.3\times10^3$ kg/m$^3$) is frequently used to fabricate high-performance magnetostrictive transducers.

Table 5.6.4 Properties of Terfenol D.

| $\mu_{r33}^T$ | $\mu_{r33}^S$ | $d_{33}$ | $k_{33}$ | Elastic Modulus $c_{33}^H$ N/mm$^2$ | Elastic Modulus $c_{33}^B$ N/mm$^2$ | Comprehensive Strength $T_t$ N/mm$^2$ | Tensile Strength $T_p$ N/mm$^2$ |
|---|---|---|---|---|---|---|---|
| 9.4 | 4.5 | $1.5\times10^{-8}$ | 0.75 | $28\times10^3$ | $52\times10^3$ | 710 | 29 |

## Reference

1. *An American National Standard, IEEE Standard on Piezoelectricity, ANSI/IEEE Standard 176-1987*, IEEE Inc., 1987.

## 5.7 FUNDAMENTALS OF MODELING OF ELECTROMAGNETIC RADIATING ENERGY MICRODEVICES

The electromagnetic power is generated and radiated by antennas. Time-varying current radiates electromagnetic waves (radiated electromagnetic fields). Radiation pattern, beam width, directivity, and other major characteristics can be studied using Maxwell's equations, see Section 2.2. We use the vectors of the electric field intensity ***E***, electric flux density ***D***, magnetic field intensity ***H***, and magnetic flux density ***B***. The constitutive equations are

$$\mathbf{D}=\varepsilon\mathbf{E} \text{ and } \mathbf{B}=\mu\mathbf{H}$$

where $\varepsilon$ is the permittivity; $\mu$ is the permiability.

It was shown in Section 2.2 that in the static (time-invariant) fields, electric and magnetic field vectors form separate and independent pairs. That is, ***E*** and ***D*** are not related to ***H*** and ***B***, and vice versa. However, for time-varying electric and magnetic fields, we have the following fundamental electromagnetic equations

$$\nabla\times\mathbf{E}(x,y,z,t)=-\mu\frac{\partial\mathbf{H}(x,y,z,t)}{\partial t},$$

$$\nabla\times\mathbf{H}(x,y,z,t)=\sigma\mathbf{E}(x,y,z,t)+\varepsilon\frac{\partial\mathbf{E}(x,y,z,t)}{\partial t}+\mathbf{J}(x,y,z,t),$$

$$\nabla\cdot\mathbf{E}(x,y,z,t)=\frac{\rho_v(x,y,z,t)}{\varepsilon},$$

$$\nabla\cdot\mathbf{H}(x,y,z,t)=0,$$

where ***J*** is the current density, and using the conductivity $\sigma$, we have $\mathbf{J}=\sigma\mathbf{E}$; $\rho_v$ is the volume charge density.

The total current density is the sum of the source current $\mathbf{J}_S$ and the conduction current density $\sigma\mathbf{E}$ (due to the field created by the source $\mathbf{J}_S$). Thus,

$$\mathbf{J}_\Sigma=\mathbf{J}_S+\sigma\mathbf{E}.$$

The equation of conservation of charge (continuity equation) is

$$\oint_s\mathbf{J}\cdot d\mathbf{s}=-\frac{d}{dt}\int_v\rho_v dv,$$

and in the point form one obtains

$$\nabla\cdot\mathbf{J}(x,y,z,t)=-\frac{\partial\rho_v(x,y,z,t)}{\partial t}.$$

Therefore, the net outflow of current from a closed surface results in decrease of the charge enclosed by the surface.

The electromagnetic waves transfer the electromagnetic power. That is, the energy is delivered by means of electromagnetic waves. Using equations

$$\nabla\times\mathbf{E}=-\mu\frac{\partial\mathbf{H}}{\partial t}\text{ and }\nabla\times\mathbf{H}=\varepsilon\frac{\partial\mathbf{E}}{\partial t}+\mathbf{J},$$

we have

$$\nabla\cdot(\mathbf{E}\times\mathbf{H})=\mathbf{H}\cdot(\nabla\times\mathbf{E})-\mathbf{E}\cdot(\nabla\times\mathbf{H})=-\mathbf{H}\cdot\mu\frac{\partial\mathbf{H}}{\partial t}-\mathbf{E}\cdot\left(\varepsilon\frac{\partial\mathbf{E}}{\partial t}+\mathbf{J}\right).$$

In a media, where the constitute parameters are constant (time-invariant), we have the so-called point-function relationship

$$\nabla\cdot(\mathbf{E}\times\mathbf{H})=-\frac{\partial}{\partial t}\left(\tfrac{1}{2}\varepsilon E^2+\tfrac{1}{2}\mu H^2\right)-\sigma E^2.$$

In integral form one obtains

$$\oint_s(\mathbf{E}\times\mathbf{H})\cdot d\mathbf{s}=\underbrace{-\frac{\partial}{\partial t}\int_v\left(\tfrac{1}{2}\varepsilon E^2+\tfrac{1}{2}\mu H^2\right)dv}_{\substack{\text{time–rate of change of energy stored in}\\ \text{the electric field }\mathbf{E}\text{ and magnetic field }\mathbf{H}}}-\underbrace{\int_v\sigma E^2 dv.}_{\substack{\text{ohmic power dissipated}\\ \text{in the presence of }\mathbf{E}}}$$

The right side of the equation derived gives the rate of decrease of the electric and magnetic energies stored minus the ohmic power dissipated as heat in the volume $v$. The pointing vector, which is a power density vector, represents the power flows per unit area, and

$$\mathbf{P}=\mathbf{E}\times\mathbf{H}.$$

Furthermore,

$$\underbrace{\oint_s(\mathbf{E}\times\mathbf{H})\cdot d\mathbf{s}=\oint_s\mathbf{P}\cdot d\mathbf{s}}_{\text{power leaving the enclosed volume}}=\frac{\partial}{\partial t}\int_v(w_E+w_H)dv+\int_v\rho_\sigma dv,$$

where $w_E=\tfrac{1}{2}\varepsilon E^2$ and $w_H=\tfrac{1}{2}\mu H^2$ are the electric and magnetic energy densities; $\rho_\sigma=\sigma E^2=\frac{1}{\sigma}J^2$ is the ohmic power density.

The important conclusion is that the total power transferred into a closed surface $s$ at any instant equals the sum of the rate of increase of the stored electric and magnetic energies and the ohmic power dissipated within the enclosed volume $v$.

If the source charge density $\rho_v(x,y,z,t)$ and the source current density $\mathbf{J}(x,y,z,t)$ vary sinusoidally, the electromagnetic field also vary sinusoidally. Hence, we have deal with the so-called time-harmonic electromagnetic fields. The sinusoidal time-varying electromagnetic fields will be studied. Hence, the phasor analysis is applied. For example,

$$\mathbf{E}(\mathbf{r})=E_x(\mathbf{r})\mathbf{a}_x+E_y(\mathbf{r})\mathbf{a}_y+E_z(\mathbf{r})\mathbf{a}_z.$$

The electric field intensity components are the complex functions, and

$$E_x(\mathbf{r})=E_{x\mathrm{Re}}+jE_{x\mathrm{Im}},\ E_y(\mathbf{r})=E_{y\mathrm{Re}}+jE_{y\mathrm{Im}},\ E_z(\mathbf{r})=E_{z\mathrm{Re}}+jE_{z\mathrm{Im}}.$$

For the real electromagnetic field, we have

$$E_x(\mathbf{r},t)=E_{x\mathrm{Re}}(\mathbf{r})\cos\omega t-E_{x\mathrm{Im}}(\mathbf{r})\sin\omega t.$$

One obtains the time-harmonic electromagnetic field equations. In particular,

- Faraday's law $\nabla\times\mathbf{E}=-j\omega\mu\mathbf{H}$,
- Generalized (by Maxwells) Amphere's law

$$\nabla\times\mathbf{H}=\sigma\mathbf{E}+j\omega\varepsilon\mathbf{E}+\mathbf{J}=j\omega\left(\frac{\sigma}{j\omega}+\varepsilon\right)\mathbf{E}+\mathbf{J},$$

- Gauss's law $\nabla\cdot\mathbf{E}=\dfrac{\rho_v}{\dfrac{\sigma}{j\omega}+\varepsilon}$,
- Continuity of magnetic flux $\nabla\cdot\mathbf{H}=0$,
- Continuity law $\nabla\cdot\mathbf{J}=-j\omega\rho_v$, (5.7.1)

where $\left(\dfrac{\sigma}{j\omega}+\varepsilon\right)$ is the complex permittivity. However, for simplicity we will use $\varepsilon$ keeping in mind that the expression for the complex permittivity $\dfrac{\sigma}{j\omega}+\varepsilon$ must be applied.

The electric field intensity ***E***, electric flux density ***D***, magnetic field intensity ***H***, magnetic flux density ***B***, and current density **J** are complex-valued functions of spatial coordinates.

From the equation (5.7.1) taking the curl of $\nabla\times\mathbf{E}=-j\omega\mu\mathbf{H}$, which is rewritten as $\nabla\times\mathbf{E}=-j\omega\mathbf{B}$, and using $\nabla\times\mathbf{H}=j\omega\mathbf{D}+\mathbf{J}$, one obtains

$$\nabla\times\nabla\times\mathbf{E}=\omega^2\mu\varepsilon\mathbf{E}=k_v^2\mathbf{E}=-j\omega\mu\mathbf{J},$$

where $k_v$ is the wave constant $k_v=\omega\sqrt{\mu\varepsilon}$, and in free space $k_{v0}=\omega\sqrt{\mu_0\varepsilon_0}=\dfrac{\omega}{c}$ because the speed of light is $c=\dfrac{1}{\sqrt{\mu_0\varepsilon_0}}$, $c=3\times10^8\ \frac{\text{m}}{\text{sec}}$.

The wavelength is found as

$$\lambda_v=\frac{2\pi}{k_v}=\frac{2\pi}{\omega\sqrt{\mu\varepsilon}},$$

and in free space

$$\lambda_{v0}=\frac{2\pi}{k_{v0}}=\frac{2\pi c}{\omega}.$$

Using the magnetic vector potential ***A***, we have $\mathbf{B}=\nabla\times\mathbf{A}$.

Hence,

$$\nabla\times(\mathbf{E}+j\omega\mathbf{A})=0,$$

and thus

$$\mathbf{E}+j\omega\mathbf{A}=-\nabla\mathbf{\Lambda},$$

where $\mathbf{\Lambda}$ is the scalar potential.

To guarantee that $\nabla\times\mathbf{H}=j\omega\mathbf{D}+\mathbf{J}$ holds, it is required that

$$\nabla\times\mu\mathbf{H}=\nabla\times\nabla\times\mathbf{A}=\nabla\nabla\cdot\mathbf{A}-\nabla^2\mathbf{A}=j\omega\mu\varepsilon\mathbf{E}+\mu\mathbf{J}.$$

Therefore, one finally finds the equation needed to be solved

$$\nabla^2\mathbf{A}+k_v^2\mathbf{A}=\nabla(\nabla\cdot\mathbf{A}+j\omega\mu\varepsilon\mathbf{\Lambda})-\mu\mathbf{J}.$$

Taking note of the Lorentz condition $\nabla\cdot\mathbf{A}=-j\omega\mu\varepsilon\mathbf{\Lambda}$, one obtains.

$$\nabla^2\mathbf{A}+k_v^2\mathbf{A}=-\mu\mathbf{J}.$$

Thus, the equation for $\mathbf{\Lambda}$ is found. In particular,

$$\nabla^2\mathbf{\Lambda}+k_v^2\mathbf{\Lambda}=-\frac{\rho_v}{\varepsilon}.$$

The equation for the magnetic vector potential is found solving the following inhomogeneous Helmholtz equation

$$\nabla^2\mathbf{A}+k_v^2\mathbf{A}=-\mu\mathbf{J}.$$

The expression for the electromagnetic field intensity, in terms of the vector potential, is

$$\mathbf{E}=-j\omega\mathbf{A}+\frac{\nabla\nabla\cdot\mathbf{A}}{j\omega\mu\varepsilon}.$$

To derive **E**, one must have **A**. The Laplacian for **A** in different coordinate systems can be found. For example, we have

$$\nabla^2A_x+k_v^2A_x=-\mu J_x,$$

$$\nabla^2A_y+k_v^2A_y=-\mu J_y,$$

$$\nabla^2A_z+k_v^2A_z=-\mu J_z.$$

It was shown that the magnetic vector potential and the scalar potential obey the time-dependent inhomogeneous wave equation

$$\left(\nabla^2-k\frac{\partial^2}{\partial t^2}\right)\Omega(\mathbf{r},t)=-F(\mathbf{r},t).$$

The solution of this equation is found using Green's function as

$$\Omega(\mathbf{r},t)=-\iiiint F(\mathbf{r}',t')G(\mathbf{r}-\mathbf{r}';t-t')dt'd\tau',$$

where $G(\mathbf{r}-\mathbf{r}';t-t')=-\frac{1}{4\pi|\mathbf{r}-\mathbf{r}'|}\delta\left(t-t'-k|\mathbf{r}-\mathbf{r}'|\right).$

The so-called retarded solution is

$$\Omega(\mathbf{r},t)=-\iiint\frac{F(\mathbf{r}',t'-k|\mathbf{r}-\mathbf{r}'|)}{|\mathbf{r}-\mathbf{r}'|}d\tau'.$$

For sinusoidal electromagnetic fields, we apply the Fourier analysis to obtain

$$\Omega(\mathbf{r}) = -\frac{1}{4\pi}\iiint \frac{e^{-jk_v|\mathbf{r}-\mathbf{r}'|}}{|\mathbf{r}-\mathbf{r}'|}F(\mathbf{r}')d\tau'.$$

Thus, we have the expressions for the phasor retarded potentials

$$\mathbf{A}(\mathbf{r}) = \frac{\mu}{4\pi}\int_v \frac{e^{-jk_v|\mathbf{r}-\mathbf{r}'|}}{|\mathbf{r}-\mathbf{r}'|}\mathbf{J}(\mathbf{r}')dv,$$

$$\Lambda(\mathbf{r}) = \frac{1}{4\pi\varepsilon}\int_v \frac{e^{-jk_v|\mathbf{r}-\mathbf{r}'|}}{|\mathbf{r}-\mathbf{r}'|}\rho(\mathbf{r}')dv.$$

*Example 5.7.1*

Consider a short (*dl*) thin filament of current located in the origin, see Figure 5.7.1. Derive the expressions for magnetic vector potential and electromagnetic field intensities.

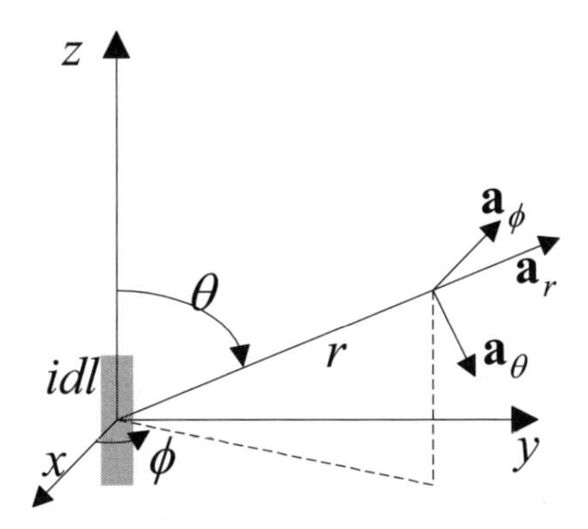


Figure 5.7.1 Current filament in the spherical coordinate system.

*Solution.*

The magnetic vector potential has only a *z* component, and thus, from

$$\nabla^2\mathbf{A} + k_v^2\mathbf{A} = -\mu\mathbf{J},$$

we have

$$\nabla^2 A_z + k_v^2 A_z = -\mu J_z = -\mu\frac{i}{ds},$$

where *ds* is the cross-sectional area of the filament.

Taking note of the spherical symmetry, we conclude that the magnetic vector potential $A_z$ is not a function of the polar and azimuth angles $\theta$ and $\phi$. In particular, the following equation results

$$\frac{1}{r^2}\frac{\partial}{\partial r}r^2\frac{\partial A_z}{\partial r} + k_v^2 A_z = 0.$$

It is well-known that the solution of equation $\frac{d^2\psi}{d\psi^2}+k_v^2\psi=0$ has two components. In particular, $e^{jk_v r}$ (outward propagation) and $e^{-jk_v r}$ (inward propagation). The inward propagation is not a part of solution for the filament located in the origin. Thus, we have

$\psi(t,r)=ae^{j\omega t-jk_v r}$ (outward propagating spherical wave).

In free space, we have

$\psi(t,r)=ae^{j\omega(t-r/c)}$.

Substituting $A_z=\frac{\psi}{r}$, one obtains $A_z(r)=\frac{a}{r}e^{-\frac{j\omega r}{c}}$.

To find the constant $a$, we use the volume integral

$$\int_v \nabla^2 A_z dv=\oint_s \nabla A_z\cdot\mathbf{a}_r r_d^2\sin\theta d\theta d\phi=-\frac{\omega^2}{c^2}\int_v A_z dv-\mu_0\int_v J_z dv,$$

where the differential spherical volume is $dv=r_d^2\sin\theta d\theta d\phi dr$; $r_d$ is the differential radius.

Making use of

$$\nabla A_z\cdot\mathbf{a}_r=\frac{\partial A_z}{\partial r}=-\left(1+j\frac{\omega}{c}r\right)\frac{a}{r^2}e^{-j\frac{\omega}{c}r},$$

we have

$$\lim_{r_d\to 0}\int_0^{2\pi}\int_0^{\pi}-\left(1+j\frac{\omega}{c}r_d\right)ae^{-j\frac{\omega}{c}r_d}\sin\theta d\theta d\phi=-4\pi a=-\mu_0 idl,$$

one has $a=\frac{\mu_0 idl}{4\pi}$.

Thus, the following expression results

$$A_z(r)=\frac{\mu_0 idl}{4\pi r}e^{-\frac{j\omega r}{c}}.$$

Therefore, the final equation for the magnetic vector potential (outward propagating spherical wave) is

$$\mathbf{A}(r)=\frac{\mu_0 idl}{4\pi r}e^{-\frac{j\omega r}{c}}\mathbf{a}_z.$$

From $\mathbf{a}_z=\mathbf{a}_r\cos\theta-\mathbf{a}_\theta\sin\theta$, we have

$$\mathbf{A}(r)=\frac{\mu_0 idl}{4\pi r}e^{-\frac{j\omega r}{c}}(\mathbf{a}_r\cos\theta-\mathbf{a}_\theta\sin\theta)$$

The magnetic and electric field intensities are found using

$$\mathbf{B} = \nabla \times \mathbf{A} \text{ and } \mathbf{E} = -j\omega\mathbf{A} + \frac{\nabla\nabla \cdot \mathbf{A}}{j\omega\mu\varepsilon}.$$

Then, one finds

$$\mathbf{H}(r) = \frac{1}{\mu_0}\nabla \times \mathbf{A}(r) = \frac{idl\sin\theta}{4\pi r}\left(\frac{j\omega}{cr} + \frac{1}{r^2}\right)e^{-\frac{j\omega r}{c}}\mathbf{a}_\phi,$$

$$\mathbf{E}(r) = \frac{j\sqrt{\frac{\mu_0}{\varepsilon_0}}cidl}{4\pi\omega}\cos\theta\left(\frac{j\omega}{cr^2} + \frac{1}{r^3}\right)e^{-\frac{j\omega r}{c}}\mathbf{a}_r$$

$$-\frac{j\sqrt{\frac{\mu_0}{\varepsilon_0}}cidl}{4\pi\omega}\sin\theta\left(-\frac{\omega^2}{c^2 r} + \frac{j\omega}{cr^2} + \frac{1}{r^3}\right)e^{-\frac{j\omega r}{c}}\mathbf{a}_\theta.$$

The intrinsic impedance is given as

$$Z_0 = \sqrt{\frac{\mu_0}{\varepsilon_0}}, \text{ and } Y_0 = \frac{1}{Z_0} = \sqrt{\frac{\varepsilon_0}{\mu_0}}.$$

Near-field and far-field electromagnetic radiation fields can be found, simplifying the expressions for $\mathbf{H}(r)$ and $\mathbf{E}(r)$.

For near-field, we have

$$\mathbf{H}(r) = \frac{1}{\mu_0}\nabla \times \mathbf{A}(r) = j\frac{idl\sin\theta\omega}{4\pi cr^2}e^{-\frac{j\omega r}{c}}\mathbf{a}_\phi$$

and $\mathbf{E}(r) = j\dfrac{\sqrt{\frac{\mu_0}{\varepsilon_0}}cidl\omega}{4\pi c^2 r}\sin\theta\mathbf{a}_\theta.$

The complex Pointing vector can be found as

$$\tfrac{1}{2}\mathbf{E}(r) \times \mathbf{H}^*(r).$$

The following expression for the complex power flowing out of a sphere of radius $r$ results

$$\tfrac{1}{2}\oint_s \left(\mathbf{E}(r) \times \mathbf{H}^*(r)\right)\cdot d\mathbf{s} = \frac{\omega^2\mu_0\sqrt{\mu_0\varepsilon_0}i^2 dl^2}{12\pi} = \frac{\omega\mu_0 k_v i^2 dl^2}{12\pi}.$$

The real quality is found, and the power dissipated in the sense that it travels away from source and cannot be recovered. □

*Example 5.7.2*

Derive the expressions for the magnetic vector potential and electromagnetic field intensities for a magnetic dipole (small current loop) which is shown in Figure 5.7.2.

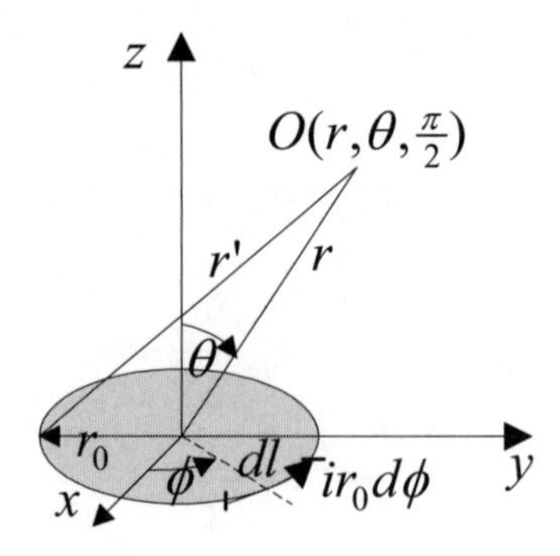


Figure 5.7.2 Current loop in the *xy* plane.

*Solution.*

The magnetic dipole moment is equal to the current loop are times current. That is,

$$\mathbf{M} = \pi r_0^2 i\mathbf{a}_z = M\mathbf{a}_z .$$

For the short current filament, it was derived in Example 5.7.1 that

$$\mathbf{A}(r) = \frac{\mu_0 idl}{4\pi r} e^{-\frac{j\omega r}{c}} \mathbf{a}_z .$$

In contrast, we have

$$\mathbf{A} = \frac{\mu_0 i}{4\pi} \oint_l \frac{1}{r'} dl .$$

The distance between the source element *dl* and point $O(r,\theta,\frac{\pi}{2})$ is denoted as *r*'. It should be emphasized that the current filament lies in the *xy* planc, and

$$dl = \mathbf{a}_\phi r_0 d\phi = (-\mathbf{a}_x \sin\phi + \mathbf{a}_y \cos\phi) r_0 d\phi .$$

Thus, due to the symmetry

$$\mathbf{A} = \mathbf{a}_\phi \frac{\mu_0 i r_0}{2\pi} \int_{-\pi/2}^{\pi/2} \frac{\sin\phi}{r'} d\phi ,$$

where using the trigonometric identities one finds

$$r'^2 = r^2 + r_0^2 - 2rr_0 \sin\theta \sin\phi .$$

Assuming that $r^2 >> r_0^2$, we have $\frac{1}{r'} \approx \frac{1}{r}\left(1 + \frac{r_0}{r} \sin\theta \sin\phi\right)$.

Therefore

$$\mathbf{A}=\mathbf{a}_\phi \frac{\mu_0 i r_0}{2\pi}\int_{-\pi/2}^{\pi/2}\frac{\sin\phi}{r'}d\phi$$

$$=\mathbf{a}_\phi \frac{\mu_0 i r_0}{2\pi r}\int_{-\pi/2}^{\pi/2}\left(1+\frac{r_0}{r}\sin\theta\sin\phi\right)\sin\phi d\phi=\mathbf{a}_\phi \frac{\mu_0 i r_0^2}{4r^2}\sin\theta.$$

Having obtained the explicit expression for the vector potential, the magnetic flux density is found. In particular,

$$\mathbf{B}=\nabla\times\mathbf{A}=\nabla\times\mathbf{a}_\phi \frac{\mu_0 i r_0^2}{4r^2}\sin\theta=\frac{\mu_0 i r_0^2}{4r^3}(2\mathbf{a}_r\cos\theta+\mathbf{a}_\theta\sin\theta).$$

Taking note of the expression for the magnetic dipole moment $\mathbf{M}=\pi r_0^2 i\mathbf{a}_z$, one has

$$\mathbf{A}=\mathbf{a}_\phi \frac{\mu_0 i r_0^2}{4r^2}\sin\theta=\frac{\mu_0}{4\pi r^2}\mathbf{M}\times\mathbf{a}_r.$$

It was shown that using $\mathbf{A}=\frac{\mu_0 i}{4\pi}\oint_l \frac{1}{r'}dl$, the desired results are obtained.

Let us apply $\mathbf{A}=\frac{\mu_0 i}{4\pi}\oint_l \frac{e^{-j\frac{\omega}{c}r'}}{r'}dl$.

From $e^{-j\frac{\omega}{c}r'}\approx\left[1-j\frac{\omega}{c}(r'-r)\right]e^{-j\frac{\omega}{c}r}$, we have

$$\mathbf{A}=\frac{\mu_0 i}{4\pi}\oint_l \frac{[1-j\frac{\omega}{c}(r'-r)]e^{-j\frac{\omega}{c}r}}{r'}dl=\mathbf{a}_\phi \frac{\mu_0 M}{4\pi r^2}\left(1+j\frac{\omega}{c}r\right)e^{-j\frac{\omega}{c}r}\sin\theta.$$

Therefore, one finds

$$E_\phi=j\frac{\mu_0\omega^3 M}{4c^2\pi}\left(\frac{1}{j\frac{\omega}{c}r}-\frac{1}{\frac{\omega^2}{c^2}r^2}\right)e^{-j\frac{\omega}{c}r}\sin\theta,$$

$$H_r=j\frac{2\mu_0\omega^3 M}{4c^2\sqrt{\frac{\mu_0}{\varepsilon_0}}\pi}\left(\frac{1}{\frac{\omega^2}{c^2}r^2}+\frac{1}{j\frac{\omega^3}{c^3}r^3}\right)e^{-j\frac{\omega}{c}r}\cos\theta,$$

$$H_\theta=-j\frac{\mu_0\omega^3 M}{4c^2\sqrt{\frac{\mu_0}{\varepsilon_0}}\pi}\left(\frac{1}{j\frac{\omega}{c}r}-\frac{1}{\frac{\omega^2}{c^2}r^2}-\frac{1}{j\frac{\omega^3}{c^3}r^3}\right)e^{-j\frac{\omega}{c}r}\sin\theta.$$

The electromagnetic fields in near- and far-fields can be straightforwardly derived, and thus, the corresponding approximations for the $E_\phi$, $H_r$ and $H_\theta$ can be obtained. □

Let the current density distribution in the volume be given as $\mathbf{J}(\mathbf{r}_0)$, and for far-field from Figure 5.7.3 one has $\mathbf{r} \approx \mathbf{r}' - \mathbf{r}_0$.

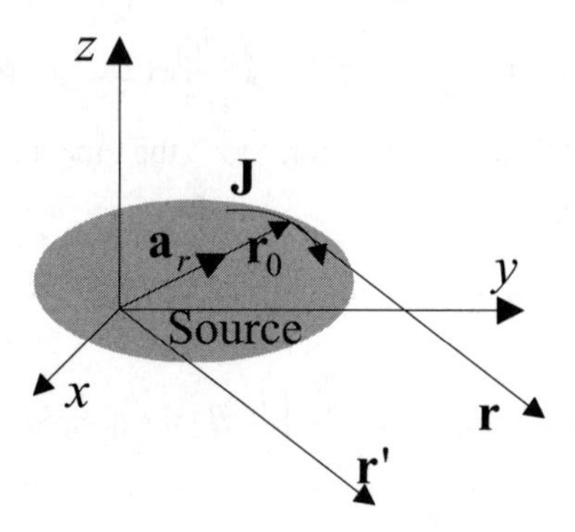


Figure 5.7.3 Radiation from volume current distribution.

The formula to calculate far-field magnetic vector potential is

$$\mathbf{A}(\mathbf{r}) = \frac{\mu}{4\pi r} e^{-jk_v r} \int_v \mathbf{J}(\mathbf{r}_0) e^{-jk_v \mathbf{r}_0} dv,$$

and the electric and magnetic field intensities are found using

$$\mathbf{E} = -j\omega \mathbf{A} + \frac{\nabla\nabla \cdot \mathbf{A}}{j\omega\mu\varepsilon}$$

and $\mathbf{B} = \nabla \times \mathbf{A}$.

We have

$$\mathbf{E}(\mathbf{r}) = \frac{jk_v Z_v}{4\pi r} e^{-jk_v r} \int_v [\mathbf{a}_r \cdot \mathbf{J}(\mathbf{r}_0)\mathbf{a}_r - \mathbf{J}(\mathbf{r}_0)] e^{-jk_v \mathbf{r}_0} dv,$$

$$\mathbf{H}(\mathbf{r}) = Y_v \mathbf{a}_r \times \mathbf{E}(\mathbf{r}).$$

*Example 5.7.3*

Consider the half-wave dipole antenna fed from a two-wire transmission line, as shown in Figure 5.7.4. The antenna is one-quarter wavelength; that is, $-\frac{1}{4}\lambda_v \le z \le \frac{1}{4}\lambda_v$. The current distribution is $i(z) = i_0 \cos k_v z$. Obtain the equations for electromagnetic field intensities and radiated power.

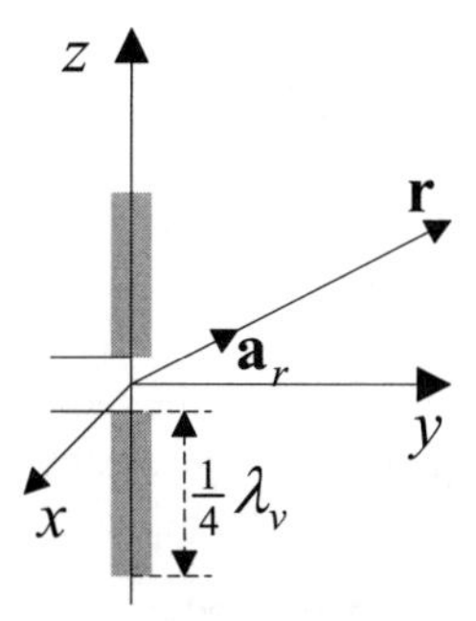


Figure 5.7.4 Half-wave dipole antenna.

*Solution.*

The wavelength is given as $\lambda_v = \dfrac{2\pi}{k_v} = \dfrac{2\pi}{\omega\sqrt{\mu\varepsilon}}$.

Thus, in free space we have $\lambda_{v0} = \dfrac{2\pi}{k_{v0}} = \dfrac{2\pi c}{\omega}$.

It was emphasized that $k_{v0} = \omega\sqrt{\mu_0\varepsilon_0}$.

Making use of

$$\mathbf{E}(\mathbf{r}) = \frac{jk_v Z_v}{4\pi r} e^{-jk_v r} \int_v \left[\mathbf{a}_r \cdot \mathbf{J}(\mathbf{r}_0)\mathbf{a}_r - \mathbf{J}(\mathbf{r}_0)\right] e^{-jk_v \mathbf{r}_0} dv,$$

we have the following line integral

$$\mathbf{E}(\mathbf{r}) = \frac{jk_v Z_v}{4\pi r} e^{-jk_v r} \oint_l \left[(\mathbf{a}_r \cdot \mathbf{a}_l)\mathbf{a}_r - \mathbf{a}_l\right] i(l) e^{-jk_v \mathbf{r}_0} dl,$$

where $\mathbf{a}_l$ is the unit vector in the current direction.

Then,

$$\mathbf{E}(\mathbf{r}) = \frac{jk_v Z_v i_0}{4\pi r} e^{-jk_v r} \int_{-\frac{1}{4}\lambda_v}^{\frac{1}{4}\lambda_v} (\mathbf{a}_r \cos\theta - \mathbf{a}_z) \cos k_v z e^{-jk_v z \cos\theta} dz$$

$$= \frac{jZ_v i_0 \cos(\frac{1}{2}\pi\cos\theta)}{2\pi r \sin\theta} e^{-jk_v r} \mathbf{a}_\theta.$$

Having found the magnetic field intensity as

$$\mathbf{H}(\mathbf{r}) = Y_v \mathbf{a}_r \times \mathbf{E}(\mathbf{r}) = H_\phi \mathbf{a}_\phi = \frac{ji_0 \cos(\frac{1}{2}\pi\cos\theta)}{2\pi r \sin\theta} e^{-jk_v r} \mathbf{a}_\phi,$$

the power flux per unit area is

$$\tfrac{1}{2}\mathrm{Re}\left(\mathbf{E}(\mathbf{r}) \times \mathbf{H}(\mathbf{r})^* \cdot \mathbf{a}_r\right) = \tfrac{1}{2} E_\phi H_\phi^* = \frac{|i_0|^2 Z_0 \cos^2(\frac{1}{2}\pi\cos\theta)}{8\pi^2 r^2 \sin^2\theta}.$$

Integrating the derived expression over the surface

$$\frac{|i_0|^2 Z_0}{8\pi^2}\int_0^{2\pi}\int_0^{\pi}\frac{\cos^2(\frac{1}{2}\pi\cos\theta)}{\sin^2\theta}\sin\theta d\theta d\phi,$$

the total radiated power is found to be $36.6|i_0|^2$. □

If the current density distribution is known, the radiation field can be found. Using Maxwell's equations, taking note of the electric and magnetic vector potentials $\mathbf{A}_E$ and $\mathbf{A}_H$, we have the following equations

$$\left(\nabla^2+k_v^2\right)\mathbf{A}_E=-\mu\mathbf{J}_E,\ \left(\nabla^2+k_v^2\right)\mathbf{A}_H=-\varepsilon\mathbf{J}_H,$$

$$\mathbf{E}=-j\omega\mathbf{A}_E+\frac{1}{j\omega\mu\varepsilon}\nabla\nabla\cdot\mathbf{A}_E-\frac{1}{\varepsilon}\nabla\times\mathbf{A}_H,$$

$$\mathbf{H}=-j\omega\mathbf{A}_H+\frac{1}{j\omega\mu\varepsilon}\nabla\nabla\cdot\mathbf{A}_H-\frac{1}{\mu}\nabla\times\mathbf{A}_E.$$

The solutions are

$$\mathbf{A}_E(\mathbf{r})=\frac{\mu}{4\pi r}e^{-jk_v r}\int_v e^{jk_v\mathbf{r}}\mathbf{J}_E(\mathbf{r})d\mathbf{r},$$

$$\mathbf{A}_H(\mathbf{r})=\frac{\varepsilon}{4\pi r}e^{-jk_v r}\int_v e^{jk_v\mathbf{r}}\mathbf{J}_H(\mathbf{r})d\mathbf{r}.$$

*Example 5.7.4*

Consider the slot (one-half wavelength long slot is dual to the half-wave dipole antenna studied in Example 5.7.3), which is exited from the coaxial line, see Figure 5.7.5. The electric field intensity in the *z*-direction is $E=E_0\sin k_v\left(l-|z|\right)$. Derive the expressions for the magnetic vector potential and electromagnetic field intensities.

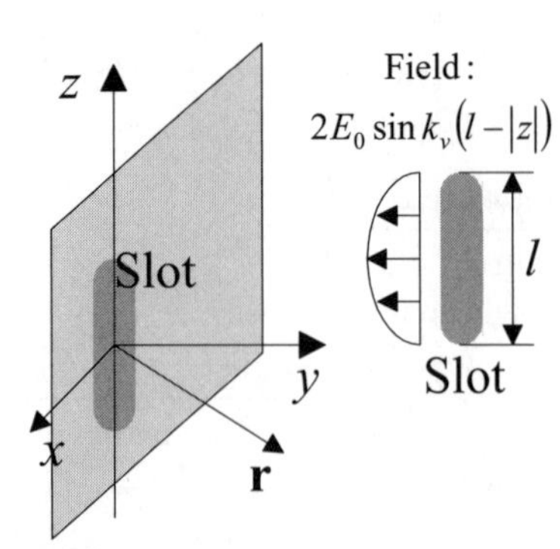


Figure 5.7.5 Slot antenna.

*Solution.*

Using the magnetic current density $\mathbf{J}_H$, from

$$\int_s \nabla\times\mathbf{E}\cdot d\mathbf{s}=\oint_l \mathbf{E}\cdot dl=-\int_s j\omega\mathbf{B}\cdot d\mathbf{s}-\int_s \mathbf{J}_H\cdot d\mathbf{s},$$

the boundary conditions for the magnetic current sheet are found as

$$\mathbf{a}_n\times\mathbf{E}_1-\mathbf{a}_n\times\mathbf{E}_2=-\mathbf{J}_H.$$

The slot antenna is exited by the magnetic current with strength $2E_0\sin k_v\left(l-|z|\right)$ in the $z$ axis.

For half-wave slot we have $i_H=i_0\sin k_v\left(l-|z|\right)$, and

$$\left(\nabla^2+k_v^2\right)\mathbf{A}_H=-\varepsilon\mathbf{J}_H,$$

$$\mathbf{H}=-j\omega\mathbf{A}_H+\frac{\nabla\nabla\cdot\mathbf{A}_H}{j\omega\mu\varepsilon}=j\frac{i_0Y_0\cos(\frac{1}{2}\pi\cos\theta)}{2\pi r\sin\theta}e^{-jk_vr}\mathbf{a}_\theta,$$

$$\mathbf{E}=-\frac{\nabla\times\mathbf{A}_H}{\varepsilon}=-j\frac{i_0\cos(\frac{1}{2}\pi\cos\theta)}{2\pi r\sin\theta}e^{-jk_vr}\mathbf{a}_\phi.,$$

$$\mathbf{A}_H(\mathbf{r})=\frac{\varepsilon}{16\pi}e^{-jk_vr}\int_s e^{jk_v\mathbf{r}}\mathbf{J}(\mathbf{r})ds.$$

The boundary condition

$$\mathbf{a}_n\times\mathbf{E}=-\tfrac{1}{2}\mathbf{J}_H=\mathbf{a}_n\times\mathbf{a}_xE_0\sin k_v\left(l-|z|\right)$$

is satisfied by the radiated electromagnetic field.

The radiation pattern of the slot antenna is the same as for the dipole antenna. □

## References

1. W. H. Hayt, *Engineering Electromagnetics*, McGraw-Hill, New York, 1989.
2. R. E. Collin, *Antennas and Radiowave propagation*," McGraw-Hill, New York, 1985.
3. C. R. Paul, K. W. Whites and S. A. Nasar, *Introduction to Electromagnetic Fields*, McGraw-Hill, New York, 1998.

## 5.8 CLASSICAL MECHANICS AND ITS APPLICATION

With advanced molecular computer-aided-design tools, one can design, analyze, and evaluate three-dimensional (3D) nanostructures in the steady-state. However, the comprehensive analysis in the time domain needs to be performed. That is, the designer must study the dynamic evolution of MEMS and NEMS. Conventional methods of molecular mechanics do not allow one to perform numerical analysis of complex MEMS and NEMS in time-domain, and even 3D modeling is restricted to simple structures. Our goal is to develop a fundamental understanding of electromechanical and electromagnetic processes in nano- and microscale structures. An addition, the basic theoretical foundations will be developed and used in analysis of MEMS and NEMS from systems standpoints. That is, we depart from the subsystem analysis, and study MEMS and NEMS as dynamics systems.

From modeling, simulation, analysis and visualization standpoints, MEMS and NEMS are very complex. In fact, MEMS and NEMS are modeled using advanced concepts of quantum mechanics, electromagnetic theory, structural dynamics, thermodynamics, thermochemistry, etc. It was illustrated that MEMS and NEMS integrate a great number of components (subsystems, devices, and structures), and the mathematical model development is an extremely challenging problem because the commonly used conventional methods, assumptions, and simplifications may not be applied to MEMS and NEMS (for example, the Newtonian mechanics is not applicable to the molecular-scale analysis, and Maxwell's equations must be contemporarily applied to study complex electromagnetic phenomena). As the result, partial differential equations describe multivariable mathematical models of MEMS and NEMS.

The visualization issues must be addressed to study the complex tensor data (tensor field). Techniques and software for visualizing scalar and vector field data are available to visualize the data in three dimensions. In contrast, techniques to visualize tensor fields are not available due to complex, multivariate nature of the data and the fact that no commonly used experimental analogy exists for visualizing tensor data. The second-order tensor fields consist of $3\times3$ matrixes defined at each node in a computational grid. Tensor field variables can include stress, viscous stress, rate of strain, and momentum (tensor variables in conventional structural dynamics include stress and strain). The tensor field can be simplified and visualized as a scalar field. Alternatively, the individual vectors that comprise the tensor field can be analyzed. However, these simplifications result in the loss of valuable information needed to analyze complex tensor fields. Vector fields can be visualized using streamlines which depict a subset of the data. Hyperstreamlines, as an extension of the streamlines to the second-order tensor fields, provide one with a continuous representation of the tensor field along a three-dimensional path. Due to obvious limitations and scope, this book does not cover the tensor field topologies, and through this brief discussion in result visualization, the author emphasizes multidisciplinary nature and complexity of the phenomena in MEMS and NEMS.

While some results have been thoroughly studied, many important aspects have not been approached and researched, primarily due to the multidisciplinary nature and complexity of MEMS and NEMS. The major objectives of this book are to study the fundamental theoretical foundations, develop innovative concepts in structural design and optimization, perform modeling and simulation, as well as solve the motion control problem, and validate the results. To develop mathematical models, we augment nano- or microactuator/sensor and circuitry dynamics (the dynamics can be studied at the nano and micro scales). Newtonian and quantum mechanics, Lagrange's and Hamilton's concepts, and other cornerstone theories are used to model MEMS and NEMS dynamics in the time domain. Taking note of these basic principles and laws, nonlinear mathematical models are found to perform comprehensive analysis and design. The control mechanisms and decision making are discussed, and control algorithms must be synthesized to attain the desired specifications and requirements imposed on the performance. It is evident that nano- and microsystem features must be thoroughly considered when approaching modeling and simulation, analysis, and design. The ability to find mathematical models is a key problem in MEMS and NEMS analysis and optimization, synthesis and control, manufacturing and commercialization. For MEMS, using electromagnetic theory and electro-mechanics, we develop adequate mathematical models to attain the design objectives. The proposed approach, which augments electromagnetics and electromechanics, allows the designer to solve a much broader spectrum of problems compared with finite-element analysis because an interactive electro-magnetic-mechanical-ICs analysis is performed.

In this book the author studies large-scale MEMS and NEMS (actuators and sensors have been primarily studied and analyzed from the fabrication standpoints) and thorough fundamental theory is developed. Applying the theoretical foundations to analyze and regulate in the desired manner the energy or information flows in MEMS and NEMS, the designer is confronted with the need to find adequate mathematical models of the phenomena and design MEMS and NEMS configurations. Mathematical models can be found using basic physical concepts. In particular, in electrical, mechanical, fluid or thermal systems, the mechanism of storing, dissipating, transforming, and transferring energies are analyzed. We will use the Lagrange equations of motion, Kirchhoff's and Newton's laws, Maxwell's equations and quantum theory to illustrate the model developments. It was emphasized that MEMS and NEMS integrate many components and subsystems. One can reduce interconnected systems to simple, idealized subsystems (components). However, this idealization is unpractical. For example, one cannot study nano- and microscale actuators and sensors without studying subsystems (devices) to actuate and control these transducers. That is, MEMS and NEMS integrate mechanical and electromechanical motion devices (actuators and sensors), power converters and antennas, processors and IO devices, etc. One of the primary objective of this book is to illustrate how one can develop comprehensive mathematical models of MEMS and NEMS using basic principles and laws. Through illustrative examples, differential equations will be found to model dynamic systems.

Based upon the synthesized MEMS and NEMS architectures, to analyze and regulate in the desired manner the energy or information flows, the designer needs to find adequate mathematical models and optimize the performance characteristics through design of control algorithms. Some mathematical models can be found using basic foundations, and mathematical theory to map the dynamics of some processes and system evolution is not developed yet. In this section we study electrical, mechanical, fluid or thermal systems, the mechanism of storing, dissipating, transforming, and transferring energies in actuators and sensors which can be manufactured using a large variety of different nano-, micro-, and miniscale technologies. In this section we will use the Lagrange equations of motion, as well as Kirchhoff's and Newton's laws to illustrate the model developments applicable to a large class of nano- and microscale transducers. It has been illustrated that one cannot reduce interconnected systems (MEMS and NEMS) to simple, idealized subsystems (components). For example, one cannot study actuators and smart structures without studying the mechanism to regulate these actuators using ICs and antennas. These ICs and antennas are controlled by the processor which received the information from sensors. The primary objective of this chapter is to illustrate how one can develop mathematical models of dynamic systems using basic principles and laws. Through illustrative examples, differential equations will be found and simulated.

Nano- and microelectromechanical systems must be studied using the fundamental laws and basic principles of mechanics and electromagnetics. Let us identify and study these key concepts to illustrate the use of cornerstone principles. The study of the motion of systems with the corresponding analysis of forces that cause motion is our interest.

### 5.8.1 Newtonian Mechanics

*Newtonian Mechanics: Translational Motion*

The equations of motion for mechanical systems can be found using Newton's second law of motion. Using the position (displacement) vector $\vec{\mathbf{r}}$, the Newton equation in the vector form is given as

$$\sum \vec{F}(t,\vec{\mathbf{r}}) = m\vec{a}\,, \tag{5.8.1}$$

where $\sum \vec{F}(t,\vec{\mathbf{r}})$ is the vector sum of all forces applied to the body ($\vec{F}$ is called the *net* force); $\vec{a}$ is the vector of acceleration of the body with respect to an inertial reference frame; $m$ is the mass of the body.

From (5.8.1), in the Cartesian system (*xyz* coordinates) we have

$$\sum \vec{F}(t,\vec{\mathbf{r}}) = m\vec{a} = m\frac{d\vec{\mathbf{r}}^2}{dt^2} = m\begin{bmatrix} \frac{d\vec{x}^2}{dt^2} \\ \frac{d\vec{y}^2}{dt^2} \\ \frac{d\vec{z}^2}{dt^2} \end{bmatrix}, \begin{bmatrix} \vec{a}_x \\ \vec{a}_y \\ \vec{a}_z \end{bmatrix} = \begin{bmatrix} \frac{d\vec{x}^2}{dt^2} \\ \frac{d\vec{y}^2}{dt^2} \\ \frac{d\vec{z}^2}{dt^2} \end{bmatrix}.$$

In the Cartesian coordinate system, Newton's second law is expressed as

$$\sum F_x = ma_x, \sum F_y = ma_y, \sum F_z = ma_z.$$

It is worth noting that $m\vec{a}$ represents the magnitude and direction of the applied net force acting on the object. Hence, $m\vec{a}$ is not a force.

A body is at equilibrium (the object is at rest or is moving with constant speed) if

$$\sum \vec{F} = 0.$$

Newton's second law in terms of the linear momentum, which is found as $\vec{p} = m\vec{v}$, is given by

$$\sum \vec{F} = \frac{d\vec{p}}{dt} = \frac{d(m\vec{v})}{dt},$$

where $\vec{v}$ is the vector of the object velocity.

Thus, the force is equal to the rate of change of the momentum. The object or particle moves uniformly if $\frac{d\vec{p}}{dt} = 0$ (thus, $\vec{p} = const$).

Newton's laws are extended to multibody systems, and the momentum of a system of N particles is the vector sum of the individual momenta. That is,

$$\vec{P} = \sum_{i=1}^{N} \vec{p}_i.$$

Consider the multibody system of N particles. The position (displacement) is represented by the vector **r** which in the Cartesian coordinate system has the components *x*, *y* and *z*. Taking note of the expression for the potential energy $\Pi(\vec{\mathbf{r}})$, one has for the conservative mechanical system

$$\sum \vec{F}(\vec{\mathbf{r}}) = -\nabla\Pi(\vec{\mathbf{r}}).$$

Therefore, the work done per unit time is

$$\frac{dW}{dt} = \sum \vec{F}(\vec{\mathbf{r}})\frac{d\vec{\mathbf{r}}}{dt} = -\nabla\Pi(\vec{\mathbf{r}})\frac{d\vec{\mathbf{r}}}{dt} = -\frac{d\Pi(\vec{\mathbf{r}})}{dt}.$$

From Newton's second law one obtains

$$m\vec{a}-\sum\vec{F}(\vec{\mathbf{r}})=m\frac{d^2\vec{\mathbf{r}}}{dt^2}-\sum\vec{F}(\vec{\mathbf{r}})=0,$$

and hence, for a conservative system we have

$$m\frac{d^2\vec{\mathbf{r}}}{dt^2}+\nabla\Pi(\vec{\mathbf{r}})=0.$$

For the system of N particles, the equations of motion are

$$m_N\frac{d^2\vec{\mathbf{r}}_N}{dt^2}+\nabla\Pi(\vec{\mathbf{r}}_N)=0,$$

or $$m_i\frac{d^2(\vec{x}_i,\vec{y}_i,\vec{z}_i)}{dt^2}+\frac{\partial\Pi(\vec{x}_i,\vec{y}_i,\vec{z}_i)}{\partial(\vec{x}_i,\vec{y}_i,\vec{z}_i)}=0,\ i=1,\ldots,N.$$

The total kinetic energy of the particle is $\Gamma=\frac{1}{2}mv^2$, and for N particles, one has

$$\Gamma\left(\frac{d\vec{x}_i}{dt},\frac{d\vec{y}}{dt}_i,\frac{d\vec{z}_i}{dt}\right)=\frac{1}{2}\sum_{i=1}^{N}m_i\left(\frac{d\vec{x}_i}{dt},\frac{d\vec{y}}{dt}_i,\frac{d\vec{z}_i}{dt}\right).$$

Furthermore, we obtain

$$m_i\frac{d(\vec{x}_i,\vec{y}_i,\vec{z}_i)}{dt}=\frac{\partial\Gamma\left(\frac{d\vec{x}_i}{dt},\frac{d\vec{y}_i}{dt},\frac{d\vec{z}_i}{dt}\right)}{\partial\left(\frac{d\vec{x}_i}{dt},\frac{d\vec{y}_i}{dt},\frac{d\vec{z}_i}{dt}\right)}.$$

Using the generilized coordinates $(q_1,\ldots,q_n)$ and generalized velocities $\left(\frac{dq_1}{dt},\ldots,\frac{dq_n}{dt}\right)$, one finds the total kinetic $\Gamma\left(q_1,\ldots,q_n,\frac{dq_1}{dt},\ldots,\frac{dq_n}{dt}\right)$ and potential $\Pi(q_1,\ldots,q_n)$ energies. Hence, using the expressions for the total kinetic and potential energies, Newton's second law of motion can be given in the following form

$$\frac{d}{dt}\left(\frac{\partial\Gamma}{\partial\dot{q}_i}\right)+\frac{\partial\Pi}{\partial q_i}=0.$$

That is, the generalized coordinates $q_i$ are used to model multibody systems, and

$$(q_1,\ldots,q_n)=(\vec{x}_1,\vec{y}_1,\vec{z}_1,\ldots,\vec{x}_N,\vec{y}_N,\vec{z}_N).$$

The obtained results are connected to the Lagrange equations of motion which will be studied later.

### *Newtonian Mechanics: Rotational Motion*

For rotational motion, the net torque and angular acceleration are used. The rotational analog of Newton's second law for a rigid body is

$$\sum \vec{T} = J\vec{\alpha},$$

where $\sum \vec{T}$ is the *net* torque; $J$ is the moment of inertia (*rotational inertia*); $\vec{\alpha}$ is the angular acceleration vector, $\vec{\alpha} = \frac{d}{dt}\frac{d\vec{\theta}}{dt} = \frac{d^2\vec{\theta}}{dt^2} = \frac{d\vec{\omega}}{dt}$; $\vec{\theta}$ is the angular displacement; $\omega$ denotes the angular velocity.

The angular momentum of the system $\vec{L}_M$ is expressed as

$$\vec{L}_M = \vec{r} \times \vec{p} = \vec{r} \times m\vec{v},$$

and $\sum \vec{T} = \frac{d\vec{L}_M}{dt} = \vec{r} \times \vec{F}$,

where $\vec{r}$ is the position vector with respect to the origin.

For the rigid body, rotating around the axis of symmetry, we have

$$\vec{L}_M = J\vec{\omega}.$$

*Example 5.8.1*

A micromotor has the equivalent moment of inertia $J = 5 \times 10^{-20}$ kg-m$^2$. Let the angular velocity of the rotor is $\omega_r = 10t^{1/5}$. Find the angular momentum and the developed electromagnetic torque as functions of time. The load and friction torques are zero.

*Solution.*

The angular momentum is found as $L_M = J\omega_r = 5 \times 10^{-19} t^{1/5}$.

The developed electromagnetic torque is $T_e = \frac{dL_M}{dt} = 1 \times 10^{-19} t^{-4/5}$. □

From Newtonian mechanics one concludes that the applied net force plays a central role in quantitatively describing the motion. An alternative analysis of motion can be performed in terms of the energy or momentum quantities, which are conserved. The principle of conservation of energy states that energy can be only converted from one form to another. Kinetic energy is associated with motion, while potential energy is associated with position. The sum of the kinetic (Γ), potential (Π), and dissipated (D) energies is called the total energy of the system ($\Sigma_T$), which is conserved, and the total amount of energy remains constant; that is,

$$\Sigma_T = \Gamma + \Pi + D = const.$$

For example, consider the translational motion of a body which is attached to an ideal spring that obeys Hooke's law. Neglecting friction, one obtains the following expression for the total energy

$$\Sigma_T = \Gamma + \Pi = \tfrac{1}{2}(mv^2 + k_s x^2) = const\,.$$

Here, the translational kinetic energy is $\Gamma = \frac{1}{2}mv^2$; the elastic potential energy is $\Pi = \frac{1}{2}k_s x^2$; $k_s$ is the force constant of the spring; $x$ is the displacement.

For rotating spring, we have

$$\Sigma_T = \Gamma + \Pi = \tfrac{1}{2}(J\omega^2 + k_s\theta^2) = const\,,$$

where the rotational kinetic energy and the elastic potential energy are

$$\Gamma = \tfrac{1}{2}J\omega^2 \text{ and } \Pi = \tfrac{1}{2}k_s\theta^2\,.$$

The kinetic energy of a rigid body having translational and rotational components of motion is found to be

$$\Gamma = \tfrac{1}{2}(mv^2 + J\omega^2)\,.$$

That is, motion of the rigid body is represented as a combination of translational motion of the center of mass and rotational motion about the axis through the center of mass. The moment of inertia depends upon how the mass is distributed with respect to the axis, and $J$ is different for different axes of rotation. If the body is uniform in density, $J$ can be easily calculated for regularly shaped bodies in terms of their dimensions. For example, a rigid cylinder with mass $m$ (which is uniformly distributed), radius $R$, and length $l$, has the following horizontal and vertical moments of inertia $J_{horizontal} = \frac{1}{2}mR^2$ and $J_{vertical} = \frac{1}{4}mR^2 + \frac{1}{12}ml^2$. The *radius of gyration* can be found for irregularly shaped objects, and the moment of inertia can be easily obtained.

In electromechanical motion devices, the force and torque are thoroughly studied. Assuming that the body is rigid and the moment of inertia is constant, one has

$$\vec{T}d\vec{\theta} = J\vec{\alpha}d\vec{\theta} = J\frac{d\vec{\omega}}{dt}d\vec{\theta} = J\frac{d\vec{\theta}}{dt}d\vec{\omega} = J\vec{\omega}d\vec{\omega}\,.$$

The total work, as given by

$$W = \int_{\theta_0}^{\theta_f}\vec{T}d\vec{\theta} = \int_{\omega_0}^{\omega_f}J\vec{\omega}d\vec{\omega} = \tfrac{1}{2}(J\omega_f^2 - J\omega_0^2)\,,$$

represents the change of the kinetic energy.

Furthermore,

$$\frac{dW}{dt} = \vec{T}\frac{d\vec{\theta}}{dt} = \vec{T}\times\vec{\omega}\,,$$

and the power is defined by

$$P = \vec{T} \times \vec{\omega}.$$

This equation is an analog of $P = \vec{F} \times \vec{v}$, which is applied for translational motion.

*Example 5.8.2*

Consider a micropositioning table actuated by a micromotor. How much work is required to accelerate a 2 mg payload ($m$ = 2 mg) from $v_0$= 0 m/sec to $v_f$ = 1 m/sec?

*Solution.*

The work needed is calculated as

$$W = \tfrac{1}{2}(mv_f^2 - mv_0^2) = \tfrac{1}{2} 2 \times 10^{-6} \times 1^2 = 1 \times 10^{-6} \text{ J.}$$ □

*Example 5.8.3*

The rated power and angular velocity of a micromotor are 0.001 W and 100 rad/sec. Calculate the rated electromagnetic torque.

*Solution.*

The electromagnetic torque is

$$T_e = \frac{P}{\omega_r} = \frac{0.001}{100} = 1 \times 10^{-5} \text{ N-m.}$$ □

*Example 5.8.4*

Consider a body of mass $m$ in the $XY$ coordinate system. The force $\vec{F}_a$ is applied in the $x$ direction. Neglecting *Coulomb* and static friction, and assuming that the viscous friction force is $F_{fr} = B_v \frac{dx}{dt}$, find the equations of motion. Here $B_v$ is the viscous friction coefficient.

*Solution.*

The free-body diagram developed is illustrated in Figure 5.8.1.

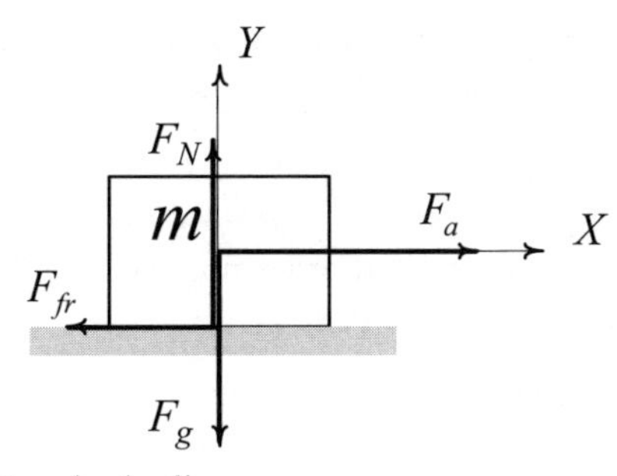


Figure 5.8.1 Free-body diagram.

The sum of the forces, acting in the $y$ direction, is expressed as

$$\sum \vec{F}_Y = \vec{F}_N - \vec{F}_g,$$

where $\vec{F}_g = mg$ is the gravitational force acting on the mass $m$; $\vec{F}_N$ is the normal force which is equal and opposite to the gravitational force.

From (5.8.1), the equation of motion in the $y$ direction is expressed as

$$\vec{F}_N - \vec{F}_g = ma_y = m\frac{d^2y}{dt^2},$$

where $a_y$ is the acceleration in the $y$ direction, $a_y = \frac{d^2y}{dt^2}$.

Making use $\vec{F}_N = \vec{F}_g$, we have

$$\frac{d^2y}{dt^2} = 0.$$

The sum of the forces acting in the $x$ direction is found using the applied force $\vec{F}_a$ and the friction force $\vec{F}_{fr}$; in particular, we have

$$\sum \vec{F}_X = \vec{F}_a - \vec{F}_{fr}.$$

The applied force can be time-invariant $\vec{F}_a = const$ or time-varying $\vec{F}_a(t) = f(t,x,y,z)$. For example,

$$\vec{F}_a(t) = x\sin(6t-4)e^{-0.5t} + \frac{dy}{dt}t^2 + z^3\cos\left(\frac{dx}{dt}t - y^2t^4\right).$$

Using (2.1), the equation motion in the $x$ direction is found to be

$$\vec{F}_a - \vec{F}_{fr} = ma_x = m\frac{d^2x}{dt^2},$$

where $a_x$ is the acceleration in the $X$ direction, $a_x = \frac{d^2x}{dt^2}$, and the velocity in the $X$ direction is $v = \frac{dx}{dt}$.

Assuming that the *Coulomb* and static friction can be neglected, the friction force, as a function of the viscous friction coefficient $B_v$ and velocity $v = \frac{dx}{dt}$, is given by $F_{fr} = B_v\frac{dx}{dt}$.

Hence, one obtains the second-order nonlinear differential equation to map the body dynamics in the $x$ direction

$$\frac{d^2x}{dt^2} = \frac{1}{m}\left(F_a - B_v\frac{dx}{dt}\right),$$

A set of two first-order linear differential equations results, and

$$\frac{dx}{dt} = v,$$

$$\frac{dv}{dt} = \frac{1}{m}\left(F_a - B_v v\right). \qquad \square$$

The application of Newton's law leads to the partial differential equations. To illustrate this, we consider two examples.

*Example 5.8.5*

The elastic membrane is illustrated in Figure 5.8.2. Derive the mathematical model to model the rectangular membrane vibration. That is, the goal is to study the time varying membrane deflection *d(t,x,y)* in the *xy* plane. The mass of the undeflected membrane per unit area $\rho$ is constant (homogeneous membrane).

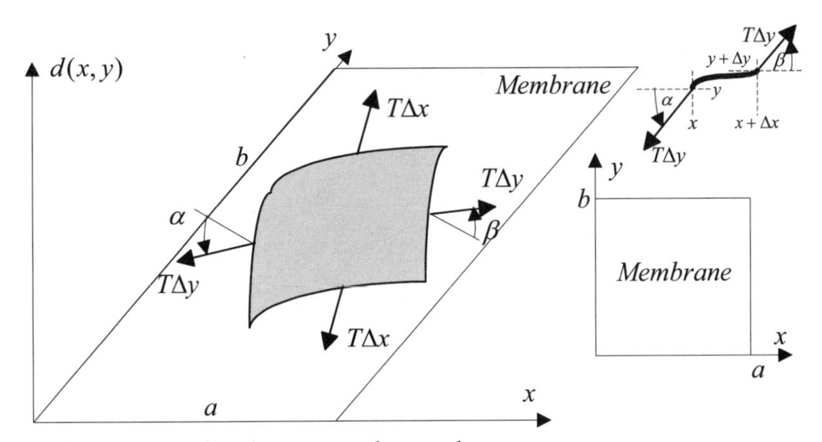


Figure 5.8.2 Vibrating rectangular membrane.

*Solution.*

Assume that the membrane is perfectly flexible. For small deflections, the tension $T$ (the force per unit length) is the same at all points in all directions, and suppose that $T$ is constant during the motion. It should be emphasized that because the deflection of the membrane is small, compared with the membrane size *ab*, the inclination angles are small.

Taking note of these assumptions, the forces acting on the sides are approximated as $F_x = T\Delta x$ and $F_y = T\Delta y$. The membrane is assumed to be perfectly flexible, therefore, forces $F_x$ and $F_y$ are tangential to the membrane.

The horizontal components of the forces are found as the cosine functions of the inclination angles. The horizontal components at the opposite sides (right and left) are equal because angles $\alpha$ and $\beta$ are small. Thus, the membrane motion in the horizontal direction can be neglected.

The vertical components of the forces are $T\Delta y \sin\beta$ and $-T\Delta y \sin\alpha$.

Using Newton's second law of motion, the net force must be found. We have the following expression

$$\sum F = T\Delta y\left(d_x(x+\Delta x, y_1) - d_x(x, y_2)\right) + T\Delta x\left(d_y(x_1, y+\Delta y) - d_y(x_2, y)\right)$$

Thus, two-dimensional partial (wave) differential equation is

$$\frac{\partial^2 d(t,x,y)}{\partial t^2} = \frac{T}{\rho}\left(\frac{\partial^2 d(t,x,y)}{\partial x^2} + \frac{\partial^2 d(t,x,y)}{\partial y^2}\right) = \frac{T}{\rho}\nabla^2 d(t,x,y).$$

Using initial and boundary conditions, the solution can be found.

Let the initial conditions are

$$d(t_0,x,y) = d_0(x,y)$$

and $\dfrac{\partial d(t_0,x,y)}{\partial t} = d_1(x,y)$.

Thus, the initial displacement $d_0(x,y)$ and initial velocity $d_1(x,y)$ are given.

Assume that the boundary conditions are

$$d(t,x_0,y_0) = 0 \text{ and } d(t,x_f,y_f) = 0.$$

The solution is found to be

$$d(t,x,y) = \sum_{i=1}^{\infty}\sum_{j=1}^{\infty} d_{ij}(t,x,y)$$

$$= \sum_{i=1}^{\infty}\sum_{j=1}^{\infty}\left(A_{ij}\cos\lambda_{ij}t + B_{ij}\sin\lambda_{ij}t\right)\sin\frac{i\pi x}{a}\sin\frac{j\pi x}{b},$$

where the characteristic eigenvalues are found as

$$\lambda_{ij} = \sqrt{\frac{T}{\rho}}\pi\sqrt{\frac{i^2}{a^2} + \frac{j^2}{b^2}}.$$

Using initial conditions, the Fourier coefficients are obtained in the form of the double Fourier series. In particular, we have

$$A_{ij} = \frac{4}{ab}\int_0^b\int_0^a d_0(x,y)\sin\frac{i\pi x}{a}\sin\frac{j\pi y}{b}\,dxdy,$$

and $B_{ij} = \dfrac{4}{ab\lambda_{ij}}\displaystyle\int_0^b\int_0^a d_1(x,y)\sin\frac{i\pi x}{a}\sin\frac{j\pi y}{b}\,dxdy$. □

*Example 5.8.6*

Derive the mathematical model of the infinitely long beam on the elastic foundation as shown in Figure 5.8.3. The load force is the square function. The modulus (the spring stiffness per unit length) of the elastic foundation is $k_s$.

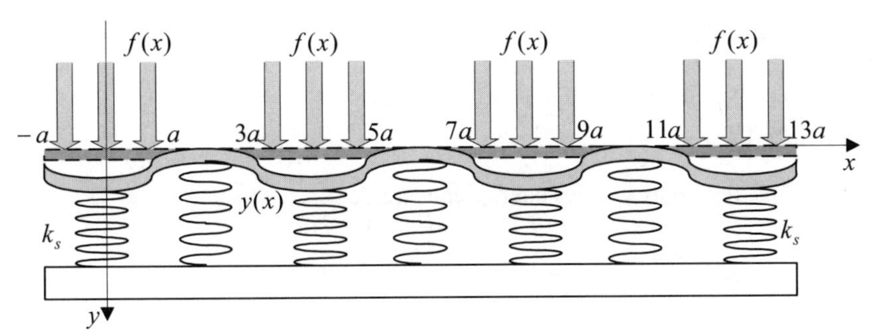


Figure 5.8.3 Beam on elastic foundation under the load force $f(x)$.

*Solution.*

Using the Euler beam theory, the deflection $y(x)$ due to the net load force $F(x)$ is modeled by the fourth-order differential equation

$$k_r \frac{d^4 y}{dt^4} = F(x),$$

where $k_r$ is the flexural rigidity constant.

Therefore, we have the following differential equation to model the infinite beam under the consideration

$$k_r \frac{d^4 y}{dt^4} + k_s y = f(x).$$

The general homogeneous solution is given by

$$y(x) = e^{\frac{1}{2}\sqrt[4]{k_r}xx}\left[k_1 \sin\left(\tfrac{1}{2}\sqrt[4]{k_r}x\right) + k_2 \sin\left(\tfrac{1}{2}\sqrt[4]{k_r}x\right)\right]$$
$$+ e^{-\frac{1}{2}\sqrt[4]{k_r}x}\left[k_3 \sin\left(\tfrac{1}{2}\sqrt[4]{k_r}x\right) + k_4 \sin\left(\tfrac{1}{2}\sqrt[4]{k_r}x\right)\right],$$

where the unknown coefficients $k_I$ can be determined using the initial and boundary conditions. The boundary-value problem can be relaxed, and the solution can be found in the series form.

The load force is the periodic function, and using the Fourier series we have

$$f(x) = \frac{f_0}{2} + \frac{2f_0}{\pi} \sum_{i=1}^{\infty} \frac{\sin\left(\frac{1}{2} i\pi\right)}{i} \cos\frac{i\pi x}{2a}.$$

The solution of the differential equation

$$k_r \frac{d^4 y}{dt^4} + k_s y = f(x)$$

is found as

$$y(x)=a_0+\sum_{i=1}^{\infty}a_i\cos\frac{i\pi x}{2a}.$$

Differentiating this equation four times gives

$$k_s a_0=\frac{f_0}{2}$$

and $$\left(k_r\frac{i^4\pi^4}{16a^4}+k_s\right)a_i=\frac{2f_0}{\pi}\frac{\sin\left(\frac{1}{2}i\pi\right)}{i}.$$

Thus, the Fourier series coefficients are

$$a_0=\frac{f_0}{2k_s}\text{ and }a_i=\frac{2f_0}{\pi}\frac{\sin\left(\frac{1}{2}i\pi\right)}{i\left[k_r\left(\frac{i\pi}{2a}\right)^4+k_s\right]},\ i\geq 1.$$

Therefore, the solution is given by

$$y(x)=\frac{f_0}{2k_s}+\frac{32f_0a^4}{\pi}\sum_{i=1}^{\infty}\frac{\sin\left(\frac{1}{2}i\pi\right)}{i\left(k_ri^4\pi^4+16a^4k_s\right)}\cos\frac{i\pi x}{2a}.$$

The first-order approximation is

$$y(x)\approx\frac{f_0}{2k}+\frac{32f_0a^4}{\pi\left(k_r\pi^4+16a^4k_s\right)}\cos\frac{\pi x}{2a}.$$ □

## *Friction Models in Electromechanical Systems*

A thorough consideration of friction is essential for understanding the operation of electromechanical systems. Friction is a very complex nonlinear phenomenon which is difficult to model. The classical *Coulomb* friction is a retarding frictional force (for translational motion) or torque (for rotational motion) that changes its sign with the reversal of the direction of motion, and the amplitude of the frictional force or torque are constant. For translational and rotational motions, the *Coulomb* friction force and torque are

$$F_{Coulomb}=k_{Fc}\,\mathrm{sgn}(v)=k_{Fc}\,\mathrm{sgn}\left(\frac{dx}{dt}\right),$$

$$T_{Coulomb}=k_{Tc}\,\mathrm{sgn}(\omega)=k_{Tc}\,\mathrm{sgn}\left(\frac{d\theta}{dt}\right),$$

where $k_{Fc}$ and $k_{Tc}$ are the Coulomb friction coefficients.

Figure 5.8.4a illustrates the *Coulomb* friction.

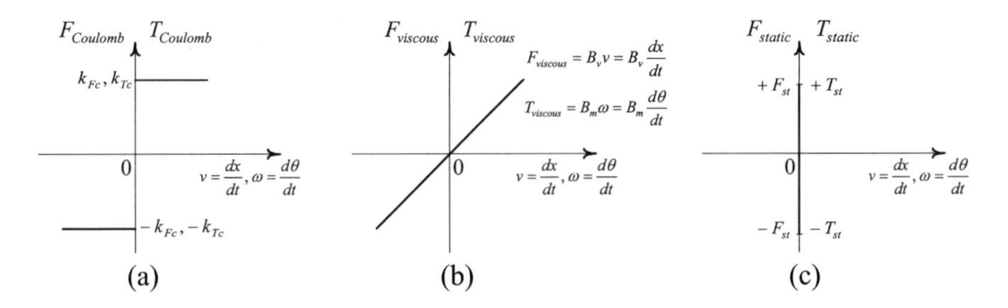


Figure 5.8.4 Functional representations of (a) *Coulomb* friction; (b) viscous friction; (c) static friction.

Viscous friction is a retarding force or torque that is a linear function of linear or angular velocity. The viscous friction force and torque versus linear and angular velocities are shown in Figure 5.8.4b. The following expressions are commonly used to model the viscous friction

$$F_{viscous}=B_v v=B_v\frac{dx}{dt} \text{ for translational motion,}$$

and $T_{viscous}=B_m\omega=B_m\frac{d\theta}{dt}$ for rotational motion,

where $B_v$ and $B_m$ are the viscous friction coefficients.

The static friction exists only when the body is stationary and vanishes as motion begins. The static friction is a force $F_{static}$ or torque $T_{static}$, and we have the following expressions

$$F_{static}=\pm F_{st}\Big|_{v=\frac{dx}{dt}=0},$$

and $T_{static}=\pm T_{st}\Big|_{\omega=\frac{d\theta}{dt}=0}$.

One concludes that the static friction is a retarding force or torque that tends to prevent the initial translational or rotational motion at beginning, see Figure 5.8.4c.

In general, the friction force and torque are nonlinear functions which must be modeled using frictional memory, presliding conditions, etc. The empirical formulas, commonly used to express $F_{static}$ and $T_{static}$, are

$$F_{fr}=\left(k_{fr1}-k_{fr2}e^{-k|v|}+k_{fr3}|v|\right)\operatorname{sgn}(v_t)=\left(k_{fr1}-k_{fr2}e^{-k\left|\frac{dx}{dt}\right|}+k_{fr3}\left|\frac{dx}{dt}\right|\right)\operatorname{sgn}\left(\frac{dx}{dt}\right)$$

and

$$T_{fr}=\left(k_{fr1}-k_{fr2}e^{-k|\omega|}+k_{fr3}|\omega|\right)\operatorname{sgn}(\omega)=\left(k_{fr1}-k_{fr2}e^{-k\left|\frac{d\theta}{dt}\right|}+k_{fr3}\left|\frac{d\theta}{dt}\right|\right)\operatorname{sgn}\left(\frac{d\theta}{dt}\right)$$

These $F_{static}$ and $T_{static}$ are shown in Figure 5.8.5.

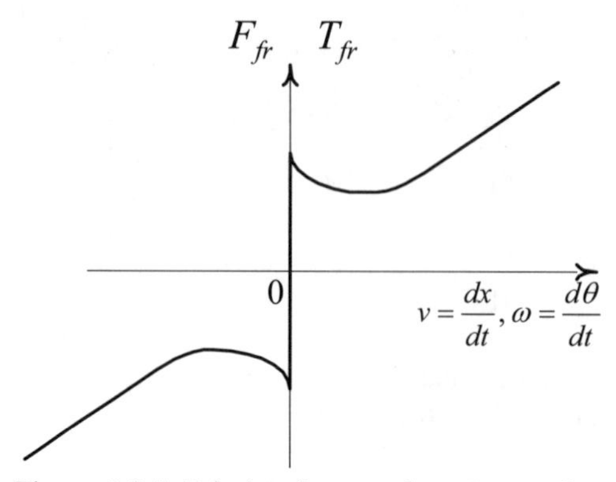


Figure 5.8.5 Friction force and torque are functions of linear and angular velocities.

*Example 5.8.7 Microtransducer Model*

Figure 5.8.6 shows a simple electromechanical device (microactuator) with a stationary member and movable plunger. Using Newton's second law, find the lumped-parameters mathematical model.

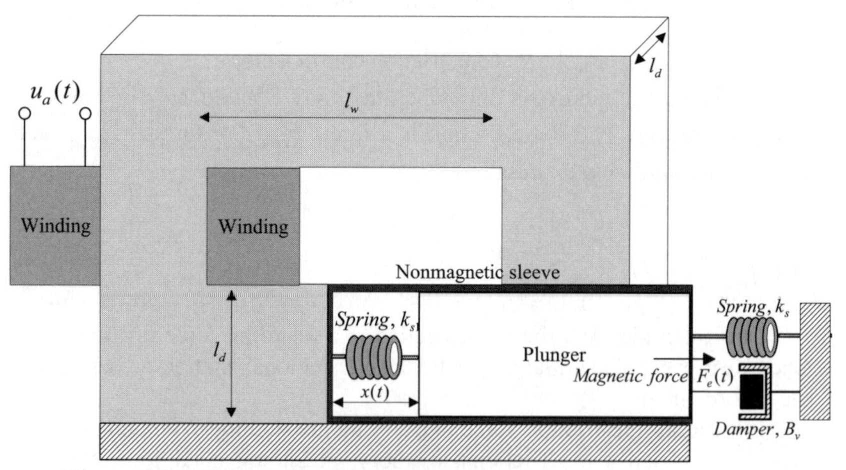


Figure 5.8.6 Schematic of a microactuator.

*Solution.*

As illustrated in the Example 5.2.8, the mathematical model was derived using the Newton's and Kirchhoff's second laws. In particular, we have

$$F(t)=m\frac{d^2x}{dt^2}+B_v\frac{dx}{dt}+(k_sx-k_{s1}x)+F_e(t)\,,\ F_e(i,x)=\frac{\partial W_c(i,x)}{\partial x},$$

$$u_a=ri+\frac{d\psi}{dt},\ \psi=L(x)i\,.$$

Using the expression for the inductance, one finally finds the following nonlinear differential equations for the microtransducer studied

$$\frac{di}{dt} = -\frac{r[A_g l_f + A_f \mu_f (x+2d)]}{N^2 \mu_f \mu_0 A_f A_g} i + \frac{\mu_f A_f}{A_g l_f + A_f \mu_f (x+2d)} iv$$
$$+ \frac{A_g l_f + A_f \mu_f (x+2d)}{N^2 \mu_f \mu_0 A_f A_g} u_a,$$

$$\frac{dx}{dt} = v,$$

$$\frac{dv}{dt} = \frac{N^2 \mu_f^2 \mu_0 A_f^{\ 2} A_g}{2m[A_g l_f + A_f \mu_f (x+2d)]^2} i^2 - \frac{1}{m}(k_s x - k_{s1} x) - \frac{B_v}{m} v.$$

□

*Newtonian Mechanics: Rotational Motion*

For one-dimensional rotational systems Newton's second law of motion is expressed as

$$M = J\alpha, \tag{5.8.2}$$

where $M$ is the sum of all moments about the center of mass of a body, (N-m); $J$ is the moment of inertia about its center of mass, (kg-m$^2$); $\alpha$ is the angular acceleration of the body, (rad/sec$^2$).

*Example 5.8.8*

Given a point mass $m$ suspended by a massless unstretchable string of length $l$, see Figure 5.8.7. Derive the equations of motion for a simple pendulum with negligible friction.

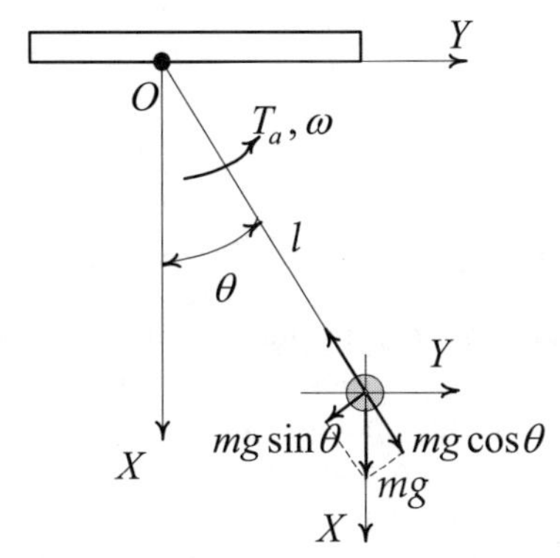


Figure 5.8.7 A simple pendulum.

*Solution.*

The restoring force, which is proportional to $\sin\theta$ and given by $-mg\sin\theta$, is the tangential component of the net force. Therefore, the sum of the moments about the pivot point $O$ is found as

$$\sum M = -mgl\sin\theta + T_a,$$

where $T_a$ is the applied torque; $l$ is the length of the pendulum measured from the point of rotation.

Using (5.8.2), one obtains the equation of motion

$$J\alpha = J\frac{d^2\theta}{dt^2} = -mgl\sin\theta + T_a,$$

where $J$ is the moment of inertial of the mass about the point $O$.

Hence, the second-order differential equation is found to be

$$\frac{d^2\theta}{dt^2} = \frac{1}{J}\left(-mgl\sin\theta + T_a\right).$$

Using the following differential equation for the angular displacement

$$\frac{d\theta}{dt} = \omega,$$

one obtains the following set of two first-order differential equations

$$\frac{d\omega}{dt} = \frac{1}{J}\left(-mgl\sin\theta + T_a\right),$$

$$\frac{d\theta}{dt} = \omega.$$

The moment of inertia is expressed by $J = ml^2$. Hence, we have the following differential equations to be used in modeling of a simple pendulum

$$\frac{d\omega}{dt} = -\frac{g}{l}\sin\theta + \frac{1}{ml^2}T_a,$$

$$\frac{d\theta}{dt} = \omega.$$

□

### 5.8.2 Lagrange Equations of Motion

Electromechanical systems, including MEMS and NEMS, augment mechanical and electronic (microelectronic) components. Therefore, one studies mechanical, electromagnetic, and circuitry transients. It was illustrated that the designer can integrate the *torsional-mechanical* dynamics and circuitry equations of motion. However, there exist general concepts to model systems. The Lagrange and Hamilton concepts are based on the energy analysis. Using the system variables, one finds the total kinetic, dissipation, and potential energies (which are denoted as $\Gamma$, $D$ and $\Pi$). Taking note of the total kinetic $\Gamma\left(t, q_1,..., q_n, \frac{dq_1}{dt},..., \frac{dq_n}{dt}\right)$, dissipation $D\left(t, q_1,..., q_n, \frac{dq_1}{dt},..., \frac{dq_n}{dt}\right)$, and potential $\Pi(t, q_1,..., q_n)$ energies, the Lagrange equations of motion are

$$\frac{d}{dt}\left(\frac{\partial\Gamma}{\partial\dot{q}_i}\right) - \frac{\partial\Gamma}{\partial q_i} + \frac{\partial D}{\partial\dot{q}_i} + \frac{\partial\Pi}{\partial q_i} = Q_i. \tag{5.8.3}$$

Here, $q_i$ and $Q_i$ are the generalized coordinates and the generalized forces (applied forces and disturbances). The generalized coordinates $q_i$ are used to derive expressions for energies $\Gamma\left(t, q_1,..., q_n, \frac{dq_1}{dt},..., \frac{dq_n}{dt}\right)$, $D\left(t, q_1,..., q_n, \frac{dq_1}{dt},..., \frac{dq_n}{dt}\right)$ and $\Pi(t, q_1,..., q_n)$.

Taking into account that for conservative (losseless) systems $D = 0$, we have the following Lagrange's equations of motion

$$\frac{d}{dt}\left(\frac{\partial\Gamma}{\partial\dot{q}_i}\right) - \frac{\partial\Gamma}{\partial q_i} + \frac{\partial\Pi}{\partial q_i} = Q_i.$$

*Example 5.8.9 Mathematical Model of a Simple Pendulum*

Derive the mathematical model for a simple pendulum using the Lagrange equations of motion.

*Solution.*

Derivation of the mathematical model for the simple pendulum, shown in Figure 5.8.7, was performed in Example 5.8.8 using the Newtonian mechanics. For the studied conservative (losseless) system we have $D = 0$. Thus, the Lagrange equations of motion are

$$\frac{d}{dt}\left(\frac{\partial\Gamma}{\partial\dot{q}_i}\right) - \frac{\partial\Gamma}{\partial q_i} + \frac{\partial\Pi}{\partial q_i} = Q_i.$$

The kinetic energy of the pendulum bob is

$$\Gamma = \tfrac{1}{2}m\left(l\dot{\theta}\right)^2.$$

The potential energy is found as

$$\Pi = mgl(1-\cos\theta).$$

As the generalized coordinate, the angular displacement is used. Thus, $q_i = \theta$.

The generalized force is the torque applied, $Q_i = T_a$.

One obtains the following expressions

$$\frac{\partial\Gamma}{\partial\dot{q}_i} = \frac{\partial\Gamma}{\partial\dot{\theta}} = ml^2\dot{\theta},$$

$$\frac{\partial\Gamma}{\partial q_i} = \frac{\partial\Gamma}{\partial\theta} = 0,$$

$$\frac{\partial\Pi}{\partial q_i} = \frac{\partial\Pi}{\partial\theta} = mgl\sin\theta .$$

Thus, the first term of the Lagrange equation is found to be

$$\frac{d}{dt}\left(\frac{\partial\Gamma}{\partial\dot{\theta}}\right) = ml^2\frac{d^2\theta}{dt^2} + 2ml\frac{dl}{dt}\frac{d\theta}{dt}.$$

Assuming that the string is unstretchable, we have

$$\frac{dl}{dt} = 0.$$

Hence,

$$ml^2\frac{d^2\theta}{dt^2} + mgl\sin\theta = T_a .$$

Thus, one obtains

$$\frac{d^2\theta}{dt^2} = \frac{1}{ml^2}\left(-mgl\sin\theta + T_a\right).$$

Recall that the equation of motion, derived by using Newtonian mechanics, is

$$\frac{d^2\theta}{dt^2} = \frac{1}{J}\left(-mgl\sin\theta + T_a\right), \text{ where } J = ml^2 .$$

One concludes that the results are the same, and the equations are

$$\frac{d\omega}{dt} = -\frac{g}{l}\sin\theta + \frac{1}{ml^2}T_a,$$

$$\frac{d\theta}{dt} = \omega . \qquad \square$$

*Example 5.8.10 Mathematical Model of a Pendulum*

Consider a double pendulum of two degrees of freedom with no external forces applied to the system, see Figure 5.8.8. Using the Lagrange equations of motion, derive the differential equations.

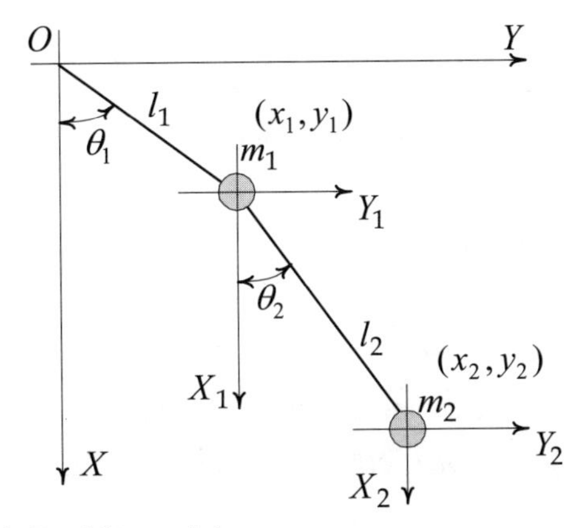


Figure 5.8.8 Double pendulum.

*Solution.*

The angular displacement $\theta_1$ and $\theta_2$ are chosen as the independent generalized coordinates. In the XY plane studied, let $(x_1, y_1)$ and $(x_2, y_2)$ be the rectangular coordinates of $m_1$ and $m_2$. Then, we obtain

$$x_1 = l_1 \cos\theta_1,\ x_2 = l_1 \cos\theta_1 + l_2 \cos\theta_2,$$

$$y_1 = l_1 \sin\theta_1,\ y_2 = l_1 \sin\theta_1 + l_2 \sin\theta_2.$$

The total kinetic energy $\Gamma$ is found to be

$$\Gamma = \frac{1}{2}m_1\left(\dot{x}_1^2 + \dot{y}_1^2\right) + \frac{1}{2}m_2\left(\dot{x}_2^2 + \dot{y}_2^2\right)$$

$$= \frac{1}{2}(m_1 + m_2)l_1^2\dot{\theta}_1^2 + m_2 l_1 l_2 \dot{\theta}_1 \dot{\theta}_2 \cos(\theta_2 - \theta_1) + \frac{1}{2}m_2 l_2^2 \dot{\theta}_2^2.$$

Then, one obtains

$$\frac{\partial\Gamma}{\partial\theta_1} = m_2 l_1 l_2 \sin(\theta_2 - \theta_1)\dot{\theta}_1\dot{\theta}_2,$$

$$\frac{\partial\Gamma}{\partial\dot{\theta}_1} = (m_1 + m_2)l_1^2\dot{\theta}_1 + m_2 l_1 l_2 \cos(\theta_2 - \theta_1)\dot{\theta}_2,$$

$$\frac{\partial\Gamma}{\partial\theta_2} = -m_2 l_1 l_2 \sin(\theta_1 - \theta_2)\dot{\theta}_1\dot{\theta}_2,$$

$$\frac{\partial\Gamma}{\partial\dot{\theta}_2} = m_2 l_1 l_2 \cos(\theta_2 - \theta_1)\dot{\theta}_1 + m_2 l_1^2 \dot{\theta}_2.$$

The total potential energy is given by

$$\Pi = m_1 g x_1 + m_2 g x_2 = (m_1 + m_2) g l_1 \cos\theta_1 + m_2 g l_2 \cos\theta_2.$$

Hence,

$$\frac{\partial\Pi}{\partial\theta_1}=-(m_1+m_2)gl_1\sin\theta_1 \text{ and } \frac{\partial\Pi}{\partial\theta_2}=-m_2gl_2\sin\theta_2 .$$

The Lagrange equations of motion are

$$\frac{d}{dt}\left(\frac{\partial\Gamma}{\partial\dot{\theta}_1}\right)-\frac{\partial\Gamma}{\partial\theta_1}+\frac{\partial\Pi}{\partial\theta_1}=0,$$

$$\frac{d}{dt}\left(\frac{\partial\Gamma}{\partial\dot{\theta}_2}\right)-\frac{\partial\Gamma}{\partial\theta_2}+\frac{\partial\Pi}{\partial\theta_2}=0.$$

Hence, the dynamic equations of the system are

$$l_1\Big[(m_1+m_2)l_1\ddot{\theta}_1+m_2l_2\cos(\theta_2-\theta_1)\ddot{\theta}_2-m_2l_2\sin(\theta_2-\theta_1)\dot{\theta}_2^{\,2}$$
$$-m_2l_2\sin(\theta_2-\theta_1)\dot{\theta}_1\dot{\theta}_2-(m_1+m_2)g\sin\theta_1\Big]=0,$$
$$m_2l_2\Big[l_2\ddot{\theta}_2+l_1\cos(\theta_2-\theta_1)\ddot{\theta}_1+l_1\sin(\theta_2-\theta_1)\dot{\theta}_1^{\,2}$$
$$l_1\sin(\theta_2-\theta_1)\dot{\theta}_1\dot{\theta}_2-g\sin\theta_2\Big]=0.$$

It should be emphasized that if the torques $T_1$ and $T_2$ are applied to the first and second joints, the following equations of motions results

$$l_1\Big[(m_1+m_2)l_1\ddot{\theta}_1+m_2l_2\cos(\theta_2-\theta_1)\ddot{\theta}_2-m_2l_2\sin(\theta_2-\theta_1)\dot{\theta}_2^{\,2}$$
$$-m_2l_2\sin(\theta_2-\theta_1)\dot{\theta}_1\dot{\theta}_2-(m_1+m_2)g\sin\theta_1\Big]=T_1,$$
$$m_2l_2\Big[l_2\ddot{\theta}_2+l_1\cos(\theta_2-\theta_1)\ddot{\theta}_1+l_1\sin(\theta_2-\theta_1)\dot{\theta}_1^{\,2}$$
$$l_1\sin(\theta_2-\theta_1)\dot{\theta}_1\dot{\theta}_2-g\sin\theta_2\Big]=T_2.$$

□

*Example 5.8.11 Mathematical Model of a Circuit Network*

Consider a two-mesh electric circuit, as shown in Figure 5.8.9. Find the circuitry dynamics.

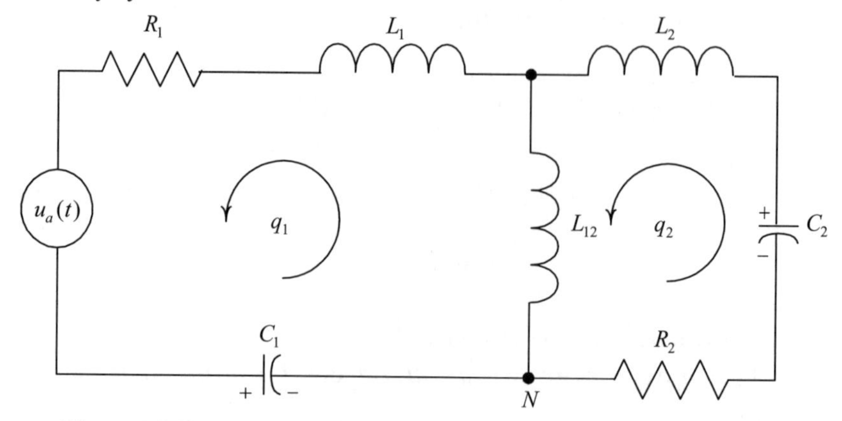


Figure 5.8.9 Two-mesh circuit network.

*Solution.*

We use $q_1$ and $q_2$ as the independent generalized coordinates, where $q_1$ is the electric charge in the first loop, $q_2$ represents the electric charge in the second loop. The generalized force, which is applied to the system, is denoted as $Q_1$. These generalized coordinates are related to the circuitry variables. In particular, the currents $i_1$ and $i_2$ are found in terms of charges as $i_1 = \dot{q}_1$ and $i_2 = \dot{q}_2$.

That is, we have

$$q_1 = \frac{i_1}{s} \text{ and } q_2 = \frac{i_2}{s}.$$

The generalized force is the applied voltage. Hence, $u_a(t) = Q_1$.

The total magnetic energy (kinetic energy) is expressed by

$$\Gamma = \tfrac{1}{2}L_1\dot{q}_1^{\,2} + \tfrac{1}{2}L_{12}(\dot{q}_1 - \dot{q}_2)^2 + \tfrac{1}{2}L_2\dot{q}_2^{\,2}.$$

By using this equation for $\Gamma$, we have

$$\frac{\partial\Gamma}{\partial q_1} = 0,\ \frac{\partial\Gamma}{\partial\dot{q}_1} = (L_1 + L_{12})\dot{q}_1 - L_{12}\dot{q}_2,$$

$$\frac{\partial\Gamma}{\partial q_2} = 0,\ \frac{\partial\Gamma}{\partial\dot{q}_2} = -L_{12}\dot{q}_1 + (L_2 + L_{12})\dot{q}_2.$$

Using the equation for the total electric energy (potential energy)

$$\Pi = \tfrac{1}{2}\frac{q_1^2}{C_1} + \tfrac{1}{2}\frac{q_2^2}{C_2},$$

one finds

$$\frac{\partial\Pi}{\partial q_1} = \frac{q_1}{C_1} \text{ and } \frac{\partial\Pi}{\partial q_2} = \frac{q_2}{C_2}.$$

The total heat energy dissipated is

$$D = \tfrac{1}{2}R_1\dot{q}_1^{\,2} + \tfrac{1}{2}R_2\dot{q}_2^{\,2}.$$

Hence,

$$\frac{\partial D}{\partial\dot{q}_1} = R_1\dot{q}_1 \text{ and } \frac{\partial D}{\partial\dot{q}_2} = R_2\dot{q}_2.$$

The Lagrange equations of motion are expressed using the independent coordinates used. We obtain

$$\frac{d}{dt}\left(\frac{\partial\Gamma}{\partial\dot{q}_1}\right) - \frac{\partial\Gamma}{\partial q_1} + \frac{\partial D}{\partial\dot{q}_1} + \frac{\partial\Pi}{\partial q_1} = Q_1,$$

$$\frac{d}{dt}\left(\frac{\partial\Gamma}{\partial\dot{q}_2}\right) - \frac{\partial\Gamma}{\partial q_2} + \frac{\partial D}{\partial\dot{q}_2} + \frac{\partial\Pi}{\partial q_2} = 0.$$

Hence, the differential equations for the circuit studied are found to be

$$(L_1 + L_{12})\ddot{q}_1 - L_{12}\ddot{q}_2 + R_1\dot{q}_1 + \frac{q_1}{C_1} = u_a,$$

$$-L_{12}\ddot{q}_1 + (L_2 + L_{12})\ddot{q}_2 + R_2\dot{q}_2 + \frac{q_2}{C_2} = 0.$$

The SIMULINK model can be built using these derived nonlinear differential equations. In particular, we have

$$\ddot{q}_1 = \frac{1}{(L_1 + L_{12})}\left(-\frac{q_1}{C_1} - R_1\dot{q}_1 + L_{12}\ddot{q}_2 + u_a\right)$$

and $$\ddot{q}_2 = \frac{1}{(L_2 + L_{12})}\left(L_{12}\ddot{q}_1 - \frac{q_2}{C_2} - R_2\dot{q}_2\right).$$

The corresponding SIMULINK diagram is shown in Figure 5.8.10.

It should be emphasized that the currents $i_1$ and $i_2$ are expressed in terms of charges as $i_1 = \dot{q}_1$ and $i_2 = \dot{q}_2$. That is, $q_1 = \frac{i_1}{s}$ and $q_2 = \frac{i_2}{s}$.

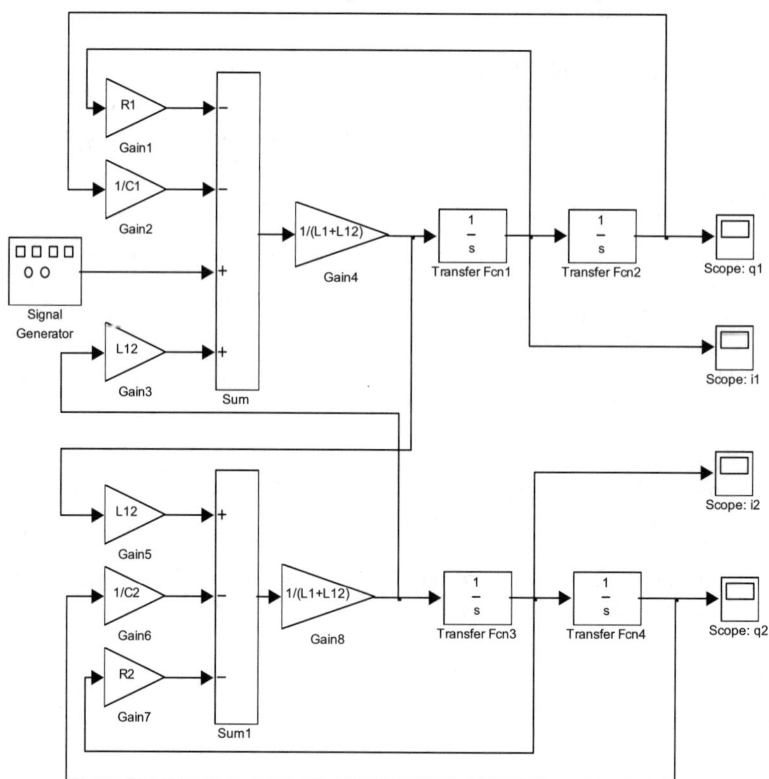


Figure 5.8.10 SIMULINK diagram.

To perform simulations and numerical analysis, the circuitry parameters are assigned to be: $L_1$=0.01 H, $L_2$=0.005 H, $L_{12}$= 0.0025 H, $C_1$=0.02 F, $C_2$=0.1 F, $R_1$=10 ohm, $R_2$= 5 ohm and $u_a = 100\sin(200t)$ V. The simulation results, which give the time history of $q_1(t), q_2(t), i_1(t)$ and $i_2(t)$, are documented in Figure 5.8.11.

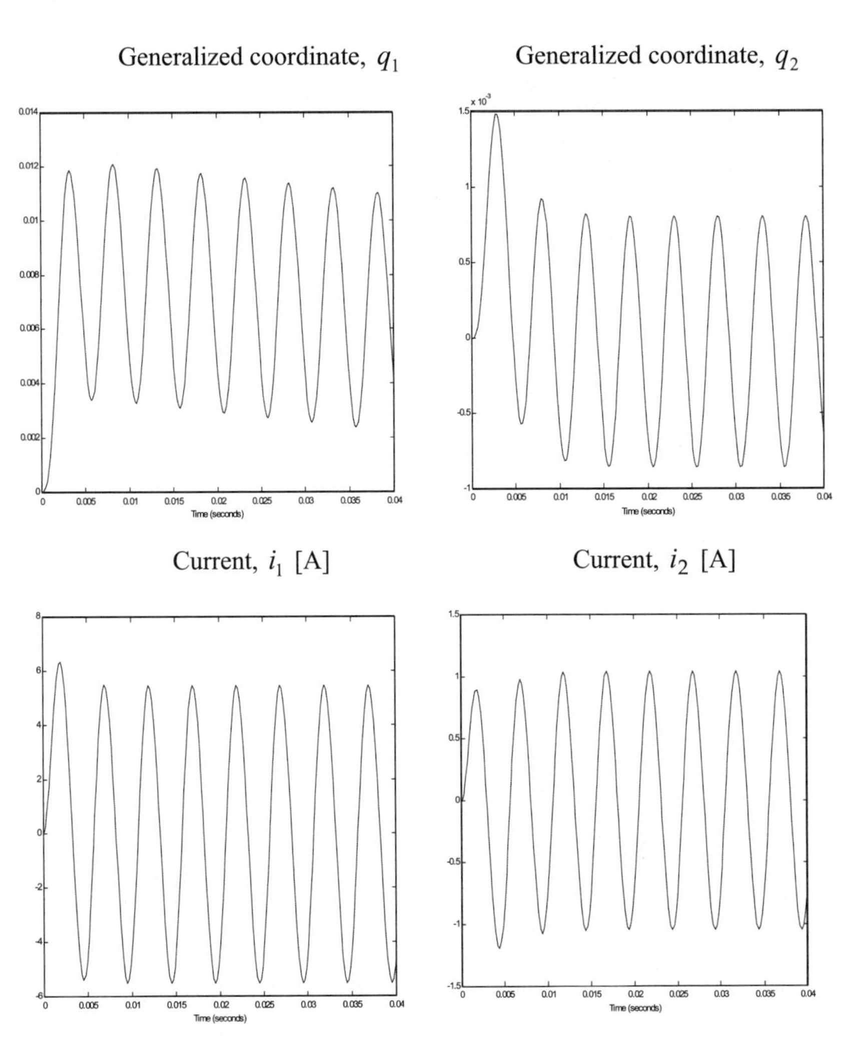


Figure 5.8.11 Circuit dynamics: evolution of the generalized coordinates and currents.

□

*Example 5.8.12 Mathematical Model of an Electric Circuit*

Using the Lagrange equations of motion develop the mathematical models for the circuit shown in the Figure 5.8.12. Prove that the model derived using the Lagrange equations of motion equivalent the model developed using Kirchhoff's law.

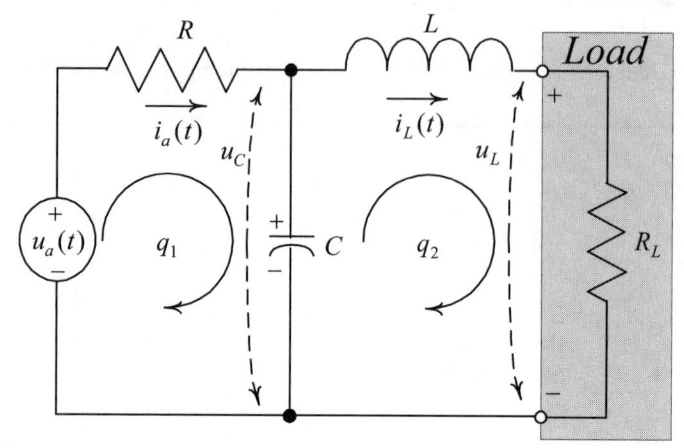


Figure 5.8.12 Electric circuit.

*Solution.*

Using $q_1$ and $q_2$ as the independent generalized coordinates, the Lagrange equations of motion can be found. Here, $q_1$ is the electric charge in the first loop and $i_a = \dot{q}_1$, and $q_2$ is the electric charge in the second loop, $i_L = \dot{q}_2$. The generalized force, applied to the system, is denoted as $Q_1$, and $u_a(t) = Q_1$.

The total kinetic energy is

$$\Gamma = \tfrac{1}{2} L\dot{q}_2^{\,2}.$$

Therefore, we have

$$\frac{\partial\Gamma}{\partial q_1} = 0, \quad \frac{\partial\Gamma}{\partial \dot{q}_1} = 0 \text{ and } \frac{d}{dt}\left(\frac{\partial\Gamma}{\partial \dot{q}_1}\right) = 0,$$

$$\frac{\partial\Gamma}{\partial q_2} = 0, \quad \frac{\partial\Gamma}{\partial \dot{q}_2} = L\dot{q}_2 \text{ and } \frac{d}{dt}\left(\frac{\partial\Gamma}{\partial \dot{q}_2}\right) = L\ddot{q}_2.$$

The total potential energy is expressed as

$$\Pi = \tfrac{1}{2}\frac{(q_1 - q_2)^2}{C}.$$

Hence, one obtains

$$\frac{\partial\Pi}{\partial q_1}=\frac{q_1-q_2}{C} \text{ and } \frac{\partial\Pi}{\partial q_2}=\frac{-q_1+q_2}{C}.$$

The total dissipated energy is

$$D=\tfrac{1}{2}R\dot{q}_1^{\,2}+\tfrac{1}{2}R_L\dot{q}_2^{\,2}.$$

Therefore

$$\frac{\partial D}{\partial\dot{q}_1}=R\dot{q}_1 \text{ and } \frac{\partial D}{\partial\dot{q}_2}=R_L\dot{q}_2.$$

The Lagrange equations of motion

$$\frac{d}{dt}\left(\frac{\partial\Gamma}{\partial\dot{q}_1}\right)-\frac{\partial\Gamma}{\partial q_1}+\frac{\partial D}{\partial\dot{q}_1}+\frac{\partial\Pi}{\partial q_1}=Q_1,$$

$$\frac{d}{dt}\left(\frac{\partial\Gamma}{\partial\dot{q}_2}\right)-\frac{\partial\Gamma}{\partial q_2}+\frac{\partial D}{\partial\dot{q}_2}+\frac{\partial\Pi}{\partial q_2}=0,$$

lead one to the following two differential equations

$$R\dot{q}_1+\frac{q_1-q_2}{C}=u_a,$$

$$L\ddot{q}_2+R_L\dot{q}_2+\frac{-q_1+q_2}{C}=0.$$

Hence, we have found a set of two differential equations. In particular,

$$\dot{q}_1=\frac{1}{R}\left(\frac{-q_1+q_2}{C}+u_a\right),$$

$$\ddot{q}_2=\frac{1}{L}\left(-R_L\dot{q}_2+\frac{q_1-q_2}{C}\right).$$

By using Kirchhoff's law, two differential equations result

$$\frac{du_C}{dt}=\frac{1}{C}\left(-\frac{u_C}{R}-i_L+\frac{u_a(t)}{R}\right),$$

$$\frac{di_L}{dt}=\frac{1}{L}\left(u_C-R_Li_L\right).$$

Taking note of $i_a=\dot{q}_1$ and $i_L=\dot{q}_2$, and making use $C\frac{du_C}{dt}=i_a-i_L$, we obtain

$$u_C=\frac{q_1-q_2}{C}.$$

The equivalence of the differential equations derived using the Lagrange equations of motion and Kirchhoff's law is proven. □

*Example 5.8.13 Mathematical Model of a Boost Converter*

A high-frequency one-quadrant *boost* (*step-up*) dc-dc switching converter is documented in Figure 5.8.13. Find the mathematical model in the form of differential equations.

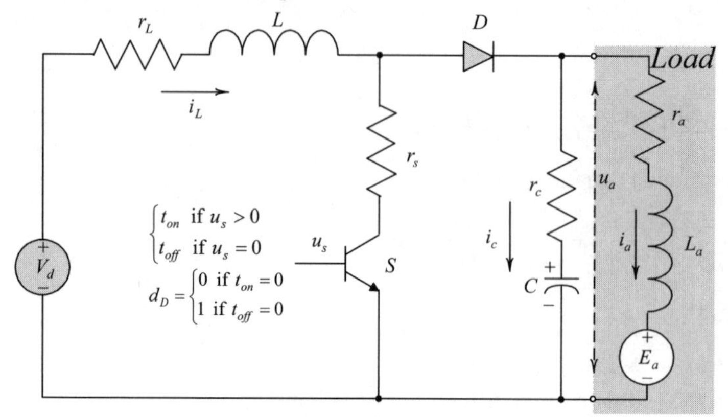


Figure 5.8.13 *Boost* converter.

*Solution.*

To solve the model development problem, we will derive the differential equations if the duty ratio $d_D$ is 1 and 0. Then, we will augment two mathematical models found to model the *boost* converter.

When the switch is closed, the diode *D* is reverse biased. For $d_D = 1$ ($t_{off} = 0$), one obtains the following set of linear differential equations

$$\frac{du_C}{dt} = -\frac{1}{C} i_a,$$

$$\frac{di_L}{dt} = \frac{1}{L}\left(-(r_L + r_s) i_L + V_d\right),$$

$$\frac{di_a}{dt} = \frac{1}{L_a}\left(u_C - (r_a + r_c) i_a - E_a\right).$$

If the switch is open ($d_D = 0$), the diode *D* is forward biased because the direction of the inductor current $i_L$ does not change instantly. Therefore, one has three linear differential equations

$$\frac{du_C}{dt} = \frac{1}{C}(i_L - i_a),$$

$$\frac{di_L}{dt} = \frac{1}{L}\left(-u_C - (r_L + r_c) i_L + r_c i_a + V_d\right),$$

$$\frac{di_a}{dt} = \frac{1}{L_a}\left(u_C + r_c i_L - (r_a + r_c) i_a - E_a\right).$$

Assuming the switching frequency is high, the *averaging* concept is applied, and we have

$$\frac{du_C}{dt}=\frac{1}{C}\left(i_L-i_a-i_Ld_D\right),$$

$$\frac{di_L}{dt}=\frac{1}{L}\left(-u_C-\left(r_L+r_c\right)i_L+r_ci_a+u_Cd_D+\left(r_c-r_s\right)i_Ld_D-r_ci_ad_D+V_d\right),$$

$$\frac{di_a}{dt}=\frac{1}{L_a}\left(u_C+r_ci_L-\left(r_a+r_c\right)i_a-r_ci_Ld_D-E_a\right).$$

Considering the duty ratio as the control input, one concludes that a set of nonlinear differential equations result. In fact, the state variables are multiplied by the control.

Let us illustrate that Lagrange's concept gives the same differential equations. We denote the electric charges in the first and the second loops as $q_1$ and $q_2$, and the generalized forces are $Q_1$ and $Q_2$. Then,

$$\frac{d}{dt}\left(\frac{\partial\Gamma}{\partial\dot{q}_1}\right)-\frac{\partial\Gamma}{\partial q_1}+\frac{\partial D}{\partial\dot{q}_1}+\frac{\partial\Pi}{\partial q_1}=Q_1,$$

$$\frac{d}{dt}\left(\frac{\partial\Gamma}{\partial\dot{q}_2}\right)-\frac{\partial\Gamma}{\partial q_2}+\frac{\partial D}{\partial\dot{q}_2}+\frac{\partial\Pi}{\partial q_2}=Q_2.$$

For the closed switch, the total kinetic, potential, and dissipated energies are

$$\Gamma=\tfrac{1}{2}\left(L\dot{q}_1^{\,2}+L_a\dot{q}_2^{\,2}\right),\Pi=\tfrac{1}{2}\frac{q_2^2}{C},D=\tfrac{1}{2}\left(\left(r_L+r_s\right)\dot{q}_1^{\,2}+\left(r_c+r_a\right)\dot{q}_2^{\,2}\right).$$

Assuming that the resistances, inductances, and capacitance are time-invariant (constant), one obtains

$$\frac{\partial\Gamma}{\partial q_1}=0,\ \frac{\partial\Gamma}{\partial q_2}=0,\ \frac{\partial\Gamma}{\partial\dot{q}_1}=L\dot{q}_1,\ \frac{\partial\Gamma}{\partial\dot{q}_2}=L_a\dot{q}_2,$$

$$\frac{d}{dt}\left(\frac{\partial\Gamma}{\partial\dot{q}_1}\right)=L\ddot{q}_1,\ \frac{d}{dt}\left(\frac{\partial\Gamma}{\partial\dot{q}_2}\right)=L_a\ddot{q}_2,$$

$$\frac{\partial\Pi}{\partial q_1}=0,\ \frac{\partial\Pi}{\partial q_2}=\frac{q_2}{C},$$

$$\frac{\partial D}{\partial\dot{q}_1}=\left(r_L+r_s\right)\dot{q}_1,\ \frac{\partial D}{\partial\dot{q}_2}=\left(r_c+r_a\right)\dot{q}_2.$$

Therefore,

$$L\ddot{q}_1+\left(r_L+r_s\right)\dot{q}_1=Q_1,$$

$$L_a\ddot{q}_2+\left(r_c+r_a\right)\dot{q}_2+\frac{1}{C}q_2=Q_2,$$

and thus,

$$\ddot{q}_1 = \frac{1}{L}\left(-\left(r_L + r_s\right)\dot{q}_1 + Q_1\right),$$

$$\ddot{q}_2 = \frac{1}{L_a}\left(-\left(r_c + r_a\right)\dot{q}_2 - \frac{1}{C}q_2 + Q_2\right),$$

The total kinetic, potential, and dissipated energies if the switch is open are found to be

$$\Gamma = \tfrac{1}{2}\left(L\dot{q}_1^{\,2} + L_a\dot{q}_2^{\,2}\right), \Pi = \tfrac{1}{2}\frac{\left(q_1 - q_2\right)^2}{C}, D = \tfrac{1}{2}\left(r_L\dot{q}_1^{\,2} + r_c\left(\dot{q}_1 - \dot{q}_2\right)^2 + r_a\dot{q}_2^{\,2}\right).$$

Thus,

$$\frac{\partial\Gamma}{\partial q_1} = 0,\ \frac{\partial\Gamma}{\partial q_2} = 0,\ \frac{\partial\Gamma}{\partial\dot{q}_1} = L\dot{q}_1,\ \frac{\partial\Gamma}{\partial\dot{q}_2} = L_a\dot{q}_2,$$

$$\frac{d}{dt}\left(\frac{\partial\Gamma}{\partial\dot{q}_1}\right) = L\ddot{q}_1,\ \frac{d}{dt}\left(\frac{\partial\Gamma}{\partial\dot{q}_2}\right) = L_a\ddot{q}_2,$$

$$\frac{\partial\Pi}{\partial q_1} = \frac{q_1 - q_2}{C},\ \frac{\partial\Pi}{\partial q_2} = -\frac{q_1 - q_2}{C},$$

$$\frac{\partial D}{\partial\dot{q}_1} = \left(r_L + r_c\right)\dot{q}_1 - r_c\dot{q}_2,\ \frac{\partial D}{\partial\dot{q}_2} = -r_c\dot{q}_1 + \left(r_c + r_a\right)\dot{q}_2.$$

Using

$$L\ddot{q}_1 + \left(r_L + r_c\right)\dot{q}_1 - r_c\dot{q}_2 + \frac{q_1 - q_2}{C} = Q_1,$$

$$L_a\ddot{q}_2 - r_c\dot{q}_1 + \left(r_c + r_a\right)\dot{q}_2 - \frac{q_1 - q_2}{C} = Q_2,$$

one has

$$\ddot{q}_1 = \frac{1}{L}\left(-\left(r_L + r_c\right)\dot{q}_1 + r_c\dot{q}_2 - \frac{q_1 - q_2}{C} + Q_1\right),$$

$$\ddot{q}_2 = \frac{1}{L_a}\left(r_c\dot{q}_1 - \left(r_c + r_a\right)\dot{q}_2 + \frac{q_1 - q_2}{C} + Q_2\right).$$

It must be emphasized that $i_L = \dot{q}_1$, $i_a = \dot{q}_2$, and $Q_1 = V_d$, $Q_2 = -E_a$.

Taking note of the differential equations when the switch is closed and open, the differential equations in Cauchy's form are found using

$$\frac{dq_1}{dt} = i_L \text{ and } \frac{dq_2}{dt} = i_a.$$

The voltage across the capacitor $u_C$ is expressed using the charges $q_1$ and $q_2$. When the switch is closed $u_C = -\frac{q_2}{C}$. If the switch is open

$u_C = \dfrac{q_1 - q_2}{C}$. The analysis of the differential equations derived using Kirchhoff's voltage law and the Lagrange equations of motion illustrates that the mathematical models are found using different state variables. In particular, $u_C, i_L, i_a$ and $q_1, i_L, q_2, i_a$ are used. However, the resulting differential equations are the same as one applies the corresponding variable transformations as given by

$$\frac{dq_1}{dt} = i_L, \ \frac{dq_2}{dt} = i_a, \ Q_1 = V_d \text{ and } Q_2 = -E_a. \qquad \square$$

*Example 5.8.14 Mathematical Model of a Microtransducer*

Consider a microtransducer with two independently excited stator and rotor windings, see Figure 5.8.14. Derive the differential equations.

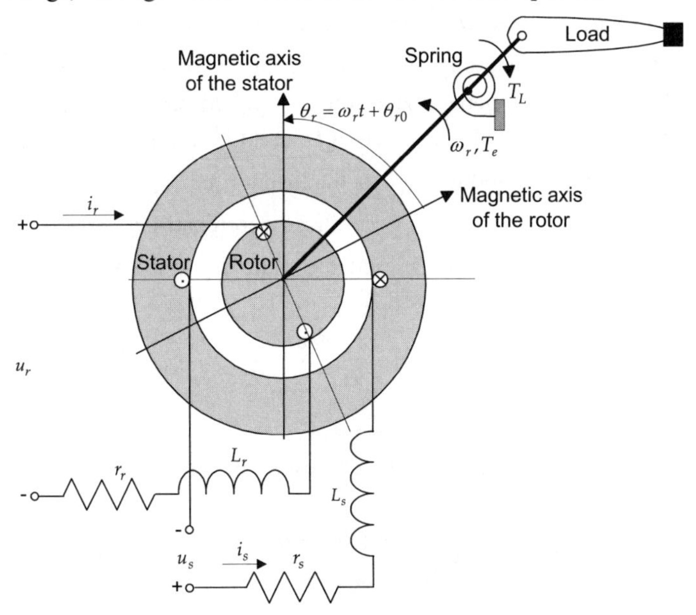


Figure 5.8.14 Microtransducer with stator and rotor windings.

*Solution.*

The following notations are used: $i_s$ and $i_r$ are the currents in the stator and rotor windings; $u_s$ and $u_r$ are the applied voltages to the stator and rotor windings; $\omega_r$ and $\theta_r$ are the rotor angular velocity and displacement; $T_e$ and $T_L$ are the electromagnetic and load torques; $r_s$ and $r_r$ are the resistances of the stator and rotor windings; $L_s$ and $L_r$ are the self-inductances of the stator and rotor windings; $L_{sr}$ is the mutual inductance of the stator and rotor windings; $\Re_m$ is the reluctance of the magnetizing path; $N_s$ and $N_r$ are the number of turns in the stator and rotor windings; $J$ is the moment of inertia of the rotor and attached load; $B_m$ is the viscous friction coefficient; $k_s$ is the spring constant.

The magnetic fluxes that cross an air gap produce a force of attraction, and the developed electromagnetic torque $T_e$ is countered by the tortional spring which causes a *counterclockwise* rotation. The load torque $T_L$ should be considered.

Our goal is to find a nonlinear mathematical model. In fact, the ability to formulate the modeling problem and find the resulting equations that describe a motion device constitute the most important issues. By using the Lagrange concept, the independent generalized coordinates must be chosen. Let us use $q_1$, $q_2$ and $q_3$, where $q_1$ and $q_2$ denote the electric charges in the stator and rotor windings; $q_3$ represents the rotor angular displacement.

We denote the generalized forces, applied to an electromechanical system, as $Q_1$, $Q_2$ and $Q_3$, where $Q_1$ and $Q_2$ are the applied voltages to the stator and rotor windings; $Q_3$ is the load torque.

The first derivative of the generalized coordinates $\dot{q}_1$ and $\dot{q}_2$ represent the stator and rotor currents $i_s$ and $i_r$, while $\dot{q}_3$ is the angular velocity of the rotor $\omega_r$. We have,

$$q_1 = \frac{i_s}{s},\ q_2 = \frac{i_r}{s},\ q_3 = \theta_r,\ \dot{q}_1 = i_s,\ \dot{q}_2 = i_r,\ \dot{q}_3 = \omega_r,$$

$$Q_1 = u_s,\ Q_2 = u_r \text{ and } Q_3 = -T_L.$$

The Lagrange equations are expressed in terms of each independent coordinate, and we have

$$\frac{d}{dt}\left(\frac{\partial \Gamma}{\partial \dot{q}_1}\right) - \frac{\partial \Gamma}{\partial q_1} + \frac{\partial D}{\partial \dot{q}_1} + \frac{\partial \Pi}{\partial q_1} = Q_1,$$

$$\frac{d}{dt}\left(\frac{\partial \Gamma}{\partial \dot{q}_2}\right) - \frac{\partial \Gamma}{\partial q_2} + \frac{\partial D}{\partial \dot{q}_2} + \frac{\partial \Pi}{\partial q_2} = Q_2,$$

$$\frac{d}{dt}\left(\frac{\partial \Gamma}{\partial \dot{q}_3}\right) - \frac{\partial \Gamma}{\partial q_3} + \frac{\partial D}{\partial \dot{q}_3} + \frac{\partial \Pi}{\partial q_3} = Q_3.$$

The total kinetic energy of electrical and mechanical systems is found as a sum of the total magnetic (electrical) $\Gamma_E$ and mechanical $\Gamma_M$ energies. The total kinetic energy of the stator and rotor circuitry is given as

$$\Gamma_E = \tfrac{1}{2} L_s \dot{q}_1^2 + L_{sr} \dot{q}_1 \dot{q}_2 + \tfrac{1}{2} L_r \dot{q}_2^2.$$

The total kinetic energy of the mechanical system, which is a function of the equivalent moment of inertia of the rotor and the payload attached, is expressed by

$$\Gamma_M = \tfrac{1}{2} J\dot{q}_3^2 .$$

Then, we have

$$\Gamma = \Gamma_E + \Gamma_M = \tfrac{1}{2} L_s \dot{q}_1^2 + L_{sr}\dot{q}_1\dot{q}_2 + \tfrac{1}{2} L_r \dot{q}_2^2 + \tfrac{1}{2} J\dot{q}_3^2 .$$

The mutual inductance is a periodic function of the angular rotor displacement, and $L_{sr}(\theta_r) = \dfrac{N_s N_r}{\Re_m(\theta_r)}$.

The magnetizing reluctance is maximum if the stator and rotor windings are not displaced, and $\Re_m(\theta_r)$ is minimum if the coils are displaced by 90 degrees. Then, $L_{sr\min} \le L_{sr}(\theta_r) \le L_{sr\max}$, where $L_{sr\max} = \dfrac{N_s N_r}{\Re_m(90^\circ)}$ and $L_{sr\min} = \dfrac{N_s N_r}{\Re_m(0^\circ)}$.

The mutual inductance can be approximated as a cosine function of the rotor angular displacement. The amplitude of the mutual inductance between the stator and rotor windings is found as $L_M = L_{sr\max} = \dfrac{N_s N_r}{\Re_m(90^\circ)}$.

Then,

$$L_{sr}(\theta_r) = L_M \cos\theta_r = L_M \cos q_3 .$$

One obtains an explicit expression for the total kinetic energy as

$$\Gamma = \tfrac{1}{2} L_s \dot{q}_1^2 + L_M \dot{q}_1\dot{q}_2 \cos q_3 + \tfrac{1}{2} L_r \dot{q}_2^2 + \tfrac{1}{2} J\dot{q}_3^2 .$$

The following partial derivatives result

$$\frac{\partial\Gamma}{\partial q_1} = 0,\quad \frac{\partial\Gamma}{\partial \dot{q}_1} = L_s\dot{q}_1 + L_M\dot{q}_2\cos q_3,$$

$$\frac{\partial\Gamma}{\partial q_2} = 0,\quad \frac{\partial\Gamma}{\partial \dot{q}_2} = L_M\dot{q}_1\cos q_3 + L_r\dot{q}_2,$$

$$\frac{\partial\Gamma}{\partial q_3} = -L_M\dot{q}_1\dot{q}_2\sin q_3,\quad \frac{\partial\Gamma}{\partial \dot{q}_3} = J\dot{q}_3 .$$

The potential energy of the spring with constant $k_s$ is

$$\Pi = \tfrac{1}{2} k_s q_3^2 .$$

Therefore, $\dfrac{\partial\Pi}{\partial q_1} = 0$, $\dfrac{\partial\Pi}{\partial q_2} = 0$ and $\dfrac{\partial\Pi}{\partial q_3} = k_s q_3$.

The total heat energy dissipated is expressed as

$$D = D_E + D_M ,$$

where $D_E$ is the heat energy dissipated in the stator and rotor windings, $D_E = \frac{1}{2} r_s \dot{q}_1^2 + \frac{1}{2} r_r \dot{q}_2^2$; $D_M$ is the heat energy dissipated by mechanical system, $D_M = \frac{1}{2} B_m \dot{q}_3^2$.

Hence,

$$D = \frac{1}{2} r_s \dot{q}_1^2 + \frac{1}{2} r_r \dot{q}_2^2 + \frac{1}{2} B_m \dot{q}_3^2 .$$

One obtains

$$\frac{\partial D}{\partial \dot{q}_1} = r_s \dot{q}_1, \quad \frac{\partial D}{\partial \dot{q}_2} = r_r \dot{q}_2 \text{ and } \frac{\partial D}{\partial \dot{q}_3} = B_m \dot{q}_3 .$$

Using

$$q_1 = \frac{i_s}{s}, \; q_2 = \frac{i_r}{s}, \; q_3 = \theta_r, \; \dot{q}_1 = i_s, \; \dot{q}_2 = i_r, \; \dot{q}_3 = \omega_r,$$

$$Q_1 = u_s, \; Q_2 = u_r \text{ and } Q_3 = -T_L,$$

we have three differential equations for a servo-system. In particular,

$$L_s \frac{di_s}{dt} + L_M \cos\theta_r \frac{di_r}{dt} - L_M i_r \sin\theta_r \frac{d\theta_r}{dt} + r_s i_s = u_s,$$

$$L_r \frac{di_r}{dt} + L_M \cos\theta_r \frac{di_s}{dt} - L_M i_s \sin\theta_r \frac{d\theta_r}{dt} + r_r i_r = u_r,$$

$$J \frac{d^2\theta_r}{dt^2} + L_M i_s i_r \sin\theta_r + B_m \frac{d\theta_r}{dt} + k_s \theta_r = -T_L .$$

The last equation should be rewritten by making use of the rotor angular velocity. In particular, $\frac{d\theta_r}{dt} = \omega_r$.

Using the stator and rotor currents, angular velocity, and displacement as the state variables, the nonlinear differential equations in Cauchy's form are

$$\frac{di_s}{dt} = \frac{-r_s L_r i_s - \frac{1}{2} L_M^2 i_s \omega_r \sin 2\theta_r + r_r L_M i_r \cos\theta_r + L_r L_M i_r \omega_r \sin\theta_r + L_r u_s - L_M \cos\theta_r u_r}{L_s L_r - L_M^2 \cos^2\theta_r},$$

$$\frac{di_r}{dt} = \frac{r_s L_M i_s \cos\theta_r + L_s L_M i_s \omega_r \sin\theta_r - r_r L_s i_r - \frac{1}{2} L_M^2 i_r \omega_r \sin 2\theta_r - L_M \cos\theta_r u_s + L_s u_r}{L_s L_r - L_M^2 \cos^2\theta_r},$$

$$\frac{d\omega_r}{dt} = \frac{1}{J}\left(-L_M i_s i_r \sin\theta_r - B_m \omega_r - k_s \theta_r - T_L\right),$$

$$\frac{d\theta_r}{dt} = \omega_r .$$

The developed nonlinear mathematical model in the form of highly coupled nonlinear differential equations cannot be linearized, and one must model the doubly exited transducer studied using the nonlinear differential equations derived. □

### 5.8.3 Hamilton Equations of Motion

The Hamilton concept allows one to model the system dynamics, and the differential equations are found using the generalized momenta $p_i$, and

$$p_i = \frac{\partial L}{\partial \dot{q}_i}.$$

As was emphasized, the generalized coordinates were used in the Lagrange equations of motion to derive the mathematical models.

The Lagrangian function

$$L\left(t, q_1, \ldots, q_n, \frac{dq_1}{dt}, \ldots, \frac{dq_n}{dt}\right)$$

for the conservative systems is the difference between the total kinetic and potential energies. In particular,

$$L\left(t, q_1, \ldots, q_n, \frac{dq_1}{dt}, \ldots, \frac{dq_n}{dt}\right) = \Gamma\left(t, q_1, \ldots, q_n, \frac{dq_1}{dt}, \ldots, \frac{dq_n}{dt}\right) - \Pi(t, q_1, \ldots, q_n).$$

Thus, $L\left(t, q_1, \ldots, q_n, \frac{dq_1}{dt}, \ldots, \frac{dq_n}{dt}\right)$ is the function of $2n$ independent variables. One has

$$dL = \sum_{i=1}^{n}\left(\frac{\partial L}{\partial q_i} dq_i + \frac{\partial L}{\partial \dot{q}_i} d\dot{q}_i\right) = \sum_{i=1}^{n}(\dot{p}_i dq_i + p_i d\dot{q}_i).$$

We define the Hamiltonian function as

$$H(t, q_1, \ldots, q_n, p_1, \ldots, p_n) = -L\left(t, q_1, \ldots, q_n, \frac{dq_1}{dt}, \ldots, \frac{dq_n}{dt}\right) + \sum_{i=1}^{n} p_i \dot{q}_i,$$

$$dH = \sum_{i=1}^{n}(-\dot{p}_i dq_i + \dot{q}_i dp_i),$$

where $\sum_{i=1}^{n} p_i \dot{q}_i = \sum_{i=1}^{n} \frac{\partial L}{\partial \dot{q}_i} \dot{q}_i = \sum_{i=1}^{n} \frac{\partial \Gamma}{\partial \dot{q}_i} \dot{q}_i = 2\Gamma$.

Thus, we have

$$H\left(t, q_1, \ldots, q_n, \frac{dq_1}{dt}, \ldots, \frac{dq_n}{dt}\right) = \Gamma\left(t, q_1, \ldots, q_n, \frac{dq_1}{dt}, \ldots, \frac{dq_n}{dt}\right) + \Pi(t, q_1, \ldots, q_n)$$

or $H(t, q_1, \ldots, q_n, p_1, \ldots, p_n) = \Gamma(t, q_1, \ldots, q_n, p_1, \ldots, p_n) + \Pi(t, q_1, \ldots, q_n)$.

One concludes that the Hamiltonian, which is equal to the total energy, is expressed as a function of the generalized coordinates and generalized momenta. The equations of motion are governed by the following equations

$$\dot{p}_i = -\frac{\partial H}{\partial q_i}, \quad \dot{q}_i = \frac{\partial H}{\partial p_i}, \tag{5.8.4}$$

which are called the Hamiltonian equations of motion.

It is evident that using the Hamiltonian mechanics, one obtains the system of $2n$ first-order partial differential equations to model the system dynamics. In contrast, using the Lagrange equations of motion, the system of $n$ second-order differential equations results. However, the derived differential equations are equivalent.

*Example 5.8.15*

Consider the harmonic oscillator. The total energy is given as the sum of the kinetic and potential energies, $\Sigma_T = \Gamma + \Pi = \frac{1}{2}(mv^2 + k_s x^2)$. Find the equations of motion using the Lagrange and Hamilton concepts.

*Solution.*

The Lagrangian function is

$$L\left(x, \frac{dx}{dt}\right) = \Gamma - \Pi = \tfrac{1}{2}(mv^2 - k_s x^2) = \tfrac{1}{2}(m\dot{x}^2 - k_s x^2).$$

Making use of the Lagrange equations of motion

$$\frac{d}{dt}\frac{\partial L}{\partial \dot{x}} - \frac{\partial L}{\partial x} = 0,$$

we have

$$m\frac{d^2x}{dt^2} + k_s x = 0.$$

From Newton's second law, the second-order differential equation of motion is

$$m\frac{d^2x}{dt^2} + k_s x = 0.$$

The Hamiltonian function is expressed as

$$H(x, p) = \Gamma + \Pi = \tfrac{1}{2}(mv^2 - k_s x^2) = \tfrac{1}{2}\left(\frac{1}{m}p^2 - k_s x^2\right).$$

From the Hamiltonian equations of motion

$$\dot{p}_i = -\frac{\partial H}{\partial q_i} \text{ and } \dot{q}_i = \frac{\partial H}{\partial p_i},$$

as given by (5.8.4), one obtains

$$\dot{p} = -\frac{\partial H}{\partial x} = -k_s x,$$

$$\dot{x} = \dot{q} = \frac{\partial H}{\partial p} = \frac{p}{m}.$$

The equivalence the results and equations of motion are obvious. □

## 5.9 THERMOANALYSIS AND HEAT EQUATION

It is known that the heat propagates (flows) in the direction of decreasing temperature, and the rate of propagation is proportional to the gradient of the temperature. Using the thermal conductivity of the media $k_t$ and the temperature $T(t,x,y,z)$, one has the following equation to calculate the velocity of the heat flow

$$\vec{\mathbf{v}}_h = -k_t \nabla T(t,x,y,z). \qquad (5.9.1)$$

Consider the region **R** and let $s$ be the boundary surface. Using the divergence theorem, from (5.9.1) one obtains the partial differential equation (heat equation) which is expressed as

$$\frac{\partial T(t,x,y,z)}{\partial t} = k^2 \nabla^2 T(t,x,y,z), \qquad (5.9.2)$$

where $k$ is the thermal diffusivity of the media.

We have $k = \dfrac{k_t}{k_h k_d}$,

where $k_h$ and $k_d$ are the specific heat and density constants.

Solving partial differential equation (5.8.2), which is subject to the initial and boundary conditions, one finds the temperature of the homogeneous media. In the Cartesian coordinate system, one has

$$\nabla^2 T(t,x,y,z) = \frac{\partial^2 T(t,x,y,z)}{\partial x^2} + \frac{\partial^2 T(t,x,y,z)}{\partial y^2} + \frac{\partial^2 T(t,x,y,z)}{\partial z^2}.$$

Using the Laplacian of $T$ in the cylindrical and spherical coordinate systems, one can reformulate the thermoanalysis problem using different coordinates in order to straightforwardly solve the problem.

It the heat flow is steady (time-invariant), then $\dfrac{\partial T(t,x,y,z)}{\partial t} = 0$.

Hence, three-dimensional heat equation (5.9.2) becomes Laplace's equation as given by $0 = k^2 \nabla^2 T(t,x,y,z)$.

The two-dimensional heat equation is

$$\frac{\partial T(t,x,y)}{\partial t} = k^2 \nabla^2 T(t,x,y) = k^2 \left( \frac{\partial^2 T(t,x,y)}{\partial x^2} + \frac{\partial^2 T(t,x,y)}{\partial y^2} \right).$$

If $\dfrac{\partial T(t,x,y)}{\partial t} = 0$, one has

$$0 = k^2 \nabla^2 T(t,x,y) = k^2 \left( \frac{\partial^2 T(t,x,y)}{\partial x^2} + \frac{\partial^2 T(t,x,y)}{\partial y^2} \right).$$

Using initial and boundary conditions, this partial differential equation can be solved using Fourier series, Fourier integrals, or Fourier transforms.

The so-called one-dimensional heat equation is

$$\frac{\partial T(t,x)}{\partial t}=k^2\frac{\partial^2 T(t,x)}{\partial x^2}$$

with initial and boundary conditions

$$T(t_0,x)=T_t(x),\ T(t,x_0)=T_0 \text{ and } T(t,x_f)=T_f.$$

A large number of analytical and numerical methods are available to solve the heat equation.

The analytic solution if

$$T(t,x_0)=0 \text{ and } T(t,x_f)=0$$

is given as

$$T(t,x)=\sum_{i=1}^{\infty}B_i\sin\frac{i\pi x}{x_f}e^{-i^2\frac{k^2\pi^2}{x_f^2}t},$$

$$B_i=\frac{2}{x_f}\int_{x_0}^{x_f}T_t(x)\sin\frac{i\pi x}{x_f}dx.$$

Assuming that $T_t(x)$ is piecewise continuous in $x\in[x_0\ \ x_f]$ and has one-sided derivatives at all interior points, one finds the coefficients of the Fourier sine series $B_i$.

*Example 5.9.1*

Consider the copper bar with length 0.1 mm. The thermal conductivity, specific heat and density constants are $k_t=1$, $k_h=0.09$ and $k_d=9$. The initial and boundary conditions are

$$T(0,x)=T_t(x)=0.2\sin\frac{\pi x}{0.001},\ T(t,0)=0 \text{ and } T(t,0.001)=0.$$

Find the temperature in the bar as a function of the position and time.

*Solution.*

From the general solution

$$T(t,x)=\sum_{i=1}^{\infty}B_i\sin\frac{i\pi x}{x_f}e^{-i^2\frac{k^2\pi^2}{x_f^2}t},$$

using the initial condition, we have

$$T(0,x)=\sum_{i=1}^{\infty}B_i\sin\frac{i\pi x}{x_f}=0.2\sin\frac{\pi x}{0.001}.$$

Thus, $B_1=0.2$ and all other $B_i$ coefficients are zero. Hence, the solution (temperature as the function of the position and time) is found to be

$$T(t,x)=\sum_{i=1}^{\infty}B_i\sin\frac{i\pi x}{x_f}e^{-i^2\frac{k^2\pi^2}{x_f^2}t}=B_1\sin\frac{\pi x}{x_f}e^{-\frac{k^2\pi^2}{x_f^2}t}=0.2\sin\frac{\pi x}{0.001}e^{-1.5\times10^7 t}.\ \square$$

# CHAPTER 6

# NANOSYSTEMS, QUANTUM MECHANICS, AND MATHEMATICAL MODELS

## 6.1 ATOMIC STRUCTURES AND QUANTUM MECHANICS

The fundamental and applied research as well as engineering developments in NEMS have undergone major developments in last years. High-performance nanostructures and nanodevices have been manufactured and implemented (accelerometers and microphones, actuators and sensors, molecular wires and transistors, etc.). Nanoengineering studies NEMS as well as their subsystems, devices, and structures which are made from atoms and molecules, and the electron is considered a fundamental particle. The students and engineers have obtained the necessary background in physics classes. The properties and performance of materials (media) is understood through the analysis of the atomic structure.

The atomic structures were studied by Rutherford and Einstein (in the 1900s), Heisenberg and Dirac (in the 1920s), Schrödinger, Bohr, Feynman, and many other scientists. For example, the theory of quantum electrodynamics studies the interaction of electrons and photons. In the 1940s, the major breakthrough appears in augmentation of the electron dynamics with electromagnetic field. One can control molecules and groups of molecules (nanostructures) applying the electromagnetic field. Micro- and nanoscale devices (e.g., actuators and sensors) have been fabricated, and some problems in structural design and optimization have been approached and solved. However, these nano- and microscale devices (which have dimensions nano- and micrometers) must be controlled, and one faces an extremely challenging problem to design NEMS integrating control and optimization, self-organization and decision making, diagnostics and self-repairing, signal processing and communication, as well as other features. In 1959, Richard Feynman gave a talk to the American Physical Society in which he emphasized the important role of nanotechnology and nanoscale organic and inorganic systems on the society and progress.

All media are made from atoms, and the medium properties depend on the atomic structure. Recalling Rutherford's structure of the atomic nuclei, we can view a very simple atomic model and omit detailed composition, because only three subatomic particles (proton, neutron and electron) have bearing on chemical behavior.

The nucleus of the atom bears the major mass. It is an extremely dense region, which contains positively charged protons and neutral neutrons. It occupies a small amount of the atomic volume compared with the virtually indistinct cloud of negatively charged electrons attracted to the positively charged nucleus by the force that exists between the particles of opposite electric charge.

For the atom of the element, the number of protons is always the same but the number of neutrons may vary. Atoms of a given element, which differ in number of neutrons (and consequently in mass), are called isotopes. For example, carbon always has 6 protons, but it may have 6 neutrons as well. In this case it is called "carbon-12" ($^{12}C$). The representation of the carbon atom is given in Figure 6.1.1.

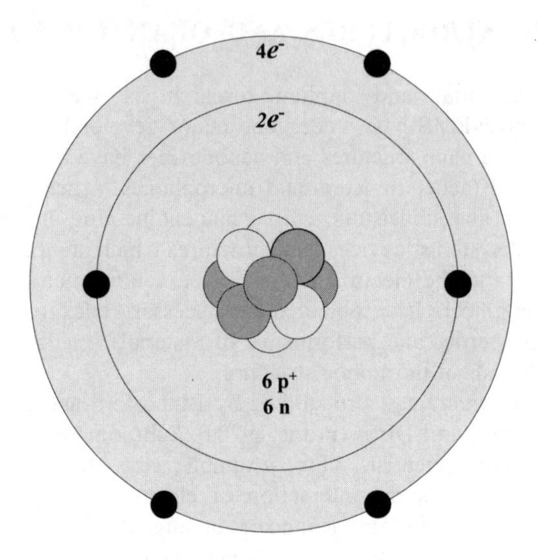


Figure 6.1.1 Simplified two-dimensional representation of carbon atom (C). Six protons (p+, dashed color) and six neutrons (n, white) are in centrally located nucleus. Six electrons ($e^-$, black), orbiting the nucleus, occupy two shells.

An atom has no net charge due to the equal number of positively charged protons in the nucleus and negatively charged electrons around it. For example, all atoms of carbon have 6 protons and 6 electrons. If electrons are lost or gained by the neutral atom due to the chemical reaction, a charged particle called an ion is formed.

When one deals with such subatomic particles as the electron, the dual nature of matter places a fundamental limitation on how accurate we can describe both location and momentum of the object. Austrian physicist Erwin Schrödinger in 1926 derived an equation that describes wave and particle natures of the electron. This fundamental equation led to the new area in physics, called quantum mechanics, which enables us to deal with subatomic particles. The complete solution to Schrödinger's equation gives a set of wave functions and set of corresponding energies. These wave functions are called orbitals. A collection of orbitals with the same principal quantum number, which describes the orbit, is called the electron shell. Each shell is divided into the number of subshells with the equal principal quantum number. Each subshell consists of number of orbitals. Each shell may contain only two electrons of the opposite spin (Pouli exclusion principle). When the electron is in the lowest energy orbital, the atom is in its ground state. When the electron enters the orbital, the atom is in an excited state. To promote the electron to the excited-state orbital, the photon of the appropriate energy should be absorbed as the energy supplement.

When the size of the orbital increases, and the electron spends more time farther from the nucleus. It possesses more energy and is less tightly bound to the nucleus. The most outer shell is called the valence shell. The electrons taht occupy it are referred to as valence electrons. Inner shell electrons are called the core electrons. There are valence electrons, which participate in the bond formation between atoms when molecules are formed, and in ion formation when the electrons are removed from the electrically neutral atom and the positively charged cation is formed. They possess the highest ionization energies (the energy which measures the ease of removing the electron from the atom), and occupy the energetically weakest orbital since it is the most remote orbital from the nucleus. The valence electrons removed from the valence shell become free electrons, transferring the energy from one atom to another. We will describe the influence of the electromagnetic field on the atom later in the text, and it is relevant to include more detailed description of the Pauli exclusion principal.

The electric conductivity of a media is predetermined by the density of free electrons, and good conductors have the free electron density in the range of $10^{23}$ free electrons per $cm^3$. In contrast, the free electron density of good insulators is in the range of 10 free electrons per $cm^3$. The free electron density of semiconductors in the range from $10^7/cm^3$ to $10^{15}/cm^3$ (for example, the free electron concentration in silicon at $25^0$C and $100^0$C are $2\times10^{10}/cm^3$ and $2\times10^{12}/cm^3$, respectively). The free electron density is determined by the energy gap between valence and conduction (free) electrons. That is, the properties of the media (conductors, semiconductors, and insulators) are determined by the atomic structure.

Using the atoms as building blocks, one can manufacture different structures using the molecular nanotechnology. There are many challenging problems needed to be solve such as mathematical modeling and analysis, simulation and design, optimization and testing, implementation and deployment, technology transfer and mass production. In addition, to build NEMS, advanced manufacturing technologies must be developed and applied. To fabricate nanoscale systems at the molecular level, the problems in atomic-scale positional assembly ("maneuvering things atom by atom" as Richard Feynman predicted) and artificial self-replication (systems are able to build copies of themselves, e.g., like the crystals growth process, complex DNA strands which copy tens of millions atoms with perfect accuracy, or self replicating tomato which has millions of genes, proteins, and other molecular components) must be solved. The author does not encourage the blind copying, and the submarine and whale are very different even though both sail. Using the Scanning or Atomic Probe Microscopes, it is possible to achieve positional accuracy in the angstrom-range. However, the atomic-scale "manipulator" (which will have a wide range of motion guaranteeing the flexible assembly of molecular components), controlled by the external source (electromagnetic field, pressure, or temperature) must be designed and used. The position control will be achieved by the molecular computer and will be based on molecular computational devices.

The quantitative explanation, analysis and simulation of natural phenomena can be approached using comprehensive mathematical models which map essential features. The Newton laws and Lagrange equations of motion, Hamilton concept and d'Alambert concept allow one to model conventional mechanical systems, and the Maxwell equations applied to model electromagnetic phenomena. In the 1920s, new theoretical developments, concepts and formulations (*quantum mechanics*) were made to develop the atomic scale theory because atomic-scale systems do not obey the classical laws of physics and mechanics. In 1900 Max Plank discovered the effect of quantization of energy, and he found that the radiated (emitted) energy is given as

$$E = nh\nu,$$

where $n$ is the nonnegative integer, $n = 0, 1, 2, \ldots$; $h$ is the Plank constant, $h = 6.626\times10^{-34}$ J-sec; $\nu$ is the frequency of radiation, $\nu = \frac{c}{\lambda}$, $c$ is the speed of light, $c = 3\times10^{8}\ \frac{\text{m}}{\text{sec}}$; $\lambda$ is the wavelength which is measured in angstroms ($\overset{o}{A} = 1\times10^{-10}$ m), $\lambda = \frac{c}{\nu}$.

The following discrete energy values result:

$E_0 = 0$, $E_1 = h\nu$, $E_2 = 2h\nu$, $E_3 = 3h\nu$, etc.

The observation of discrete energy spectra suggests that each particle has the energy $h\nu$ (the radiation results due to N particles), and the particle with the energy $h\nu$ is called *photon*.

The *photon* has the momentum as expressed as

$$p = \frac{h\nu}{c} = \frac{h}{\lambda}.$$

Soon, Einstein demonstrated the discrete nature of light, and Niels Bohr develop the model of the hydrogen atom using the planetary system analog, see Figure 6.1.2. It is clear that if the electron has planetary-type orbits, it can be excited to an outer orbit and can "fall" to the inner orbits. Therefore, to develop the model, Bohr postulated that the electron has the certain stable circular orbit (that is, the orbiting electron does not produces the radiation because otherwise the electron would lose the energy and change the path); the electron changes the orbit of higher or lower energy by receiving or radiating a discrete amount of energy; the angular momentum of the electron is $p = nh$.

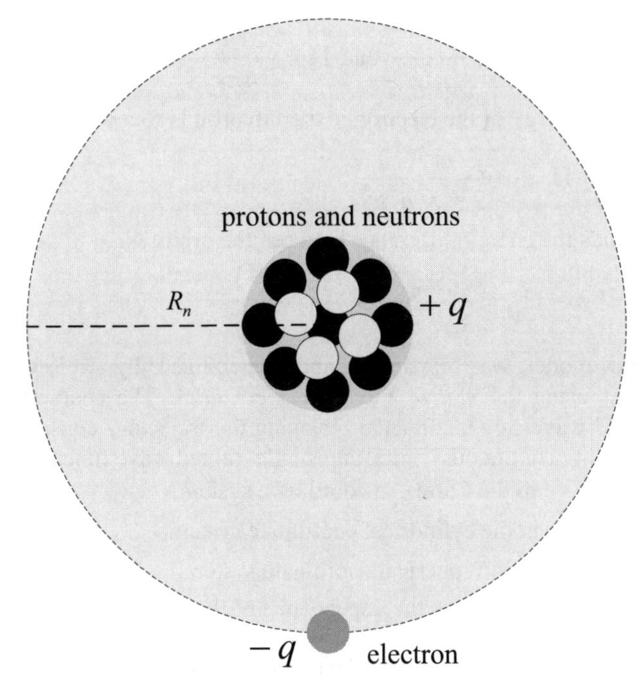


Figure 6.1.2 Hydrogen atom: uniform circular motion.

To attain the uniform circular motion, from Newton's law, the electrostatic (Coulomb) force must be equal to the radial force. For radii $R_1$ and $R_2$ we have

$$\frac{q^2}{4\pi\varepsilon_0 R_1^2}=\frac{mv^2}{R_1} \text{ and } \frac{q^2}{4\pi\varepsilon_0 R_2^2}=\frac{mv^2}{R_2}.$$

That is, in general

$$\frac{q^2}{4\pi\varepsilon_0 R_n^2}=\frac{mv^2}{R_n},$$

where $R_n$ is the radius of the $n$ orbit, and $R_n=\dfrac{4\pi\varepsilon_0 n^2 h^2}{mq^2}$.

Applying the expression for the angular momentum

$p = nh = mvR_n$,

one obtains

$$\frac{q^2}{4\pi\varepsilon_0 R_n^2}=\frac{m}{R_n}\left(\frac{nh}{mR_n}\right)^2=\frac{1}{mR_n}\frac{n^2h^2}{R_n^2}.$$

The kinetic and potential energies are

$$\Gamma = \tfrac{1}{2}mv^2 = \frac{mq^4}{32\pi^2\varepsilon_0^2 n^2 h^2} \text{ and } \Pi = -\frac{q^2}{4\pi\varepsilon_0 R_n} = -\frac{mq^4}{16\pi^2\varepsilon_0^2 n^2 h^2}.$$

The total energy of the electron in the *n*th orbit is found to be

$$E_n = \Gamma + \Pi = -\frac{mq^4}{32\pi^2\varepsilon_0^2 n^2 h^2}.$$

One finds the energy difference between the orbits as

$$\Delta E = E_{n2} - E_{n1} = \frac{mq^4}{32\pi^2\varepsilon_0^2 h^2}\left(\frac{1}{n_1^2} - \frac{1}{n_2^2}\right).$$

Bohr's model was expanded and generalized by Heisenberg and Schrödinger using the *matrix* and *wave mechanics*. The characteristics of particles and waves are augmented replacing the trajectory consideration by the waves using continuous, finite, and single-valued wave function

- $\Psi(x,y,z,t)$ in the Cartesian coordinate system,
- $\Psi(r,\phi,z,t)$ in the cylindrical coordinate system,
- $\Psi(r,\theta,\phi,t)$ in the spherical coordinate system.

The wavefunction gives the dependence of the wave amplitude on space coordinates and time.

Using the classical mechanics, for a particle of mass *m* with energy *E* moving in the Cartesian coordinate system one has

$$\underbrace{E(x,y,z,t)}_{\textit{total energy}} = \underbrace{\Gamma(x,y,z,t)}_{\textit{kinetic energy}} + \underbrace{\Pi(x,y,z,t)}_{\textit{potential energy}}$$

$$= \frac{p^2(x,y,z,t)}{2m} + \Pi(x,y,z,t) = \underbrace{H(x,y,z,t)}_{\textit{Hamiltonian}}.$$

Thus, we have

$$p^2(x,y,z,t) = 2m\left[E(x,y,z,t) - \Pi(x,y,z,t)\right].$$

Using the formula for the wavelength (Broglie's equation)

$$\lambda = \frac{h}{p} = \frac{h}{mv},$$

one finds

$$\frac{1}{\lambda^2} = \left(\frac{p}{h}\right)^2 = \frac{2m}{h^2}\left[E(x,y,z,t) - \Pi(x,y,z,t)\right].$$

This expression is substituted in the *Helmholtz* equation

$$\nabla^2\Psi + \frac{4\pi^2}{\lambda^2}\Psi = 0$$

which gives the evolution of the wavefunction.

We obtain the Schrödinger equation as

$$E(x,y,z,t)\Psi(x,y,z,t) = -\frac{\hbar^2}{2m}\nabla^2\Psi(x,y,z,t) + \Pi(x,y,z,t)\Psi(x,y,z,t)$$

or

$$E(x,y,z,t)\Psi(x,y,z,t)$$
$$= -\frac{\hbar^2}{2m}\left(\frac{\partial^2\Psi(x,y,z,t)}{\partial x^2} + \frac{\partial^2\Psi(x,y,z,t)}{\partial y^2} + \frac{\partial^2\Psi(x,y,z,t)}{\partial z^2}\right)$$
$$+ \Pi(x,y,z,t)\Psi(x,y,z,t).$$

Here, the modified Plank constant is $\hbar = \frac{h}{2\pi} = 1.055\times 10^{-34}$ J-sec.

In 1926, Erwine Schrödinger derived the following equation

$$-\frac{\hbar^2}{2m}\nabla^2\Psi + \Pi\Psi = E\Psi$$

which can be related to the Hamiltonian function

$$H = -\frac{\hbar^2}{2m}\nabla + \Pi.$$

Thus,

$$H\Psi = E\Psi.$$

For different coordinate systems we have

- Cartesian system

$$\nabla^2\Psi(x,y,z,t)$$
$$= \frac{\partial^2\Psi(x,y,z,t)}{\partial x^2} + \frac{\partial^2\Psi(x,y,z,t)}{\partial y^2} + \frac{\partial^2\Psi(x,y,z,t)}{\partial z^2};$$

- Cylindrical system

$$\nabla^2\Psi(r,\phi,z,t)$$
$$= \frac{1}{r}\frac{\partial}{\partial r}\left(r\frac{\partial\Psi(r,\phi,z,t)}{\partial r}\right) + \frac{1}{r^2}\frac{\partial^2\Psi(r,\phi,z,t)}{\partial\phi^2} + \frac{\partial^2\Psi(r,\phi,z,t)}{\partial z^2};$$

- Spherical system

$$\nabla^2\Psi(r,\theta,\phi,t) =$$
$$\frac{1}{r^2}\frac{\partial}{\partial r}\left(r^2\frac{\partial\Psi(r,\theta,\phi,t)}{\partial r}\right) + \frac{1}{r^2\sin\theta}\frac{\partial}{\partial\theta}\left(\sin\theta\frac{\partial\Psi(r,\theta,\phi,t)}{\partial\theta}\right)$$
$$+ \frac{1}{r^2\sin^2\theta}\frac{\partial^2\Psi(r,\theta,\phi,t)}{\partial\phi^2}.$$

The Schrödinger partial differential equation must be solved, and the wavefunction is normalized using the probability density

$$\iint|\Psi|^2 d\varsigma = 1.$$

*Example 6.1.1*

Let us illustrate the application of the Schrödinger equation. Assume that the particle moves in the $x$ direction (translational motion). We have,

$$-\frac{\hbar^2}{2m}\frac{d^2\Psi(x)}{dx^2}+\Pi(x)\Psi(x)=E(x)\Psi(x).$$

The Hamiltonian function is given as

$$H(x,p)=\frac{p^2(x)}{2m}+\Pi(x)=-\frac{\hbar^2}{2m}\frac{d^2}{dx^2}+\Pi(x).$$

Let the particle moves from $x = 0$ to $x = x_f$, and the potential energy is

$$\Pi(x)=\begin{cases}0, & 0\le x\le x_f\\ \infty, & x<0 \text{ and } x>x_f\end{cases}.$$

Thus, the motion of the particle is bounded in the "potential wall", and

$$\Psi(x)=\begin{cases}\text{continuous if } 0\le x\le x_f\\ 0 \text{ if } x<0 \text{ and } x>x_f\end{cases}.$$

If $0\le x\le x_f$, the potential energy is zero, and we have

$$-\frac{\hbar^2}{2m}\frac{d^2\Psi(x)}{dx^2}=E\Psi(x),\ 0\le x\le x_f.$$

The solution of the resulting second-order differential equation

$$\frac{d^2\Psi(x)}{dx^2}+k^2\Psi(x)=0,\ k=\sqrt{\frac{2mE}{\hbar^2}}$$

is

$$\Psi(x)=ae^{ikx}+be^{-ikx}=a(\cos kx+i\sin kx)+b(\cos kx-i\sin kx)$$
$$=c\sin kx+d\cos kx.$$

The solution can be easily verified by plugging the solution in the left-side of the differential equation

$$-\frac{\hbar^2}{2m}\frac{d^2\Psi(x)}{dx^2}=E\Psi(x),$$

and we have

$$E\Psi(x)=E\Psi(x).$$

It should be emphasized that the kinetic energy of the particle is $\frac{p^2}{2m}$, where $p = kh$.

It is obvious that one must use the boundary conditions.

We have $\Psi(x)\big|_{x=0}=\Psi(0)=0$, and therefore $d = 0$.

From $\Psi(x)\big|_{x=x_f}=\Psi(x_f)=0$ using $c\sin kx_f=0$ one must find the constant $c$ and the expression for $kx_f$.

Assuming that $c \neq 0$ from $c \sin kx_f = 0$, we have

$kx_f = n\pi$,

where $n$ is the positive or negative integer (if $n$ = 0, the wavefunction vanishes everywhere, and thus, $n \neq 0$).

From $k = \sqrt{\frac{2mE}{\hbar^2}}$ and making use of $kx_f = n\pi$ we have the expression for the energy (discrete values of the energy which allow of solution of the Schrödinger equation) as

$$E_n = \frac{\hbar^2 \pi^2}{2mx_f^2} n^2, n = 1, 2, 3, \ldots,$$

where the integer $n$ designates the allowed energy level ($n$ is called the quantum number).

For example, if $n$ = 1 and $n$ = 2, we have $E_1 = \frac{\hbar^2 \pi^2}{2mc^2}$ (the lowest possible energy which is called the ground state) and $E_2 = \frac{2\hbar^2 \pi^2}{mc^2}$.

Thus, we have illustrated that the energy of the particle is quantized.
The expression for the wavefunction is found to be

$$\Psi_n(x) = c \sin kx + d \cos kx = c \sin \frac{n\pi}{x_f} x.$$

Using the probability density, we normalize the wavefunction, and

$$\int_0^{x_f} \Psi_n^2(x) dx = c^2 \int_0^{x_f} \sin^2 \frac{n\pi}{x_f} x dx = c^2 \frac{x_f}{n\pi} \int_0^{n\pi} \sin^2 g dg$$

$$c^2 \frac{x_f}{n\pi} \frac{n\pi}{2} = c^2 \frac{x_f}{2} = 1, \ g = \frac{n\pi}{x_f} x.$$

Hence, $c = \sqrt{\frac{2}{x_f}}$, and one obtains

$$\Psi_n(x) = \sqrt{\frac{2}{x_f}} \sin \frac{n\pi}{x_f} x, \ 0 \le x \le x_f.$$

For $n = 1$ and $n = 2$, we have

$$\Psi_1(x) = \sqrt{\frac{2}{x_f}} \sin \frac{\pi}{x_f} x \text{ and } \Psi_2(x) = \sqrt{\frac{2}{x_f}} \sin \frac{2\pi}{x_f} x.$$

Using the formula for the probability density $\rho = \Psi^T \Psi$, one has

$$\rho_n(x) = \frac{2}{x_f} \sin^2 \frac{n\pi}{x_f} x.$$

□

It was shown that

$$H\Psi = E\Psi ,$$

$$H = -\frac{\hbar^2}{2m}\nabla + \Pi .$$

Using the CGS (centimeter/gram/second) units, when the electromagnetic field is quantized, the potential can be used instead of wavefunction. In particular, using the momentum operator due to electron orbital angular momentum **L**, the classical Hamiltonian for electrons in electromagnetic field is

$$H = \frac{1}{2m}\left(\mathbf{p} + \frac{e}{c}\mathbf{A}\right)^2 - e\phi .$$

From the Hamilton equations

$$\dot{q} = \frac{\partial H}{\partial p},$$

$$\dot{p} = -\frac{\partial H}{\partial q},$$

by making use of

$$\frac{d\mathbf{r}}{dt} = \frac{1}{m}\left(\mathbf{p} + \frac{e}{c}\mathbf{A}\right),$$

$$\mathbf{p} = m\mathbf{v} - \frac{e}{c}\mathbf{A},$$

$$\dot{\mathbf{p}} = -\frac{e}{mc}\left(\mathbf{p} + \frac{e}{c}\mathbf{A}\right)\cdot\frac{\partial \mathbf{A}}{\partial \mathbf{x}} + e\frac{\partial \phi}{\partial \mathbf{x}},$$

one finds the Lorentz force equation as

$$\mathbf{F} = -\frac{e}{c}\mathbf{v}\times\mathbf{B} - e\mathbf{E} .$$

This equation gives the force due to motion in a magnetic field and the force due to electric field.

It is important to emphasize that the following equation results

$$-\frac{\hbar^2}{2m}\nabla^2\Psi + \frac{e}{2mc}\mathbf{B}\cdot\mathbf{L}\Psi + \frac{e^2}{8mc^2}\left(r^2B^2 - (\mathbf{r}\cdot\mathbf{B})^2\right)\Psi = (E + e\phi)\Psi$$

to study the quantized Hamilton equation, where the dominant term due to magnetic field is

$$\frac{e}{2mc}\mathbf{B}\cdot\mathbf{L} = -\boldsymbol{\mu}\cdot\mathbf{B},$$

where $\boldsymbol{\mu}$ the magnetic momentum due to the electron orbital angular momentum (the so-called Zeeman effect) is $\boldsymbol{\mu} = -\frac{e}{2mc}\mathbf{L}$.

## 6.2 MOLECULAR AND NANOSTRUCTURE DYNAMICS

Conventional, mini- and microscale electromechanical systems can be modeled, simulated and analyzed using electromagnetic and circuitry theories, classical mechanics and thermodynamics. Other fundamental concepts are applied as well. The complexity of mathematical models of mini- and microelectromechanical systems (nonlinear ordinary and partial differential equations explicitly describe the spectrum of electromagnetic and electromechanical phenomena, processes and effects) is not ambiguous, and numerical algorithms to solve the equations have been derived. Computationally efficient software and environments to support heterogeneous simulation and data-intensive analysis are available. Nanoscale structures, devices and systems, in general, cannot be studied using the conventional concepts.

The ability to find equations (mathematical models), which adequately describe nanosystems properties, phenomena and effects, is a key problem in modeling, analysis, synthesis, optimization, control, fabrication, manufacturing, and implementation of NEMS. In this section, using classical and quantum mechanics, functional density concept, and electromagnetic theory we further contribute to the nanoelectromechanical theory to model, analyze, and simulate nanosystems. The reported developments support the existing paradigms in modeling, analysis, and design of NEMS. The proposed fundamental results allow the designer to solve a broad spectrum of problems compared with currently existing methods. The reported theoretical and applied results are verified and demonstrated. We study novel phenomena in nanosystems using quantum, classical, and optimization theories. This research is critical to overcome current obstacles in complete understanding of processes and phenomena in nanoscale. In particular, the long-standing goal is developing of fundamental and experimental tools to design and fabricate nanostructures using synthesis methods devised.

The fundamental and applied research in molecular nanotechnology and nanosystems is concentrated on design, modeling, simulation, and fabrication of molecular scale structures and devices. The design, modeling, and simulation of NEMS, MEMS, and their components can be attacked using advanced theoretical developments and simulation concepts. Comprehensive analysis must be performed before the designer embarks in costly fabrication (a wide range of nanoscale structures and devices, molecular machines and subsystems, can be fabricated with atomic precision) because through modeling and simulation the rapid evaluation and prototyping can be performed facilitating significant advantages and manageable perspectives to attain the desired objectives. With advanced computer-aided-design tools, complex large-scale nanostructures, nanodevices, and nanosystems can be designed, analyzed, and evaluated.

Classical quantum mechanics does not allow the designer to perform analytical and numerical analysis even for simple nanostructures which consist of a couple of molecules. Steady-state three-dimensional modeling and simulation are also restricted to simple nanostructures. Our goal is to develop a fundamental understanding of phenomena and processes in nanostructures with emphasis on their further applications in nanodevices, nanosubsystems, NEMS, and MEMS. The objective is the development of theoretical fundamentals (theory of nanoelectromechanics) to perform 3D+ (three-dimensional dynamics in the time domain) modeling and simulation.

The atomic level dynamics can be studied using the wave function. The Schrödinger equation for N-electron systems must be solved. However, this problem cannot be solved even for simple nanostrustures. In [1-5], the density functional theory was developed, and the charge density is used rather than the electron wavefunctions. In particular, the N-electron problem is formulated as N one-electron equations where each electron interacts with all other electrons via an effective exchange-correlation potential. These interactions are augmented using the charge density. Plane wave sets and total energy pseudo-potential methods can be used to solve the Kohn-Sham one electron equations [1-3]. The Hellmann-Feynman theory can be applied to calculate the forces solving the molecular dynamics problem [4, 6].

### 6.2.1 Schrödinger Equation and Wavefunction Theory

For two point charges, Coulomb's law is given as

$$\mathbf{F}=\frac{q_1 q_2}{4\pi\varepsilon d^2}\mathbf{a}_r=\frac{q_1 q_2}{4\pi\varepsilon}\frac{(\mathbf{r}-\mathbf{r}')}{|\mathbf{r}-\mathbf{r}'|^3},$$

and in the Cartesian coordinate systems one has

$$\mathbf{F}=\frac{q_1 q_2}{4\pi\varepsilon d^2}\mathbf{a}_r=\frac{q_1 q_2}{4\pi\varepsilon d^2}\frac{(x-x')\mathbf{a}_x+(y-y')\mathbf{a}_y+(z-z')\mathbf{a}_z}{\sqrt{(x-x')^2+(y-y')^2+(z-z')^2}}.$$

In the case of charge distribution, using the volume charge density $\rho_v$, the net force exerted on $q_1$ by the entire volume charge distribution is the vector sum of the contribution from all differential elements of charge within this distribution. In particular, $\mathbf{F}=\frac{q_1}{4\pi\varepsilon}\int_v \rho_v \frac{(\mathbf{r}-\mathbf{r}')}{|\mathbf{r}-\mathbf{r}'|^3}dv$, see Figure 6.2.1.

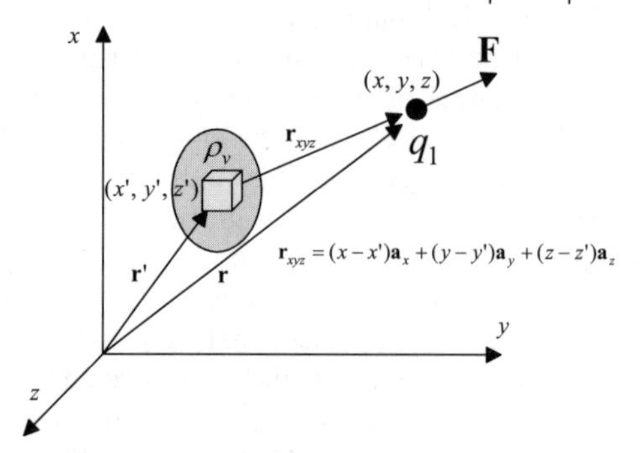


Figure 6.2.1 Coulomb's law.

In the electrostatic field, the potential energy stored in a region of continuous charge distribution is found as

$$\Pi_V = \tfrac{1}{2}\int_v \mathbf{D}\cdot\mathbf{E}dv = \tfrac{1}{2}\int_v \varepsilon\mathbf{E}^2 dv = \tfrac{1}{2}\int_v \rho_v(\mathbf{r})V(\mathbf{r})dv,$$

where $V(\mathbf{r})$ is the potential; $v$ is the volume containing $\rho_v$.

The charge distribution can be given in terms of volume, surface, and line charges. In particular, we have

$$V(\mathbf{r}) = \int_v \frac{\rho_v(\mathbf{r}')}{4\pi\varepsilon|\mathbf{r}-\mathbf{r}'|}dv',$$

$$V(\mathbf{r}) = \int_s \frac{\rho_s(\mathbf{r}')}{4\pi\varepsilon|\mathbf{r}-\mathbf{r}'|}ds',$$

and $V(\mathbf{r}) = \int_l \frac{\rho_l(\mathbf{r}')}{4\pi\varepsilon|\mathbf{r}-\mathbf{r}'|}dl'$.

It should be emphasized that that the electric field intensity is found as

$$\mathbf{E}(\mathbf{r}) = \int_v \frac{\rho_v(\mathbf{r}')}{4\pi\varepsilon}\frac{(\mathbf{r}-\mathbf{r}')}{|\mathbf{r}-\mathbf{r}'|^3}dv'.$$

Thus, the energy of an electric field or a charge distribution is stored in the field.

The energy, stored in the steady magnetic field is

$$\Pi_M = \tfrac{1}{2}\int_v \mathbf{B}\cdot\mathbf{H}dv.$$

## *Mathematical Models: Energy-Based Quantum and Classical Mechanics*

To perform the comprehensive modeling and analysis of nanostructures in the time domain, there is a critical need to develop and apply advanced theories using fundamental physical laws. Classical and quantum mechanics are widely used, and this section illustrates that the Schrödinger equation can be found using Hamilton's concept (it is well known that the Euler-Largange equations, given in terms of the generalized coordinates and forces, can be straightforwardly derived applying the variational principle). The quantum mechanics gives the system evolution in the form of the Schrödinger equations.

Newton's second law $\sum \vec{F}(t,\vec{\mathbf{r}}) = m\vec{a}$ in terms of the linear momentum $\vec{p} = m\vec{v}$, is given by

$$\sum \vec{F} = \frac{d\vec{p}}{dt} = \frac{d(m\vec{v})}{dt}.$$

Using the potential energy $\Pi(\vec{\mathbf{r}})$, for the conservative mechanical system we have

$$\sum \vec{F}(\vec{\mathbf{r}}) = -\nabla\Pi(\vec{\mathbf{r}}).$$

Hence, $m\dfrac{d^2\vec{\mathbf{r}}}{dt^2} + \nabla\Pi(\vec{\mathbf{r}}) = 0$.

For the system of N particles, the equations of motion are

$$m_N \frac{d^2\vec{\mathbf{r}}_N}{dt^2} + \nabla\Pi(\vec{\mathbf{r}}_N) = 0,$$

$$m_i \frac{d^2(\vec{x}_i, \vec{y}_i, \vec{z}_i)}{dt^2} + \frac{\partial\Pi(\vec{x}_i, \vec{y}_i, \vec{z}_i)}{\partial(\vec{x}_i, \vec{y}_i, \vec{z}_i)} = 0, i = 1,\ldots, N.$$

The total kinetic energy of the particle is $\Gamma = \frac{1}{2}mv^2$. For N particles, one has

$$\Gamma\left(\frac{d\vec{x}_i}{dt}, \frac{d\vec{y}}{dt}_i, \frac{d\vec{z}_i}{dt}\right) = \frac{1}{2}\sum_{i=1}^{N} m_i \left(\frac{d\vec{x}_i}{dt}, \frac{d\vec{y}}{dt}_i, \frac{d\vec{z}_i}{dt}\right).$$

Using the generilized coordinates $(q_1,\ldots, q_n)$ and generalized velocities $\left(\dfrac{dq_1}{dt},\ldots,\dfrac{dq_n}{dt}\right)$, one finds the total kinetic $\Gamma\left(q_1,\ldots, q_n, \dfrac{dq_1}{dt},\ldots,\dfrac{dq_n}{dt}\right)$ and potential $\Pi(q_1,\ldots, q_n)$ energies. Thus, Newton's second law of motion can be given as

$$\frac{d}{dt}\left(\frac{\partial\Gamma}{\partial\dot{q}_i}\right) + \frac{\partial\Pi}{\partial q_i} = 0.$$

That is, the generalized coordinates $q_i$ are used to model multibody systems, and $(q_1,\ldots, q_n) = (\vec{x}_1, \vec{y}_1, \vec{z}_1, \ldots, \vec{x}_N, \vec{y}_N, \vec{z}_N)$.

The obtained results are connected to the Lagrange equations of motion. Using the total kinetic $\Gamma\left(t, q_1,\ldots, q_n, \dfrac{dq_1}{dt},\ldots,\dfrac{dq_n}{dt}\right)$, dissipation $D\left(t, q_1,\ldots, q_n, \dfrac{dq_1}{dt},\ldots,\dfrac{dq_n}{dt}\right)$, and potential $\Pi(t, q_1,\ldots, q_n)$ energies, the Lagrange equations of motion are

$$\frac{d}{dt}\left(\frac{\partial\Gamma}{\partial\dot{q}_i}\right) - \frac{\partial\Gamma}{\partial q_i} + \frac{\partial D}{\partial\dot{q}_i} + \frac{\partial\Pi}{\partial q_i} = Q_i.$$

Here, $q_i$ and $Q_i$ are the generalized coordinates and the generalized forces (applied forces and disturbances).

The Hamilton concept allows one to model the system dynamics, and the differential equations are found using the generalized momenta $p_i$, $p_i = \frac{\partial L}{\partial \dot{q}_i}$. The Lagrangian function $L\left(t, q_1, \ldots, q_n, \frac{dq_1}{dt}, \ldots, \frac{dq_n}{dt}\right)$ for the conservative systems is the difference between the total kinetic and potential energies. We have

$$L\left(t, q_1, \ldots, q_n, \frac{dq_1}{dt}, \ldots, \frac{dq_n}{dt}\right) = \Gamma\left(t, q_1, \ldots, q_n, \frac{dq_1}{dt}, \ldots, \frac{dq_n}{dt}\right) - \Pi(t, q_1, \ldots, q_n).$$

Thus, $L\left(t, q_1, \ldots, q_n, \frac{dq_1}{dt}, \ldots, \frac{dq_n}{dt}\right)$ is the function of $2n$ independent variables, and

$$dL = \sum_{i=1}^{n}\left(\frac{\partial L}{\partial q_i} dq_i + \frac{\partial L}{\partial \dot{q}_i} d\dot{q}_i\right) = \sum_{i=1}^{n}\left(\dot{p}_i dq_i + p_i d\dot{q}_i\right).$$

Define the Hamiltonian function as

$$H(t, q_1, \ldots, q_n, p_1, \ldots, p_n) = -L\left(t, q_1, \ldots, q_n, \frac{dq_1}{dt}, \ldots, \frac{dq_n}{dt}\right) + \sum_{i=1}^{n} p_i \dot{q}_i,$$

$$dH = \sum_{i=1}^{n}\left(-\dot{p}_i dq_i + \dot{q}_i dp_i\right),$$

where $\sum_{i=1}^{n} p_i \dot{q}_i = \sum_{i=1}^{n} \frac{\partial L}{\partial \dot{q}_i}\dot{q}_i = \sum_{i=1}^{n} \frac{\partial \Gamma}{\partial \dot{q}_i}\dot{q}_i = 2\Gamma$.

Thus,

$$H\left(t, q_1, \ldots, q_n, \frac{dq_1}{dt}, \ldots, \frac{dq_n}{dt}\right) = \Gamma\left(t, q_1, \ldots, q_n, \frac{dq_1}{dt}, \ldots, \frac{dq_n}{dt}\right) + \Pi(t, q_1, \ldots, q_n)$$

or $H(t, q_1, \ldots, q_n, p_1, \ldots, p_n) = \Gamma(t, q_1, \ldots, q_n, p_1, \ldots, p_n) + \Pi(t, q_1, \ldots, q_n)$.

One concludes that the Hamiltonian function, which is equal to the total energy, is expressed as a function of the generalized coordinates and generalized momenta. The equations of motion are governed by the following equations

$$\dot{p}_i = -\frac{\partial H}{\partial q_i},$$

$$\dot{q}_i = \frac{\partial H}{\partial p_i},$$

which are the Hamiltonian equations of motion.

The Hamiltonian function

$$H = \underbrace{-\frac{\hbar^2}{2m}\nabla^2}_{\text{one-electron kinetic energy}} + \underbrace{\Pi}_{\text{potential energy}}$$

can be used to derive the one-electron Schrödinger equation. To describe the behavior of electrons in a media, one must use N-dimensional Schrödinger equation to obtain the N-electron wavefunction $\Psi(t, \mathbf{r}_1, \mathbf{r}_2, ..., \mathbf{r}_{N-1}, \mathbf{r}_N)$.

The Hamiltonian for an isolated *N*-electron atomic system is

$$H = -\frac{\hbar^2}{2m}\sum_{i=1}^{N}\nabla_i^2 - \frac{\hbar^2}{2M}\nabla^2 - \sum_{i=1}^{N}\frac{1}{4\pi\varepsilon}\frac{e_i q}{\left|\mathbf{r}_i - \mathbf{r}_n^{'}\right|} + \sum_{i \neq j}^{N}\frac{1}{4\pi\varepsilon}\frac{e^2}{\left|\mathbf{r}_i - \mathbf{r}_j^{'}\right|},$$

where $q$ is the potential due to nucleus; $e = 1.6 \times 10^{-19}$ C.

For an isolated *N*-electron, *Z*-nucleus molecular system, the Hamiltonian function (Hamiltonian operator) is found to be

$$H = -\frac{\hbar^2}{2m}\sum_{i=1}^{N}\nabla_i^2 - \sum_{k=1}^{Z}\frac{\hbar^2}{2m_k}\nabla_k^2$$

$$-\sum_{i=1}^{N}\sum_{k=1}^{Z}\frac{1}{4\pi\varepsilon}\frac{e_i q_k}{\left|\mathbf{r}_i - \mathbf{r}_k^{'}\right|} + \sum_{i \neq j}^{N}\frac{1}{4\pi\varepsilon}\frac{e^2}{\left|\mathbf{r}_i - \mathbf{r}_j^{'}\right|} + \sum_{k \neq m}^{Z}\frac{1}{4\pi\varepsilon}\frac{q_k q_m}{\left|\mathbf{r}_k - \mathbf{r}_m^{'}\right|},$$

where $q_k$ are the potentials due to nuclei.

The first and second terms of the Hamiltonian function

$$-\frac{\hbar^2}{2m}\sum_{i=1}^{N}\nabla_i^2 \text{ and } -\sum_{k=1}^{Z}\frac{\hbar^2}{2m_k}\nabla_k^2$$

are the multibody kinetic energy operators.

The term

$$-\sum_{i=1}^{N}\sum_{k=1}^{Z}\frac{1}{4\pi\varepsilon}\frac{e_i q_k}{\left|\mathbf{r}_i - \mathbf{r}_k^{'}\right|}$$

maps the interaction of the electrons with the nuclei at **R** (the electron-nucleus attraction energy operator).

In the Hamiltonian, the fourth term

$$\sum_{i \neq j}^{N}\frac{1}{4\pi\varepsilon}\frac{e^2}{\left|\mathbf{r}_i - \mathbf{r}_j^{'}\right|}$$

gives the interactions of electrons with each other (the electron-electron repulsion energy operator).

Term $\sum_{k \neq m}^{Z}\frac{1}{4\pi\varepsilon}\frac{q_k q_m}{\left|\mathbf{r}_k - \mathbf{r}_m^{'}\right|}$ describes the interaction of the Z nuclei at **R** (the nucleus-nucleus repulsion energy operator).

For an isolated *N*-electron *Z*-nucleus atomic or molecular systems in the Born-Oppenheimer nonrelativistic approximation, we have

$H\Psi = E\Psi$.

The Schrödinger equation is

$$\left[-\frac{\hbar^2}{2m}\sum_{i=1}^{N}\nabla_i^2-\sum_{k=1}^{Z}\frac{\hbar^2}{2m_k}\nabla_k^2\right.$$

$$\left.-\sum_{i=1}^{N}\sum_{k=1}^{Z}\frac{1}{4\pi\varepsilon}\frac{e_i q_k}{\left|\mathbf{r}_i-\mathbf{r}_k^{'}\right|}+\sum_{i\neq j}^{N}\frac{1}{4\pi\varepsilon}\frac{e^2}{\left|\mathbf{r}_i-\mathbf{r}_j^{'}\right|}+\sum_{k\neq m}^{Z}\frac{1}{4\pi\varepsilon}\frac{q_k q_m}{\left|\mathbf{r}_k-\mathbf{r}_m^{'}\right|}\right]$$

$$\times\Psi(t,\mathbf{r}_1,\mathbf{r}_2,...,\mathbf{r}_{N-1},\mathbf{r}_N)=E(t,\mathbf{r}_1,\mathbf{r}_2,...,\mathbf{r}_{N-1},\mathbf{r}_N)\Psi(t,\mathbf{r}_1,\mathbf{r}_2,...,\mathbf{r}_{N-1},\mathbf{r}_N).$$

(6.2.1)

The total energy $E(t,\mathbf{r}_1,\mathbf{r}_2,...,\mathbf{r}_{N-1},\mathbf{r}_N)$ must be found using the nucleus-nucleus Coulomb repulsion energy as well as the electron energy.

It is very difficult, or impossible, to solve analytically or numerically the nonlinear partial differential equation (6.2.1). Taking into account only the Coulomb force (electrons and nuclei are assumed to interact due to the Coulomb force only), the Hartree approximation is applied. In particular, the N-electron wavefunction $\Psi(t,\mathbf{r}_1,\mathbf{r}_2,...,\mathbf{r}_{N-1},\mathbf{r}_N)$ is expressed as a product of N one-electron wavefunctions as

$$\Psi(t,\mathbf{r}_1,\mathbf{r}_2,...,\mathbf{r}_{N-1},\mathbf{r}_N)=\psi_1(t,\mathbf{r}_1)\psi_2(t,\mathbf{r}_2)...\psi_{N-1}(t,\mathbf{r}_{N-1})\psi_N(t,\mathbf{r}_N).$$

The one-electron Schrödinger equation for *j*th electron is

$$\left(-\frac{\hbar^2}{2m}\nabla_j^2+\Pi(t,\mathbf{r})\right)\psi_j(t,\mathbf{r})=E_j(t,\mathbf{r})\psi_j(t,\mathbf{r}). \qquad (6.2.2)$$

In equation (6.2.2), the first term $-\frac{\hbar^2}{2m}\nabla_j^2$ is the one-electron kinetic energy, and $\Pi(t,\mathbf{r}_j)$ is the total potential energy. The potential energy includes the potential that *j*th electron feels from the nucleus (considering the ion, the repulsive potential in the case of anion, or attractive in the case of cation). It is obvious that *j*th electron feels the repulsion (repulsive forces) from other electrons. Assumed that the negative electrons' charge density $\rho(\mathbf{r})$ is smoothly distributed in **R**. Hence, the potential energy due interaction (repulsion) of an electron in **R** is

$$\Pi_{Ej}(t,\mathbf{r})=\int_{\mathbf{R}}\frac{e\rho(\mathbf{r}')}{4\pi\varepsilon\left|\mathbf{r}-\mathbf{r}'\right|}d\mathbf{r}'.$$

We made some assumptions, and the results derived contradict with some fundamental principles. The Pauli exclusion principle requires that the multisystem wavefunction is an antisymmetric under the interchange of electrons. For two electrons, we have,

$$\Psi(t,\mathbf{r}_1,\mathbf{r}_2,...,\mathbf{r}_j,...,\mathbf{r}_{j+i},...,\mathbf{r}_{N-1},\mathbf{r}_N)=-\Psi(t,\mathbf{r}_1,\mathbf{r}_2,...,\mathbf{r}_{j+i},...,\mathbf{r}_j,...,\mathbf{r}_{N-1},\mathbf{r}_N).$$

This principle is not satisfied, and the generalization is needed to integrate the asymmetry phenomenon using the asymmetric coefficient $\pm 1$. The Hartree-Fock equation is

$$\left[-\frac{\hbar^2}{2m}\nabla_j^2+\Pi(t,\mathbf{r})\right]\psi_j(t,\mathbf{r})$$
$$-\sum_i\int_{\mathbf{R}}\frac{\psi_i^*(t,\mathbf{r}')\psi_j(t,\mathbf{r}')\psi_i(t,\mathbf{r})\psi_j^*(t,\mathbf{r})}{|\mathbf{r}-\mathbf{r}'|}d\mathbf{r}'=E_j(t,\mathbf{r})\psi_j(t,\mathbf{r}). \quad (6.2.3)$$

The so-called Hartree-Fock nonlinear partial differential equation (6.2.3), which is difficult to solve, is the approximation because the multi-body electron interactions should be considered in general. Thus, the explicit equation for the total energy must be used. This phenomenon can be integrated using the charge density function.

### 6.2.2 Density Functional Theory

There is a critical need to develop computationally efficient and accurate procedures to perform quantum modeling of nanoscale structures. This section reports the related results and gives the formulation of the modeling problem to avoid the complexity associated with many-electron wavefunctions which result if the classical quantum mechanics formulation is used. The complexity of the Schrödinger equation is enormous even for very simple molecules. For example, the carbon atom has 6 electrons. Can one visualize six-dimensional space? Furthermore, the simplest carbon nanotube molecule has 6 carbon atoms. That is, one has 36 electrons, and a 36-dimensional problem results. The difficulties associated with the solution of the Schrödinger equation drastically limit the applicability of the conventional quantum mechanics. The analysis of properties, processes, phenomena, and effects in even the simplest nanostructures cannot be studied and comprehended. The problems can be solved applying the Hohenberg-Kohn density functional theory.

The statistical consideration, proposed by Thomas and Fermi in 1927, gives the distribution of electrons in atoms. The following assumptions were used: electrons are distributed uniformly, and there is an effective potential field that is determined by the nuclei charge and the distribution of electrons. Considering electrons distributed in a three-dimensional box, the energy analysis can be performed. Summing all energy levels, one finds the energy. Thus, one can relate the total kinetic energy and the electron charge density. The statistical consideration can be used in order to approximate the distribution of electrons in an atom. The relation between the total kinetic energy of $N$ electrons $E$, and the electron density was derived using the local density approximation concept.

The Thomas-Fermi kinetic energy functional is

$$\Gamma_F(\rho_e(\mathbf{r}))=2.87\int_{\mathbf{R}}\rho_e^{5/3}(\mathbf{r})d\mathbf{r},$$

and the exchange energy is found to be

$$E_F(\rho_e(\mathbf{r})) = 0.739 \int_{\mathbf{R}} \rho_e^{4/3}(\mathbf{r})d\mathbf{r}.$$

For homogeneous atomic systems, applying of the electron charge density $\rho_e(\mathbf{r})$, Thomas and Fermi derived the following energy functional

$$E_F(\rho_e(\mathbf{r})) = 2.87 \int_{\mathbf{R}} \rho_e^{5/3}(\mathbf{r})d\mathbf{r} - q\int_{\mathbf{R}} \frac{\rho_e(\mathbf{r})}{r} d\mathbf{r} + \int_{\mathbf{R}}\int_{\mathbf{R}} \frac{1}{4\pi\varepsilon} \frac{\rho_e(\mathbf{r})\rho_e(\mathbf{r}')}{|\mathbf{r}-\mathbf{r}'|} d\mathbf{r}d\mathbf{r}'$$

considering electrostatic electron-nucleus attraction and electron-electron repulsion

Following this idea, instead of the many-electron wavefunctions, the charge density for N-electron systems can be used. Only the knowledge of the charge density is needed to perform analysis of molecular dynamics. The charge density is the function that describes the number of electrons per unit volume (function of three spatial variables $x$, $y$ and $z$ in the Cartesian coordinate system). The quantum mechanics and quantum modeling must be applied to understand and analyze nanostructures and nanodevices because they operate under the quantum effects.

The total energy of $N$-electron system under the external field is defined in terms of the three-dimensional charge density $\rho(\mathbf{r})$. The complexity is significantly decreased because the problem of modeling of $N$-electron $Z$-nucleus systems becomes equivalent to the solution of the equation for one electron. The total energy is given as

$$E(t,\rho(\mathbf{r})) = \underbrace{\Gamma_1(t,\rho(\mathbf{r})) + \Gamma_2(t,\rho(\mathbf{r}))}_{\text{kinetic energy}} + \underbrace{\int_{\mathbf{R}} \frac{e\rho(\mathbf{r}')}{4\pi\varepsilon|\mathbf{r}-\mathbf{r}'|} d\mathbf{r}'}_{\text{potential energy}}, \tag{6.2.4}$$

where $\Gamma_1(t,\rho(\mathbf{r}))$ and $\Gamma_2(t,\rho(\mathbf{r}))$ are the interacting (exchange) and non-interacting kinetic energies of a single electron in $N$-electron $Z$-nucleus system,

$$\Gamma_1(t,\rho(\mathbf{r})) = \int_{\mathbf{R}} \gamma(t,\rho(\mathbf{r}))\rho(\mathbf{r})d\mathbf{r}, \Gamma_2(t,\rho(\mathbf{r})) = -\frac{\hbar^2}{2m}\sum_{j=1}^{N}\int_{\mathbf{R}} \psi_j^*(t,\mathbf{r})\nabla_j^2\psi_j(t,\mathbf{r})d\mathbf{r};$$

$\gamma(t,\rho(\mathbf{r}))$ is the parameterization function.

It should be emphasized that the Kohn-Sham electronic orbitals are subject to the following orthogonal condition

$$\int_{\mathbf{R}} \psi_i^*(t,\mathbf{r})\psi_j(t,\mathbf{r})d\mathbf{r} = \delta_{ij}.$$

The state of substance (media) depends largely on the balance between the kinetic energies of the particles and the interparticle energies of attraction.

The expression for the total potential energy is easily justified.

The term $\int\limits_{\mathbf{R}} \frac{e\rho(\mathbf{r}')}{4\pi\varepsilon|\mathbf{r}-\mathbf{r}'|} d\mathbf{r}'$ represents the Coulomb interaction in **R**, and the total potential energy is a function of the charge density $\rho(\mathbf{r})$.

The total kinetic energy (interactions of electrons and nuclei, and electrons) is integrated into the equation for the total energy. The total energy, as given by (6.2.4), is stationary with respect to variations in the charge density. The charge density is found taking note of the Schrödinger equation. The first-order Fock-Dirac electron charge density matrix is

$$\rho_e(\mathbf{r}) = \sum_{j=1}^{N} \psi_j^*(t,\mathbf{r})\psi_j(t,\mathbf{r}). \tag{6.2.5}$$

The three-dimensional electron charge density is a function in three variables ($x$, $y$ and $z$ in the Cartesian coordinate system). Integrating the electron charge density $\rho_e(\mathbf{r})$, one obtains the charge of the total number of electrons $N$. Thus,

$$\int\limits_{\mathbf{R}} \rho_e(\mathbf{r})d\mathbf{r} = Ne.$$

Hence, $\rho_e(\mathbf{r})$ satisfies the following properties

$$\rho_e(\mathbf{r}) > 0, \ \int\limits_{\mathbf{R}} \rho_e(\mathbf{r})d\mathbf{r} = Ne,$$

$$\int\limits_{\mathbf{R}} \left|\sqrt{\nabla\rho_e(\mathbf{r})}\right|^2 d\mathbf{r} < \infty,$$

$$\int\limits_{\mathbf{R}} \nabla^2\rho_e(\mathbf{r})d\mathbf{r} = \infty.$$

For the nuclei charge density, we have

$$\rho_n(\mathbf{r}) > 0 \ \text{and} \ \int\limits_{\mathbf{R}} \rho_n(\mathbf{r})d\mathbf{r} = \sum_{k=1}^{Z} q_k.$$

There exist an infinite number of antisymmetric wavefunctions that give the same $\rho(\mathbf{r})$. The minimum-energy concept (energy-functional minimum principle) is applied. The total energy is a function of $\rho(\mathbf{r})$, and the so-called ground state $\Psi$ must minimize the expectation value $\langle E(\rho)\rangle$.

The searching density functional $F(\rho)$, which searches all $\Psi$ in the N-electron Hilbert space $H$ to find $\rho(\mathbf{r})$ and guarantee the minimum to the energy expectation value, is expressed as

$$F(\rho) \le \min_{\substack{\Psi\to\rho \\ \Psi\in H_\Psi}} \langle\Psi| E(\rho) |\Psi\rangle,$$

where $H_\Psi$ is any subset of the N-electron Hilbert space.

Using the variational principle, we have

$$\frac{\Delta E(\rho)}{\Delta f(\rho)} = \int_{\mathbf{R}} \frac{\Delta E(\rho)}{\Delta \rho(\mathbf{r}')} \frac{\Delta \rho(\mathbf{r}')}{\Delta f(\mathbf{r})} d\mathbf{r}' = 0,$$

where $f(\rho)$ is the nonnegative function.

Thus, $\left. \frac{\Delta E(\rho)}{\Delta f(\rho)} \right|_N = \text{const}$.

The solution to the system of equations (6.2.2) is found using the charge density (6.2.5).

To perform the analysis of nanostructure dynamics, one studies the molecular dynamics. The force and displacement must be found. Substituting the expression for the total kinetic and potential energies in (6.2.4), where the charge density is given by (6.2.5), the total energy $E(t, \rho(\mathbf{r}))$ results.

The external energy is supplied to control nanoscale actuators, and one has

$$E_\Sigma(t,\mathbf{r}) = E_{external}(t,\mathbf{r}) + E(t, \rho(\mathbf{r})).$$

Then, the force at position $\mathbf{r}_r$ is

$$\mathbf{F}_r(t,\mathbf{r}) = -\frac{dE_\Sigma(t,\mathbf{r})}{d\mathbf{r}_r}$$
$$= -\frac{\partial E_\Sigma(t,\mathbf{r})}{\partial \mathbf{r}_r} - \sum_j \frac{\partial E(t,\mathbf{r})}{\partial \psi_j(t,\mathbf{r})} \frac{\partial \psi_j(t,\mathbf{r})}{\partial \mathbf{r}_r} - \sum_j \frac{\partial E(t,\mathbf{r})}{\partial \psi_j^*(t,\mathbf{r})} \frac{\partial \psi_j^*(t,\mathbf{r})}{\partial \mathbf{r}_r}. \tag{6.2.6}$$

Taking note of

$$\sum_j \frac{\partial E(t,\mathbf{r})}{\partial \psi_j(t,\mathbf{r})} \frac{\partial \psi_j(t,\mathbf{r})}{\partial \mathbf{r}_r} + \sum_j \frac{\partial E(t,\mathbf{r})}{\partial \psi_j^*(t,\mathbf{r})} \frac{\partial \psi_j^*(t,\mathbf{r})}{\partial \mathbf{r}_r} = 0,$$

the expression for the force is found from (6.2.6). In particular, one finds

$$\mathbf{F}_r(t,\mathbf{r}) = -\frac{\partial E_{external}(t,\mathbf{r})}{\partial \mathbf{r}_r}$$
$$- \int_{\mathbf{R}} \rho(t,\mathbf{r}) \frac{\partial[\Pi_r(t,\mathbf{r}) + \Gamma_r(t,\mathbf{r})]}{\partial \mathbf{r}_r} d\mathbf{r} - \int_{\mathbf{R}} \frac{\partial E_\Sigma(t,\mathbf{r})}{\partial \rho(t,\mathbf{r})} \frac{\partial \rho(t,\mathbf{r})}{\partial \mathbf{r}_r} d\mathbf{r}.$$

As the wavefunctions converge (the conditions of the Hellmann-Feynman theorem are satisfied), we have

$$\int_{\mathbf{R}} \frac{\partial E(t,\mathbf{r})}{\partial \rho(t,\mathbf{r})} \frac{\partial \rho(t,\mathbf{r})}{\partial \mathbf{r}_r} d\mathbf{r} = 0.$$

One can deduce the expression for the wavefunctions, find the charge density, calculate the forces, and study processes and phenomena in nanoscale. The displacement is found using the following equation of motion

$$m\frac{d^2\mathbf{r}}{dt^2} = \mathbf{F}_r(t,\mathbf{r}) \text{ or } m\frac{d^2(\vec{x},\vec{y},\vec{z})}{dt^2} = \mathbf{F}_r(\vec{x},\vec{y},\vec{z}).$$

### 6.2.3 Nanostructures and Molecular Dynamics

Atomistic modeling can be performed using the force field method. The effective interatomic potential for a system of N particles is found as the sum of the second-, third-, fourth-, and higher-order terms as

$$\Pi(\mathbf{r}_1,...,\mathbf{r}_N)=\sum_{i,j=1}^{N}\Pi^{(2)}(\mathbf{r}_{ij})+\sum_{i,j,k=1}^{N}\Pi^{(3)}(\mathbf{r}_i,\mathbf{r}_j,\mathbf{r}_k)+\sum_{i,j,k,l=1}^{N}\Pi^{(4)}(\mathbf{r}_i,\mathbf{r}_j,\mathbf{r}_k,\mathbf{r}_l)+...$$

Usually, the interatomic effective pair potential $\sum_{i,j=1}^{N}\Pi^{(2)}(\mathbf{r}_{ij})$, which depends on the interatomic distance $r_{ij}$ between the nuclei $i$ and $j$, dominates. For example, the three-body interconnection terms cannot be omitted only if the angle-dependent potentials are considered. Using the effective ionic charges $Q_i$ and $Q_j$, we have

$$\Pi^{(2)}=\underbrace{\frac{Q_iQ_j}{4\pi\varepsilon r_{ij}}}_{electrostatic}+\underbrace{\phi(\mathbf{r}_{ij})}_{short-range},$$

where $\phi(\mathbf{r}_{ij})$ is the short-range interaction energy due to the repulsion between electron charge clouds, Van der Waals attraction, bond bending and stretching phenomena.

For ionic and partially ionic media we have

$$\phi(r_{ij})=k_{1ij}e^{-k_{2ij}r_{ij}}-k_{3ij}r_{ij}^{-6}+k_{4ij}r_{ij}^{-12},$$

where $k_{1ij}=\sqrt{k_{1i}k_{1j}}$, $k_{2ij}=\sqrt{k_{2i}k_{2j}}$, $k_{3ij}=\sqrt{k_{3i}k_{3j}}$ and $k_{4ij}=\sqrt{k_{4i}k_{4j}}$; $k_i$ are the bond energy constants (for example, for *Si* we have $Q$ = 2.4, $k_3$ = 0.00069 and $k_4$ = 104, for *Al* one has $Q$ = 1.4, $k_3$ = 1690 and $k_4$ = 278, and for $Na^+$ we have $Q$ = 1, $k_3$ = 0.00046 and $k_4$ = 67423).

Another, commonly used approximation is $\phi(r_{ij})=k_{5ij}(r_{ij}-r_{Eij})$, where $r_{ij}$ is the bond length, $r_{ij}=|\mathbf{r}_j-\mathbf{r}_i|$; $r_{Eij}$ is the equilibrium bond distance.

Performing the summations in the studied **R**, one finds the potential energy, and the force results. The position (displacement) is represented by the vector **r** which in the Cartesian coordinate system has the components $x$, $y$ and $z$. Taking note of the expression for the potential energy $\Pi(\vec{\mathbf{r}})=\Pi(\mathbf{r}_1,...,\mathbf{r}_N)$, one has

$$\sum\vec{F}(\vec{\mathbf{r}})=-\nabla\Pi(\vec{\mathbf{r}}).$$

From Newton's second law for the system of $N$ particles, we have the following equation of motion

$$m_N\frac{d^2\vec{\mathbf{r}}_N}{dt^2}+\nabla\Pi(\vec{\mathbf{r}}_N)=0,$$

or

$$m_i \frac{d^2(\vec{x}_i, \vec{y}_i, \vec{z}_i)}{dt^2} + \frac{\partial \Pi(\vec{x}_i, \vec{y}_i, \vec{z}_i)}{\partial(\vec{x}_i, \vec{y}_i, \vec{z}_i)} = 0, i = 1, \ldots, N .$$

To perform molecular modeling one applies the energy-based methods. It was shown that electrons can be considered explicitly. However, it can be assumed that electrons will obey the optimum distribution once the positions of the nuclei in **R** are known. This assumption is based on the Born-Oppenheimer approximation of the Schrödinger equation. This approximation is satisfied because nuclei mass is much greater then electron mass, and thus, nuclei motions (vibrations and rotations) are slow compared with the electrons' motions. Therefore, nuclei motions can be studied separately from electrons' dynamics. Molecules can be studied as $Z$-body systems of elementary masses (nuclei) with springs (bonds between nuclei). The molecule potential energy (potential energy equation) is found using the number of nuclei and bond types (bending, stretching, lengths, geometry, angles, and other parameters), van der Waals radius, parameters of media, etc. The molecule potential energy surface is

$$E_T = E_{bs} + E_b + E_{sb} + E_{ts} + E_W + E_{dd} .$$

Here, the energy due to bond stretching is found using the equation similar to Hook's law. In particular,

$$E_{bs} = k_{bs1}(l - l_0) + k_{bs3}(l - l_0)^3 ,$$

where $k_{bs1}$ and $k_{bs3}$ are the constants; $l$ and $l_0$ are the actual and natural bond length (displacement).

The equations for energies due to bond angle bending $E_b$, stretch-bend interactions $E_{sb}$, torsion strain $E_{ts}$, van der Waals interactions $E_W$, and dipole-dipole interactions $E_{dd}$ are well known and can be readily applied.

### 6.2.4 Electromagnetic Fields and Their Quantization

The mathematical models for energy conversion (energy storage, transport, and dissipation) and electromagnetic field (propagation, radiation and other major characteristics) in media are found using Maxwell's equations. The vectors of electric field intensity $\boldsymbol{E}$, electric flux density $\boldsymbol{D}$, magnetic field intensity $\boldsymbol{H}$, and magnetic flux density $\boldsymbol{B}$ are used as the cornerstone variables. The finite element analysis concept cannot be viewed as a meaningful paradigm because it provides one with the steady-state solution which does not describe the important electromagnetic phenomena and effects even for conventional systems. The time-independent and frequency-domain Maxwell's and Schrödinger equations also have serious limitations in nano- and microscale system analysis. Therefore, complete mathematical models must be developed without simplifications and assumptions to understand, analyze, and comprehend a wide spectrum of phenomena and effects.

Maxwell's equations can be quantized. This concept provides a meaningful means for interpreting, understanding, predicting, and analysis of complex time-dependent behavior at nanoscale without invoking all the intricacy and difficulty of quantum electrodynamics, electromagnetics, and mechanics.

The Lorentz force on the charge $q$ moving at the velocity $\mathbf{v}$ is found as

$$\mathbf{F}(t,\mathbf{r}) = q\mathbf{E}(t,\mathbf{r}) + q\mathbf{v}\times\mathbf{B}(t,\mathbf{r}).$$

Using the electromagnetic potential $\mathbf{A}$ we have

$$\mathbf{F} = q\left(-\frac{\partial \mathbf{A}}{\partial \mathbf{t}} - \nabla V + \mathbf{v}\times(\nabla\times\mathbf{A})\right),$$

$$\mathbf{v}\times(\nabla\times\mathbf{A}) = \nabla(\mathbf{v}\cdot\mathbf{A}) - \frac{d\mathbf{A}}{dt} + \frac{\partial\mathbf{A}}{\partial t},$$

where $V(t, \mathbf{r})$ is the scalar electrostatic potential function (potential difference).

Four Maxwell's equations in the time domain are

$$\nabla\times\mathbf{E}(t,\mathbf{r}) = -\mu\frac{\partial\mathbf{H}(t,\mathbf{r})}{\partial t} - \mu\frac{\partial\mathbf{M}(t,\mathbf{r})}{\partial t},$$

$$\nabla\times\mathbf{H}(t,\mathbf{r}) = \mathbf{J}(t,\mathbf{r}) + \varepsilon\frac{\partial\mathbf{E}(t,\mathbf{r})}{\partial t} + \frac{\partial\mathbf{P}(t,\mathbf{r})}{\partial t},$$

$$\nabla\cdot\mathbf{E}(t,\mathbf{r}) = \frac{\rho_v(t,\mathbf{r})}{\varepsilon} - \frac{\nabla\mathbf{P}(t,\mathbf{r})}{\varepsilon},$$

$$\nabla\cdot\mathbf{H}(t,\mathbf{r}) = 0,$$

where $\mathbf{J}$ is the current density, and using the conductivity $\sigma$, we have $\mathbf{J} = \sigma\mathbf{E}$; $\varepsilon$ is the permittivity; $\mu$ is the permeability; $\rho_v$ is the volume charge density.

Using the electric $\mathbf{P}$ and magnetic $\mathbf{M}$ polarizations (dipole moment per unit volume) of the medium, one obtains the constitutive equations as

$$\mathbf{D}(t,\mathbf{r}) = \varepsilon\mathbf{E}(t,\mathbf{r}) + \mathbf{P}(t,\mathbf{r}) \text{ and } \mathbf{B}(t,\mathbf{r}) = \mu\mathbf{H}(t,\mathbf{r}) + \mu\mathbf{M}(t,\mathbf{r}).$$

The electromagnetic waves transfer the electromagnetic power, and we have

$$\underbrace{\int_v \nabla\cdot(\mathbf{E}\times\mathbf{H})dv - \oint_s(\mathbf{E}\times\mathbf{H})\cdot d\mathbf{s}}_{\text{total power flowing into volume bounded by s}} =$$

$$-\underbrace{\int_v\frac{\partial}{\partial t}\left(\tfrac{1}{2}\varepsilon\mathbf{E}\cdot\mathbf{E} + \tfrac{1}{2}\mu\mathbf{H}\cdot\mathbf{H}\right)dv}_{\text{rate of change of the electromagetic stored energy in electromagetic fields}} - \underbrace{\int_v\mathbf{E}\cdot\mathbf{J}dv}_{\text{power expended by the field on moving charges}} - \underbrace{\int_v\mathbf{E}\cdot\frac{\partial\mathbf{P}}{\partial t}dv}_{\text{power expended by the field on electric dipoles}} - \int_v\mu\mathbf{H}\cdot\frac{\partial\mathbf{M}}{\partial t}dv.$$

The pointing vector $\mathbf{E}\times\mathbf{H}$, which is a power density vector, represents the power flows per unit area. Furthermore, the electromagnetic momentum is found as $\mathbf{M} = \frac{1}{c^2}\int_v\mathbf{E}\times\mathbf{H}dv$.

The electromagnetic field can be studied using the magnetic vector potential $\mathbf{A}$. In particular,

$$\mathbf{B}(t,\mathbf{r}) = \nabla\times\mathbf{A}(t,\mathbf{r}),\ \mathbf{E}(t,\mathbf{r}) = -\frac{\partial\mathbf{A}(t,\mathbf{r})}{\partial t} - \nabla V(t,\mathbf{r}).$$

Making use of the *coulomb gauge* $\nabla\cdot\mathbf{A}(t,\mathbf{r})=0$, for the free electromagnetic field (***A*** is determined by the transverse current density), we have

$$\nabla^2\mathbf{A}(t,\mathbf{r})-\frac{1}{c^2}\frac{\partial^2\mathbf{A}(t,\mathbf{r})}{\partial t^2}=0,$$

where $c$ is the speed of light, $c=\frac{1}{\sqrt{\mu_0\varepsilon_0}}$, $c=3\times10^8\ \frac{\mathrm{m}}{\mathrm{sec}}$.

The solution of the partial differential equation is

$$\mathbf{A}(t,\mathbf{r})=\frac{1}{2\sqrt{\varepsilon}}\sum_s a_s(t)\mathbf{A}_s(\mathbf{r}),$$

and using the separation of variables technique we have

$$\nabla^2\mathbf{A}_s(\mathbf{r})+\frac{\omega_s}{c^2}\mathbf{A}_s(\mathbf{r})=0,\quad \frac{d^2a_s(t)}{dt^2}+\omega_s a_s(t)=0,$$

where $\omega_s$ is the separation constant which determines the eigenfunctions.

The stored electromagnetic energy $\langle W(t)\rangle=\frac{1}{2v}\int_v(\varepsilon\mathbf{E}\cdot\mathbf{E}+\mu\mathbf{H}\cdot\mathbf{H})dv$ is given by

$$\langle W(t)\rangle=\frac{1}{4v}\int_v\left(\omega_s\omega_{s'}\mathbf{A}_s\cdot\mathbf{A}^*_{s'}+c^2\nabla\times\mathbf{A}_s\cdot\nabla\times\mathbf{A}^*_{s'}\right)a_s(t)a^*_{s'}(t)dv$$

$$=\frac{1}{4v}\sum_{s,s'}\left(\omega_s\omega_{s'}+\omega^2_{s'}\right)a_s(t)a^*_{s'}(t)\int_v\mathbf{A}_s\cdot\mathbf{A}^*_{s'}dv=\tfrac{1}{2}\sum_s\omega_s^2a_s(t)a^*_s(t).$$

The Hamiltonian is $H=\frac{1}{2v}\int_v(\varepsilon\mathbf{E}\cdot\mathbf{E}+\mu\mathbf{H}\cdot\mathbf{H})dv$

Let us apply the quantum mechanics to examine very important features.

The Hamiltonian function is found using the kinetic and potential energies $\Gamma$ and $\Pi$.

For a particle of mass $m$ with energy $E$ moving in the Cartesian coordinate system one has

$$\underbrace{E(x,y,z,t)}_{\text{total energy}}=\underbrace{\Gamma(x,y,z,t)}_{\text{kinetic energy}}+\underbrace{\Pi(x,y,z,t)}_{\text{potential energy}}=\frac{p^2(x,y,z,t)}{2m}+\Pi(x,y,z,t)$$

$$=\underbrace{H(x,y,z,t)}_{\text{Hamiltonian}}.$$

Thus, $p^2(x,y,z,t)=2m[E(x,y,z,t)-\Pi(x,y,z,t)]$.

Using the formula for the wavelength (Broglie's equation) $\lambda=\frac{h}{p}=\frac{h}{mv}$, one finds

$$\frac{1}{\lambda^2}=\left(\frac{p}{h}\right)^2=\frac{2m}{h^2}\left[E(x,y,z,t)-\Pi(x,y,z,t)\right].$$

This expression is substituted in the *Helmholtz* equation $\nabla^2\Psi+\frac{4\pi^2}{\lambda^2}\Psi=0$ which gives the evolution of the wavefunction.

We obtain the Schrödinger equation as

$$E(x,y,z,t)\Psi(x,y,z,t)=-\frac{\hbar^2}{2m}\nabla^2\Psi(x,y,z,t)+\Pi(x,y,z,t)\Psi(x,y,z,t)$$

or

$$E(x,y,z,t)\Psi(x,y,z,t)=-\frac{\hbar^2}{2m}\left(\frac{\partial^2\Psi(x,y,z,t)}{\partial x^2}+\frac{\partial^2\Psi(x,y,z,t)}{\partial y^2}+\frac{\partial^2\Psi(x,y,z,t)}{\partial z^2}\right)$$
$$+\Pi(x,y,z,t)\Psi(x,y,z,t).$$

Here, the modified Plank constant is $\hbar=\frac{h}{2\pi}=1.055\times10^{-34}$ J-sec.

The Schrödinger equation

$$-\frac{\hbar^2}{2m}\nabla^2\Psi+\Pi\Psi=E\Psi$$

is related to the Hamiltonian $H=-\frac{\hbar^2}{2m}\nabla+\Pi$, and

$$H\Psi=E\Psi.$$

The Schrödinger partial differential equation must be solved, and the wavefunction is normalized using the probability density $\int|\Psi|^2 d\varsigma=1$.

Let us illustrate the application of the Schrödinger equation.

*Example 6.2.1*

Consider a simple one-degree-of-freedom harmonic oscillator.

We have,

$$-\frac{\hbar^2}{2m}\frac{d^2\Psi(q)}{dq^2}+\Pi(q)\Psi(q)=E(q)\Psi(q),$$

and the Hamiltonian function is

$$H(q,p)=\frac{p^2(q)}{2m}+\Pi(q)=\frac{1}{2m}\left(-\hbar^2\frac{d^2}{dq^2}+m^2\omega^2q^2\right)=-\frac{\hbar^2}{2m}\frac{d^2}{dq^2}+\frac{m\omega^2q^2}{2}$$

The Hamilton equations of motion relating $\frac{dq}{dt}$ to $p$, and $\frac{dp}{dt}$ to $q$, are

$$\frac{dq}{dt} = \frac{\partial H}{\partial p} = \frac{p}{m}$$

and $\dfrac{dp}{dt} = -\dfrac{\partial H}{\partial q} = -m\omega^2 q$.

Therefore, one has the following second-order homogeneous equation of motion

$$\ddot{q} = -\omega^2 q^2 .$$

Making use of the initial conditions, the solution is found to be [7, 8]

$$q(t) = q_0 \cos \omega t + \frac{p_0}{m\omega} \sin \omega t ,$$

$$p(t) = -q_0 m\omega \sin \omega t + p_0 \cos \omega t .$$

The quantization is performed by considering $p$ and $q$ equivalent to the momentum and coordinate operators.

Applying the variables

$$a(t) = \sqrt{\frac{m}{2\omega}}\left(\omega q + i\frac{p}{m}\right) \text{ and } a^+(t) = \sqrt{\frac{m}{2\omega}}\left(\omega q - i\frac{p}{m}\right),$$

one has

$$q = \frac{1}{\sqrt{2m\omega}}\left(a + a^+\right),$$

$$p = -i\sqrt{\tfrac{1}{2} m\omega}\left(a - a^+\right).$$

Thus, the Hamiltonian is given by

$$H = \omega a^+ a .$$

Therefore Hamilton equations

$$\frac{dq}{dt} = \frac{\partial H}{\partial p} = \frac{p}{m},$$

$$\frac{dp}{dt} = -\frac{\partial H}{\partial q} = -m\omega^2 q$$

are written in the decoupled form as

$$i\frac{da}{dt} = \frac{\partial H}{\partial a^+} = \omega a \text{ and } i\frac{da^+}{dt} = -\frac{\partial H}{\partial a} = -\omega a^+ . \qquad \square$$

The $q$ and $p$, as well as $a$ and $a^+$, were used as the variables. We apply the Hermitian operators $\boldsymbol{q}$ and $\boldsymbol{p}$ which satisfy the commutative relations

$$[\mathbf{q}, \mathbf{q}] = 0,\ [\mathbf{p}, \mathbf{p}] = 0 \text{ and } [\mathbf{q}, \mathbf{p}] = i\hbar\delta .$$

The Schrödinger representation of the energy eigenvector $\Psi_n(q) = \langle q | E_n \rangle$ satisfies the following equations

$$\langle q|H|E_n\rangle = E_n\langle q|E_n\rangle,\ \left\langle q\left|\frac{1}{2m}\left(\mathbf{p}^2+m^2\omega^2\mathbf{q}^2\right)\right|E_n\right\rangle = E_n\langle q|E_n\rangle,$$

and $\left(-\frac{\hbar^2}{2m}\frac{d^2}{dq^2}+\frac{m\omega^2q^2}{2}\right)\Psi_n(q)=E_n\Psi_n(q)$.

The solution is

$$\Psi_n(q)=\sqrt{\frac{1}{2^n n!}\sqrt{\frac{\omega}{\pi\hbar}}}H_n\left(\sqrt{\frac{m\omega}{\hbar}}q\right)e^{-\frac{m\omega}{2\hbar}q^2},$$

where $H_n\left(\sqrt{\frac{m\omega}{\hbar}}q\right)$ is the Hermite polynomial, and the energy eigenvalues which correspond to the given eigenstates are $E_n=\hbar\omega_n=\hbar\omega\left(n+\frac{1}{2}\right)$.

The eigenfunctions can be generated using the following procedure. Using non-Hermitian operators

$$\mathbf{a}=\sqrt{\frac{m}{2\hbar\omega}}\left(\omega\mathbf{q}+i\frac{\mathbf{p}}{m}\right) \text{ and } \mathbf{a}^+=\sqrt{\frac{m}{2\hbar\omega}}\left(\omega\mathbf{q}-i\frac{\mathbf{p}}{m}\right),$$

we have, $\mathbf{q}=\sqrt{\frac{\hbar}{2m\omega}}\left(\mathbf{a}+\mathbf{a}^+\right)$ and $\mathbf{p}=-i\sqrt{\frac{m\hbar\omega}{2}}\left(\mathbf{a}-\mathbf{a}^+\right)$.

The commutation equation is $[\mathbf{a},\mathbf{a}^+]=1$, and one obtains

$$H=\frac{\hbar\omega}{2}\left(\mathbf{a}\mathbf{a}^+ +\mathbf{a}^+\mathbf{a}\right)=\hbar\omega\left(\mathbf{a}\mathbf{a}^+ +\tfrac{1}{2}\right).$$

The Heisenberg equations of motion are

$$\frac{d\mathbf{a}}{dt}=\frac{1}{i\hbar}[\mathbf{a},H]=-i\omega\mathbf{a} \text{ and } \frac{d\mathbf{a}^+}{dt}=\frac{1}{i\hbar}[\mathbf{a}^+,H]=i\omega\mathbf{a}^+$$

with solutions

$$\mathbf{a}=\mathbf{a}_s e^{-i\omega t} \text{ and } \mathbf{a}^+=\mathbf{a}_s^+e^{i\omega t},$$

and $\mathbf{a}|0\rangle=0$, $\mathbf{a}|n\rangle=\sqrt{n}\,|n-1\rangle$ and $\mathbf{a}^+|n\rangle=\sqrt{n+1}\,|n+1\rangle$.

Using the *state vector generating rule* $|n\rangle=\frac{\mathbf{a}^{+n}}{\sqrt{n!}}|0\rangle$, one has the eigenfuction generator

$$\langle q|n\rangle=\frac{1}{\sqrt{n!}}\left(\sqrt{\frac{m}{2\hbar\omega}}\right)^n\left(\omega q-\frac{\hbar}{m}\frac{d}{dq}\right)^n\langle q|0\rangle$$

for the equation $\Psi_n(q)=\sqrt{\frac{1}{2^n n!}\sqrt{\frac{\omega}{\pi\hbar}}}H_n\left(\sqrt{\frac{m\omega}{\hbar}}q\right)e^{-\frac{m\omega}{2\hbar}q^2}$.

Compare the equation for the stored electromagnetic energy $\langle W(t)\rangle = \frac{1}{2}\sum_s \omega_s^2 a_s(t) a_s^*(t)$ and the Hamiltonian $H = \omega a^+ a$, we have

$$\sqrt{\frac{\omega_s}{2}} a_s \Rightarrow a = \sqrt{\frac{m}{2\omega}}\left(\omega q + i\frac{p}{m}\right) \text{ and } \sqrt{\frac{\omega_s}{2}} a_s^* \Rightarrow a^+ = \sqrt{\frac{m}{2\omega}}\left(\omega q - i\frac{p}{m}\right).$$

Therefore, to perform the canonical quantization, the electromagnetic field variables are expressed as the field operators using

$$\sqrt{\frac{\omega_s}{2}} a_s \Rightarrow \sqrt{\frac{m}{2\omega}}\left(\omega \mathbf{q} + i\frac{\mathbf{p}}{m}\right) = \sqrt{\hbar}\mathbf{a}$$

and $$\sqrt{\frac{\omega_s}{2}} a_s^* \Rightarrow \sqrt{\frac{m}{2\omega}}\left(\omega \mathbf{q} - i\frac{\mathbf{p}}{m}\right) = \sqrt{\hbar}\mathbf{a}^+.$$

The following equations finally result

$$H = \sum_s \frac{\hbar\omega_s}{2}\left[\mathbf{a}_s(t)\mathbf{a}_s^+(t) + \mathbf{a}_s^+(t)\mathbf{a}_s(t)\right],$$

$$\mathbf{E}(t,\mathbf{r}) = \sum_s \sqrt{\frac{\hbar\omega_s}{2\varepsilon}}\left[\mathbf{a}_s(t)\mathbf{A}_s(\mathbf{r}) - \mathbf{a}_s^+(t)\mathbf{A}_{s'}^*(\mathbf{r})\right],$$

$$\mathbf{H}(t,\mathbf{r}) = c\sum_s \sqrt{\frac{\hbar}{2\mu\omega_s}}\left[\mathbf{a}_s(t)\nabla\times\mathbf{A}_s(\mathbf{r}) + \mathbf{a}_s^+(t)\nabla\times\mathbf{A}_{s'}^*(\mathbf{r})\right],$$

$$\mathbf{A}(t,\mathbf{r}) = \sum_s \sqrt{\frac{\hbar}{2\varepsilon\omega_s}}\left[\mathbf{a}_s(t)\mathbf{A}_s(\mathbf{r}) + \mathbf{a}_s^+(t)\mathbf{A}_{s'}^*(\mathbf{r})\right].$$

The derived expressions can be straightforwardly applied. For example, for a single mode field we have

$$\mathbf{E}(t,\mathbf{r}) = i\mathbf{e}\sqrt{\frac{\hbar\omega}{2\varepsilon v}}\left(\mathbf{a}(t)e^{i\mathbf{k}\cdot\mathbf{r}-i\omega t} - \mathbf{a}^+(t)e^{-i\mathbf{k}\cdot\mathbf{r}-i\omega t}\right)$$

and $$\mathbf{H}(t,\mathbf{r}) = i\sqrt{\frac{\hbar\omega}{2\mu v}}\mathbf{k}\times\mathbf{e}\left(\mathbf{a}(t)e^{i\mathbf{k}\cdot\mathbf{r}-i\omega t} - \mathbf{a}^+(t)e^{-i\mathbf{k}\cdot\mathbf{r}-i\omega t}\right),$$

$$\langle n|\,\mathbf{E}\,|n\rangle = 0,\ \langle n|\,\mathbf{H}\,|n\rangle = 0,$$

$$\Delta\mathbf{E} = \sqrt{\frac{\hbar\omega}{\varepsilon v}\left(n+\tfrac{1}{2}\right)},\ \Delta\mathbf{H} = \sqrt{\frac{\hbar\omega}{\mu v}\left(n+\tfrac{1}{2}\right)},$$

$$\Delta\mathbf{E}\Delta\mathbf{H} = \sqrt{\frac{1}{\varepsilon\mu}}\frac{\hbar\omega}{v}\left(n+\tfrac{1}{2}\right),$$

where $E_p = \sqrt{\frac{\hbar\omega}{2\varepsilon v}}$ is the electric field per photon.

The complete Hamiltonian of a coupled system is $H_\Sigma = H + H_{ex}$, where $H_{ex}$ is the interaction Hamiltonian. For example, the photoelectric interaction Hamiltonian is found as

$$H_{exp} = -e\sum_n \mathbf{r}_n \cdot \mathbf{E}(t,\mathbf{R}),$$

where $\boldsymbol{r}_n$ are the relative spatial coordinates of the electrons bound to a nucleus located at $\boldsymbol{R}$.

Taking note of the quantum electrodynamics, electromagnetics, and mechanics, as well as the reported results in quantization and relationship between Maxwell's and Schrödinger equations, the documented paradigm is illustrated is Figure 6.2.2.

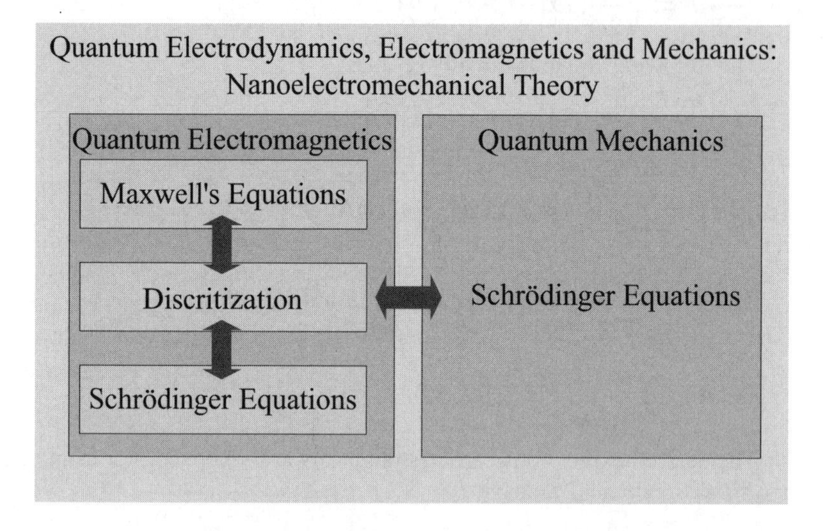


Figure 6.2.2 Nanoelectromechanics augments quantum electrodynamics, electromagnetics, and mechanics.

## 6.3 MOLECULAR WIRES AND MOLECULAR CIRCUITS

Different molecular wires have been devised and tested. For example, the molecular wire can consist of the single molecule chain with its end adsorbed to the surface of the gold lead that can cover monolayers of other molecules. Molecular wires connect the nanoscale structures and devices. The current density of carbon nanotubes, 1,4-dithiol benzene (molecular wire) and copper are $10^{11}$, $10^{12}$ and $10^{6}$ electroncs/sec-nm$^2$, respectively [9]. The current technology allows one to fill carbon nanotubes with other media (metals, organic and inorganic materials). That is, to connect nanostructures, as shown in Figure 6.3.1, it is feasible to use molecular wires which can be synthesized through the organic synthesis.

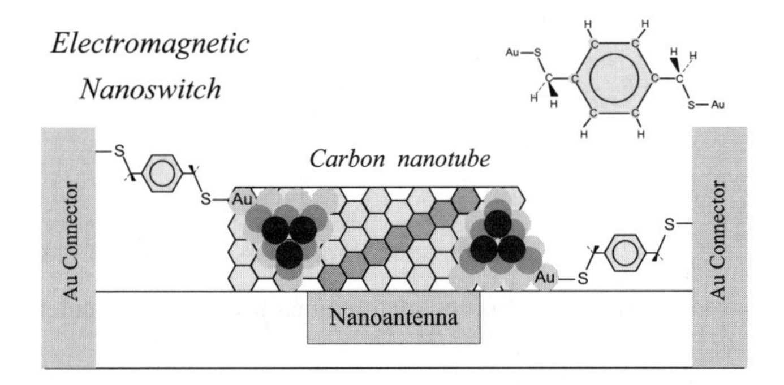


Figure 6.3.1 Nanoswitch with carbon nanotube, molecular wire (1,4-dithiol benzene) and nanoantenna.

Consider covalent bonds. These bonds occur from sharing the electrons between two atoms. Covalent bonds represent the interactions of two nonmetallic elements, or metallic and nonmetallic elements. Let us study the electron density around the nuclei of two atoms. If the electron cloud's overlap region passes through on the line joining two nuclei, the bond is called $\sigma$ bond, see Figure 6.3.2. The overlap may occur between orbitals perpendicularly oriented to the internuclear axis. The resulting covalent bond produces overlap above and below the internuclear axis. Such bond is called $\pi$ bond. There is no probability of finding the electron on the internuclear axis in a $\pi$ bond, and the overlap in it is less than in the $\sigma$ bond. Therefore, $\pi$ bonds are generally weaker than $\sigma$ bonds.

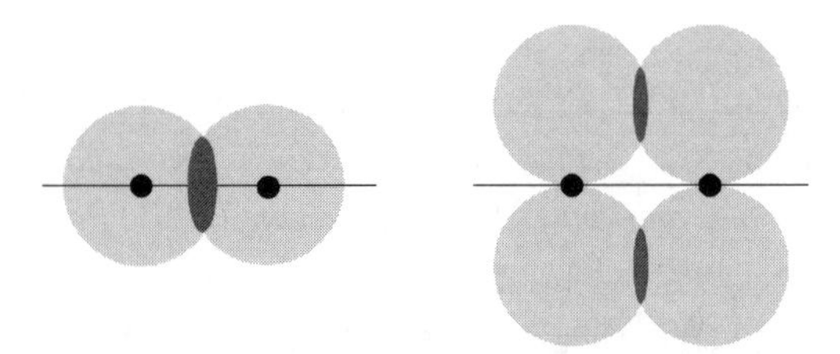

Figure 6.3.2 $\sigma$ and $\pi$ covalent bonds.

Single bonds are usually $\sigma$ bonds. Double bonds, which are much stronger, consist of one $\sigma$ bond and one $\pi$ bond, and the triple bond (the strongest one) consists of one $\sigma$ bond and two $\pi$ bonds. In the case of carbon nanotubes, the strong interaction among the carbon atoms is guaranteed by the strength of the C-C single bond which holds carbon atoms together in the honeycomb-like hexagon unit (open-ended nanotube).

In molecular wires, the current $i_m$ is a function of the applied voltage $u_m$, and Landauer's formula is [10]

$$i_m = \frac{2e}{h}\int_{-\infty}^{+\infty} T(E_m, u_m)\left(\frac{1}{e^{\frac{E_m-\mu_{p1}}{k_BT}}+1} - \frac{1}{e^{\frac{E_m-\mu_{p2}}{k_BT}}+1}\right)dE_m,$$

where $\mu_{p1}$ and $\mu_{p2}$ are the electrochemical potentials, $\mu_{p1} = E_F + \frac{1}{2}eu_m$ and $\mu_{p2} = E_F - \frac{1}{2}eu_m$; $E_F$ is the equilibrium Fermi energy of the source; $T(E_m, u_m)$ is the transmission function obtained using the molecular energy levels and coupling.

We have [10]

$$i_m = \frac{2e}{h}\int_{\mu_{p1}}^{\mu_{p2}} T(E_m, u_m)\frac{1}{4k_BT}\operatorname{sech}^2\left(\frac{E_m}{2k_BT}\right)dE_m,\ k_BT = 26\text{meV}.$$

Thus, the molecular wire conductance is found as

$$c_m = \frac{\partial i_m}{\partial u_m} \approx \frac{e^2}{h}\left[T(\mu_{p1}) + T(\mu_{p2})\right].$$

## References

1. P. Hohenberg and W. Kohn, "Inhomogeneous electron gas," *Phys. Rev.*, vol. 136, pp. B864-B871, 1964.
2. W. Kohn and R. M. Driezler, "Time-dependent density-fuctional theory: conceptual and practical aspects," *Phys. Rev. Letters*, vol. 56, pp. 1993 - 1995, 1986.
3. W. Kohn and L. J. Sham, "Self-consistent equations including exchange and correlation effects," *Phys. Rev.*, vol. 140, pp. A1133 - A1138, 1965.
4. S. E. Lyshevski, *Nano- and Microelectromechanical Systems: Fundamentals of Nano- and Microengineering*," CRC Press, Boca Raton, FL, 2000.
5. R. G. Parr and W. Yang, *Density-Functional Theory of Atoms and Molecules*, Oxford University Press, New York, NY, 1989.
6. E. R. Davidson, *Reduced Density Matrices in Quantum Chemistry,* Academic Press, New York, NY, 1976.
7. A. Yariv, *Quantum Electronics*, John Wiley and Sons, New York, 1989.
8. V. R. Jones, *Optical Physics and Quantum Electronics*, Lecture Notes, Harvard University, 2000.
9. J. C. Ellenbogen and J. C. Love, *Architectures for molecular electronic computers*, MP 98W0000183, MITRE Corporation, 1999.
10. W. T. Tian, S. Datta, S. Hong, R. Reifenberger, J. I. Henderson, and C. P. Kubiak, "Conductance spectra of molecular wires," *Int. Journal Chemical Physics*, vol. 109, no. 7, pp. 2874-2882, 1998.

# CHAPTER 7

# CONTROL OF MICROELECTROMECHANICAL SYSTEMS

## 7.1 INTRODUCTION TO MICROELECTROMECHANICAL SYSTEMS CONTROL

The solution of a spectrum of problems in nonlinear analysis, structural synthesis, modeling, and optimization of MEMS leads to the development of superior high-performance MEMS. In this chapter, we address and solve control problems. It must be emphasized that the reported paradigms and methods are applicable to nanoscale devices and structures which can be modeled by differential equations.

Mathematical models of MEMS in the form of nonlinear differential equations were derived in previous chapters. Microelectromechanical systems augment a great number of subsystems and components. It was illustrated that MEMS must be controlled. To control motion microdevices (for example translational and rotational microtransducers and micromachines), ICs regulate the voltage supplied to the windings. These ICs are controlled based upon the reference (command), output, state, decision making, event, and other variables. Studying the end-to-end MEMS behavior in the actuation application, the outputs are tracking errors, linear and angular displacements, and velocities. There exist infinite numbers of possible MEMS configurations, and it is impossible to cover all possible scenarios. Therefore, our efforts will be concentrated on the generic results which can be obtained describing MEMS by differential equations. That is, using the mathematical model, as given by differential equations, our goal is to design control algorithms to guarantee the desired performance characteristics (settling time, accuracy, overshoot, controllability, stability, disturbance attenuation, etc.) addressing the motion control problem. The application of the Lyapunov theory, Hamilton-Jacobi concept, and intelligent control methods is studied as applied to solve the control problems.

Several methods have been developed to address and solve nonlinear design and motion control problems for multi-input/multi-output dynamic systems. In particular, the Hamilton-Jacobi and Lyapunov theories are found to be the most meaningful in the analytic design of control laws [1]. For complex multi-input/multi-output MEMS, the neural network paradigm [2] and the concepts based upon evolutionary learning and adaptation [3] can be efficiently and straightforwardly applied.

## 7.2 LYAPUNOV STABILITY THEORY

The MEMS dynamics are described by nonlinear differential equations. In particular,

$$\dot{x}(t) = F(x,r,d) + B(x)u \,,\, y = H(x) \,,\, u_{\min} \le u \le u_{\max} \,,\, x(t_0) = x_0 \,, \tag{7.2.1}$$

where $x \in X \subset \mathbb{R}^c$ is the state vector (displacement, position, velocity, current, etc.); $u \in U \subset \mathbb{R}^m$ is the bounded control vector (voltage); $r \in R \subset \mathbb{R}^b$ and $y \in Y \subset \mathbb{R}^b$ are the measured reference and output vectors; $d \in D \subset \mathbb{R}^s$ is the disturbance vector (load and friction torques, etc.); $F(\cdot):\mathbb{R}^c \times \mathbb{R}^b \times \mathbb{R}^s \to \mathbb{R}^c$ and $B(\cdot):\mathbb{R}^c \to \mathbb{R}^{c\times m}$ are jointly continuous and Lipschitz; $H(\cdot):\mathbb{R}^c \to \mathbb{R}^b$ is the smooth map defined in the neighborhood of the origin, $H(0) = 0$.

In (7.2.1), the state-space equation is $\dot{x}(t) = F(x,r,d) + B(x)u$, while the output equation is $y = H(x)$.

The control bounds are represented as $u_{\min} \le u \le u_{\max}$.

Before engaging in the design of closed-loop systems, which in this section will be based upon the Lyapunov stability theory, let us study stability of time-varying nonlinear dynamic systems described by

$$\dot{x}(t) = F(t,x) \,,\; x(t_0) = x_0 \,,\; t \ge 0 \,.$$

The following Theorem is formulated.

**Theorem.**

*Consider the microelectromechanical system described by nonlinear differential equations*

$$\dot{x}(t) = F(t,x) \,,\; x(t_0) = x_0 \,,\; t \ge 0 \,.$$

*If there exists a positive-definite scalar function $V(t,x)$, called the Lyapunov function, with continuous first-order partial derivatives with respect to t and x*

$$\frac{dV}{dt} = \frac{\partial V}{\partial t} + \left(\frac{\partial V}{\partial x}\right)^T \frac{dx}{dt} = \frac{\partial V}{\partial t} + \left(\frac{\partial V}{\partial x}\right)^T F(t,x) \,,$$

*then*

- *the equilibrium state of $\dot{x}(t) = F(t,x)$ is stable if the total derivative of the positive-definite function $V(t,x) > 0$ is $\frac{dV}{dt} \le 0$;*
- *the equilibrium state of $\dot{x}(t) = F(t,x)$ is uniformly stable if the total derivative of the positive-definite decreasing function $V(t,x) > 0$ is $\frac{dV}{dt} \le 0$;*

- *the equilibrium state of* $\dot{x}(t) = F(t,x)$ *is uniformly asymptotically stable in the large if the total derivative of* $V(t,x) > 0$ *is negative definite; that is,* $\frac{dV}{dt} < 0$ ;
- *the equilibrium state of* $\dot{x}(t) = F(t,x)$ *is exponentially stable in the large if the exist the* $K_\infty$*-functions* $\rho_1(\cdot)$ *and* $\rho_2(\cdot)$, *and K-function* $\rho_3(\cdot)$ *such that*

$$\rho_1(\|x\|) \le V(t,x) \le \rho_2(\|x\|) \text{ and } \frac{dV(x)}{dt} \le -\rho_3(\|x\|). \qquad \blacksquare$$

Examples are studied to illustrate how this theorem can be straightforwardly applied.

*Example 7.2.1*

Study stability of a microstructure which is modeled by $\dot{x}(t) = F(x)$.

In particular, two nonlinear time-invariant differential equations

$\dot{x}_1(t) = -x_1^5 - x_1^3 x_2^4$, $\dot{x}_2(t) = -x_2^9$, $t \ge 0$,

describe the microstructure end-to-end unforced dynamics.

*Solution.*

A scalar positive-definite function is expressed in the quadratic form as $V(x_1, x_2) = \frac{1}{2}(x_1^2 + x_2^2)$.

The total derivative is found to be

$$\frac{dV(x_1,x_2)}{dt} = \left(\frac{\partial V}{\partial x}\right)^T \frac{dx}{dt} = \left(\frac{\partial V}{\partial x}\right)^T F(x)$$

$$= \frac{\partial V}{\partial x_1}\left(-x_1^5 - x_1^3 x_2^4\right) + \frac{\partial V}{\partial x_2}\left(-x_2^9\right) = -x_1^6 - x_1^4 x_2^4 - x_2^{10}.$$

Thus, we have $\frac{dV(x_1,x_2)}{dt} < 0$.

The total derivative of $V(x_1, x_2) > 0$ is negative definite.

Therefore, the equilibrium state of the microstructure is uniformly asymptotically stable. $\square$

*Example 7.2.2*

Study stability of a microstructure if the mathematical model of the end-to-end behavior is given by the time-varying nonlinear differential equations $\dot{x}(t) = F(t,x)$. In particular,

$$\dot{x}_1(t) = -x_1 + x_2^3, \; \dot{x}_2(t) = -e^{-10t} x_1 x_2^2 - 5x_2 - x_2^3, \; t \ge 0.$$

*Solution.*

A scalar positive-definite function is chosen in the quadratic form as $V(t, x_1, x_2) = \frac{1}{2}\left(x_1^2 + e^{10t} x_2^2\right)$, $V(t, x_1, x_2) > 0$.

The total derivative is given by

$$\frac{dV(t, x_1, x_2)}{dt}$$

$$= \frac{\partial V}{\partial t} + \frac{\partial V}{\partial x_1}\left(-x_1 + x_2^3\right) + \frac{\partial V}{\partial x_2}\left(-e^{-10t} x_1 x_2^2 - 5x_2 - x_2^3\right)$$

$$= -x_1^2 - e^{10t} x_2^4.$$

Therefore, the total derivative is negative definite, $\frac{dV(x_1, x_2)}{dt} < 0$.

Hence, making use if the Theorem, one concludes that the equilibrium state is uniformly asymptotically stable. □

*Example 7.2.3*

Study stability of the ICs which is described by the differential equations

$$\dot{x}_1(t) = -x_1 + x_2,$$

$$\dot{x}_2(t) = -x_1 - x_2 - x_2|x_2|,\ t \ge 0.$$

*Solution.*

The positive-definite scalar Lyapunov candidate is chosen in the following form

$$V(x_1, x_2) = \tfrac{1}{2}\left(x_1^2 + x_2^2\right).$$

Thus, $V(x_1, x_2) > 0$.

The total derivative is

$$\frac{dV(x_1, x_2)}{dt} = x_1\dot{x}_1 + x_2\dot{x}_2 = -x_1^2 - x_2^2\left(1 + |x_2|\right).$$

Therefore, $\frac{dV(x_1, x_2)}{dt} < 0$.

Hence, the equilibrium state of ICs is uniformly asymptotically stable, and the quadratic function $V(x_1, x_2) = \frac{1}{2}\left(x_1^2 + x_2^2\right)$ is the Lyapunov function. □

*Example 7.2.4 Stability of Synchronous Micromotors*

Consider a microdrive actuated by the permanent-magnet synchronous micromotor if $T_L$=0. Using (5.4.12), three nonlinear differential equations in the rotor reference frame are given as

$$\frac{di_{qs}^r}{dt} = -\frac{r_s}{L_{ls} + \frac{3}{2}\overline{L}_m} i_{qs}^r - \frac{\psi_m}{L_{ls} + \frac{3}{2}\overline{L}_m} \omega_r - i_{ds}^r \omega_r + \frac{1}{L_{ls} + \frac{3}{2}\overline{L}_m} u_{qs}^r,$$

$$\frac{di_{ds}^r}{dt} = -\frac{r_s}{L_{ls} + \frac{3}{2}\overline{L}_m} i_{ds}^r + i_{qs}^r \omega_r + \frac{1}{L_{ls} + \frac{3}{2}\overline{L}_m} u_{ds}^r,$$

$$\frac{d\omega_r}{dt} = \frac{3P^2\psi_m}{8J} i_{qs}^r - \frac{B_m}{J}\omega_r .$$

Study the microdrive stability letting for the following cases:

1. $u_{qs}^r = 0$ and $u_{ds}^r = 0$ (open-loop microsystem),
2. $u_{qs}^r \neq 0$, $u_{qs}^r = -k_\omega \omega_r$ and $u_{ds}^r = 0$ (closed-loop microsystem).

*Solution.*

For open-loop microdrive, we have $u_{qs}^r = 0$ and $u_{ds}^r = 0$.

Hence, the differential equations $\dot{x}(t) = F(x)$ are rewritten as

$$\frac{di_{qs}^r}{dt} = -\frac{r_s}{L_{ls} + \frac{3}{2}\overline{L}_m} i_{qs}^r - \frac{\psi_m}{L_{ls} + \frac{3}{2}\overline{L}_m} \omega_r - i_{ds}^r \omega_r ,$$

$$\frac{di_{ds}^r}{dt} = -\frac{r_s}{L_{ls} + \frac{3}{2}\overline{L}_m} i_{ds}^r + i_{qs}^r \omega_r ,$$

$$\frac{d\omega_r}{dt} = \frac{3P^2\psi_m}{8J} i_{qs}^r - \frac{B_m}{J}\omega_r .$$

The state-space and output equations (7.2.1) model MEMS. Let us illustrate the application of the matrix notations for the permanent-magnet synchronous micromotor.

Using three differential equations given above, rewritten in the state-space form

$$\dot{x}(t) = F(x) = Ax + F_N(x),$$

one obtains

$$\dot{x}(t) = \begin{bmatrix} -\frac{r_s}{L_{ls} + \frac{3}{2}\overline{L}_m} & 0 & -\frac{\psi_m}{L_{ls} + \frac{3}{2}\overline{L}_m} \\ 0 & -\frac{r_s}{L_{ls} + \frac{3}{2}\overline{L}_m} & 0 \\ \frac{3P^2\psi_m}{8J} & 0 & -\frac{B_m}{J} \end{bmatrix} \begin{bmatrix} i_{qs}^r \\ i_{ds}^r \\ \omega_r \end{bmatrix} + \begin{bmatrix} -i_{ds}^r \omega_r \\ i_{qs}^r \omega_r \\ 0 \end{bmatrix},$$

where

$$A=\begin{bmatrix} -\frac{r_s}{L_{ls}+\frac{3}{2}\overline{L}_m} & 0 & -\frac{\psi_m}{L_{ls}+\frac{3}{2}\overline{L}_m} \\ 0 & -\frac{r_s}{L_{ls}+\frac{3}{2}\overline{L}_m} & 0 \\ \frac{3P^2\psi_m}{8J} & 0 & -\frac{B_m}{J} \end{bmatrix},\ F_N(x)=\begin{bmatrix} -i^r_{ds}\omega_r \\ i^r_{qs}\omega_r \\ 0 \end{bmatrix}.$$

Using the quadratic positive-definite Lyapunov function

$$V(i^r_{qs},i^r_{ds},\omega_r)=\tfrac{1}{2}(i^{r\,2}_{qs}+i^{r\,2}_{ds}+\omega_r^2),$$

the expression for the total derivative is found to be

$$\frac{dV(i^r_{qs},i^r_{ds},\omega_r)}{dt}=$$

$$-\frac{r_s}{L_{ss}}\left(i^{r\,2}_{qs}+i^{r\,2}_{ds}\right)-\frac{B_m}{J}\omega_r^{\,2}-\frac{8J\psi_m-3P^2L_{ss}\psi_m}{8JL_{ss}}i^r_{qs}\omega_r.$$

Thus,

$$\frac{dV\left(i^r_{qs},i^r_{ds},\omega_r\right)}{dt}<0.$$

One concludes that the equilibrium state of a microdrive is uniformly asymptotically stable.

Consider the closed-loop microsystem.

To guarantee the balanced operation (see Section 5.4.2), we set the following *quadrature* and *direct* voltages

$$u^r_{qs}=-k_\omega\omega_r$$

and $u^r_{ds}=0$.

That is, the *quadrature* voltage is proportional to the angular velocity, and $k_\omega$ is the proportional feedback gain.

Thus, the following differential equations result for the closed-loop microsystem

$$\frac{di^r_{qs}}{dt}=-\frac{r_s}{L_{ls}+\frac{3}{2}\overline{L}_m}i^r_{qs}-\frac{\psi_m}{L_{ls}+\frac{3}{2}\overline{L}_m}\omega_r-i^r_{ds}\omega_r-\frac{1}{L_{ls}+\frac{3}{2}\overline{L}_m}k_\omega\omega_r,$$

$$\frac{di^r_{ds}}{dt}=-\frac{r_s}{L_{ls}+\frac{3}{2}\overline{L}_m}i^r_{ds}+i^r_{qs}\omega_r,$$

$$\frac{d\omega_r}{dt}=\frac{3P^2\psi_m}{8J}i^r_{qs}-\frac{B_m}{J}\omega_r.$$

In the state-space form one has

$$\dot{x}(t)=\begin{bmatrix} -\dfrac{r_s}{L_{ls}+\frac{3}{2}\overline{L}_m} & 0 & -\dfrac{\psi_m+k_\omega}{L_{ls}+\frac{3}{2}\overline{L}_m} \\ 0 & -\dfrac{r_s}{L_{ls}+\frac{3}{2}\overline{L}_m} & 0 \\ \dfrac{3P^2\psi_m}{8J} & 0 & -\dfrac{B_m}{J} \end{bmatrix}\begin{bmatrix} i^r_{qs} \\ i^r_{ds} \\ \omega_r \end{bmatrix}+\begin{bmatrix} -i^r_{ds}\omega_r \\ i^r_{qs}\omega_r \\ 0 \end{bmatrix}.$$

Taking note of the quadratic positive-definite Lyapunov function

$$V(i^r_{qs}, i^r_{ds}, \omega_r)=\tfrac{1}{2}(i^{r\,2}_{qs}+i^{r\,2}_{ds}+\omega_r^2),$$

we obtain

$$\frac{dV(i^r_{qs}, i^r_{ds}, \omega_r)}{dt}=$$

$$-\frac{r_s}{L_{ss}}\left(i^{r\,2}_{qs}+i^{r\,2}_{ds}\right)-\frac{B_m}{J}\omega_r^{\,2}-\frac{8J(\psi_m+k_\omega)-3P^2L_{ss}\psi_m}{8JL_{ss}}i^r_{qs}\omega_r.$$

Hence, $V\left(i^r_{qs}, i^r_{ds}, \omega_r\right)>0$ and $\dfrac{dV\left(i^r_{qs}, i^r_{ds}, \omega_r\right)}{dt}<0$.

Therefore, the conditions for asymptotic stability are guaranteed.

It is obvious that the rate of decreasing of $\dfrac{dV\left(i^r_{qs}, i^r_{ds}, \omega_r\right)}{dt}$ affects the microsystem dynamics. The derived expression for $\dfrac{dV\left(i^r_{qs}, i^r_{ds}, \omega_r\right)}{dt}$ clearly illustrates the role of the proportional feedback gain $k_\omega$. □

In Example 7.2.4 it was shown that dynamic systems can be controlled to attain the desired transient dynamics, stability margins, etc. Let us study how to solve the motion control problem with the ultimate goal to synthesize tracking controllers applying Lyapunov's stability theory.

Using the reference (command) vector $r(t)$ and the system output $y(t)$, the tracking error (which ideally must be zero) is

$$e(t)=Nr(t)-y(t). \tag{7.2.2}$$

For example, one assigns the reference angular velocity, and the tracking error is found as the difference of the command and actual angular velocities, e.g.,

$$e(t)=r(t)-y(t)=\omega_{ref}(t)-\omega_r(t).$$

The Lyapunov theory is applied to derive the *admissible* control laws $u_{\min}\le u\le u_{\max}$ (voltages and currents are bounded, and therefore the saturation effect is always the reality). That is, the *admissible* bounded controller should be designed as a continuous function within the constrained control set

$$U=\{u\in\mathbb{R}^m: u_{\min}\le u\le u_{\max}, u_{\min}<0, u_{\max}>0\}\subset\mathbb{R}^m.$$

Making use of the Lyapunov candidate $V(t,x,e)$, the bounded proportional-integral controller with the state feedback extension is expressed as

$$u = \text{sat}_{u_{\min}}^{u_{\max}}\left( G_x(t)B(x)^T \frac{\partial V(t,x,e)}{\partial x} + G_e(t)B_e^T \frac{\partial V(t,x,e)}{\partial e} + G_i(t)B_e^T \frac{1}{s}\frac{\partial V(t,x,e)}{\partial e} \right) \tag{7.2.3}$$

where $G_x(\cdot):\mathbb{R}_{\geq 0}\to\mathbb{R}^{m\times m}$, $G_e(\cdot):\mathbb{R}_{\geq 0}\to\mathbb{R}^{m\times m}$ and $G_i(\cdot):\mathbb{R}_{\geq 0}\to\mathbb{R}^{m\times m}$ are the bounded symmetric weighting matrix-functions defined on $[t_0,\infty)$, $G_x>0$, $G_e>0$, $G_i>0$; $V(\cdot):\mathbb{R}_{\geq 0}\times\mathbb{R}^c\times\mathbb{R}^b\to\mathbb{R}_{\geq 0}$ is the continuously differentiable real-analytic $C^\kappa$ ($\kappa\geq 1$) function with respect to $x\in X$ and $e\in E$ on $[t_0,\infty)$.

It was emphasized that the control signal is saturated, and this feature is documented in Figure 7.2.1.

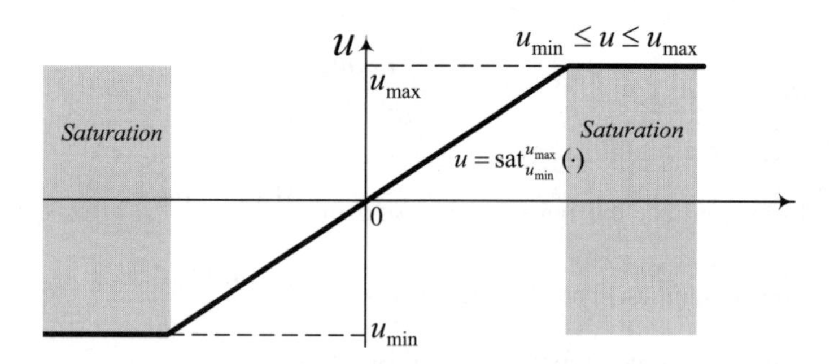


Figure 7.2.1 Bounded control, $u_{\min} \leq u \leq u_{\max}$.

For closed-loop MEMS (7.2.1)–(7.2.3) with $X_0=\{x_0\in\mathbb{R}^c\}\subseteq X\subset\mathbb{R}^c$, $u\in U\subset\mathbb{R}^m$, $r\in R\subset\mathbb{R}^b$ and $d\in D\subset\mathbb{R}^s$, it is straightforward to find the evolution set $X(X_0, U, R, D)\subset\mathbb{R}^c$. Furthermore, using the output equation, one obtain $X \xrightarrow{H} Y$. Thus, the system (7.2.1)-(7.2.3) evolves in

$XY(X_0, U, R, D)=\{(x,y)\in X\times Y: x_0\in X_0, u\in U, r\in R, d\in D, t\in[t_0,\infty)\}\subset\mathbb{R}^c\times\mathbb{R}^b$.

The tracking error vector

$e(t) = Nr(t) - y(t)$, $e(\cdot):[t_0,\infty)\to\mathbb{R}^b$

gives the difference between the reference input $r(\cdot):[t_0,\infty)\to\mathbb{R}^b$ and system output $y(\cdot):[t_0,\infty)\to\mathbb{R}^b$. Our goal is to find the feedback coefficients of controller (7.2.3) to guarantee that the closed-loop MEMS will evolve in the desired manner. The following Lyapunov-based Lemma is formulated to study the stability of closed-loop dynamic systems as well as to find the feedback coefficients to guarantee the criteria imposed on the Lyapunov pair [1].

**Lemma.**

*Consider the closed-loop microelectromechanical systems (7.2.1)–(7.2.3).*

1. *Solutions of the closed-loop system are uniformly ultimately bounded;*
2. *equilibrium point is exponentially stable in the convex and compact state evolution set* $X(X_0, U, R, D)\subset\mathbb{R}^c$;
3. *tracking is ensured and disturbance attenuation is guaranteed in the state-error evolution set* $XE(X_0, E_0, U, R, D)\subset\mathbb{R}^c\times\mathbb{R}^b$,

*if there exists a* $C^\kappa$ *function* $V(t,x,e)$ *in XE such that for all* $x\in X$, $e\in E$, $u\in U$, $r\in R$ *and* $d\in D$ *on* $[t_0,\infty)$

*(i)* $\rho_1\|x\|+\rho_2\|e\|\le V(t,x,e)\le\rho_3\|x\|+\rho_4\|e\|$, (7.2.4)

*(ii) along (7.2.1) with (7.2.3), the following inequality holds*

$$\frac{dV(t,x,e)}{dt}\le-\rho_5\|x\|-\rho_6\|e\|. \quad (7.2.5)$$

*Here,* $\rho_1(\cdot):\mathbb{R}_{\ge0}\to\mathbb{R}_{\ge0}$, $\rho_2(\cdot):\mathbb{R}_{\ge0}\to\mathbb{R}_{\ge0}$, $\rho_3(\cdot):\mathbb{R}_{\ge0}\to\mathbb{R}_{\ge0}$ *and* $\rho_4(\cdot):\mathbb{R}_{\ge0}\to\mathbb{R}_{\ge0}$ *are the* $K_\infty$*-functions;* $\rho_5(\cdot):\mathbb{R}_{\ge0}\to\mathbb{R}_{\ge0}$ *and* $\rho_6(\cdot):\mathbb{R}_{\ge0}\to\mathbb{R}_{\ge0}$ *are the K-functions.* ■

The major problem is to design the Lyapunov candidate functions.

Let us apply a family of nonquadratic Lyapunov candidates

$$V(t,x,e)=\sum_{i=0}^{\eta}\frac{2\gamma+1}{2(i+\gamma+1)}\left(x^{\frac{i+\gamma+1}{2\gamma+1}}\right)^T K_{xi}(t)x^{\frac{i+\gamma+1}{2\gamma+1}}+\sum_{i=0}^{\varsigma}\frac{2\beta+1}{2(i+\beta+1)}\left(e^{\frac{i+\beta+1}{2\beta+1}}\right)^T K_{ei}(t)e^{\frac{i+\beta+1}{2\beta+1}}$$
$$+\sum_{i=0}^{\sigma}\frac{2\mu+1}{2(i+\mu+1)}\left(e^{\frac{i+\mu+1}{2\mu+1}}\right)^T K_{si}(t)e^{\frac{i+\mu+1}{2\mu+1}}. \quad (7.2.6)$$

To design the Lyapunov functions, the nonnegative integers were used.

In particular, $\eta=0,1,2,\ldots$, $\gamma=0,1,2,\ldots$, $\varsigma=0,1,2,\ldots$, $\beta=0,1,2\ldots$, $\sigma=0,1,2,\ldots$, and $\mu=0,1,2,\ldots$.

From (7.2.3) and (7.2.6), one obtains a bounded *admissible* controller as

$$u=\mathbf{sat}_{u_{\min}}^{u_{\max}}\left(G_x(t)B(x)^T\sum_{i=0}^{\eta}\operatorname{diag}\left[x^{\frac{i-\gamma}{2\gamma+1}}\right]K_{xi}(t)x^{\frac{i+\gamma+1}{2\gamma+1}}\right.$$
$$\left.+G_e(t)B_e^T\sum_{i=0}^{\varsigma}\operatorname{diag}\left[e^{\frac{i-\beta}{2\beta+1}}\right]K_{ei}(t)e^{\frac{i+\beta+1}{2\beta+1}}+G_i(t)B_e^T\frac{1}{s}\sum_{i=0}^{\sigma}\operatorname{diag}\left[e^{\frac{i-\mu}{2\mu+1}}\right]K_{si}(t)e^{\frac{i+\mu+1}{2\mu+1}}\right). \quad (7.2.7)$$

Here, $K_{xi}(\cdot):\mathbb{R}_{\ge0}\to\mathbb{R}^{c\times c}$, $K_{ei}(\cdot):\mathbb{R}_{\ge0}\to\mathbb{R}^{b\times b}$ and $K_{si}(\cdot):\mathbb{R}_{\ge0}\to\mathbb{R}^{b\times b}$ are the matrix-functions.

It is evident that assigning the integers to be zero, the well-known quadratic Lyapunov candidate results, and

$$V(t,x,e)=\tfrac{1}{2}x^TK_{x0}(t)x+\tfrac{1}{2}e^TK_{e0}(t)e+\tfrac{1}{2}e^TK_{s0}(t)e.$$

The bounded controller is found to be

$$u=\mathbf{sat}_{u_{\min}}^{u_{\max}}\left(G_x(t)B(x)^T K_{x0}(t)x+G_e(t)B_e^T K_{e0}(t)e+G_i(t)B_e^T K_{s0}(t)\frac{1}{s}e\right).$$

Substituting (7.2.7) into (7.2.1), the total derivative of the Lyapunov candidate $V(t,x,e)$ is obtained. Solving (7.2.5), the feedback coefficients are obtained.

*Example 7.2.5*

Consider a microdrive actuated by a permanent-magnet DC micromotor with IC (*step-down* converter), see Figure 7.2.2. Find the control algorithm.

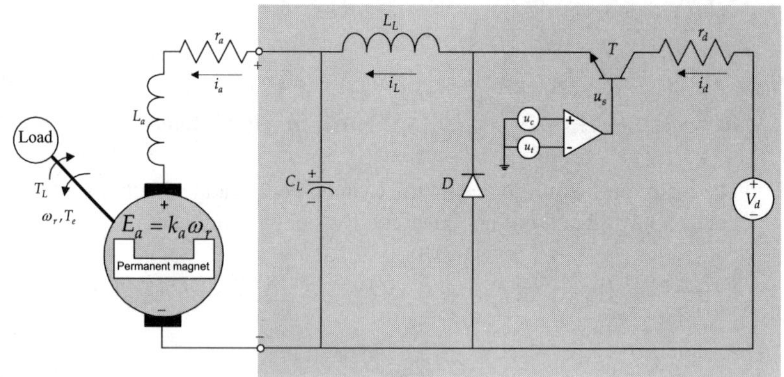


Figure 7.2.2 Permanent-magnet DC micromotor with *step-down* converter.

*Solution.*

Using the Kirchhoff laws and the *averaging* concept, we have the following nonlinear state-space model with bounded control

$$\begin{bmatrix}\frac{du_a}{dt}\\ \frac{di_L}{dt}\\ \frac{di_a}{dt}\\ \frac{d\omega_r}{dt}\end{bmatrix}=\begin{bmatrix}0 & \frac{1}{C_L} & -\frac{1}{C_L} & 0\\ -\frac{1}{L_L} & 0 & 0 & 0\\ \frac{1}{L_a} & 0 & -\frac{r_a}{L_a} & -\frac{k_a}{L_a}\\ 0 & 0 & \frac{k_a}{J} & -\frac{B_m}{J}\end{bmatrix}\begin{bmatrix}u_a\\ i_L\\ i_a\\ \omega_r\end{bmatrix}+\begin{bmatrix}0\\ \left(\frac{V_d}{L_L u_{t\max}}-\frac{r_d}{L_L u_{t\max}}i_L\right)\\ 0\\ 0\end{bmatrix}u_c-\begin{bmatrix}0\\ 0\\ 0\\ \frac{1}{J}\end{bmatrix}T_L,$$

$u_c \in [0 \quad 10]$ V.

A bounded control law should be synthesized.

From (7.2.6), letting $\varsigma=\sigma=1$ and $\beta=\mu=\eta=\gamma=0$, one finds the nonquadratic function $V(e,x)$. In particular, we apply the following Lyapunov candidate

$$V(e,x) = \tfrac{1}{2}k_{e0}e^2 + \tfrac{1}{4}k_{e1}e^4 + \tfrac{1}{2}k_{ei0}e^2 + \tfrac{1}{4}k_{ei1}e^4 + \tfrac{1}{2}[u_a \;\; i_L \;\; i_a \;\; \omega_r]K_{x0}\begin{bmatrix} u_a \\ i_L \\ i_a \\ \omega_r \end{bmatrix},$$

where $K_{x0} \in \mathbb{R}^{4\times4}$.

Therefore, from (7.2.7), one obtains

$$u_c = \begin{cases} 10 \text{ for } u \ge 10, \\ u \text{ for } 0 < u < 10, \\ 0 \text{ for } u \le 0, \end{cases}$$

$$u = k_1 e + k_2 e^3 + k_3 \int e dt + k_4 \int e^3 dt - k_5 u_a - k_6 i_L - k_7 i_a - k_8 \omega_r .$$

If the criteria imposed on the Lyapunov pair are guaranteed, one concludes that the stability conditions are satisfied. The positive-definite nonquadratic function $V(e,x)$ was used. The feedback gains must be found by solving inequality $dV(e,x)/dt < 0$ which represent the second condition imposed on the Lyapunov function.

The following inequality is solved

$$\frac{dV(e,x)}{dt} \le -\tfrac{1}{2}\|e\|^2 - \tfrac{1}{4}\|e\|^4 - \tfrac{1}{2}\|x\|^2$$

to find the feedback coefficients in the controller designed.

Making use of

$$V(e,x) > 0 \text{ and } \frac{dV(e,x)}{dt} < 0,$$

one concludes that stability is guaranteed. □

Many examples in design of tracking controllers for electromechanical systems are reported in [1].

*Example 7.2.6*

Study the flip-chip MEMS: eight-layered lead magnesium niobate actuator (3 mm diameter, 0.25 mm thickness), actuated by a monolithic high-voltage switching regulator, $-1 \le u \le 1$. Find the control law. A set of differential equations to model the microactuator dynamics is

$$\frac{dF_y}{dt} = -9472F_y + 13740F_y u + 48593u ,$$

$$\frac{dv_y}{dt} = 947F_y - 94100v_y - 2609v_y^{1/3} - 2750x_y ,$$

$$\frac{dx_y}{dt} = v_y .$$

*Solution.*

The control is bounded. In particular, we have

$-1 \le u \le 1$.

The error is the difference between the reference and microactuator position. That is,

$e(t) = r(t) - y(t)$,

where $y(t) = x_y$ and $r(t) = r_y(t)$.

Using (7.2.6) and setting the nonnegative integers to be $\varsigma = \sigma = 1$ and $\beta = \mu = \eta = \gamma = 0$, we have

$$V(e,x) = \tfrac{1}{2}k_{e0}e^2 + \tfrac{1}{4}k_{e1}e^4 + \tfrac{1}{2}k_{ei0}e^2 + \tfrac{1}{4}k_{ei1}e^4 + \tfrac{1}{2}[F_y \ v_y \ x_y]K_{xo}\begin{bmatrix} F_y \\ v_y \\ x_y \end{bmatrix}.$$

Applying the design procedure, a bounded control law is synthesized. By making use of (7.2.7), one has

$$u = \mathrm{sat}_{-1}^{+1}\left(94827e + 2614e^3 + 4458\int e dt + 817\int e^3 dt\right).$$

The feedback gains were found by solving inequality

$$\frac{dV\,(e,x)}{dt} \le -\|e\|^2 - \|e\|^4 - \|x\|^2 .$$

The criteria imposed on the Lyapunov pair are satisfied. In fact,

$V(e,x) > 0$

and $\dfrac{dV(e,x)}{dt} \le 0$.

Hence, the bounded control law guarantees stability and ensures tracking. The experimental validation of stability and tracking is important. The controller is tested and examined. Figure 7.2.3 illustrates the transient dynamics for the position if the reference signal (desired position) is assigned to be $r_y(t) = 4\times10^{-6}\sin 1000t$.

Figure 7.2.4 illustrates the microactuator position dynamics if the reference is $r_y(t) = const = 4\times10^{-6}$.

From these end-to-end transient dynamics it is evident that the stability is guaranteed, desired performance is achieved, and the output precisely follows the reference position $r_y(t)$. In general, the sensor accuracy must be studied implementing control laws. In fact, the tracking error $e(t)$ is used to derive the control efforts (duty cycle).

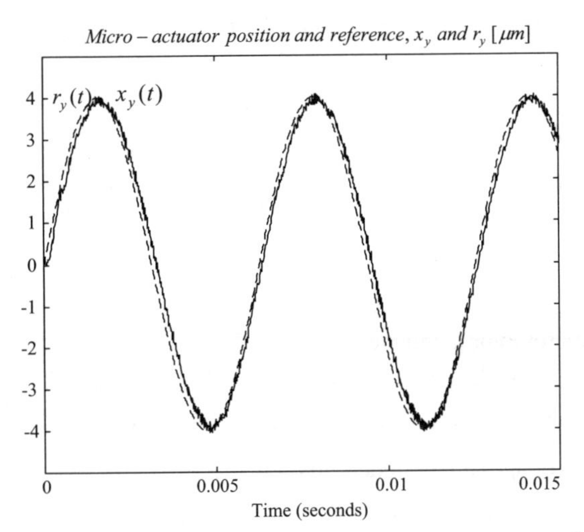


Figure 7.2.3 Transient output dynamics if $r_y(t) = 4\times10^{-6}\sin 1000t$.

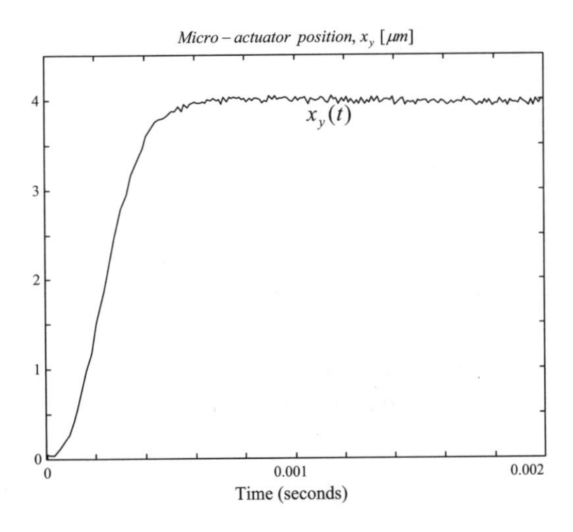


Figure 7.2.4 Actuator position, $r_y(t) = const = 4\times10^{-6}$.

□

*Example 7.2.7*

Consider a flip-chip MEMS (microservo) with permanent-magnet stepper micromotor controlled by ICs. Design the tracking control algorithm making use the control bounds. The mathematical model in the *ab* variables is given in the form of nonlinear differential equations. In particular,

$$\frac{di_{as}}{dt}=-\frac{r_s}{L_{ss}}i_{as}+\frac{RT\psi_m}{L_{ss}}\omega_{rm}\sin(RT\theta_{rm})+\frac{1}{L_{ss}}u_{as},$$

$$\frac{di_{bs}}{dt}=-\frac{r_s}{L_{ss}}i_{bs}-\frac{RT\psi_m}{L_{ss}}\omega_{rm}\cos(RT\theta_{rm})+\frac{1}{L_{ss}}u_{bs},$$

$$\frac{d\omega_{rm}}{dt}=-\frac{RT\psi_m}{J}\left[i_{as}\sin(RT\theta_{rm})+i_{bs}\cos(RT\theta_{rm})\right]-\frac{B_m}{J}\omega_{rm}-\frac{1}{J}T_L,$$

$$\frac{d\theta_{rm}}{dt}=\omega_{rm}.$$

Stepper micromotor parameters are:

$RT=6$, $r_s=60$ ohm, $\psi_m=0.0064$ N-m/A, $L_{ss}=0.05$ H,

$B_m=1.3\times10^{-7}$ N-m-sec/rad and $J=1.8\times10^{-8}$ kg-m$^2$.

The phase voltages are bounded. In particular,

$u_{min}\le u_{as}\le u_{max}$ and $u_{min}\le u_{bs}\le u_{max}$,

where $u_{min}$ = - 12 V and $u_{max}$ = 12 V.

*Solution.*

The nonlinear controller is given as

$$u=\begin{bmatrix}u_{as}\\ u_{bs}\end{bmatrix}=\begin{bmatrix}-\sin(RT\theta_{rm}) & 0\\ 0 & \cos(RT\theta_{rm})\end{bmatrix}$$

$$\times\,\mathbf{sat}_{u_{min}}^{u_{max}}\left(G_x(t)B^T\frac{\partial V(t,x,e)}{\partial x}+G_e(t)B_e^T\frac{\partial V(t,x,e)}{\partial e}+\frac{1}{s}G_i(t)B_e^T\frac{\partial V(t,x,e)}{\partial e}\right).$$

The rotor displacement is denoted as $\theta_{rm}(t)$, and the output is $y(t)=\theta_{rm}(t)$. Thus, the tracking error is

$e(t)=r(t)-y(t)$

The Lyapunov candidate is found using (7.2.6).

Choosing a candidate Lyapunov function to be (letting $\eta=\gamma=0$ and $\varsigma=\beta=\sigma=\mu=1$)

$$V(e,x)=\tfrac{3}{4}K_{e0}e^{4/3}+\tfrac{1}{2}K_{e1}e^2+\tfrac{3}{4}K_{ei0}e^{4/3}+\tfrac{1}{2}K_{ei1}e^2+\tfrac{1}{2}\left[i_{as}\ \ i_{bs}\ \ \omega_{rm}\ \ \theta_{rm}\right]K_{x0}\begin{bmatrix}i_{as}\\ i_{bs}\\ \omega_{rm}\\ \theta_{rm}\end{bmatrix},$$

and solving

$$\frac{dV(e,x)}{dt}\le-\|e\|^{4/3}-\|e\|^2-\|x\|^2,$$

a bounded controller is found as

$$u_{as} = -\sin(RT\theta_{rm})\text{sat}_{-12}^{+12}\left(14e + 2.9e^{1/3} + \frac{1}{s}6.1e + \frac{1}{s}4.3e^{1/3}\right),$$

$$u_{bs} = \cos(RT\theta_{rm})\text{sat}_{-12}^{+12}\left(14e + 2.9e^{1/3} + \frac{1}{s}6.1e + \frac{1}{s}4.3e^{1/3}\right).$$

The sufficient conditions for robust stability are satisfied because

$$V(e,x) > 0 \text{ and } \frac{dV(e,x)}{dt} < 0.$$

Figures 7.2.5 and 7.2.6 document the dynamic if the reference (command) displacement was assigned to be 0.5 and 1 radians, respectively. From analytical and experimental results one concludes that the robust stability and tracking are guaranteed.

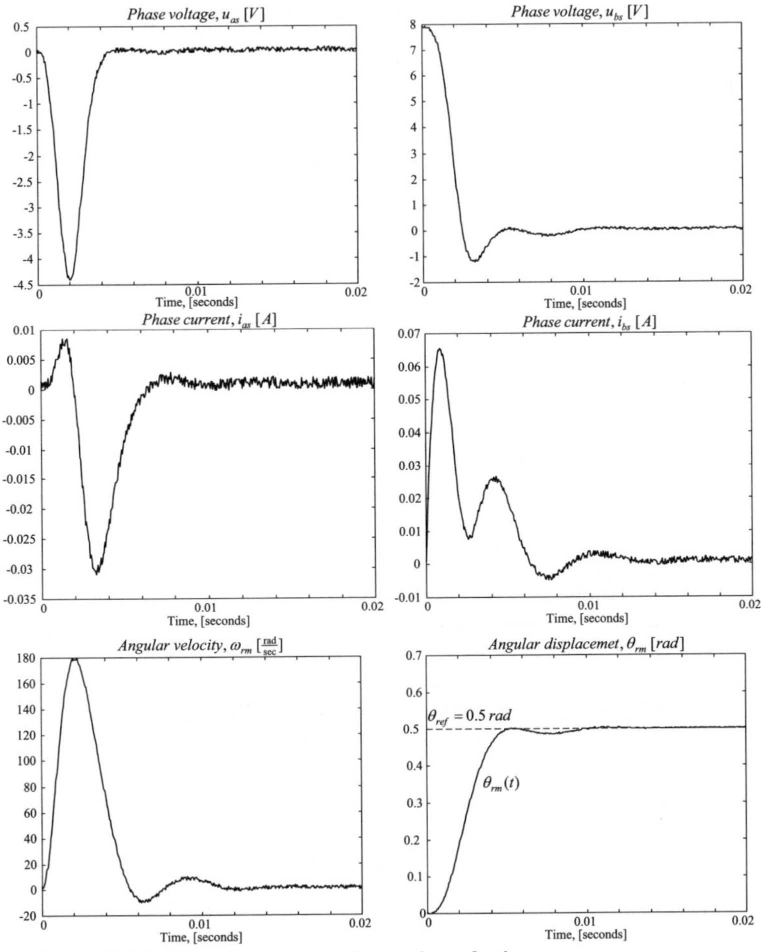


Figure 7.2.5 Transient output dynamics of microservo.

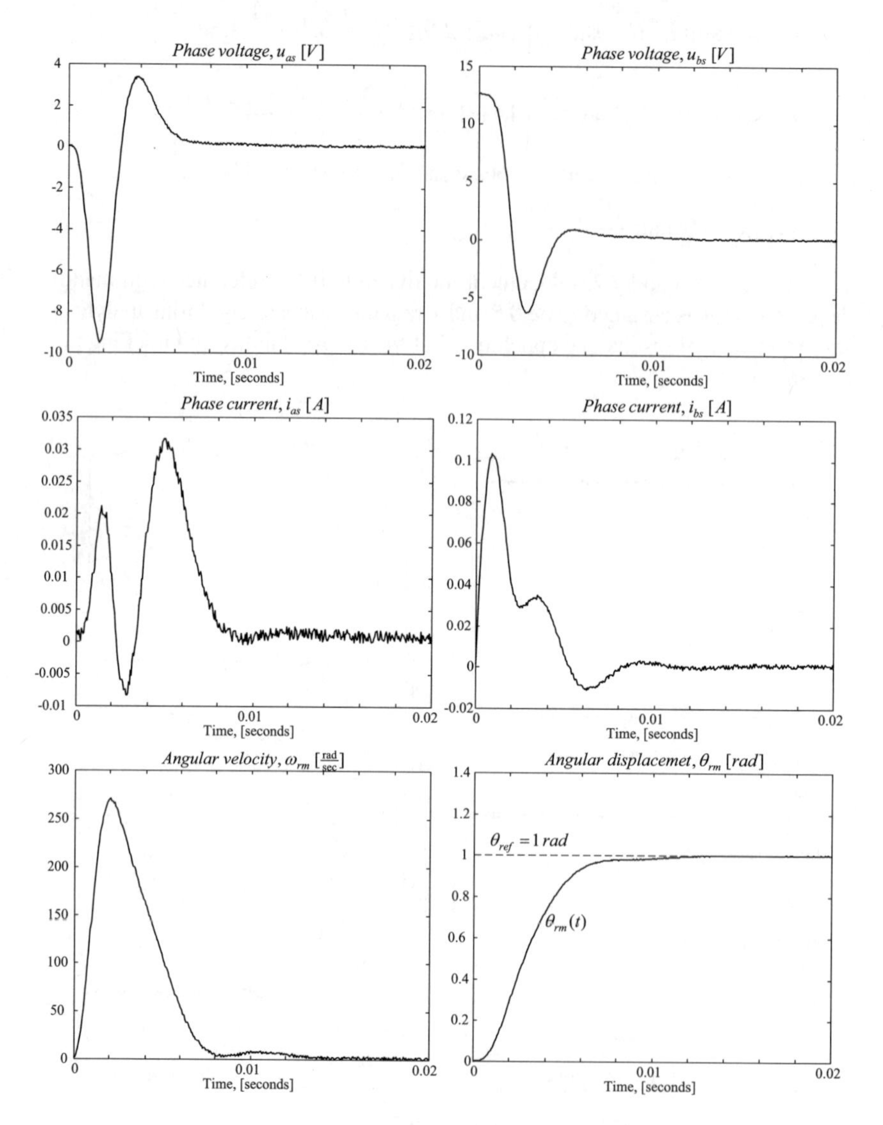


Figure 7.2.6 Transient output dynamics of microservo.

□

## 7.3 CONTROL OF MICROELECTROMECHANICAL SYSTEMS

It was illustrated that mathematical models of MEMS can be developed with different degrees of complexity. In addition to the models of motion and radiating energy microdevices, the fast dynamics of ICs should be examined. Due to the complexity of complete mathematical models of ICs, impracticality of the developed high-order partial differential equations, and very fast dynamics, the IC dynamics can be modeled using reduced-order differential equation or as unmodeled dynamics. For MEMS, modeled using linear and nonlinear differential equations

$$\dot{x}(t) = Ax + Bu\,,\ u_{\min} \le u \le u_{\max}\,, y=Hx, \tag{7.3.1}$$

$$\dot{x}(t) = F_z(t,x,r,z) + B_p(t,x,p)u\,, u_{\min} \le u \le u_{\max}\,, y = H(x)\,, \tag{7.3.2}$$

different control algorithms can be designed.

Here, as in the previous section, the state, control, output, and reference (command) vectors are denoted as $x$, $u$, $y$ and $r$.

The parameter uncertainties (e.g., time-varying coefficients, unmodeled dynamics, unpredicted changes, etc.) are modeled using $z$ and $p$ vectors.

The matrices of coefficients are $A$, $B$ and $H$.

The smooth mapping fields of the nonlinear model are denoted as $F_z(\cdot)$, $B_p(\cdot)$ and $H(\cdot)$.

It should be emphasized that the control is bounded. For example, using the IC duty ratio $d_D$ as the control signal, we have $0 \le d_D \le 1$ or $-1 \le d_D \le +1$. Four-quadrant ICs are used due to superior performance, and $-1 \le d_D \le +1$. Hence, we have $-1 \le u \le +1$. However, in general,

$$u_{\min} \le u \le u_{\max}\,.$$

### 7.3.1 Proportional-Integral-Derivative Controllers

Many MEMS can be controlled by the proportional-integral-derivative (PID) controllers which taking note of control bounds are given as [1]

$$u(t) = \mathbf{sat}_{u_{\min}}^{u_{\max}}\left(e, \int edt, \frac{de}{dt}\right)$$

$$= \mathbf{sat}_{u_{\min}}^{u_{\max}}\left(\underbrace{\sum_{j-0}^{\varsigma} k_{pj} e^{\frac{2j+1}{2\beta+1}}}_{proportion\ al} + \underbrace{\sum_{j=0}^{\sigma} k_{ij} \int e^{\frac{2j+1}{2\mu+1}} dt}_{integral} + \underbrace{\sum_{j=0}^{\alpha} k_{dj} \dot{e}^{\frac{2j+1}{2\gamma+1}}}_{differenti\ al}\right),$$

$$u_{\min} \le u \le u_{\max}\,,$$

where $k_{pj}$, $k_{ij}$ and $k_{dj}$ are the matrices of the proportional, integral and derivative feedback gains; $\varsigma, \beta, \sigma$, $\mu$, $\alpha$ and $\gamma$ are the nonnegative integers.

In the nonlinear PID controllers, the tracking error is used. In particular,

$$e(t) = \underbrace{r(t)}_{reference/command} - \underbrace{y(t)}_{output}.$$

Linear bounded controllers can be straightforwardly designed. For example, letting $\varsigma = \beta = \sigma = \mu = 0$, we have the following linear PID control law

$$u(t) = \mathbf{sat}_{u_{\min}}^{u_{\max}}\left(k_{p0}e(t) + k_{i0}\int e(t)dt\right).$$

The PID controllers with the state feedback extension can be synthesized as

$$u(t) = \mathbf{sat}_{u_{\min}}^{u_{\max}}(e, x)$$

$$= \mathbf{sat}_{u_{\min}}^{u_{\max}}\left(\underbrace{\sum_{j=0}^{\varsigma} k_{pj} e^{\frac{2j+1}{2\beta+1}}}_{proportional} + \underbrace{\sum_{j=0}^{\sigma} k_{ij} \int e^{\frac{2j+1}{2\mu+1}} dt}_{integral} + \underbrace{\sum_{j=0}^{\alpha} k_{dj} \dot{e}^{\frac{2j+1}{2\gamma+1}}}_{differential} + G(t)B\frac{\partial V(e,x)}{\partial \begin{bmatrix} e \\ x \end{bmatrix}}\right),$$

$$u_{\min} \le u \le u_{\max},$$

where $V(e,x)$ is the function which satisfies the general requirements imposed on the Lyapunov pair [1]. Thus, the sufficient conditions for stability are used.

It is evident that nonlinear feedback mappings result, and the nonquadratic function $V(e,x)$ can be synthesized and used to obtain the control algorithm and feedback gains.

### 7.3.2 Tracking Control

*Integral Controller Design*

Tracking control is designed for the augmented systems which is modeled using the state variables and the reference dynamics. In particular, using (7.3.1). from

$$\dot{x}(t) = Ax + Bu, \ \dot{x}^{ref}(t) = r(t) - y(t) = r(t) - Hx(t),$$

one finds

$$\dot{x}_\Sigma(t) = A_\Sigma x_\Sigma + B_\Sigma u + N_\Sigma r, \ y = Hx, \ x_\Sigma = \begin{bmatrix} x \\ x^{ref} \end{bmatrix},$$

$$A_\Sigma = \begin{bmatrix} A & 0 \\ -H & 0 \end{bmatrix}, B_\Sigma = \begin{bmatrix} B \\ 0 \end{bmatrix}, \ N_\Sigma = \begin{bmatrix} 0 \\ I \end{bmatrix}.$$

Minimizing the quadratic performance functional

$$J = \frac{1}{2}\int_{t_0}^{t_f}\left(x_\Sigma{}^T Q x_\Sigma + u^T G u\right)dt ,$$

one finds the control law using the first-order necessary condition for optimality.

Necessary conditions that the control function $u(\cdot)$ guarantees a minimum to the Hamiltonian

$$H = \frac{1}{2}x_\Sigma^T Q x_\Sigma + \frac{1}{2}u^T G u + \frac{\partial V(x)}{\partial x}^T\left(A_\Sigma x_\Sigma + B_\Sigma u\right)$$

are the first-order necessary condition $(n1)$ $\frac{\partial H}{\partial u} = 0$, and the second-order necessary condition $(n2)$ $\frac{\partial^2 H}{\partial u \times \partial u^T} > 0$.

Using $\frac{\partial H}{\partial u} = 0$, we have

$$u = -G^{-1}B_\Sigma^T\frac{\partial V}{\partial x_\Sigma} = -G^{-1}\begin{bmatrix} B \\ 0 \end{bmatrix}^T\frac{\partial V}{\partial x_\Sigma}.$$

Here, $Q$ is the positive semi-definite constant-coefficient matrix, and $G$ is the positive weighting constant-coefficient matrix. Since $G > 0$, the second-order necessary condition for optimality $(n2)$ is guaranteed,

$$\frac{\partial^2 H}{\partial u \times \partial u^T} = G > 0$$

The solution of the Hamilton-Jacobi equation

$$-\frac{\partial V}{\partial t} = \frac{1}{2}x_\Sigma^T Q x_\Sigma + \left(\frac{\partial V}{\partial x_\Sigma}\right)^T A x_\Sigma - \frac{1}{2}\left(\frac{\partial V}{\partial x_\Sigma}\right)^T B_\Sigma G^{-1} B_\Sigma^T\frac{\partial V}{\partial x_\Sigma}$$

is satisfied by the quadratic return function $V = \frac{1}{2}x_\Sigma^T K x_\Sigma$.

Here, $K$ is the symmetric matrix which must be found by solving the nonlinear differential equation

$$-\dot{K} = Q + A_\Sigma^T K + K^T A_\Sigma - K^T B_\Sigma G^{-1} B_\Sigma^T K,\ K(t_f) = K_f.$$

The controller is given as

$$u = -G^{-1}B_\Sigma^T K x_\Sigma = -G^{-1}\begin{bmatrix} B \\ 0 \end{bmatrix}^T K x_\Sigma .$$

From $\dot{x}_{ref}(t) = e(t)$, one has $x_{ref}(t) = \int e(t)dt$.

Therefore, we obtain the integral control law

$$u(t) = -G^{-1}\begin{bmatrix} B \\ 0 \end{bmatrix}^T K \begin{bmatrix} x(t) \\ \int e(t)dt \end{bmatrix}.$$

In this control algorithm, the error vector is used in addition to the state feedback.

As was illustrated, the bounds are imposed on the control, and $u_{\min} \le u \le u_{\max}$. Therefore, the bounded controllers must be designed. Using the nonquadratic performance functional [1]

$$J = \int_{t_0}^{t_f} \left( x_\Sigma{}^T Q x_\Sigma + G \int \tan^{-1} u \, du \right) dt,$$

with positive semi-definite constant-coefficient matrix $Q$ and positive-definite matrix $G$, one finds

$$u(t) = -\tanh\left( G^{-1}\begin{bmatrix} B \\ 0 \end{bmatrix}^T K \begin{bmatrix} x(t) \\ \int e(t)dt \end{bmatrix}\right)$$

$$\approx -\text{sat}_{-1}^{+1}\left( G^{-1}\begin{bmatrix} B \\ 0 \end{bmatrix}^T K \begin{bmatrix} x(t) \\ \int e(t)dt \end{bmatrix}\right), -1 \le u \le 1.$$

This controller is obtained assuming that the solution of the functional partial differential equation can be approximated by the return function

$V = \frac{1}{2} x_\Sigma^T K x_\Sigma$,

where $K$ is the symmetric matrix.

*Proportional-Integral Control Laws Design*

We define the tracking error vector as

$e(t) = Nr(t) - y(t) = Nr(t) - Hx^{sys}(t)$.

Then, for linear systems we have

$\dot{e}(t) = N\dot{r}(t) - \dot{y}(t) = N\dot{r}(t) - H\dot{x}^{sys}(t) = N\dot{r}(t) - HA^{sys}x^{sys} - HB^{sys}u$.

Using the expanded state vector $x(t) = \begin{bmatrix} x^{sys}(t) \\ e(t) \end{bmatrix}$, one finds

$$\dot{x}(t) = \begin{bmatrix} \dot{x}^{sys}(t) \\ \dot{e}(t) \end{bmatrix} = \begin{bmatrix} A^{sys} & 0 \\ -HA^{sys} & 0 \end{bmatrix}\begin{bmatrix} x^{sys} \\ e \end{bmatrix} + \begin{bmatrix} B^{sys} \\ -HB^{sys} \end{bmatrix} u + \begin{bmatrix} 0 \\ N \end{bmatrix}\dot{r}$$

$$= Ax + Bu + \begin{bmatrix} 0 \\ N \end{bmatrix}\dot{r}, \; y = Hx^{sys}.$$

Let us illustrate the *space transformation* method. We introduce the following $z$ and $v$ vectors

$$z=\begin{bmatrix} x \\ u \end{bmatrix} \text{ and } v=\dot{u}.$$

Therefore, using $z$ and $v$, for linear systems, one obtains

$$\dot{z}(t)=\begin{bmatrix} A & B \\ 0 & 0 \end{bmatrix} z+\begin{bmatrix} 0 \\ I \end{bmatrix} v=A_z z+B_z v,\ y=Hx^{sys},\ z(t_0)=z_0.$$

Minimizing the functional

$$J=\int_{t_0}^{t_f}\left(z^T Q_z z+v^T G_z v\right)dt,\ Q_z\in\mathbb{R}^{(n+m)\times(n+m)},\ Q_z\geq 0,\ G\in\mathbb{R}^{m\times m},\ G>0,$$

the application of the first-order necessary condition for optimality gives

$$v=-G_z^{-1}B_z^T Kz.$$

The Riccati equation to find the unknown matrix $K\in\mathbb{R}^{(n+m)\times(n+m)}$ is

$$-\dot{K}=KA_z+A_z^T K-KB_z G_z^{-1}B_z^T K+Q_z,\ K(t_f)=K_f.$$

Hence, one has

$$\dot{u}(t)=-G_z^{-1}B_z^T Kz=-G_z^{-1}\begin{bmatrix} 0 \\ I \end{bmatrix}^T \begin{bmatrix} K_{11} & K_{21}^T \\ K_{21} & K_{22} \end{bmatrix}\begin{bmatrix} x \\ u \end{bmatrix}$$

$$=-G_z^{-1}K_{21}x-G_z^{-1}K_{22}u=K_{f1}x+K_{f2}u.$$

From $\dot{x}(t)=Ax+Bu$, we have $u=B^{-1}\left(\dot{x}(t)-Ax\right)$. Thus,

$$u=B^{-1}\left(\dot{x}(t)-Ax\right)=\left(B^T B\right)^{-1}B^T\left(\dot{x}(t)-Ax\right).$$

One obtains

$$\begin{aligned}\dot{u}(t)&=K_{f1}x+K_{f2}u=K_{f1}x+K_{f2}\left(B^T B\right)^{-1}B^T\left(\dot{x}(t)-Ax\right)\\ &=\left[K_{f1}-K_{f2}\left(B^T B\right)^{-1}B^T A\right]x(t)+K_{f2}\left(B^T B\right)^{-1}B^T\dot{x}(t)\\ &=\left(K_{f1}-K_{F1}A\right)x(t)+K_{F1}\dot{x}(t)=K_{F2}x(t)+K_{F1}\dot{x}(t).\end{aligned}$$

The controller is derived the following form

$$u(t)=K_{F1}x(t)-K_{F1}x_0+\int K_{F2}x(\tau)d\tau+u_0.$$

It is obvious that the designed controller is the proportional-integral control law with state feedback because $x(t)=\begin{bmatrix} x^{sys}(t) \\ e(t) \end{bmatrix}$.

For nonlinear systems, the proposed procedure can be straightforwardly used. In particular, we have the following proportional-integral controller

$$\dot{u}(t)=-G_z^{-1}B_z^T\frac{\partial V}{\partial z}=-G_z^{-1}\begin{bmatrix} 0 \\ I \end{bmatrix}^T\frac{\partial V(x,u)}{\partial[x\,u]^T},$$

where $V(x,u)$ is the return function.

It should be emphasized that nonquadratic functionals

$$J = \int_{t_0}^{t_f} \left( \sum_{i=0}^{\varsigma} \frac{2\eta+1}{2(\kappa i+\eta+1)} \begin{bmatrix} x^{sys \frac{\kappa i+\eta+1}{2\eta+1}} \\ e^{\frac{\kappa i+\eta+1}{2\eta+1}} \end{bmatrix}^T Q_{zi} \begin{bmatrix} x^{sys \frac{\kappa i+\eta+1}{2\eta+1}} \\ e^{\frac{\kappa i+\eta+1}{2\eta+1}} \end{bmatrix} + v^T G_z v \right) dt$$

can be used to guarantee the desired tracking performance.

*Example 7.3.1*

Design the tracking controller for a PZT microactatuator controlled by changing the applied voltage ($V$) using ICs. The equation of motion of PZT microactuator is

$$m_e \frac{d^2 y}{dt^2} + b\frac{dy}{dt} + ky = kd_e V,$$

where $y$ is the microactuator displacement (output).

*Solution.*

The second-order differential equation of the microactuator behavior is rewritten as

$$\frac{dy}{dt} = v,$$

$$\frac{dv}{dt} = -\frac{k}{m_e} y - \frac{b}{m_e} v + \frac{kd_e}{m_e} V.$$

Consider the reference input (desired microactuator displacement) $r(t)$. The tracking error between the reference and the output is

$e(t) = r(t) - y(t)$, $N$=1.

Thus, we have

$$\dot{x}^{sys} = A^{sys} x^{sys} + B^{sys} u,$$

$$\dot{e} = N\dot{r} - HA^{sys} x^{sys} - HB^{sys} u,$$

where

$$A^{sys} = \begin{bmatrix} 0 & 1 \\ -\frac{k}{m_e} & -\frac{b}{m_e} \end{bmatrix}, \quad B^{sys} = \begin{bmatrix} 0 \\ \frac{kd_e}{m_e} \end{bmatrix} \text{ and } H = \begin{bmatrix} 1 & 0 \end{bmatrix}.$$

One obtains

$$\dot{x} = \begin{bmatrix} \dot{x}^{sys} \\ \dot{e} \end{bmatrix} = \begin{bmatrix} A^{sys} & 0 \\ -HA^{sys} & 0 \end{bmatrix} \begin{bmatrix} x^{sys} \\ e \end{bmatrix} + \begin{bmatrix} B^{sys} \\ -HB^{sys} \end{bmatrix} u + \begin{bmatrix} 0 \\ N \end{bmatrix} \dot{r},$$

$$y = \begin{bmatrix} H & 0 \end{bmatrix} \begin{bmatrix} x^{sys} \\ e \end{bmatrix}.$$

Following the design procedure reported, using

$$z = \begin{bmatrix} x^{sys} \\ e \\ u \end{bmatrix},$$

we have

$$\dot{u}(t) = -K_f z(t) = -K_f \begin{bmatrix} y(t) \\ v(t) \\ e(t) \\ V(t) \end{bmatrix}.$$

The proportional-integral tracking controller is derived as

$$u(t) = K_{F1} \begin{bmatrix} y(t) \\ v(t) \\ e(t) \end{bmatrix} + \int K_{F2} \begin{bmatrix} y(\tau) \\ v(\tau) \\ e(\tau) \end{bmatrix} d\tau .$$

Using the following microactuator parameters $k$=3000, $b$=1, $d_e$=0.000001 and $m_e$=0.02, the tracking controller is designed using the weighting matrices $Q_z = \begin{bmatrix} 1 & 0 & 0 & 0 \\ 0 & 1 & 0 & 0 \\ 0 & 0 & 1\times10^{10} & 0 \\ 0 & 0 & 0 & 1 \end{bmatrix}$ and $G_z$=10.

The closed-loop microactuator dynamics is documented in Figure 7.3.1.

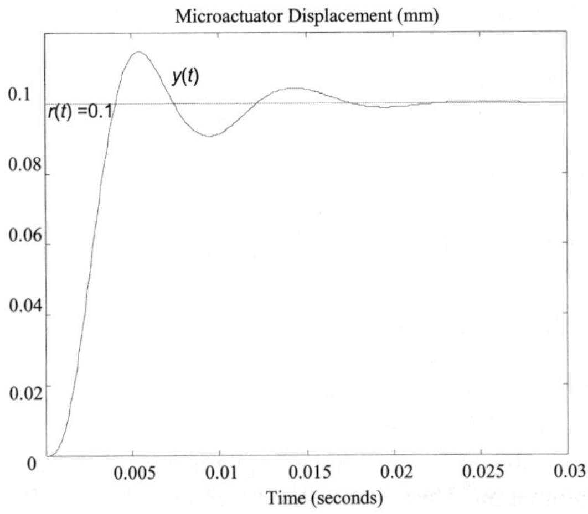


Figure 7.3.1 Closed-loop microactuator displacement if $r(t)$=0.1.

It should be emphasized that the differential equations which model the microactuator dynamics

$$m_e \frac{d^2 y}{dt^2} + b\frac{dy}{dt} + ky = k(d_e V - z),$$

must be integrated with the hysteresis model

$$\dot{z} = \alpha d_e \dot{V} - \beta |\dot{V}| z - \gamma \dot{V} |z|.$$

Thus, we have

$$m_e \frac{d^2 y}{dt^2} + b\frac{dy}{dt} + ky = k(d_e V - z),$$

$$\dot{z} = \alpha d_e \dot{V} - \beta |\dot{V}| z - \gamma \dot{V} |z|.$$

The state-space nonlinear model of PZT microactuators is

$$\frac{dy}{dt} = v,$$

$$\frac{dv}{dt} = -\frac{k}{m_e} y - \frac{b}{m_e} v - \frac{k}{m_e} z + \frac{k d_e}{m_e} V,$$

$$\frac{dz}{dt} = -\beta |\dot{V}| z + \alpha d_e \dot{V} - \gamma |z| \dot{V}.$$

□

### 7.3.3 Time-Optimal Control

Time-optimal controllers can be designed using the functional

$$J = \frac{1}{2} \int_{t_0}^{t_f} \left( x_\Sigma{}^T Q x_\Sigma \right) dt .$$

Taking note of the Hamilton-Jacobi equation

$$-\frac{\partial V}{\partial t} = \min_{-1 \le u \le 1} \left[ \frac{1}{2} x_\Sigma^T Q x_\Sigma + \left( \frac{\partial V}{\partial x_\Sigma} \right)^T \left( A x_\Sigma + B_\Sigma u \right) \right],$$

and using the optimality conditions, the relay-type controller is found to be

$$u = -\operatorname{sgn}\left( B_\Sigma^T \frac{\partial V}{\partial x_\Sigma} \right), \ -1 \le u \le 1 .$$

This control algorithm cannot be implemented in practice due to the chattering phenomenon. Therefore, relay-type control laws with dead zone

$$u = -\operatorname{sgn}\left( B_\Sigma^T \frac{\partial V}{\partial x_\Sigma} \right)\Bigg|_{\text{dead zone}}, \ -1 \le u \le 1$$

are commonly used.

### 7.3.4 Sliding Mode Control

Soft-switching sliding mode control laws are synthesized in [1]. Sliding mode soft switching algorithms provide superior performance, and the chattering effect is eliminated.

To design controllers, we model the states and errors dynamics as

$$\dot{x}(t) = Ax + Bu, \ -1 \le u \le 1,$$

$$\dot{e}(t) = N\dot{r}(t) - HAx - HBu.$$

The smooth sliding manifold is

$$M = \{(t,x,e) \in R_{\ge 0} \times X \times E \mid \upsilon(t,x,e) = 0\}$$

$$= \bigcap_{j=1}^{m} \{(t,x,e) \in R_{\ge 0} \times X \times E \mid \upsilon_j(t,x,e) = 0\}$$

The time-varying nonlinear switching surface is

$$\upsilon(t,x,e) = K_{\upsilon xe}(t,x,e) = 0.$$

The soft switching control law is given as

$$u(t,x,e) = -G\phi(\upsilon), \ -1 \le u \le 1, \ G>0,$$

where $\phi(\cdot)$ is the continuous real-analytic function of class $C^{\epsilon}$ ($\epsilon \ge 1$), for example, `tanh` and `erf`.

### 7.3.5 Constrained Control of Nonlinear MEMS: Hamilton-Jacobi Method

Constrained optimization of MEMS is a topic of great practical interest. We consider the systems modeled by nonlinear differential equations (7.3.2). Using the Hamilton-Jacobi theory, the bounded controllers can be synthesized for continuous-time systems given in the following form

$$\dot{x}^{MEMS}(t) = F_s(x^{MEMS}) + B_s(x^{MEMS})u^{2w+1}, \ y = H(x^{MEMS}), u_{\min} \le u \le u_{\max},$$

$$x^{MEMS}(t_0) = x_0^{MEMS}.$$

Here, $x^{MEMS} \in X_s$ is the state vector; $u \in U$ is the vector of control inputs; $y \in Y$ is the measured output; $F_s(\cdot)$, $B_s(\cdot)$ and $H(\cdot)$ are the smooth mappings, $F_s(0)=0$, $B_s(0)=0$ and $H(0)=0$; $w$ is the nonnegative integer.

To design the tracking controller, we augment the MEMS dynamics

$$\dot{x}^{MEMS}(t) = F_s(x^{MEMS}) + B_s(x^{MEMS})u^{2w+1}, \ y = H(x^{MEMS}), u_{\min} \le u \le u_{\max},$$

$$x^{MEMS}(t_0) = x_0^{MEMS}.$$

with the *exogenous* dynamics $\dot{x}^{ref}(t) = Nr - y = Nr - H(x^{system})$.

Using the augmented state vector $x = \begin{bmatrix} x^{MEMS} \\ x^{ref} \end{bmatrix} \in X$, one obtains

$$\dot{x}(t)=F(x,r)+B(x)u^{2w+1},\ u_{\min}\le u\le u_{\max},\ x(t_0)=x_0,\ x=\begin{bmatrix} x^{MEMS} \\ x^{ref}\end{bmatrix},$$

$$F(x,r)=\begin{bmatrix} F_s(x^{MEMS}) \\ -H(x^{MEMS})\end{bmatrix}+\begin{bmatrix} 0 \\ N\end{bmatrix}r,\ B(x)=\begin{bmatrix} B_s(x^{MEMS}) \\ 0\end{bmatrix}.$$

The set of admissible control $U$ consists of the Lebesgue measurable function $u(\cdot)$, and a bounded controller should be designed within the constrained control set

$$U=\{u\in\mathbb{R}^m \mid u_{i\min}\le u_i\le u_{i\max},\ i=1,\ldots,m\}.$$

We map the control bounds imposed by a bounded, integrable, one-to-one globally Lipschitz, vector-valued continuous function $\Phi\in C^{\epsilon}$ ($\epsilon\ge 1$). Our goal is to analytically design the bounded admissible state-feedback controller in the closed form as

$$u=\Phi(x).$$

The most common $\Phi$ are the algebraic and transcendental (exponential, hyperbolic, logarithmic, trigonometric) continuously differentiable, integrable, one-to-one functions. For example, the odd one-to-one integrable function tanh with domain $(-\infty, +\infty)$ maps the control bounds. This function has the corresponding inverse function $\tanh^{-1}$ with range $(-\infty, +\infty)$.

The performance cost to be minimized is given as

$$J=\int_{t_0}^{\infty}\left[W_x(x)+W_u(u)\right]dt$$

$$=\int_{t_0}^{\infty}\left[W_x(x)+(2w+1)\int\left(\Phi^{-1}(u)\right)^T G^{-1}\mathrm{diag}\left(u^{2w}\right)du\right]dt,$$

where $G^{-1}\in\mathbb{R}^{m\times m}$ is the positive-definite diagonal matrix.

Performance integrands $W_x(\cdot)$ and $W_u(\cdot)$ are real-valued, positive-definite, and continuously differentiable integrand functions. Using the properties of $\Phi$ one concludes that inverse function $\Phi^{-1}$ is integrable. Hence, the integral

$$\int\left(\Phi^{-1}(u)\right)^T G^{-1}\mathrm{diag}\left(u^{2w}\right)du$$

exists.

*Example 7.3.2*

Consider a nonlinear dynamic microsystem described by the differential equation

$$\frac{dx}{dt}=ax+bu^3,\ u_{\min}\le u\le u_{\max}.$$

Taking note of

$$W_u(u)=(2w+1)\int\left(\Phi^{-1}(u)\right)^T G^{-1}\mathrm{diag}\left(u^{2w}\right)du,$$

one has the following positive-definite integrand

$$W_u(u)=3\int \tanh^{-1}uG^{-1}u^2du=\tfrac{1}{3}u^3\tanh^{-1}u+\tfrac{1}{6}u^2+\tfrac{1}{6}\ln\left(1-u^2\right),\ G^{-1}=\tfrac{1}{3}.$$

In general, if the hyperbolic tangent is used to map the saturation effect, for the single-input case, one has

$$W_u(u)=(2w+1)\int u^{2w}\tanh^{-1}\frac{u}{k}du=u^{2w+1}\tanh^{-1}\frac{u}{k}-k\int\frac{u^{2w+1}}{k^2-u^2}du.$$

□

Necessary conditions that the control function $u(\cdot)$ guarantees a minimum to the Hamiltonian

$$H=W_x(x)+(2w+1)\int\left(\Phi^{-1}(u)\right)^T G^{-1}\mathrm{diag}\left(u^{2w}\right)du$$

$$+\frac{\partial V(x)}{\partial x}^T\left[F(x,r)+B(x)u^{2w+1}\right]$$

are:

first-order necessary condition ($n1$) $\dfrac{\partial H}{\partial u}=0$

second-order necessary condition ($n2$) $\dfrac{\partial^2 H}{\partial u\times\partial u^T}>0$.

The positive-definite return function $V(\cdot)$, $V\in C^{\kappa}, \kappa\geq 1$, is

$$V(x_0)=\inf_{u\in U}J(x_0,u)=\inf J\left(x_0,\Phi(\cdot)\right)\geq 0.$$

The Hamilton-Jacobi-Bellman equation is given as

$$-\frac{\partial V}{\partial t}$$

$$=\min_{u\in U}\left\{W_x(x)+(2w+1)\int\left(\Phi^{-1}(u)\right)^T G^{-1}\mathrm{diag}\left(u^{2w}\right)du+\frac{\partial V(x)}{\partial x}^T\left[F(x,r)+B(x)u^{2w+1}\right]\right\}.$$

The controller should be derived by finding the control value which attains the minimum to nonquadratic functional. The first-order necessary condition (*n1*) leads us to an admissible bounded control law. In particular,

$$u=-\Phi\left(GB(x)^T\frac{\partial V(x)}{\partial x}\right),\ u\in U.$$

The second-order necessary condition for optimality (*n2*) is met because the matrix $G^{-1}$ is positive-definite. Hence, a unique, bounded, real-analytic, and continuous control candidate is designed.

If there exists a proper function $V(x)$ which satisfies the Hamilton-Jacobi equation, then, the resulting closed-loop system is robustly stable in the specified state $X$ and control $U$ sets, and robust tracking is ensured in the convex and compact set $XY(X_0, U, R, E_0)$.

The solution of the functional equation should be found using nonquadratic return functions. To obtain $V(\cdot)$, the performance cost must be evaluated at the allowed values of the states and control. Linear and nonlinear functionals admit the final values, and the minimum value of the nonquadratic cost is given by power-series forms [1]. That is,

$$J_{\min} = \sum_{i=0}^{\eta} v(x_0)^{\frac{2(i+\gamma+1)}{2\gamma+1}}, \eta = 0,1,2,\ldots, \gamma = 0,1,2,\ldots$$

The solution of the partial differential equation is satisfied by a continuously differentiable positive-definite return function

$$V(x) = \sum_{i=0}^{\eta} \frac{2\gamma+1}{2(i+\gamma+1)} \left( x^{\frac{i+\gamma+1}{2\gamma+1}} \right)^T K_i x^{\frac{i+\gamma+1}{2\gamma+1}},$$

where matrices $K_i$ are found by solving the Hamilton-Jacobi equation.

The quadratic return function

$$V(x) = \tfrac{1}{2} x^T K_0 x$$

is found by letting $\eta=\gamma=0$. This quadratic candidate may be employed only if the designer enables to neglect the high-order terms in Taylor's series expansion.

Using $\eta=1$ and $\gamma=0$, one obtains

$$V(x) = \tfrac{1}{2} x^T K_0 x + \tfrac{1}{4} \left( x^2 \right)^T K_1 x^2.$$

For $\eta=4$ and $\gamma=1$, we have the following function

$$V(x) = \tfrac{3}{4} \left( x^{\frac{2}{3}} \right)^T K_0 x^{\frac{2}{3}} + \tfrac{1}{2} x^T K_1 x + \tfrac{3}{8} \left( x^{\frac{4}{3}} \right)^T K_2 x^{\frac{4}{3}} + \tfrac{3}{10} \left( x^{\frac{5}{3}} \right)^T K_3 x^{\frac{5}{3}} + \tfrac{1}{4} \left( x^2 \right)^T K_4 x^2.$$

The nonlinear bounded controller is given as

$$u = -\Phi\left( GB(x)^T \sum_{i=0}^{\eta} \operatorname{diag}\left[ x(t)^{\frac{i-\gamma}{2\gamma+1}} \right] K_i(t) x(t)^{\frac{i+\gamma+1}{2\gamma+1}} \right),$$

$$\operatorname{diag}\left[ x(t)^{\frac{i-\gamma}{2\gamma+1}} \right] = \begin{bmatrix} x_1^{\frac{i-\gamma}{2\gamma+1}} & 0 & \cdots & 0 & 0 \\ 0 & x_2^{\frac{i-\gamma}{2\gamma+1}} & \cdots & 0 & 0 \\ \vdots & \vdots & \ddots & \vdots & \vdots \\ 0 & 0 & \cdots & x_{c-1}^{\frac{i-\gamma}{2\gamma+1}} & 0 \\ 0 & 0 & \vdots & 0 & x_c^{\frac{i-\gamma}{2\gamma+1}} \end{bmatrix}.$$

### 7.3.6 Constrained Control of Nonlinear Uncertain MEMS: Lyapunov Method

Over the horizon $[t_0,\infty)$ we consider the dynamics of MEMS modeled as

$$\dot{x}(t)=F_z(t,x,r,z)+B_p(t,x,p)u\,,\ y=H(x)\,,$$

$$u_{\min}\le u\le u_{\max}\,,x(t_0)=x_0\,,$$

where $z\ Z$ and $p\ P$ are the parameter uncertainties, functions $z(\cdot)$ and $p(\cdot)$ are Lebesgue measurable and known within bounds; $Z$ and $P$ are the known non-empty compact sets; $F_z(\cdot)$, $B_p(\cdot)$ and $H(\cdot)$ are the smooth mapping fields.

Let us formulate and solve the motion control problem by synthesizing robust controllers that guarantee stability and robust tracking. Our goal is to design control laws which robustly stabilize nonlinear systems with uncertain parameters and drive the tracking error $e(t)=r(t)-y(t)$, $e\in E$ robustly to the compact set. For MEMS modeled by nonlinear differential equations with parameter variations, the robust tracking of the measured output vector $y\in Y$ must be accomplished with respect to the measured uniformly bounded reference input vector $r\in R$.

The *nominal* and uncertain dynamics are mapped by $F(\cdot)$, $B(\cdot)$ and $\Xi(\cdot)$. Hence, the system evolution is described as

$$\dot{x}(t)=F(t,x,r)+B(t,x)u+\Xi(t,x,u,z,p)\,,\ y=H(x)\,,$$

$$u_{\min}\le u\le u_{\max}\,,x(t_0)=x_0\,.$$

There exists a norm of $\Xi(t,x,u,z,p)$. In particular,

$$\left\|\Xi(t,x,u,z,p)\right\|\le\rho(t,x)\,,$$

where $\rho(\cdot)$ is the continuous Lebesgue measurable function.

Our goal is to solve the motion control problem, and tracking controllers must be synthesized using the tracking error vector and the state variables. Furthermore, to guarantee robustness and to expand stability margins, to improve dynamic performance and to meet other requirements, nonquadratic Lyapunov functions $V(t,e,x)$ will be used in stability analysis and design of robust tracking control laws.

It was demonstrated that the Hamilton-Jacobi theory can be used to find control laws, and the minimization of nonquadratic performance functionals leads one to the bounded controllers.

Letting $u=\Phi(t,e,x)$, one obtains a set of admissible controllers. Applying the error and state feedback we define a family of tracking controllers as

$$u=\Omega(x)\Phi(t,e,x)$$

$$=-\Omega(x)\Phi\left(G_E(t)B_E(t,x)^T\frac{1}{s}\frac{\partial V(t,e,x)}{\partial e}+G_X(t)B(t,x)^T\frac{\partial V(t,e,x)}{\partial x}\right),$$

where $\Omega(\cdot)$ is the nonlinear function; $G_E(\cdot)$ and $G_X(\cdot)$ are the diagonal matrix-functions defined on $[t_0,\infty)$; $B_E(\cdot)$ is the matrix-function; $V(\cdot)$ is the continuous, differentiable and real-analytic function.

Let us design the Lyapunov function. The quadratic Lyapunov candidates can be used. However, for uncertain nonlinear systems, nonquadratic functions $V(t,e,x)$ allow one to realize the full potential of the Lyapunov-based theory and leads us to the nonlinear feedback maps which are needed to achieve conflicting design objectives. We introduce the following family of Lyapunov candidates

$$V(t,e,x)=\sum_{i=0}^{\varsigma}\frac{2\beta+1}{2(i+\beta+1)}\left(e^{\frac{i+\beta+1}{2\beta+1}}\right)^T K_{Ei}(t)e^{\frac{i+\beta+1}{2\beta+1}}+\sum_{i=0}^{\eta}\frac{2\gamma+1}{2(i+\gamma+1)}\left(x^{\frac{i+\gamma+1}{2\gamma+1}}\right)^T K_{Xi}(t)x^{\frac{i+\gamma+1}{2\gamma+1}}$$

where $K_{Ei}(\cdot)$ and $K_{Xi}(\cdot)$ are the symmetric matrices; $\zeta, \beta, \eta$ and $\gamma$ are the nonnegative integers, $\zeta=0,1,2,..$, $\beta=0,1,2,..$, $\eta=0,1,2,..$ and $\gamma=0,1,2,..$

The quadratic form of $V(t,e,x)$ is found by letting $\zeta=\beta=\eta=\gamma=0$. We have $V(t,e,x)=\frac{1}{2}e^T K_{E0}(t)e+\frac{1}{2}x^T K_{X0}(t)x$.

By using $\zeta=1,\beta=0,\eta=1$ and $\gamma=0$, one obtains a nonquadratic candidate

$$V(t,e,x)=\tfrac{1}{2}e^T K_{E0}(t)e+\tfrac{1}{4}e^{2^T} K_{E1}(t)e^2+\tfrac{1}{2}x^T K_{X0}(t)x+\tfrac{1}{4}x^{2^T} K_{X1}(t)x^2.$$

One obtains the following tracking control law

$$u=-\Omega(x)\Phi\left(G_E(t)B_E(t,x)^T\sum_{i=0}^{\varsigma}\operatorname{diag}\left[e(t)^{\frac{i-\beta}{2\beta+1}}\right]K_{Ei}(t)\frac{1}{s}e(t)^{\frac{i+\beta+1}{2\beta+1}}\right.$$

$$\left.+G_X(t)B(t,x)^T\sum_{i=0}^{\eta}\operatorname{diag}\left[x(t)^{\frac{i-\gamma}{2\gamma+1}}\right]K_{Xi}(t)x(t)^{\frac{i+\gamma+1}{2\gamma+1}}\right),$$

where

$$\operatorname{diag}\left[e(t)^{\frac{i-\beta}{2\beta+1}}\right]=\begin{bmatrix} e_1^{\frac{i-\beta}{2\beta+1}} & 0 & \cdots & 0 & 0\\ 0 & e_2^{\frac{i-\beta}{2\beta+1}} & \cdots & 0 & 0\\ \vdots & \vdots & \ddots & \vdots & \vdots\\ 0 & 0 & \cdots & e_{b-1}^{\frac{i-\beta}{2\beta+1}} & 0\\ 0 & 0 & \vdots & 0 & e_b^{\frac{i-\beta}{2\beta+1}}\end{bmatrix}$$

and

$$\operatorname{diag}\left[x(t)^{\frac{i-\gamma}{2\gamma+1}}\right]=\begin{bmatrix} x_1^{\frac{i-\gamma}{2\gamma+1}} & 0 & \cdots & 0 & 0\\ 0 & x_2^{\frac{i-\gamma}{2\gamma+1}} & \cdots & 0 & 0\\ \vdots & \vdots & \ddots & \vdots & \vdots\\ 0 & 0 & \cdots & x_{n-1}^{\frac{i-\gamma}{2\gamma+1}} & 0\\ 0 & 0 & \vdots & 0 & x_n^{\frac{i-\gamma}{2\gamma+1}}\end{bmatrix}.$$

If matrices $K_{Ei}$ and $K_{Xi}$ are diagonal, we have

$$u=-\Omega(x)\Phi\left(G_E(t)B_E(t,x)^T\sum_{i=0}^{\varsigma}K_{Ei}(t)\frac{1}{s}e(t)^{\frac{2i+1}{2\beta+1}}+G_X(t)B(t,x)^T\sum_{i=0}^{\eta}K_{Xi}(t)x(t)^{\frac{2i+1}{2\gamma+1}}\right)$$

A closed-loop uncertain system is robustly stable in $X(X_0,U,Z,P)$ and robust tracking is guaranteed in the convex and compact set $E(E_0,Y,R)$ if for reference inputs $r\in R$ and uncertainties in $Z$ and $P$ there exists a $C^{\kappa}$ ($\kappa\geq 1$) function $V(\cdot)$, as well as $K_\infty$-functions $\rho_{X1}(\cdot)$, $\rho_{X2}(\cdot)$, $\rho_{E1}(\cdot)$, $\rho_{E2}(\cdot)$ and $K$-functions $\rho_{X3}(\cdot)$, $\rho_{E3}(\cdot)$, such that the following sufficient conditions

$$\rho_{X1}(\|x\|)+\rho_{E1}(\|e\|)\leq V(t,e,x)\leq\rho_{X2}(\|x\|)+\rho_{E2}(\|e\|),$$

$$\frac{dV(t,e,x)}{dt}\leq-\rho_{X3}(\|x\|)-\rho_{E3}(\|e\|),$$

are guaranteed in an invariant domain of stability $S$, and $XE(X_0,E_0,U,R,Z,P)\subseteq S$.

The sufficient conditions under which the robust control problem is solvable were given. Computing the derivative of the $V(t,e,x)$, the unknown coefficients of $V(t,e,x)$ can be found. That is, matrices $K_{Ei}(\cdot)$ and $K_{Xi}(\cdot)$ are obtained. This problem is solved using the nonlinear inequality concept [1].

## 7.4 INTELLIGENT CONTROL OF MEMS

Hierarchical distributed closed-loop systems must be designed for large-scale multinode MEMS in order to perform a number of complex functions and tasks in dynamic and uncertain environments. In particular, the goal is the synthesis of control algorithms and architectures which maximize performance and efficiency minimizing system complexity through:

- Intelligence, learning, evolution, and organization,
- Adaptive decision making and control,
- Coordination and autonomy of multinode MEMS through tasks and functions generation, organization, and decomposition,
- Performance analysis with outcomes prediction and assessment,
- Real-time diagnostics, health monitoring, and estimation,
- Real-time adaptation and reconfiguration,
- Fault tolerance and robustness, etc.

Control theory and engineering practice in the design and implementation of hierarchical hybrid (digital- and continuous-time subsystems are integrated, discrete and continuous events are augmented) real-time large-scale closed-loop systems have not matured. Synthesis of optimal controllers for elementary (single-input/single-output) single node MEMS can be performed using conventional methods such as the Hamilton-Jacobi and Lyapunov theories. However, these methods do not allow the designer to attain the desired features for complex multinode MEMS even though some methods (e.g., adaptive control, fuzzy logic, and neural networks) ensure performance assessment, diagnostics, adaptation, and reconfiguration. In fact, hierarchical architectures need to be designed and optimized to achieve intelligence, evolution, adaptive decision making, and performance analysis with outcome prediction. The design of intelligent systems can be mathematically formulated as a search problem in high-dimensional space, and the performance criteria form hypersurfaces. Efficient and robust search algorithms are used to perform optimization. Due to the complexity of large-scale systems and uncertainties, it is difficult to develop accurate analytic models, explicitly formulate performance specifications, derive regret functionals and performance indexes, design optimal architectures, synthesize hierarchical structures, as well as design control algorithms. The situation becomes much more complex in the synthesis of robust closed-loop systems under uncertainties in dynamic environments.

Intelligence can be defined as the ability of closed-loop MEMS to achieve the desired goals (for example, maximize safety, stability, robustness, controllability, efficiency, reliability, and survivability, while minimizing failures, electromagnetic interference, and losses) in dynamic and uncertain environments through the MEMS abilities to sense the environment, learn and evolve, perform adaptive decision making with performance analysis and outcome prediction, and control.

Let us discuss the design of a control algorithm for *j*th level of *k*-level hierarchical MEMS. The controller for *j*th level of the *N*-layer hierarchical closed-loop system can be synthesized in the following form

$$u_j(t) = f_j\left(\mathbf{P}, \sum_{i=0}^{N} e_i(t), \sum_{i=0}^{N} x_i(t), \sum_{i=0}^{N} s_i(t), \sum_{i=0}^{N} p_i(t), p_j(t)\right),$$

where $u_j(t)$ is the control vector for *j*th level; $f_j$ is the nonlinear map; **P** is the system performance variables, e.g., efficiency, robustness, stability, controllability, etc.; $\sum_{i=0}^{N} e_i(t)$ is the error vector which represents the difference between the reference commands-events $r_i(t)$ and outputs $y_i(t)$, and the end-to-end behavior error is $e(t) = r(t) - y(t)$; $\sum_{i=0}^{N} x_i(t)$ is the state and decision variable vector; $\sum_{i=0}^{N} s_i(t)$ is the sensed information vector (inputs, outputs, state variables, events, disturbances, noise, parameters, etc.) measured by *j*th and lower level sensors; $\sum_{i=0}^{N} p_i(t)$ is the parameter vector (for example, time-varying system coefficients, MEMS parameters, adjustable feedback gains changed through the decision making, learning, evolution, intelligence, control, adaptation, and reconfiguration); $p_j(t)$ is the parameter vector of the *j*th layer.

The simplest control algorithms are the proportional-integral-derivative (PID) controllers which can be designed using

$$u(t)=\underbrace{\sum_{j=1}^{2N_p-1}k_{p(2j-1)}e^{2j-1}(t)}_{proportional}+\underbrace{\sum_{j=1}^{2N_i-1}k_{i(2j-1)}\frac{e^{2j-1}(t)}{s}}_{integral}+\underbrace{\sum_{j=1}^{2N_d-1}k_{d(2j-1)}\frac{de^{2j-1}(t)}{dt}}_{derivative},$$

where $N_p, N_i$ and $N_d$ are the positive integers assigned by the designer; $k_{p(2j-1)}, k_{i(2j-1)}$ and $k_{d(2j-1)}$ are the proportional, integral, and derivative feedback coefficients.

For $N_p = 1$, $N_i = 1$ and $N_d = 1$, one has

$$u(t)=k_p e(t)+k_i\int e(t)dt+k_d\frac{de(t)}{dt}.$$

The feedback gains $k_p$, $k_i$ and $k_d$ can be time-varying. In addition, $k_p$, $k_i$ and $k_d$ can be nonlinear functions of the state variables $x(t)$, tracking error $e(t)$, disturbances, events, etc. For example,

$$u(t)=-\left(2+e^{-10t}\right)e(t)-\left(3+4e^{-20t}\right)\tanh\left(|e(t)|\right)\int e(t)dt,$$

where $k_p(t)=-2-e^{-10t}$ and $k_i\left(e(t)\right)=-\left(3+4e^{-20t}\right)\tanh\left(|e(t)|\right)$.

It is obvious that for multivariable systems the tuning problem cannot be performed through the empirical procedure, and the search for the feedback gains must be made through optimization in the *performance domain*, e.g., time domain cost functionals can be synthesized and used. It must be emphasized that simple matching of the PID-type control laws to optimal controllers (synthesized using the performance functionals) can be achieved only for the centralized controllers synthesis. The performance functionals in the time domain can be given as

$$J(k_j)=\int_{t_0}^{t_f}\left[f_P(\mathbf{P})+\sum_{i=1}^{n}te^{2i}(t,k_j)\right]dt,\ J(k_j)=\int_{t_0}^{t_f}\left[f_P(\mathbf{P})+\sum_{i=1}^{n}t^{2i}\left|e^{i}(t,k_j)\right|\right]dt,$$

$$J(k_j)=\int_{t_0}^{t_f}\left[f_P(\mathbf{P})+\sum_{i=1}^{n}\left|e^{i}(t,k_j)\right|\right]dt,$$

where $f_P$ (**P**) is the integrand which measures the microscale systems, subsystems, devices, and structures performance; $k_j$ is the vector of the controller feedback coefficients which must be found to maximize the performance.

Assume that in multinode MEMS, which have thousands of nodes (MEMS with microdevices and ICs), one sensor and actuator failed. These types of failures must be identified in real-time (through diagnostics and health monitoring), and closed-loop MEMS must be reconfigurated through intelligence and adaptive decision making with performance analysis, outcome prediction, and assessment. Hierarchically distributed closed-loop systems must be designed for large-scale multinode MEMS using hierarchical layers. For example, for three-layer configuration, the possible architecture consists of

- High-level layer (intelligent augmented/coordinated control with intelligence and adaptive decision making with performance analysis, outcome prediction, and assessment analysis capabilities),
- Medium-level layer (intelligent coordinated or autonomous control for subsystems),
- Low-level layer (sensor processing, data acquisition, simple feedback, e.g., single-input/single-output device/structure for single node with state, event, and decision variables).

Thus, the problem can be decomposed into subproblems performed at different layers (which can operate at different sampling rates) with synthesized layouts and decomposed tasks and functions. The system architecture must be synthesized, and the tasks are decomposed by the high-level layer into narrow tasks/functions and fed (with or without defined adaptation, decision making, diagnostics, estimation, implementation, performance analysis, realization, as well as other details) to the medium-level layer, which further decomposes the tasks and supervises the low-level layer. This hierarchically distributed standardized control architecture is shown in Figure 7.4.1.

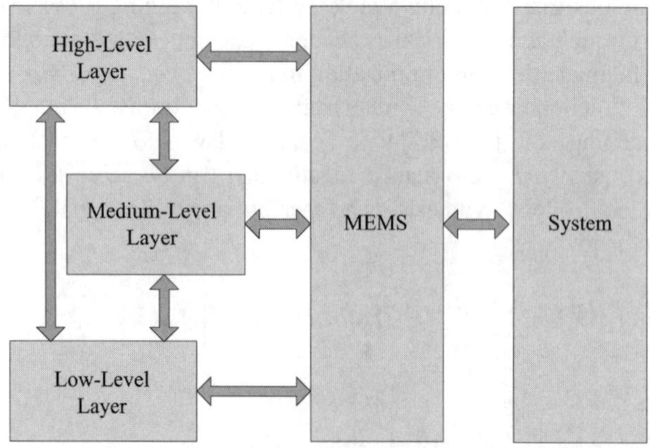


Figure 7.4.1 Three-layer hierarchically distributed architecture for large-scale multinode MEMS.

Different operating systems, interfaces, and platforms should be supported by advanced software, and there is a critical need for novel high-performance robust software.

The designer can:

- Lay out and support hierarchical controllers in `if-then-else` execution format,
- Generate codes for different platforms,

- Add and remove layers,
- Set up communication and networks based upon timing requirements (write data to the shared memory buffers and read data from the buffers, protocols development, code and encode data from the buffers using different file formats),
- Perform diagnostics and decision-making, etc.

To perform these tasks, novel design tools are needed. At high-level, intelligence, evolution, coordination and autonomy through tasks generation/organization and decomposition, adaptive decision making with performance analysis and outcome prediction, diagnostics and estimation, adaptation and reconfiguration, fault tolerance and robustness, as well as other functions must be performed through sensing-actuation, learning, evolution, analysis, evaluation, behavioral (dynamic and steady-state performance) optimization and adaptation, etc.

Architectures for hierarchically distributed complex closed-loop systems can be synthesized based upon the decomposition of tasks and functions. The analysis of complexity, hierarchy, data flow (sensing and actuation), and controllers design, allows the designer to synthesize architectures starting from lowest structural level and then governing and augmenting lower levels to upper levels based on physical relationships, functional correlation, order, sequence, and arrays to attain the desired performance, capability, and efficiency.

Consider the closed-loop system to displace (move) the flight control surfaces as shown in Figure 7.4.2. Ailerons, elevators, canards, flaps, rudders, stabilizers and tips are controlled by MEMS-based microactuators.

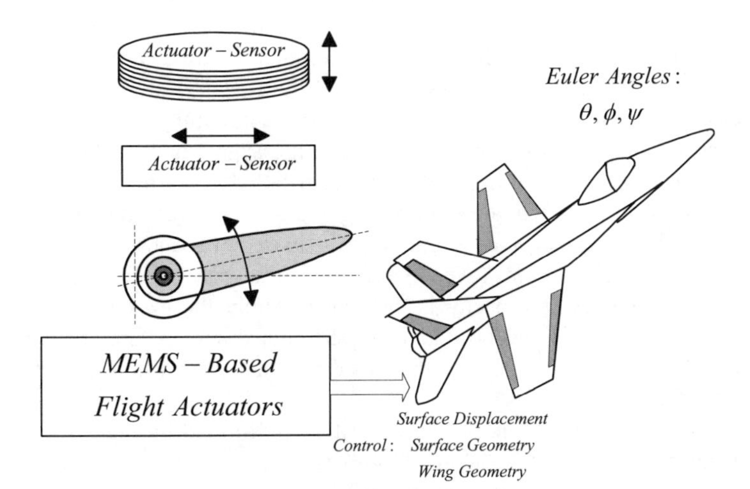


Figure 7.4.2 Aircraft with MEMS-based translational and rotational flight microactuators.

In micro air vehicles, the deflections and torques are small, and MEMS technology is uniquely suitable to attain active aerodynamic control. Smart flight surfaces can be designed using the array of MEMS nodes controlled by the distributed controller. The MEMS node integrates electromechanical microstructure or microdevice (actuator–sensor) and ICs. These MEMS can be fabricated using batch-fabrication processes described in Chapter 3. The integration of MEMS is performed through energy, communication, and control busses and IO devices. Point-to-point actuation or sensing, performed by microstructures is not sufficient in advanced systems, and, real-time intelligent coordinated motion with complex surface geometry synthesis is required. Smart control surfaces can be built using thousands of MEMS nodes, and these nodes must be interfaced and controlled. Decentralized control systems with increased levels of integration and functionality must be designed. The application of multiscalar digital signal processors (DSPs) allows one to design distributed control systems to control MEMS arrays. In addition to challenges in analysis and design of high-performance distributed systems and MEMS, other critical issues are wireless communication, high-bandwidth networking, software developments, etc. Distributed embedded control systems offer several advantages because affordable systems can be designed using low-cost, robust, and high-performance MEMS nodes. Energy and control signals are transmitted and distributed within the microstructures or microdevices (actuators–sensors) through energy, communication, and control channels. The distributed systems must have distributed processor capacity. The communication between different nodes in distributed control systems consists of data and administrative messages. Therefore, the complexity of distributed systems is higher than the complexity of centralized control systems (centralized systems are controlled by a central processor, and the processor is connected to the IO devices; the communication between the processor and IO devices consists of data messages, and, in general, the status messages are not used). In distributed control systems, the communication between the supervising high-level layer and lower level layers is based upon access messages, set-up messages, requests and responses, as well as status (events) and error messages. Hence, in distributed systems, the processor capacity is distributed within the system. Each node performs the specific function (task), and distributed systems are organized in different ways as the designer synthesizes the overall system as well as defines the node functions and tasks.

Control of smart flight surfaces must be performed by the hierarchical distributed management system to properly displace microstructures, and thus to change the geometry of control surfaces in order to guarantee the desired moments, control the aerodynamic flows, and reduce the drag. The design method and system architecture are different compared with conventional systems because the benefits and capabilities of smart flight surfaces can be utilized through hierarchical decentralized control and adaptive decision making with performance analysis and outcomes prediction. The integration and interfacing of high-, medium-, and low-level layers are critical to attaining the specified performance characteristics.

The local aerodynamic flow can be actively controlled and sensed by the specific microstructure (MEMS node), and MEMS-based actuator–sensor arrays are designed in order to deflect and change the geometry of smart flight surfaces. Different MEMS configurations and architectures can be designed for different flight control surfaces (ailerons, elevators, fins, flaps, stabilizers, and tips) by analyzing the forces and moments needed to be developed, deflection angle, deflection rate, surface geometry, aerodynamic loads, drag, as well as other specifications. However, for all control surfaces, hierarchical distributed closed-loop systems need to be designed in order to guarantee the coordinated longitudinal and lateral vehicle control as well as to ensure the optimal active aerodynamic flow control. Despite the complexity of management system architectures and control algorithms, the hierarchical decentralized management systems can be designed to guarantee the specifications and requirements imposed on flight vehicles. Ailerons, elevators, canards, fins, flaps, rudders, stabilizers and tips can be controlled by MEMS. Consider the smart flight surfaces (two flaps) which are designed using MEMS-based actuator-sensor arrays (with translational and rotational microstructures [4, 5]) in order to perform coordinated vehicle control by displacing surfaces, changing the surfaces geometry, as well as sensing the aerodynamic loads, see Figure 7.4.3.

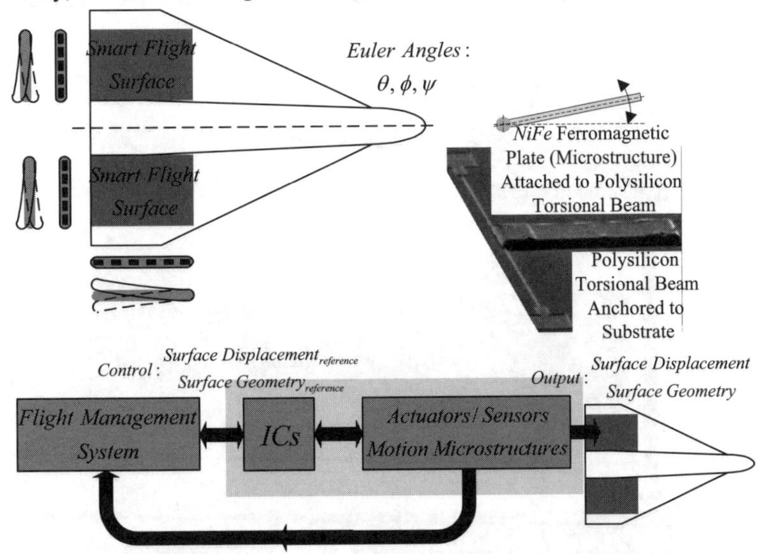


Figure 7.4.3 Flight vehicle with MEMS-based smart flight surfaces.

The MEMS array can be built using MEMS nodes, and a single node is illustrated in Figure 7.4.4.

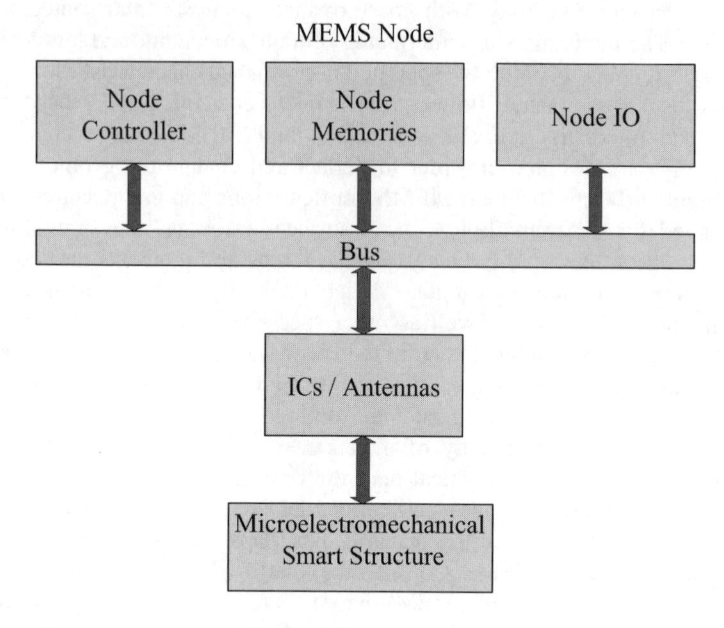


Figure 7.4.4 Elementary MEMS.

Three-layer hierarchically distributed closed-loop system architecture for a flight vehicle is documented in Figure 7.4.5.

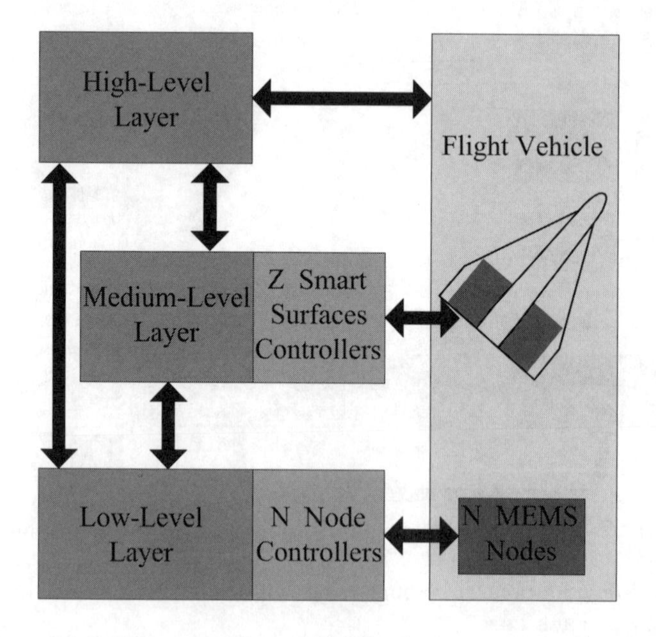


Figure 7.4.5 Three-layer hierarchical distributed closed-loop system.

To fabricate motion microstructures, micromachining, high-aspect-ratio, and CMOS technologies with photolithography process are applied. The continuous batch-fabrication process, which combines surface micromachining and thin film deposition has been widely used. The electrodeposit is patterned by selective deposition (additive process), not through etching. The electrodeposition rate is proportional to the current density, and an uniform current density over the entire seed layer is needed to guarantee a uniform thickness. To attain selective deposition, portions of the seed layer are covered with an insulating mask material that makes the current density in its proximity nonuniform. The *NiFe* thin film deposition (plating) is performed using the process reported in [4]. The ferromagnetic materials do not adhere well to silicon, and adhesion-promoting layers of titanium or chromium are deposited on the silicon surface prior to the ferromagnetic material. The need for adhesion layers, which lead to complexity, can be eliminated using a mechanical form of adhesion.

Using the desired thickness $T$, the density constant $d_c$, the Faraday constant $F_c$, the number of electrons $n_e$, the current density $J$ (which typically in the range from 2 to 20 mA/cm$^2$), and the molecular weight $m_w$, one calculates the time needed for electroplating as

$$\text{time} = \frac{T \times d_c \times F_c \times n_e}{J \times m_w}.$$

The current is found using the exposed seed layer area $A$. In particular,

$$i = A \times J$$

The layout and photolithography masks for ICs and microdevices are designed using the computer-aided design software (E-edit). Different microstructure shapes (plate, rectangular, spherical, etc.) can be synthesized. The sizing, spacing, accuracy, and other specifications are specified and relayed on the fabrication facilities used.

Results, reported in [4], demonstrate that the ferromagnetic microstructure attached to the polysicicon tensional beam can develop the electromagnetic torque and deflection angle up to 3 nN-m and 90 degrees, respectively. These electrostatic microstructures with ICs can be used in rotational-based smart flight control surfaces.

Alternatively, translational motion microstructures can be used. Thin film ceramic materials are increasingly being used in electromechanical motion devices, and lead-zirconate-titanate (PZT) thin films (affordable high torque density actuators and high-performance sensors) should be deposited on the silicon or electrode. The PZT thin films obtained from deposition can have degraded characteristics compared with the bulk PZT. In general, the designer attempts to augment motion microstructures with ICs, and therefore, a micromachined silicon technology can be the preferable solution. The MEMS node is fabricated on the Si substrate. The simplest solution is: a platinum (Pt) electrode is deposited on the substrate, followed by a piezoresistive layer of PZT, and the second Pt electrode is deposited on top. To actuate PZT actuators, one applies the voltage to the electrodes. The PZT-based microstructures can be used as actuators and sensors in smart flight control surfaces.

Affordable fabrication technologies were developed to manufacture PZT thin films with properties comparable to conventional piezoelectric ceramics. Many fabrication methods, e.g., evaporation, sputtering, sol-gel, and chemical vapor deposition were used to fabricate PZT thin films. The texture control of PZT thin films becomes increasingly important since their properties depend on the orientation. The orientation of $PbZr_{1-x}Ti_xO_3$ thin films affects the electromagnetic and mechanical properties such as: polarization-electric field (voltage), charge-displacement and strain (displacement)- electric field (voltage) curves, piezoelectric and dielectric constants, electromechanical coupling, bandwidth, capacitances, admittances, losses, elastic compliance, etc. The orientation of sol-gel PZT thin films is strongly affected by the orientation of the Pt electrode [5]. The $PbTiO_3$ thin films can be deposited on (111) or (100) textured Pt films through dc magnetron sputtering using $Ar/O_2$ gas and subsequent controlled annealing. The (111) and (100)/(101) $PbTiO_3$ film textures were obtained on (111)- and (200)-textured $Pt/SiO_2/Si$ substrates, respectively. Thus, the Pt electrode orientation is the major factor to control the orientation of PZT thin films. Oxygen incorporation into the Pt grain boundaries and lattice change deposition rate, resistivity, stress, and film orientation. Alternatively, Pt thin films with good adhesion can be deposited on $SiO_2/Si$ substrates by a two-step magnetron sputtering (in the first sputtering step, $Ar/O_2$ gas is used, and in the second step, Ar gas is applied). After two-step deposition, the annealing process can be performed at 600-1000$^\circ$C. During the first step, oxygen-containing Pt films were deposited. Oxygen incorporated in the Pt films diffused out during the high temperature annealing. After the annealing process, the films became dense without failures (hillock, pinhole, buckling, etc.). In general, sol-gel PZT films on $Pt(111)/Ti/SiO_2/Si$ are (100)-oriented, since the (100) plane is the lowest-energy surface. However, (111)-oriented PZT films on the same substrate obtained when the films were pyrolyzed at low temperatures (~350°C) for a short time (less than 5 min). The $PbZr_{.53}Ti_{.47}O_3$ thin films were deposited on Ti-free $Pt(111)/SiO_2/Si$ and $Pt(200)/SiO_2/Si$ substrates through the modified sol-gel process: $Pb(CH_3CO_2)_2H_2O$, $Zr(OC_4H_9)_4$, and $Ti(i\text{-}OC_3H_7)_4$ in 2-methoxyethanol. Pyrolysis temperature was varied from 320°C to 420°C. After the pyrolysis (takes from 1 to 30 min), the samples were annealed in the furnace at 670°C during 18 min. The films' texture and microstructure were examined using X-ray diffraction and scanning electron microscopy. Most PZT films showed (100) orientation on Pt(111) substrates, while (111) orientation was observed when the film was pyrolyzed for 1 min. The PZT films with (100) orientation were observed on Pt(200) substrates [5].

The surface-micromachined PZT-based microstructures, fabricated on silicon wafers, can be deposited using the sol-gel method. Sol-gel method offers precise control of the composition at the molecular level and low temperature processing. High-yield PZT films with the 100 to 4000 Å thickness per coating are made using methanol or isopropyl alcohol (nontoxic solvents). This fabrication procedure allows one to obtain the desired electromagnetic and mechanical properties: specified polarization-electric field (voltage) and strain (displacement)-electric field (voltage) curves, piezoelectric and dielectric constants, capacitances, admittances, low operating voltage, robustness, and reliability) of PZT thin films. Highly crystalline PZT thin films can be obtained by inserting the perovskite titanate as the seeding layers between the PZT layers. These seeding layers strongly affect the crystal growth and lattice matching. High-performance rotational and translational microstructures can be synthesized and optimized.

Describing the microstructures' fabrication processes, let us return to the control problem. Three-layer architecture of the hierarchically distributed flight management system integrates three layers: high, medium, and low. At the low level, the single MEMS-based node (actuator/sensor) is controlled using sensing data. The control can be generated using the PID or fuzzy logic control laws (single-input/single-output system) which are based on mathematical and logical (linguistic rules to describe the system operation) reasoning.

The simple tasks that can be performed by a low-level layer are

- Displace in the clockwise or counterclockwise directions (microactuator is actuated by supplying the voltage which is controlled by ICs, and the duty ratio of high-frequency transistors is controlled using the tracking error, decision variables, and events);
- Measure the displacement,
- Compare the command and actual displacements (obtain the tracking error),
- Diagnose and detect failures (For example, positive or negative values of the duty ratio, which is bounded by $\pm 1$, correspond to the clockwise or counterclockwise angular deflections, respectively; an increase of the duty ratio must lead to an increase of the current and electromagnetic field intensity).

Thus, we have a set of commands to attain the desired tasks and functions. The low-level layer is primarily responsible for actuation, sensing, simple analysis, diagnostics, and decision-making. The internal decision making mechanism and local diagnostics can be performed at a low level. The medium-level layer (which controls all control surfaces, e.g., left and right horizontal stabilizers, right and left leading and trailing flaps, etc.) coordinate the actions of hundreds of MEMS-based nodes (actuators/sensors). Control of all control surfaces is performed by a supervisory management system (high-level layer), which, in addition to microactuators management, integrates many other functions. The task analysis is accomplished by the high-level layer. In particular, based upon the information obtained from medium- and low-level layers, the high-level layer defines tasks (through intelligence, learning, evolution, analysis, adaptation, coordination, organization, decomposition, adaptive decision making with performance analysis and outcome prediction, diagnostics, etc.) to guarantee safety, flying and handling requirements, effectiveness, etc. The commands to displace the control surfaces are generated by the high-level layer based upon the overall analysis and high-level decision making. It must be emphasized that high-, medium-, and low-level layers communicate with each other, and the high-level layer possesses a key role.

Decision-making theory must be applied to develop and integrate key enabling methods, algorithms, and tools for the use in intelligent multinode MEMS. These intelligent systems must make optimal (robust) decision based upon the evolution strategies using specified requirements and priorities, monitoring (sensing) the external environment for entities of interest, recognizing those entities and then infer high-level attributes about those entities, etc. The closed-loop systems use the data from different sensors, feedback commands (controls) are generated and executed, and intelligent updates and evolution are performed. The feedback for sensor and control mechanisms are integrated, and particular emphasis is concentrated to gather the critical and essential data from the agents (nodes) of a greatest interest. Extensive information data must be constantly updated to guarantee a complete situation awareness, graduate evolution, and intelligence using performance analysis, outcome evaluation, prediction, and assessment. Thus, qualitative and quantitative analysis is performed to study the overall system evolution. To perform the inferences required, to develop an assessment of the current situation, and to predict performance and outcomes, extensive knowledge and information about MEMS and environment are needed.

A multiple learning concept can be implemented to design high-, medium-, and low-level closed-loop systems. Agents (nodes) exhibit complex behavior which can be optimized using low-level evolutionary decision-making subsystems which use learning algorithms. Reinforcement learning can be performed based upon the prioritized objectives through upper level decision-making. The agents' behavior and performance are analyzed by the high-level layer to collect and assess the evidence data. Decision trees are commonly used to provide a comprehensive set of strategies, simplify and improve (optimize) them, attain robustness and comprehensibility, and make the final decision. The low-level subsystems can perform the following functions: sensing, actuation, recognition, local diagnostics, local assessment, and local prediction with decision making.

Hierarchical distributed closed-loop systems can have different organization (architecture), and the number of layers is based upon the complexity of multinode MEMS to be controlled. The level of control hierarchy is defined by hardware and software complexity (rate of tasks completion, rate of continuous/discrete events, bandwidth, sampling time, update rate, etc.), as well as by the overall specifications and requirements imposed. Complex problems and tasks can be logically decomposed into simpler subproblems and subtasks which are easy to understand, support, and implement using *the state table of rules* (for each rule, the actuator or sensor action and operation are determined using *the system state, decision, event, and performance variables*). These subtasks must be performed in the defined sequence scenarios that lead to the desired operation, and the architecture is synthesized. Usually, a low-level subsystem is designed for each MEMS node (actuator/sensor) at the lower level of the hierarchy, and the layer level is defined based upon the overall objectives, analytical and numerical complexity of problems, information flow, etc. Thus, the complexity gradually arises form subsystems design (to control single node), to the synthesis of closed-loop architecture (layers).

Different intelligent concepts can be applied. For example, neural networks allow the designer to:

- Approximate unknown functions (function approximator neural networks);
- Generalize control vector (control neural networks).

The backpropagation method is applied for training multilayer perceptron neural networks [6, 7].

The multi-input/single-output neuron output is given by

$$u = f(Wv + B_1),$$

where $u$ is the neuron output, $u \in \mathbb{R}^1$; $f$ is the nonlinear function (*transfer function*); $W$ is the weighting matrix, $W = \lfloor w_{11}, w_{12}, \ldots, w_{1k-1}, w_{1k} \rfloor$, $W \in \mathbb{R}^{1\times k}$; $v$ is the input vector (performance variables), $v \in \mathbb{R}^k$; $B_1$ is the bias variable.

It should be emphasized that $W$ and $B_1$ are adjusted through the training (learning) mechanism.

For a single-layer neural network of $z$ neurons, one has

$$u = f(Wv + B),$$

where the weighting matrix and bias vector are $W \in \mathbb{R}^{z\times k}$ and $B \in \mathbb{R}^z$.

For a multilayer neural network of $z$ neurons, one can find the following expression for the $(i + 1)$ network outputs

$$u_{i+1} = f_{i+1}(W_{i+1}u_i + B_{i+1}), i = 0, 1, \ldots, M-2, M-1,$$

where $M$ is the number of layers in the neural network.

For example, for three-layer network, we have

$$u_3 = f_3(W_3u_2 + B_3), i = 2, \ u_2 = f_2(W_2u_1 + B_2), i = 1,$$

and $u_1 = f_1(W_1v + B_1), i = 0$.

Hence, one obtains

$$u_3 = f_3(W_3u_2 + B_3) = f_3[W_3f_2(W_2f_1(W_1v + B_1) + B_2) + B_3],$$

where the corresponding subscripts 1, 2 and 3 are used to denote the layer variables.

To approximate the unknown functions, weighting matrix $W$ and the bias vector $B$ must be determined, and the procedure for selecting $W$ and $B$ is called the network training. Many concepts are available to attain training, and backpropagation, which is based upon the gradient descent optimization methods, is commonly used. Applying the gradient descent optimization procedure, one minimizes a mean square error performance index using the end-to-end neural network behavior. That is, using the inputs vector $v$ and the output vector $c$, $c \in \mathbb{R}^k$, the quadratic performance functional is given as [6, 7]

$$J = \sum_{j=1}^{p} (c_j - u_j)^T Q (c_j - u_j) = \sum_{j=1}^{p} e_j^T Q e_j ,$$

where $e_j = c_j - u_j$ is the error vector; $Q \in \mathbb{R}^{p \times p}$ is the diagonal weighting matrix.

The steepest descent algorithm is applied to approximate the mean square errors, and the learning rate and sensitivity have been widely studied for the quadratic performance indexes.

## 7.5 HAMILTON-JACOBI THEORY AND QUANTUM MECHANICS

This section studies several key problems in modeling, analysis, control, and optimization of nanosystems. From the optimization perspectives, it is illustrated that the Schrödinger equation can be derived using Hamilton-Jacobi principle [8]. The importance of the results is that the Schrödinger equation was obtained from the closed-loop solution through optimization of the functional. This establishes the relationship between the Hamilton-Jacobi theory and quantum mechanics.

The Hamiltonian for a particle is

$$\underset{v,p}{\Delta} \int_{t_0}^{t_f} [pv - H(t,x,p)] dt = 0 ,$$

where $\underset{v,p}{\Delta}$ is the variation of the succeeding expression with respect to $v$ and $p$.

Using the optimal (stationary) value of the integral

$$\int_{t_0}^{t_f} [pv - H(t,x,p)] dt ,$$

as denoted by $V(t,x)$, one has

$$\underset{v,p}{S} \left[ \frac{dV}{dt} + pv - H(t,x,p) \right] = \underset{v,p}{S} \left[ \frac{\partial V}{\partial t} + \frac{\partial V}{\partial x} v + pv - H(t,x,p) \right] = 0 ,$$

where $\underset{v,p}{S}$ is the optimal (stationary) value obtained by varying $v$ and $p$.

Therefore, we have

$$p = -\frac{\partial V}{\partial x},$$

$$\dot{p} = -\frac{\partial V}{\partial x}\frac{\partial V}{\partial t} - v\frac{\partial^2 V}{\partial x^2} = -\frac{\partial H}{\partial x} + \frac{\partial H}{\partial p}\frac{\partial p}{\partial x},$$

$$v = \frac{\partial H}{\partial p},$$

$$H = \frac{\partial V}{\partial t}.$$

The variables $x$ and $p$ are independent, and therefore, the Hamiltonian equations of motion

$$\dot{p} = -\frac{\partial H}{\partial x},$$

$$\dot{x} = \frac{\partial H}{\partial p}$$

results.

The complex functions are used in quantum mechanics, and we replace the displacement $x$ by the complex variable $q$, and instead of velocity $v$ we have $dq = vdt + ndz$, where $n$ is complex, $n = -i\hbar\frac{1}{m}$, and $z$ is the white noise (normalized Wiener process).

Therefore, one obtains

$$\underset{v,p}{\Delta}\, E\int_{t_0}^{t_f}[pv - H(t,q,p)]dt = 0,$$

where $E$ denotes the expectation.

Making use of the variational principle, from

$$Edz = 0,$$

$$Edz^2 = dt,$$

$$Edq^2 = n^2 dt,$$

we have

$$\underset{v,p}{S}\left[\frac{dV}{dt} + pv + v\frac{\partial V}{\partial q} + \tfrac{1}{2}n^2\frac{\partial^2 V}{\partial q^2} - H(t,q,p)\right] = 0.$$

The minimization gives the following Hamiltonian equations

$$p = -\frac{\partial V}{\partial q},$$

$$v = \frac{\partial H}{\partial p},$$

$$H = \frac{\partial V}{\partial t} - \frac{i\hbar}{2m}\frac{\partial^2 V}{\partial q^2}.$$

Letting

$$V = i\hbar \log \Psi$$

and taking note of

$$p = -i\hbar \frac{\partial}{\partial q},$$

one finds

$$p = -\frac{\partial V}{\partial q} = \Psi^{-1} p \Psi,$$

$$i\hbar \frac{\partial^2 V}{\partial q^2} = \Psi^{-1} p^2 \Psi - p^2.$$

Thus, we obtain the Schrödinger equation

$$E\Psi = i\hbar \frac{\partial}{\partial t}\Psi(t,q) = \left[-\frac{\hbar^2}{2m}\frac{\partial^2}{\partial q^2} + \Pi(q)\right]\Psi(t,q)$$

$$= \left[\frac{p^2}{2m} + \Pi(q)\right]\Psi(t,q) = H\Psi.$$

The wavefunction $\Psi = e^{\frac{V}{i\hbar}}$ is the solution of the Schrödinger equation. The Schrödinger equation was derived using Hamilton-Jacobi principle. Furthermore, it was illustrated that the Schrödinger equation was found minimizing the functional. It is very important that the Schrödinger equation was derived from the closed-loop solution through optimization of the functional because it establishes the relationship between the Hamilton-Jacobi theory and quantum mechanics. The Schrödinger equation leads to the solution in the form of wavefunctions, while the Hamiltonian-Jacobi concept results in the optimal cost function. Furthermore, it was shown that the Schrödinger equation results form a closed-loop solution of $\underset{v,p}{\Delta}\, E \int_{t_0}^{t_f} [pv - H(t,q,p)]dt = 0$, and the closed-loop solution represents the goal-seeking behavior (dynamics) of nature.

## References

1. S. E. Lyshevski, *Control Systems Theory With Engineering Applications*, Birkhäuser, Boston, MA, 2001.
2. S. E. Lyshevski, *Nano- and Microelectromechanical System: Fundamentals of Nano- and Microengineering*, CRC Press, Boca Raton, FL, 2000.
3. S. E. Lyshevski and M. G. Safonov, "Intelligent motion control for electromechanical servos using evolutionary learning and adaptation mechanisms," *Proc. American Control Conf*, Arlington, VA, 2001.
4. J. W. Judy and R. S. Muller, "Magnetically actuated, addressable microstructures," *J. Microelectromechanical Systems,* vol. 6, no. 3, pp. 249-256, 1997.
5. D. S. Lee, D. Y. Park, H. J. Woo, J. Ha and E. Yoon, "Preferred orientation control of platinum thin films deposited by dc magnetron sputtering using $Ar/N_2$ gas mixture," *Proc. MRS Fall Meeting*, Boston, MA, 1999.
6. S. Haykin, *Neural Networks: A Comprehensive Foundation*, Prentice Hall, Upper Saddle River, NJ, 1999.
7. K. S. Narendra, "Neural networks for control: theory and practice," *Proc. IEEE*, vol. 84, no. 10, pp. 1385-1406, 1385.
8. H. H. Rosenbrock, "A stochastic variational principle for quantum mechanics," *Phys. Letters*, vol. 100A, pp. 343-346, 1986.

# CHAPTER 8

# CASE STUDIES: SYNTHESIS, ANALYSIS, FABRICATION, AND COMPUTER-AIDED-DESIGN OF MEMS

## 8.1 INTRODUCTION

In many applications (from medicine and biotechnology to aerospace and security), the use of nano- and microscale structures, devices and systems is very important [1-4]. This chapter discusses the analysis, modeling, design, and fabrication of electromagnetic-based microscale structures, devices and MEMS (microtransducers controlled by ICs). It is obvious that to attain our objectives and goals, the synergy of multidisciplinary engineering, science, and technology must be utilized. In particular, electromagnetic theory and mechanics comprise the fundamentals for analysis, modeling, simulation, design, and optimization, while fabrication is based on the micromachining and high-aspect-ratio techniques and processes. As was illustrated in Chapter 3, micromachining is the extension of the CMOS technologies developed to fabricate ICs. For many years, the developments in MEMS have been concentrated on the fabrication of microstructures adopting, modifying, and redesigning silicon-based technologies commonly used in integrated microelectronics. The reason for refining of conventional microelectronic technologies, development of novel techniques, design of new processes, as well as application of new materials and chemicals is simple: in general, microstructures are three-dimensional with high aspect ratios and large structural heights in contrast to two-dimensional (planar) microelectronic devices.

It was documented in Chapter 3 that various silicon structures can be made through bulk silicon micromachining (using wet or dry processes) or through surface micromachining. Metallic micromolding techniques, based upon photolithographic processes, are also widely used to fabricate microstructures. Molds are created in polymer films (usually photoresist) on planar surfaces, and then filled by electrodepositing metal (electrodeposition plays a key role in the fabrication of the microstructures and microdevices which are the components of MEMS). High aspect ratio technologies use optical, e-beam and x-ray lithography to create the millimeter-range high trenches made deep in polymethylmethacrylate resist on the electroplating base (called seed layer). Electrodeposition of magnetic materials and conductors, plating, etching and lift-off are extremely important processes to fabricate microscale structures and devices.

Although it is recognized that the ability to use and refine existing microelectronics fabrication technologies, techniques and materials is very important, and the development of novel high-yield processes to fabricate MEMS is a key factor in the rapid growth of affordable MEMS, other emerging areas arise. In particular, devising (synthesis), design, modeling, analysis, and optimization of novel MEMS are extremely important. Therefore, recently, the MEMS theory and microengineering fundamentals have been expanded to thoroughly study other critical problems such as system-level synthesis, computer-aided design, integration, synergetic classification, analysis, modeling, prototyping, optimization, and simulation. This chapter covers the fabrication, analysis, and design problems for electromagnetic microstructures, microdevices and MEMS (micro-transducers with ICs). The descriptions of the fabrication processes are given, modeling and analysis issues are emphasized, and the design is performed. Finally, the computer-aided design issues are addressed.

## 8.2 DESIGN AND FABRICATION

In MEMS, the fabrication of thin film magnetic microstructures require deposition of conductors, insulators, magnetic and other materials with the specified properties needed to attain the goals, e.g., design efficient and reliable MEMS capable to perform the desired functions and tasks. To fabricate MEMS and NEMS, different materials are used. Some bulk material constants (conductivity $\sigma$, resistivity $\rho$ at $20^0$C, relative permeability $\mu_r$, thermal expansion $t_e$ and dielectric constant–relative permittivity $\varepsilon_r$) in the SI units are given in Table 8.1.

Table 8.1 Material constants.

| Material | Silver | Copper | Gold | Al | Tungsten | Zinc | Nickel | Iron |
|---|---|---|---|---|---|---|---|---|
| $\sigma$ | $6.2\times10^7$ | $5.8\times10^7$ | $4.1\times10^7$ | $3.8\times10^7$ | $1.82\times10^7$ | $1.67\times10^7$ | $1.45\times10^7$ | $1\times10^7$ |
| $\rho$ | $1.6\times10^{-8}$ | $1.7\times10^{-8}$ | $2.4\times10^{-8}$ | $2.6\times10^{-8}$ | $5.5\times10^{-8}$ | $6\times10^{-8}$ | $6.9\times10^{-8}$ | $1\times10^{-7}$ |
| $\mu_r$ | 0.9999998 | 0.99999 | 0.99999 | 1.000001 | N/a | N/a | 600 nonlinear | 3000-5000 nonlinear |
| $t_e\times10^{-6}$ | N/a | 16.7 | 14 | 25 | N/a | N/a | N/a | N/a |

| Material | Si | $SiO_2$ | $Si_3N_4$ | SiC | GaAs | Ge |
|---|---|---|---|---|---|---|
| $\varepsilon_r$ | 11.8 | 3.8 | 7.6 | 6.5 | 13 | 16.1 |
| $t_e\times10^{-6}$ | 2.65 | 0.51 | 2.7 | 3.0 | 6.9 | 2.2 |

Although MEMS and microdevices topologies and configurations vary (see the MEMS classification concept documented in Chapter 4), in general, electromagnetic microtransducers have closed-ended, open-ended, and integrated electromagnetic systems. As examples, Figure 8.1 illustrates the microtoroid and the linear micromotor with the closed-ended and open-ended electromagnetic systems, respectively. The copper microwindings and ferromagnetic cores (microstructures made using different magnetic materials) can be fabricated through electroplating, patterning, planarization, and other fabrication processes. Figure 8.1 depicts the electroplated circular copper conductors which form the windings (10 μm wide and thick with 10 μm spacing) deposited on the insulated layer of the ferromagnetic core.

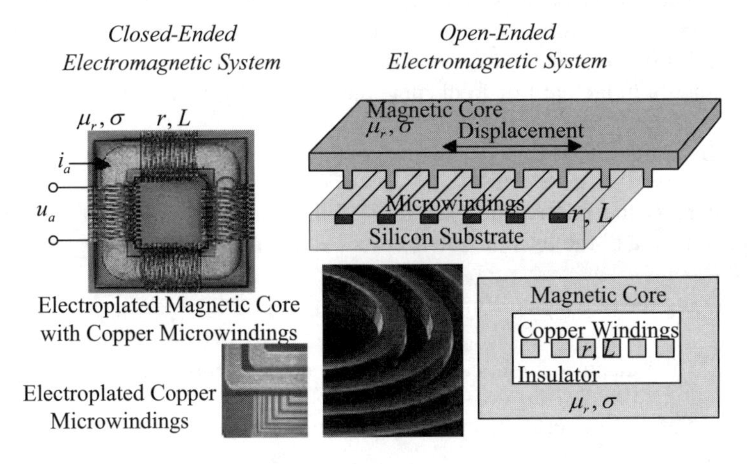


Figure 8.1 Closed-ended and open-ended electromagnetic systems in microtransducers (toroidal microstructure with the insulated copper circular conductors wound around the magnetic material and linear micromotor) with ferromagnetic cores (electroplated thin films).

The comprehensive electromagnetic analysis must be performed for microscale structures, devices, and systems. For example, the torque (force) developed and the voltage induced by microtransducers depend upon the inductance, and the microdevice efficiency is a function of the winding resistance (resistivity of the coils deposited vary), eddy currents, hysteresis, etc. Studying the microtoroid, consider a circular path of radius $R$ in a plane normal to the axis. The magnetic flux intensity is calculated using the following formula

$$\oint_s \mathbf{H} \cdot d\mathbf{s} = 2\pi RH = Ni,$$

where $N$ is the number of turns.

Thus, one has $H = \dfrac{Ni}{2\pi R}$.

The value of $H$ is a function of the $R$, and therefore, the field is not uniform.

Microwindings must guarantee the adequate inductance in the limited footprint area with the minimal resistance. For example, in the microscale transducers and power converters, 0.5 μH (or higher) inductance is required at high frequency (1–10 MHz). Compared with the conventional minidevices, thin film electromagnetic microtransducers have lower efficiency due to higher resistivity of microcoils (thin films), eddy currents, hysteresis, fringing effect, and other undesirable phenomena which usually have the secondary (negligible) effects in the miniscale and conventional electromechanical devices. The inductance can be increased by ensuring a large number of turns, using core magnetic materials with high relative permeability, increasing the cross-sectional core area, and decreasing the path length. In fact, at low frequency, the formula for inductance is

$$L = \frac{\mu_0 \mu_r N^2 A}{l},$$

where $\mu_r$ is the relative permeability of the core material; $A$ is the cross-sectional area of the ferromagnetic core; $l$ is the magnetic path length.

Using the reluctance $\Re = \dfrac{l}{\mu_0 \mu_r A}$, one has $L = \dfrac{N^2}{\Re}$.

For the electromagnetic microtransducers, the flux is a very important variable of our interest. Using the *net* current, one has $\Phi = \dfrac{Ni}{\Re}$.

It is important to recall that the inductance is related to the energy stored in the magnetic field, and $L = \dfrac{2W_m}{i^2} = \dfrac{1}{i^2}\int_v \mathbf{B}\cdot\mathbf{H}dv$. Thus, one has

$$L = \frac{1}{i^2}\int_v \mathbf{B}\cdot\mathbf{H}dv = \frac{1}{i^2}\int_v \mathbf{H}\cdot(\nabla\times\mathbf{A})dv = \frac{1}{i^2}\int_v \mathbf{A}\cdot\mathbf{J}dv = \frac{1}{i}\oint_l \mathbf{A}\cdot d\mathbf{l}$$

$$= \frac{1}{i}\oint_s \mathbf{B}\cdot d\mathbf{s} = \frac{\Phi}{i},$$

or $L = \dfrac{N\Phi}{i}$.

We found that the inductance is the function of the number of turns, flux, and current.

Making use of the equation $L = \dfrac{\mu_0 \mu_r N^2 A}{l}$, one concludes that the inductance increases as a function of the squared number of turns. However, a large number of turns requires the high turn density fabrication (small track width and spacing in order to place many turns in a given footprint area). However, reducing the track width leads to increase of the conductor resistance decreasing the efficiency. Hence, the trade-off between inductance and winding resistance must be studied. The dc resistance is found as

$$R = \rho_c \frac{l_c}{A_c},$$

where $\rho_c$ is the conductor resistivity which is a nonlinear function of thickness; $l_c$ is the conductor length; $A_c$ is the conductor cross-sectional area.

To achieve low resistance, one must deposit thick conductors with the thickness in the order of tens of micrometers. Therefore, the most feasible process for deposition of conductors is electroplating. High-aspect-ratio processes ensure thick conductors and small track widths and spaces (high-aspect-ratio conductors have a high thickness to width ratio). However, the footprint area is limited not allowing one to achieve a large conductor cross-sectional area. High inductance value can also be achieved increasing the ferromagnetic core cross-sectional area using thick magnetic cores with large $A$. However most magnetic materials used are thin film metal alloys, which generally have characteristics not as good as the bulk ferromagnetic materials commonly used in electromechanical and electromagnetic devices. These result in the eddy current and undesirable hysteresis effects which increase the core losses and decreases the inductance. It should be emphasized that eddy currents must be minimized. In addition, the size of MEMS is usually specified, and therefore, the limits on the maximum cross-sectional area are imposed.

As illustrated, ferromagnetic cores and microwindings are key components of microstructures, and different magnetic and conductor materials and processes to fabricate microtransducers are employed. Commonly, the *permalloy* (nickel $Ni_{80\%}$ – iron $Fe_{20\%}$ alloy) thin films are used. It should be emphasized that *permalloy* as well as other alloys (e.g., $Ni_{x\%}Fe_{100-x\%}$, $Ni_{x\%}Co_{100-x\%}$,, $Ni_{x\%}Fe_{y\%}Mo_{100-x-y\%}$, amorphous cobalt-phosphorous, etc.) are soft magnetic materials that can be made through electrodeposition and other deposition processes. In general, the deposits have nonuniform composition and nonuniform thickness due to the electric current nonuniformity over the electrodeposition area and other phenomena. For example, hydrodynamic effects in the electrolyte increase nonuniformity (these nonuniformities are reduced by applying specific electrochemicals). The inductance and losses remain constant up to a certain frequency (which is a function of the layer thickness, materials, fabrication processes, etc.). In the high frequency operating regimes, the inductance rapidly decreases and the losses increase due to the eddy current and hysteresis effects. For example, for the microinductor fabricated using *permalloy* ($Ni_{80\%}Fe_{20\%}$) thin film ferromagnetic core with copper microwindings, the inductance decreases rapidly above 1 MHz, 3 MHz and 6 MHz for the 10 μm, 8 μm and 5 μm thick layers, respectively. It should be emphasized that the skin depth of the ferromagnetic core thin films is a function of the magnetic properties and the frequency $f$. In particular, the skin depth is found as

$$\delta = \sqrt{\frac{1}{\pi f \mu \sigma}},$$

where $\mu$ and $\sigma$ are the permeability and conductivity of the ferromagnetic core material.

The total power losses are found using the Pointing vector $\mathbf{\Xi} = \mathbf{E} \times \mathbf{H}$, and the total power loss can be approximately derived using the expression for the power crossing the conductor surface within the area, e.g.,

$$P_{\text{average}} = \int_s \Xi_{\text{average}} ds = \tfrac{1}{4} \int_s \sigma \delta E_o^2 e^{-2/\delta \sqrt{\pi f \mu \sigma}} ds .$$

It is important to emphasize that the skin depth (depth of penetration) is available. For the bulk copper as well as for copper microcoils with thickness greater than 10 μm deposited using optimized processes, we have

$$\delta_{Cu} = \frac{0.066}{\sqrt{f}} .$$

In general, the inductance begins to decrease when the ratio of the lamination thickness to skin depth is greater than one. Thus, the lamination thickness must be less than skin depth at the operating frequency $f$ to attain the high inductance value. In order to illustrate the need to comprehensively study microinductors, we analyze the toroidal microinductor (1 mm by 1 mm, 3 μm core thickness, 2000 permeability). The inductance and winding resistance are analyzed as the functions of the operating frequency. Modeling results indicate that the inductance remains constant up to 100 kHz and decreases for the higher frequency. The resistance increases significantly at the frequencies higher than 150 kHz (the copper microconductor thickness is 2 μm, and the dc winding resistance is 10 ohm). The decreased inductance and increased resistance at high frequency are due to hysteresis and eddy current effects. Therefore, computer-aided design (CAD) packages and MEMS computational environments must integrate nonlinear phenomena and secondary effects to attain high-fidelity analysis and simulation.

The skin depth of the magnetic core materials depend on permeability and conductivity. The $Ni_{x\%}Fe_{100-x\%}$ thin films have a relative permeability in the range from 600 to 2000, and the resistivity is from 18 to 23 μohm-cm. It should be emphasized that the materials with high resistivity have low eddy current losses and allow one to deposit thicker layers as the skin depth is high. Therefore, high resistivity magnetic materials are under consideration, and the electroplated FeCo thin films have 100-130 μohm-cm resistivity. Other high resistivity materials (deposited by sputtering) are FeZrO and CoHfTaPd. In general, sputtering has advantages for the deposition of laminated layers of magnetic and insulating materials because magnetic and insulating composites can be deposited in the same process step. Electroplating, as a technique for deposition of laminated multilayered structures, requires different processes to deposit magnetic and insulating materials (layers).

Three elements that are ferromagnetic at room temperature are iron, nickel, and cobalt. Ferromagnetic materials lose their ferromagnetic characteristics above the Curi temperature, which for iron is 770$^0$C (1043$^0$K). Correspondingly, alloys of these metals (and other compounds) are also ferromagnetic, e.g., alnico (aluminum–nickel–cobalt alloy with small amount of copper). Some alloys of nonferromagnetic metals are ferromagnetic, for example, bismuth–manganese and copper–manganese–tin alloys. At low temperature, some rare-earth elements (gadolinium, dysprosium, and others) are ferromagnetic.

Ferrimagnetic materials (which have smaller response on an external magnetic field than ferromagnetic materials) are also examined. Ferrites (group of ferromagnetic materials) have low conductivity. To fabricate microstructures and devices with low eddy currents, iron oxide magnetite ($Fe_3O_4$) and nickel ferrite ($NiFe_2O_4$) are used.

The fabrication of ICs, which are one of the major functional components of MEMS, are reported in the literature [1]. The goal of this chapter is to examine microscale structures and devices (microtransducers) controlled by ICs. Therefore, the fabrication of microstructures and microtransducers should be covered. Microtransducers (which integrate stationary and movable ferromagnetic cores, windings, bearing, etc.) are much more complex compared with microstructures. Therefore, the fabrication of microinductors are covered first. The major processes involved in the electromagnetic microtransducers and microinductors fabrication are etching and electroplating magnetic vias and through-holes, and then fabricating the inductor-type microstructures on top of the through-hole wafer using multi-layer thick photoresist processes [5-7]. For example, let us use the silicon substrate (100-oriented n-type double-sided polished silicon wafers) with a thin layer of thermally grown silicon dioxide ($SiO_2$). Through-holes are patterned on the topside of the Si-$SiO_2$ wafer (photolithography process), and then, etched in the KOH system (different etch rate can be achieved based upon the concentration and temperature). The wafer then is removed from the KOH system with 20-30 μm of silicon remaining to be etched. A Ti-Cu or Cr-Cu seed layer (20-40 nm and 400-500 nm thickness, respectively) is deposited on the backside of the wafer using electron beam evaporation. The copper acts as the electroplating seed layer, while a titanium (or chromium) layer is used to improve adhesion of the copper layer to the silicon wafer. On the copper seed layer, a protective NiFe thin film layer is electroplated directly above the through-holes to attain protection and stability. The through-holes are fully etched again (in the KOH system). Then, the remaining $SiO_2$ is stripped (using the BHF solution) to reveal the backside metal layers. Then, the titanium adhesion layer is etched in the HF solution (chromium, if the Cr-Cu seed layer is used, can be removed using the $K_3Fe(CN)_6$–NaOH solution). This allows the electroplating of through-holes from the exposed copper seed layer. The empty through-holes are electroplated with the NiFe thin film. This forms the magnetic vias. Because the KOH-based etching process is crystallographically dependent, the sidewalls of the electroplating mold are the 111-oriented crystal planes ($54.7^0$ angular orientation to the surface). As a result of these $54.7^0$-agularly oriented sidewalls, the electroplating can be nonuniform. To overcome this problem, the through-holes can be over-plated and polished to the surface level [5-7]. After the through-hole overplating and polishing, the seed layer is removed, and 10-20 μm coat (e.g., polyimide PI2611) is spun on the backside and cured at $300^0$C to cover the protective NiFe layer. At this time, the microinductor can be fabricated on the topside of the wafer. In particular, the microcoils are fabricated on top of the through-hole wafer with the specified ferromagnetic core geometry (e.g., plate, toroidal, horseshoe or other shapes) parallel to the surface of the wafer. The microcoils must be wound around the ferromagnetic core to form the electromagnetic system. Therefore, the additional structural layers are needed (for example, the first level is the conductors that are the bottom segments of each microcoil turn, the second level includes the ferromagnetic core and vertical conductors which connect the top and bottom of each microcoil turn segment, and the third level consists the top conductors that are connected to the electrical vias, and thus form microcoil turns wound around the ferromagnetic core). It is obvious that the insulating (dielectric) layers are required to insulate the core and microcoils. The fabrication can be performed through the electron beam evaporation of the Ti-Cu seed layer, and then, 25-35 μm electroplating molds are formed (AZ-4000 photoresist can be used). The copper microcoils are electroplated on the top of the mold through electroplating. After electroplating is completed, the photoresist is removed with acetone. Then, the seed layer is removed (copper is etched in the $H_2SO_4$ solution, while the titanium adhesion layer is etched by the HF solution). A new layer of the AZ-4000 photoresist is spun on the wafer to insulate the bottom conductors from the core. The vias openings are patterned at the ends of the conductors, and the photoresist is cured forming the insulating layer. In addition to insulation, the hard curing leads to re-flow of the photoresist serving the planarization purpose needed to pattern additional layers. Another seed layer is deposited from which electrical vias and ferromagnetic core are patterned and electroplated. This leads to two lithography sequential steps, and the electrical vias (electroplated Cu) and ferromagnetic core (NiFe thin film) are electroplated using the same seed layer. After the vias and ferromagnetic core are completed, the photoresist and seed layers are removed. Then, the hard curing is performed. The top microconductors are patterned and deposited from another seed layer using the same process as explained above for the bottom microconductors. The detailed description of the processes described and the fabricated microtransducers are available in [5-7]. We have outlined the fabrication of microinductors because this technique can be adopted and used to fabricate microtransducers. It also must be emphasized that the analysis and design must be performed using the equations given. That is, mathematical modeling and analysis should be performed before fabrication.

## 8.3 ANALYSIS OF TRANSLATIONAL MICROTRANSDUCERS

Figure 8.2 illustrates a microelectromechanical device (translational microtransducer) with a stationary member (ferromagnetic core with windings) and moveable member (microplunger) which can be fabricated using the micromachining technologies. Our goal is to perform the analysis and modeling of the microtransducer developing the lumped-parameter mathematical model. That is, the goal is to derive the differential equations which model the microtransducer steady-state and dynamic behavior.

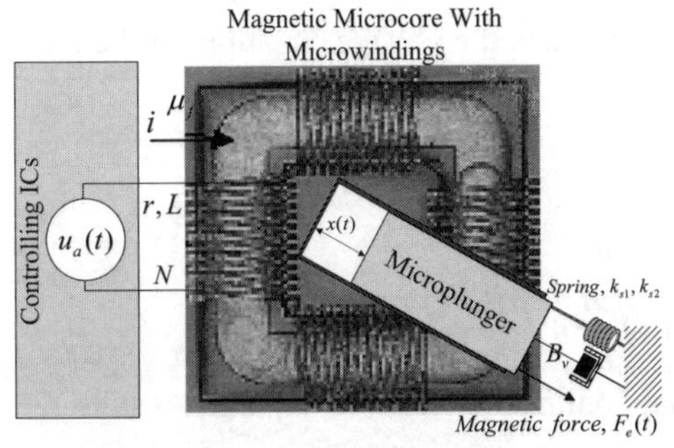


Figure 8.2 Schematic of the microtransducer with controlling ICs.

Applying Newton's second law for translational motion, we have

$$F(t) = m\frac{d^2x}{dt^2} + B_v\frac{dx}{dt} + (k_{s1}x + k_{s2}x^2) + F_e(t),$$

where $x$ denotes the microplunger displacement; $m$ is the mass of a movable member (microplunger); $B_v$ is the viscous friction coefficient; $k_{s1}$ and $k_{s2}$ are the spring constants; $F_e(t)$ is the magnetic force, $F_e(i,x) = \frac{\partial W_c(i,x)}{\partial x}$.

It should be emphasized that the restoring/stretching force exerted by the spring is given by $(k_{s1}x + k_{s2}x^2)$.

Assuming that the magnetic system is linear, the coenergy is

$W_c(i,x) = \frac{1}{2}L(x)i^2$.

Thus, the electromagnetic force developed is found to be

$$F_e(i,x) = \frac{1}{2}i^2\frac{dL(x)}{dx}.$$

In this formula, the analytic expression for the term $dL(x)/dx$ must be derived. The inductance, as a nonlinear function of the displacement, is

$$L(x) = \frac{N^2}{\Re_f + \Re_g} = \frac{N^2\mu_f\mu_0A_fA_g}{A_gl_f + A_f\mu_f(x+2d)},$$

where $\Re_f$ and $\Re_g$ are the reluctances of the magnetic material and air gap; $A_f$ and $A_g$ are the cross-sectional areas; $l_f$ and $(x + 2d)$ are the lengths of the magnetic material and the air gap.

Therefore, $\frac{dL}{dx} = -\frac{N^2\mu_f^2\mu_0A_f^2A_g}{[A_gl_f + A_f\mu_f(x+2d)]^2}$, and hence

$$F_e(i,x) = -\frac{N^2 \mu_f^2 \mu_0 A_f^2 A_g}{2[A_g l_f + A_f \mu_f (x+2d)]^2} i^2 .$$

Using Kirchhoff's law, the voltage equation for the electric circuit is

$$u_a = ri + \frac{d\psi}{dt},$$

where the flux linkage $\psi$ is given as $\psi = L(x)i$ .

Thus, one obtains

$$u_a = ri + L(x)\frac{di}{dt} + i\frac{dL(x)}{dx}\frac{dx}{dt}.$$

The following nonlinear differential equation results

$$\frac{di}{dt} = -\frac{r}{L(x)} i + \frac{1}{L(x)} \frac{N^2 \mu_f^2 \mu_0 A_f^2 A_g}{[A_g l_f + A_f \mu_f (x+2d)]^2} iv + \frac{1}{L(x)} u_a .$$

Augmenting this equation with the differential equation with the *torsional-mechanical* dynamics

$$F(t) = m\frac{d^2 x}{dt^2} + B_v \frac{dx}{dt} + (k_{s1}x + k_{s2}x^2) + F_e(t),$$

three nonlinear differential equations for the considered translational microtransducer are found to be

$$\frac{di}{dt} = -\frac{r[A_g l_f + A_f \mu_f (x+2d)]}{N^2 \mu_f \mu_0 A_f A_g} i + \frac{\mu_f A_f}{A_g l_f + A_f \mu_f (x+2d)} iv + \frac{A_g l_f + A_f \mu_f (x+2d)}{N^2 \mu_f \mu_0 A_f A_g} u_a,$$

$$\frac{dx}{dt} = v,$$

$$\frac{dv}{dt} = \frac{N^2 \mu_f^2 \mu_0 A_f^2 A_g}{2m[A_g l_f + A_f \mu_f (x+2d)]^2} i^2 - \frac{1}{m}(k_{s1}x + k_{s2}x^2) - \frac{B_v}{m} v.$$

The derived differential equations represent the lumped-parameter mathematical model of the microtransducer. In general, the high-fidelity mathematical modeling and analysis must be performed integrating nonlinearities (for example nonlinear magnetic characteristics and hysteresis) and secondary effects. However, the lumped-parameter mathematical models as given in the form of nonlinear differential equations have been validated for microtransducers. It is found that the major phenomena and effects are modeled for the current, velocity and displacement (secondary effects such as Coulomb friction, hysteresis and eddy currents, fringing effect and other phenomena have not been modeled and analyzed). At low frequency (up to 1 MHz), the lumped-parameter modeling provides one with the capabilities to attain reliable preliminary steady-state and dynamic analysis using primary circuitry and mechanical variables. It is also important to emphasize that the voltage, applied to the microwinding, is regulated by ICs. The majority of ICs to control microtransducers are designed using the pulse-width-modulation topologies. The switching frequency of ICs is usually 1 MHz or higher. Therefore, as was shown, it is very important to study the microtransducer performance at the high operating frequency. This can be performed using Maxwell's equations which will lead to the high-fidelity mathematical models [3].

## 8.4 SINGLE-PHASE RELUCTANCE MICROMOTORS: MODELING, ANALYSIS, AND CONTROL

Consider the single-phase reluctance micromachined micromotors as illustrated in Figure 8.3.

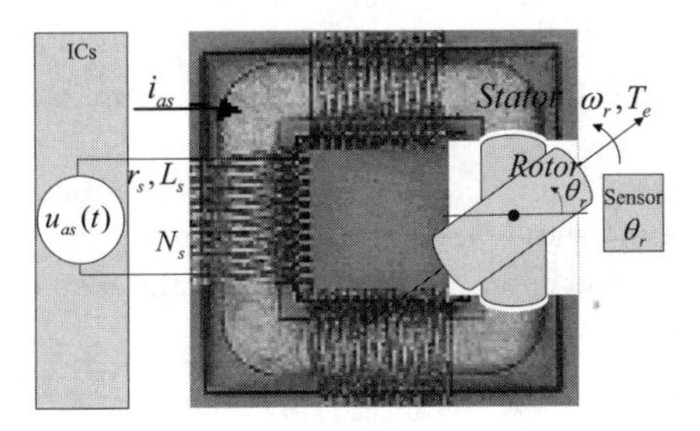


Figure 8.3 Single-phase reluctance micromotor with ICs and Hall-sensor to measure the rotor displacement.

The emphasis is concentrated on the analysis, modeling, and control of reluctance micromotors in the rotational microtransducer applications. Therefore, mathematical models must be found. The lumped-parameter modeling paradigm is based upon the use of the circuitry (voltage and current) and mechanical (velocity and displacement) variables to derive the differential equations using Newton's and Kirchhoff's laws. In these differential equations, the micromotor parameters are used. In particular, for the studied micromotor, the parameters are the stator resistance $r_s$, the magnetizing inductances in the *quadrature* and *direct* axes $L_{mq}$ and $L_{md}$ (it is evident that the air gap depends upon the rotor position, and thus, the magnetizing inductance varies as the function of the rotor displacement, see Figure 8.3), the average magnetizing inductance $\bar{L}_m$, the leakage inductance $L_{ls}$, the moment of inertia *J*, and the viscous friction coefficient $B_m$.

The expression for the electromagnetic torque was derived in [8]. In particular,

$$T_e = L_{\Delta m} i_{as}^2 \sin 2\theta_r ,$$

where $L_{\Delta m}$ is the half-magnitude of the sinusoidal magnetizing inductance $L_m$ variations, $L_m(\theta_r) = \bar{L}_m - L_{\Delta m} \cos 2\theta_r$.

Thus, to develop the electromagnetic torque, the current $i_{as}$ must be fed as a function of the rotor angular displacement $\theta_r$. For example, if

$$i_{as} = i_M \operatorname{Re}\left(\sqrt{\sin 2\theta_r}\right),$$

then,

$$T_{eaverage} = \frac{1}{\pi}\int_0^{\pi} L_{\Delta m} i_{as}^2 \sin 2\theta_r d\theta_r = \tfrac{1}{4} L_{\Delta m} i_M^2 .$$

The micromotor under our consideration is the synchronous micromachine. The obtained expression for the phase current is very important to control the synchronous microtransducers. In particular, the Hall-effect position sensor should be used to measure the rotor displacement, and the ICs must feed the phase current as a nonlinear function of $\theta_r$. Furthermore, the electromagnetic torque is controlled by changing the current magnitude $i_M$.

The mathematical model of the single-phase reluctance micromotor is found using Kirchhoff's and Newton's second laws.

In particular, we have the following differential equations

$$u_{as} = r_s i_{as} + \frac{d\psi_{as}}{dt}, \qquad \text{(circuitry equation – Kirchhoff's law )}$$

$$T_e - B_m \omega_r - T_L = J\frac{d^2\theta_r}{dt^2} \text{ (\textit{torsional-mechanical} equation–Newton's law).}$$

Here, the electrical angular velocity $\omega_r$ and displacement $\theta_r$ are used as the mechanical system variables.

From

$$u_{as} = r_s i_{as} + \frac{d\psi_{as}}{dt},$$

by using the flux linkage equation

$$\psi_{as} = \left(L_{ls} + \bar{L}_m - L_{\Delta m} \cos 2\theta_r\right) i_{as},$$

one obtains the circuitry dynamics equation with the mechanical variable $\theta_r$. In addition, the *torsional-mechanical* equation integrates the electromagnetic torque, which is the function of the electrical variable $i_{as}$. Making use the circuitry and *torsional-mechanical* dynamics, one obtains a set of three first-order nonlinear differential equations which models single-phase reluctance micromotors. In particular, we have

$$\frac{di_{as}}{dt}=-\frac{r_s}{L_{ls}+\bar{L}_m-L_{\Delta m}\cos 2\theta_r}i_{as}-\frac{2L_{\Delta m}}{L_{ls}+\bar{L}_m-L_{\Delta m}\cos 2\theta_r}i_{as}\omega_r\sin 2\theta_r$$

$$+\frac{1}{L_{ls}+\bar{L}_m-L_{\Delta m}\cos 2\theta_r}u_{as},$$

$$\frac{d\omega_r}{dt}=\frac{1}{J}\left(L_{\Delta m}i_{as}^2\sin 2\theta_r-B_m\omega_r-T_L\right),$$

$$\frac{d\theta_r}{dt}=\omega_r.$$

As the mathematical model is found and the micromotor parameters are measured, nonlinear simulation and analysis can be straightforwardly performed to study the dynamic responses and analyze the micromotor performance characteristics. In particular, the resistance, inductances, moment of inertia, viscous friction coefficient, and other parameters can be directly measured or identified based upon micromotor testing. The steady-state and dynamic analysis based upon the lumped-parameter mathematical model is straightforward. However, the lumped-parameter mathematical models simplify the analysis, and thus, these models must be compared with the experimental data to validate the results. Microfabrication processes to make reluctance micromotors are described latter in this chapter.

The disadvantage of single-phase reluctance micromotors are high torque ripple, vibration, noise, low reliability, etc. Therefore, let us study three-phase synchronous reluctance micromotors.

## 8.5 THREE-PHASE SYNCHRONOUS RELUCTANCE MICROMOTORS

Our goal is to address and solve a spectrum of problems in analysis, modeling, and control of synchronous reluctance micromachines. The electromagnetic features must be thoroughly analyzed before attempting to control micromotors. In fact, all micromotors must be controlled. That is, the angular velocity or displacement must be regulated by changing the phase voltages applied or phase currents fed to the microwindings. The electromagnetic features significantly restrict the control algorithms to be applied. Depending upon the conceptual methods employed to analyze synchronous reluctance micromachines, different control laws can be designed and implemented using ICs. Analysis and control of synchronous reluctance micromotors can be performed using different modeling, analysis, and optimization concepts. Complete lumped-parameter mathematical models of synchronous reluctance micromotors in the *machine* (*abc*) and in the *quadrature-direct-zero* (*qd0*) variables should be developed in the form of nonlinear differential equations. In particular, the lumped-parameter mathematical model for circuitry is found using the Kirchhoff's voltage law. The synchronous reluctance micromotor is illustrated in Figure 8.4.

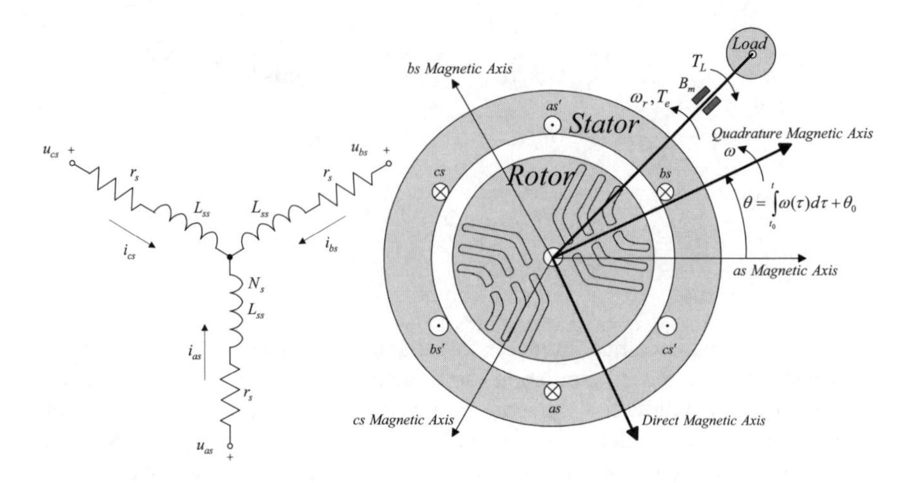


Figure 8.4 Three-phase synchronous reluctance micromotor.

The micromachine parameters are the stator resistance $r_s$ (it is assumed that the phase resistances are equal), the magnetizing inductances in the *quadrature* and *direct* axes $L_{mq}$ and $L_{md}$, the average magnetizing inductance $\overline{L}_m$, the leakage inductance $L_{ls}$, the moment of inertia *J*, and the viscous friction coefficient $B_m$.

The circuitry dynamics, is modeled as

$$\mathbf{u}_{abcs} = \mathbf{r}_s \mathbf{i}_{abcs} + \frac{d\boldsymbol{\psi}_{abcs}}{dt},$$

where $u_{as}$, $u_{bs}$ and $u_{cs}$ are the phase voltages; $i_{as}$, $i_{bs}$ and $i_{cs}$ are the phase currents; $\psi_{as}$, $\psi_{bs}$ and $\psi_{cs}$ are the flux linkages, $\boldsymbol{\psi}_{abcs} = \mathbf{L}_s \mathbf{i}_{abcs}$;

$$\mathbf{r}_s = \begin{bmatrix} r_s & 0 & 0 \\ 0 & r_s & 0 \\ 0 & 0 & r_s \end{bmatrix};$$

$$\mathbf{L}_s = \begin{bmatrix} L_{ls} + \overline{L}_m - L_{\Delta m}\cos(2\theta_r) & -\frac{1}{2}\overline{L}_m - L_{\Delta m}\cos 2\left(\theta_r - \frac{1}{3}\pi\right) & -\frac{1}{2}\overline{L}_m - L_{\Delta m}\cos 2\left(\theta_r + \frac{1}{3}\pi\right) \\ -\frac{1}{2}\overline{L}_m - L_{\Delta m}\cos 2\left(\theta_r - \frac{1}{3}\pi\right) & L_{ls} + \overline{L}_m - L_{\Delta m}\cos 2\left(\theta_r - \frac{2}{3}\pi\right) & -\frac{1}{2}\overline{L}_m - L_{\Delta m}\cos 2\left(\theta_r + \pi\right) \\ -\frac{1}{2}\overline{L}_m - L_{\Delta m}\cos 2\left(\theta_r + \frac{1}{3}\pi\right) & -\frac{1}{2}\overline{L}_m - L_{\Delta m}\cos 2\left(\theta_r + \pi\right) & L_{ls} + \overline{L}_m - L_{\Delta m}\cos 2\left(\theta_r + \frac{2}{3}\pi\right) \end{bmatrix}$$

$\overline{L}_m = \frac{1}{3}\left(L_{mq} + L_{md}\right)$ and $L_{\Delta m} = \frac{1}{3}\left(L_{md} - L_{mq}\right)$.

The expressions for inductances are nonlinear functions of the electrical angular displacement $\theta_r$. Hence, the *torsional-mechanical* dynamics must be used. Taking note of the Newton's second law of rotational motion, and using $\omega_r$ and $\theta_r$ (electrical angular velocity and displacement) as the state variables (mechanical variables), one obtains

$$T_e - B_m \frac{2}{P}\omega_r - T_L = J\frac{2}{P}\frac{d\omega_r}{dt},$$

$$\frac{d\theta_r}{dt} = \omega_r,$$

where $T_e$ and $T_L$ are the electromagnetic and load torques.

*Torque Production Analysis*

Using the coenergy, the electromagnetic torque, which is a nonlinear function of the micromotor variables (phase currents and electrical angular position) and micromotor parameters (number of poles $P$ and inductance $L_{\Delta m}$), is found to be [8]

$$T_e = \frac{P}{2}L_{\Delta m}\Big[i_{as}^2 \sin 2\theta_r + 2i_{as}i_{bs}\sin 2\left(\theta_r - \tfrac{1}{3}\pi\right) + 2i_{as}i_{cs}\sin 2\left(\theta_r + \tfrac{1}{3}\pi\right)$$

$$+ i_{bs}^2 \sin 2\left(\theta_r - \tfrac{2}{3}\pi\right) + 2i_{bs}i_{cs}\sin 2\theta_r + i_{cs}^2\sin 2\left(\theta_r + \tfrac{2}{3}\pi\right)\Big]$$

To control the angular velocity, the electromagnetic torque must be regulated. To maximize the electromagnetic torque, ICs must feed the following phase currents as functions of the angular displacement measuring or observing (sensorless control) the rotor displacement $\theta_r$

$$i_{as} = \sqrt{2}i_M \sin\left(\theta_r + \tfrac{1}{3}\varphi_i\pi\right),$$

$$i_{bs} = \sqrt{2}i_M \sin\left(\theta_r - \tfrac{1}{3}(2-\varphi_i)\pi\right),$$

$$i_{cs} = \sqrt{2}i_M \sin\left(\theta_r + \tfrac{1}{3}(2+\varphi_i)\pi\right).$$

Thus, for $\varphi_i = 0.3245$, one obtains

$$T_e = \sqrt{2}PL_{\Delta m}i_M^2.$$

That is, $T_e$ is maximized and controlled by changing the magnitude of the phase currents $i_M$. Furthermore, in theory, using the equations used, there is no torque ripple (in practice, based upon the experimental results, and performing the high-fidelity modeling integrating nonlinear electromagnetics using Maxwell's equations, one finds that there exists the torque ripple which is due to the cogging torque, eccentricity, bearing, pulse-width-modulation, electromagnetic field nonuniformity, and other phenomena).

The majority of ICs are designed to control the phase voltages $u_{as}$, $u_{bs}$ and $u_{cs}$. Therefore, the three-phase balance voltage set is needed to be introduced. We have

$$u_{as} = \sqrt{2}u_M \sin\left(\theta_r + \tfrac{1}{3}\varphi_i\pi\right),$$

$$u_{bs} = \sqrt{2}u_M \sin\left(\theta_r - \tfrac{1}{3}(2-\varphi_i)\pi\right),$$

$$u_{cs} = \sqrt{2}u_M \sin\left(\theta_r + \tfrac{1}{3}(2+\varphi_i)\pi\right),$$

where $u_M$ is the magnitude of the supplied voltages.

*Lumped-Parameter Mathematical Models*

The mathematical model of synchronous reluctance micromotors in the *abc* variables is found to be

$$\frac{di_{as}}{dt}=\frac{1}{L_D}\Big[\left(r_s i_{as}-u_{as}\right)\left(4L_{ls}^2+3\bar{L}_m^2-3L_{\Delta m}^2+8\bar{L}_m L_{ls}-4L_{ls}L_{\Delta m}\cos 2\theta_r\right)$$

$$+\left(r_s i_{bs}-u_{bs}\right)\left(3\bar{L}_m^2-3L_{\Delta m}^2+2\bar{L}_m L_{ls}+4L_{ls}L_{\Delta m}\cos 2\left(\theta_r-\tfrac{1}{3}\pi\right)\right)$$

$$+\left(r_s i_{cs}-u_{cs}\right)\left(3\bar{L}_m^2-3L_{\Delta m}^2+2\bar{L}_m L_{ls}+4L_{ls}L_{\Delta m}\cos 2\left(\theta_r+\tfrac{1}{3}\pi\right)\right)$$

$$+6\sqrt{3}L_{\Delta m}^2 L_{ls}\omega_r\left(i_{cs}-i_{bs}\right)$$

$$+\left(8L_{\Delta m}L_{ls}^2\omega_r+12L_{\Delta m}\bar{L}_m L_{ls}\omega_r\right)\left(\sin 2\theta_r i_{as}+\sin 2\left(\theta_r-\tfrac{1}{3}\pi\right)i_{bs}+\sin 2\left(\theta_r+\tfrac{1}{3}\pi\right)i_{cs}\right)\Big],$$

$$\frac{di_{bs}}{dt}=\frac{1}{L_D}\Big[\left(r_s i_{as}-u_{as}\right)\left(3\bar{L}_m^2-3L_{\Delta m}^2+2\bar{L}_m L_{ls}+4L_{ls}L_{\Delta m}\cos 2\left(\theta_r-\tfrac{1}{3}\pi\right)\right)$$

$$+\left(r_s i_{bs}-u_{bs}\right)\left(4L_{ls}^2+3\bar{L}_m^2-3L_{\Delta m}^2+8\bar{L}_m L_{ls}-4L_{ls}L_{\Delta m}\cos 2\left(\theta_r+\tfrac{1}{3}\pi\right)\right)$$

$$+\left(r_s i_{cs}-u_{cs}\right)\left(3\bar{L}_m^2-3L_{\Delta m}^2+2\bar{L}_m L_{ls}+4L_{ls}L_{\Delta m}\cos 2\theta_r\right)$$

$$+6\sqrt{3}L_{\Delta m}^2 L_{ls}\omega_r\left(i_{as}-i_{cs}\right)$$

$$+\left(8L_{\Delta m}L_{ls}^2\omega_r+12L_{\Delta m}\bar{L}_m L_{ls}\omega_r\right)\left(\sin 2\left(\theta_r-\tfrac{1}{3}\pi\right)i_{as}+\sin 2\left(\theta_r+\tfrac{1}{3}\pi\right)i_{bs}+\sin 2\theta_r i_{cs}\right)\Big],$$

$$\frac{di_{cs}}{dt}=\frac{1}{L_D}\Big[\left(r_s i_{as}-u_{as}\right)\left(3\bar{L}_m^2-3L_{\Delta m}^2+2\bar{L}_m L_{ls}+4L_{ls}L_{\Delta m}\cos 2\left(\theta_r+\tfrac{1}{3}\pi\right)\right)$$

$$+\left(r_s i_{bs}-u_{bs}\right)\left(3\bar{L}_m^2-3L_{\Delta m}^2+2\bar{L}_m L_{ls}+4L_{ls}L_{\Delta m}\cos 2\theta_r\right)$$

$$+\left(r_s i_{cs}-u_{cs}\right)\left(4L_{ls}^2+3\bar{L}_m^2-3L_{\Delta m}^2+8\bar{L}_m L_{ls}-4L_{ls}L_{\Delta m}\cos 2\left(\theta_r-\tfrac{1}{3}\pi\right)\right)$$

$$+6\sqrt{3}L_{\Delta m}^2 L_{ls}\omega_r\left(i_{bs}-i_{as}\right)$$

$$+\left(8L_{\Delta m}L_{ls}^2\omega_r+12L_{\Delta m}\bar{L}_m L_{ls}\omega_r\right)\left(\sin 2\left(\theta_r+\tfrac{1}{3}\pi\right)i_{as}+\sin 2\theta_r i_{bs}+\sin 2\left(\theta_r-\tfrac{1}{3}\pi\right)i_{cs}\right)\Big],$$

$$\frac{d\omega_r}{dt}=\frac{P^2}{4J}L_{\Delta m}\Big(i_{as}^2\sin 2\theta_r+2i_{as}i_{bs}\sin 2\left(\theta_r-\tfrac{1}{3}\pi\right)+2i_{as}i_{cs}\sin 2\left(\theta_r+\tfrac{1}{3}\pi\right)$$

$$+i_{bs}^2\sin 2\left(\theta_r-\tfrac{2}{3}\pi\right)+2i_{bs}i_{cs}\sin 2\theta_r+i_{cs}^2\sin 2\left(\theta_r+\tfrac{2}{3}\pi\right)\Big)-\frac{B_m}{J}\omega_r-\frac{P}{2J}T_L,$$

$$\frac{d\theta_r}{dt}=\omega_r\,.$$

In these differential equations, the following notations are used $\bar{L}_m=\tfrac{1}{3}\left(L_{mq}+L_{md}\right), L_{\Delta m}=\tfrac{1}{3}\left(L_{md}-L_{mq}\right),\ L_D=L_{ls}\left(9L_{\Delta m}^2-4L_{ls}^2-12\bar{L}_m L_{ls}-9\bar{L}_m^2\right)$.

That is, we have a set of five highly coupled nonlinear differential equations. Though analytic solution of these equations is virtually impossible, nonlinear modeling can be straightforwardly performed in the MATLAB environment which includes the SIMULINK toolbox.

The mathematical model can be simplified. In particular, in the rotor reference frame, we apply the Park transformation [8]

$$\mathbf{u}^r_{qd0s} = \mathbf{K}^r_s \mathbf{u}_{abcs},\ \mathbf{i}^r_{qd0s} = \mathbf{K}^r_s \mathbf{i}_{abcs},\ \boldsymbol{\psi}^r_{qd0s} = \mathbf{K}^r_s \boldsymbol{\psi}_{abcs},$$

$$\mathbf{K}^r_s = \frac{2}{3}\begin{bmatrix} \cos\theta_r & \cos\left(\theta_r - \frac{2}{3}\pi\right) & \cos\left(\theta_r + \frac{2}{3}\pi\right) \\ \sin\theta_r & \sin\left(\theta_r - \frac{2}{3}\pi\right) & \sin\left(\theta_r + \frac{2}{3}\pi\right) \\ \frac{1}{2} & \frac{1}{2} & \frac{1}{2} \end{bmatrix},$$

where $u_{qs}, u_{ds}, u_{0s}, i_{qs}, i_{ds}, i_{0s}$ and $\psi_{qs}, \psi_{ds}, \psi_{0s}$ are the *qd0* voltages, currents, and flux linkages.

Using the circuitry and *torsional-mechanical* dynamics, one finds the following nonlinear differential equations to model synchronous reluctance micromotors in the rotor reference frame

$$\frac{di^r_{qs}}{dt} = -\frac{r_s}{L_{ls}+L_{mq}} i^r_{qs} - \frac{L_{ls}+L_{md}}{L_{ls}+L_{mq}} i^r_{ds}\omega_r + \frac{1}{L_{ls}+L_{mq}} u^r_{qs},$$

$$\frac{di^r_{ds}}{dt} = -\frac{r_s}{L_{ls}+L_{md}} i^r_{ds} + \frac{L_{ls}+L_{mq}}{L_{ls}+L_{md}} i^r_{qs}\omega_r + \frac{1}{L_{ls}+L_{md}} u^r_{ds},$$

$$\frac{di^r_{0s}}{dt} = -\frac{r_s}{L_{ls}} i^r_{0s} + \frac{1}{L_{ls}} u^r_{0s},$$

$$\frac{d\omega_r}{dt} = \frac{3P^2}{8J}\left(L_{md} - L_{mq}\right) i^r_{qs} i^r_{ds} - \frac{B_m}{J}\omega_r - \frac{P}{2J} T_L,$$

$$\frac{d\theta_r}{dt} = \omega_r .$$

One can easily observe that this model is much simpler compared with the lumped-parameter mathematical model derived using the *abc* variables.

To attain the balanced operation, the *quadrature* and *direct* currents and voltages must be derived using the *direct* Park transformation

$$\mathbf{i}^r_{qd0s} = \mathbf{K}^r_s \mathbf{i}_{abcs},\ \mathbf{u}^r_{qd0s} = \mathbf{K}^r_s \mathbf{u}_{abcs} .$$

The *qd0* voltages $u^r_{qs}$, $u^r_{ds}$ and $u^r_{0s}$ are found using the three-phase balance voltage set. In particular, we have

$$u^r_{qs} = \sqrt{2}u_M,\ u^r_{ds} = 0,\ u^r_{0s} = 0 .$$

We derived the lumped-parameter mathematical models of three-phase synchronous reluctance micromotors. Based upon the differential equations obtained, nonlinear analysis can be performed, and the phase currents and voltages needed to guarantee the balance operating conditions were found. In particular, taking note of electromagnetic features, we found the phase currents and voltages needed to be applied to guarantee the desired operating features. The results reported can be straightforwardly used in nonlinear simulation.

## 8.6 MICROFABRICATION

Electromechanical microstructures and microtransducers can be fabricated through deposition of the conductors (coils and windings), ferromagnetic core, insulation layers, as well as other microstructures (movable and stationary members and their components including bearing). The order of the processes, materials, and sequential steps are different depending on the MEMS which must be devised, designed, analyzed, and optimized first. This section is aimed to provide the reader with the basic fabrication features and processes involved in the MEMS fabrication.

### 8.6.1 Microcoils/Microwindings Fabrication Through the Copper, Nickel and Aluminum Electrodeposition

Among many important processes to fabricate MEMS, the deposition is the most critical one. Therefore, let us focus our attention on the deposition through electroplating. The reader should be familiar with the basics of electrochemistry where the electrolytic processes are covered. In particular, purification (copper, zinc, cobalt, nickel and other metals) and electroplating processes are discussed and documented in the undergraduate introductory chemistry textbooks. Let us recall the major principles. When aqueous solutions are electrolyzed using metal electrodes, an electrode will be oxidized if its oxidation potential is greater than for water. As examples, nickel and copper are oxidized more readily than water, and the reactions are:

- nickel

  $Ni\ (s) \rightarrow Ni^{2+}\ (aq) + 2e^-,\ E_{ox} = 0.28\ V,$

  $2H_2O\ (l) \rightarrow 4H^+\ (aq) + O_2\ (g)\ + 4e^-,\ E_{ox} = -\ 1.23\ V;$

- copper

  $Cu\ (s) \rightarrow Cu^{2+}\ (aq) + 2e^-,\ E_{ox} = -\ 0.34\ V,$

  $2H_2O\ (l) \rightarrow 4H^+\ (aq) + O_2\ (g)\ + 4e^-,\ E_{ox} = -\ 1.23\ V,$

where s, aq, l and g denote the solid, aqueous, liquid and gas states.

If the anode is made from nickel in an electrolytic cell, nickel metal is oxidized as the anode reaction. If $Ni^{2+}$ (aq) is the solution, it is reduced at the cathode in the preference to reduction of water. As current flows, nickel dissolves from the anode and deposits on the cathode. The reactions are

$Ni\ (s) \rightarrow Ni^{2+}\ (aq) + 2e^-$ (anode),

and $Ni^{2+}\ (aq) + 2e^- \rightarrow Ni\ (s)$ (cathode).

Most metals (chromium, iron, cobalt, nickel, copper, zinc, silver, gold, and others) used to fabricate MEMS are the transition metals which occupy the "d" block of the periodic table. Important physical properties of these metals were listed in Table 8.1. The chemistry of transition metals (reactions involved and chemicals used) is very important due to the significance of electroplating, etching and other fabrication processes. In this book, most practical and effective processes, techniques, and materials (chemicals) are covered.

The conductors (microcoils to make windings) in microstructures and microtransducers can be fabricated by electrodepositing the copper and other low resistivity metals. Electrodeposition of metals is made by immersing a conductive surface in a solution containing ions of the metal to be deposited. The surface is electrically connected to an external power supply, and current is fed through the surface into the solution. In general, the reaction of the metal ions ($Metal^{x+}$) with x electrons ($xe^-$) to form metal (Metal) is

$Metal^{x+} + xe^- = Metal$.

To electrodeposit copper on the silicon wafer, the wafer is typically coated with a thin conductive layer of copper (seed layer) and immersed in a solution containing cupric ions. Electrical contact is made to the seed layer, and current is flow (passed) such that the reaction $Cu^{2+}+2e^-\rightarrow Cu$ occurs at the wafer surface. The wafer, which is electrically interacted such that the metal ions are changed to metal atoms, is the cathode. Another electrically active surface (anode) is the conductive solution to make the electrical path. At the anode, the oxidation reaction occurs that balances the current flow at the cathode, thus maintaining the electric neutrality. In the case of copper electroplating, all cupric ions removed from solution at the wafer cathode are replaced by dissolution from the copper anode. According to the Faraday law of electrolysis, in the absence of secondary reactions, the current delivered to a conductive surface during electroplating is proportional to the quantity of the metal deposited. Thus, the metal deposited can be controlled varying the electroplating current (current density) and the electrodeposition time. If the applied voltage potential is zero, there is an equilibrium. Once potential is changed by the external power source, the current flows and electroplating results. For electroplating, the current is approximated by an exponential Tafel equation. Figure 8.5 illustrates a current-voltage curve for a copper electrodeposition process. The electrodeposition rate is controlled by changing the reaction rate kinetics. The nonlinear current-voltage dependence results in the need for special electroplating cell design to attain the uniform voltage potentials across the wafer surface.

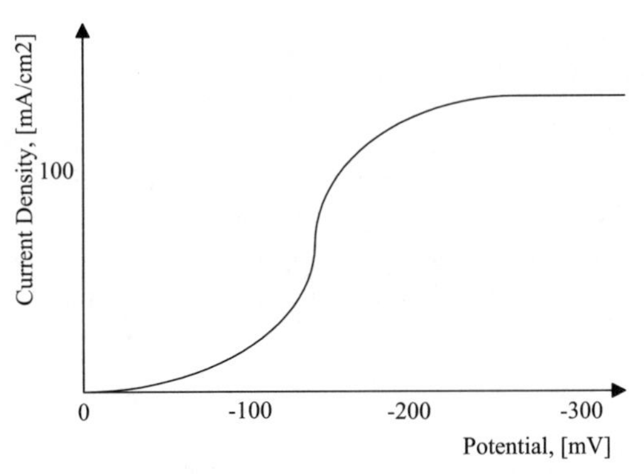


Figure 8.5 Current-voltage curve for a copper electroplating process.

As the voltage potential increases, the mass transfer effects become predominant, and the current saturation (current density limit) is reached (the elements reacting the cathode $Cu^{2+}$ do not reach the interface at a rate sufficient to sustain the high rate of reaction). Usually, electroplating processes are performed at the 30-50% current density limit to avoid undesirable electrodeposition effects. To ensure that the rate of mass transfer of electroactive elements to the interface is large compared to the reaction rate and the desired uniformity is achieved, the rates of migration diffusion and convection must be controlled. Convection is the most important phase of the mass transfer and can vary from stagnant to laminar or turbulent flow. It includes impinging flow caused by the solution pumping, undesirable flows due to substrate movement, and flows resulting from density variations. Electroplating can be carried out using a constant current, a constant voltage, or variable waveforms of current or voltage at different temperature. Using a constant current, accurate control of the mass of the electrodeposited metal can be obtained. The electroplating using variable waveforms requires more complex equipment to control, but is meaningful in maintaining the specific thickness distributions and desired film properties. The waveforms are periodic (rectangular, sinusoidal, trapezoidal, or triangular) forward and reverse pulses with variable (peak, high, average, low, or controlled) magnitude of the forward and reverse current or voltage. In addition, the duty cycle (the ratio of the forward and reverse time periods) can be controlled.

From the electrochemistry viewpoint the commonly used solutions for the copper electroplating are acidic (copper sulfate in fluoborate pyrophosphate bath) and alkaline (cyanide in not-cynaide bath), and different additives can be added to achieved the desired characteristics.

The hydrated Cu ions reaction is

$$Cu^{2+} \rightarrow Cu(H_2O)_6^{2+},$$

and the cathode reactions are

$$Cu^{2+}+2e^- \rightarrow Cu,$$

$$Cu^{2+}+e^- \rightarrow Cu^+,$$

$$Cu^+ + e^- \rightarrow Cu,$$

$$2Cu^+ \rightarrow Cu^{2+}+Cu,$$

$$H^+ + e^- \rightarrow \tfrac{1}{2} H_2.$$

The copper electroplating solution commonly used is

$CuSO_4$–$5H_2O$ (250 g/l) and $H_2SO_4$ (25 ml/l).

The basic processes are shown in Figure 8.6, and the brief description of the sequential steps and equipment which can be used are given.

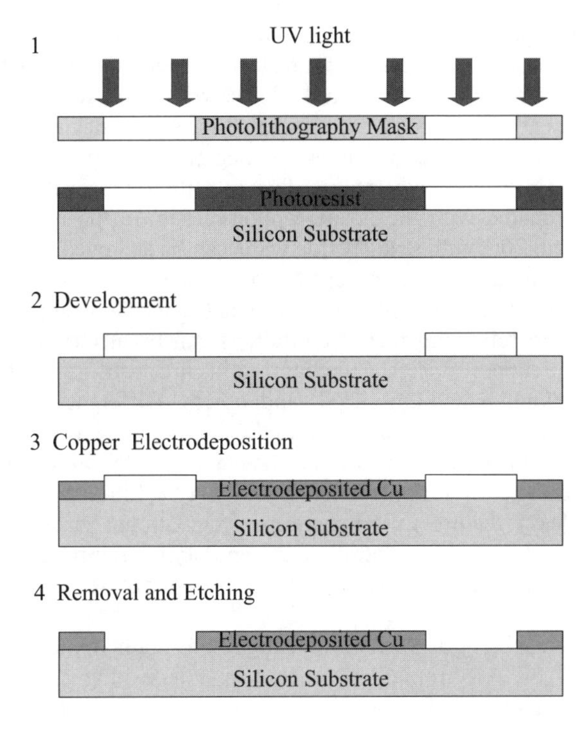


Figure 8.6 Electrodeposition of copper and basic processes: Silicon, Kapton, and other substrates can be used. After clearing, the silicon substrate is covered with a 5-10 nm chromium or titanium and 100-200 nm copper seed layer by sputtering. The copper microcoils (microstructures) are patterned using the UV photolitography. The AZ-4562 photoresist can be spincoated and prebaked on a ramped hot plate at 90-100$^0$C (ramp 30-40% with initial temperature 20-25$^0$C) for one hour. Then, the photoresist is exposed in the Karl Suss Contact Masker with the energy 1200-1800 mJ-cm$^2$. The development is released in 1:4 diluted alkaline solution (AZ-400) for 4-6 min. This gives the photoresist thickness 15-25 μm. Copper is electroplated with a three-electrode system with a copper anode and a saturated calomel reference electrode (the current power supply is the Perkin Elmer Current Source EG&G 263). The Shipley sulfate bath with the 5-10 ml/l brightener to smooth the deposit can be used. The electrodeposition is performed at the 20-25$^0$C with the magnetic stirring and the dc current density 40-60 mA/cm$^2$ (this current density leads to smooth copper thin films with the 5-10 nm rms roughness for the 10 μm thickness of the deposited copper thin film). The resistivity of the electrodeposited copper thin film (microcoils) is 1.6-1.8 μohm-cm (close to the bulk copper resistivity). After the deposition, the photoresist is removed.

As was emphasized, to reduce the losses in the copper microwindings, it is desirable to deposit the thicker conductors. Therefore, a thick photoresist (tens of μm) is required instead of the standard photoresist. For the positive photoresists (AZ-4000), 20-30 μm thickness with good structural resolution can be obtained. When thicker layers are desired, the negative photoresists (e.g., EPON Resin SU-8 which is the negative epoxy-based photoresist, Microchem Co.) have advantages because higher aspect ratio can be achieved compared with the positive photoresists. In the one-step-process, the aspect ratio 20:1 with straight side walls can be achieved.

Different industrial electrodeposition systems are available to perform electroplating. Usually, the systems are available for the wafer size is up to 6 inches, and the following metals and alloys can be electroplated using the same station: copper, nickel, palladium, gold, tin, lead, iron, silver, nickel-iron (*permalloy*), palladium-nickel, and others. Different rated current is used (for example, from –1 to 1 A), and the maximum peak current can be 3-5 times greater than the rated current. The current is stabilized usually within 0.001 A, and the current ripple is less than 0.5-1 %. Filtration (cartridge with 1μm or less features), temperature regulation, current waveform specification, duty cycle control, and ventilation feature are guaranteed. Furthermore, the industrial electrodeposition systems are computerized, see Figure 8.7.

The electroplating is performed using anode, cathode, and solution. Let us describe the electroplating process by depositing the nickel. The electroplating bath and electrodeposition process are illustrated in Figure 8.7.

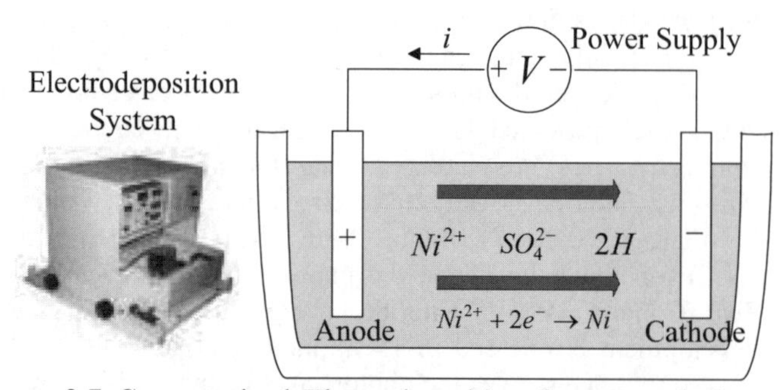


Figure 8.7 Computerized Electrodeposition System and electroplating bath and nickel electrodeposition.

The anode is made from nickel, and the cathode is made from the conductive material. The solution consists of the nickel $Ni^{2+}$, hydrogen $H^+$, and sulfate $SO_4^{2-}$ ions. As the voltage is applied, the positive ions in the solution are attracted to the negative cathode. The nickel ions $Ni^{2+}$ that reach the cathode gain electrons and are deposited (electroplated) on the surface of the cathode forming the nickel electrodeposit. At the same time, nickel is electrochemically etched from the nickel anode to produce ions for the aqueous solution and electrons for the power source. The hydrogen ions $H^+$

gain electrons from the cathode and form hydrogen gas leading to bubbles. The resulting hydrogen gas formation is undesirable because it slows the electroplating process (current is not fully used for electrodeposition). Furthermore, bubbles degrade the electrodeposits uniformity. To fabricate Ni-based microstructures by electrodeposition, a conductive plating base (seed layer is the sputtered nickel) and electrodeposit patterning are involved (this was discussed explaining the copper electrodeposition). Usually, the electrodeposit is patterned by an additive process (selective electrodeposition) instead of a subtractive process (etching). The recipes (chemicals) for chemical baths to electrodeposit materials are different. For example, for nickel deposition, the solution can be made using:
nickel sulfate (250 g/l), nickel chloride (10 g/l), boric acid (30 g/l), ferrous sulfate (10 g/l) and saccharin (5 g/l).

The solutions for deposition of magnetic thin films alloys, layers of which can be made through electrodeposition, will be given later.

It must be emphasized that commonly used magnetic materials and conductors do not adhere well to silicon. Therefore, as was described, the adhesion layers (e.g., titanium Ti or chromium Cr) are deposited on the silicon surface prior to the magnetic material electroplating.

The electrodeposition rate is proportional to the current density, and therefore, the uniform current density at the substrate seed layer is needed to attain the uniform thickness of the electrodeposit. To achieve the selective electrodeposition, portions of the seed layer are covered with the resist (the current density at the mask edges is nonuniform degrading electroplating). In addition to the current density, the deposition rate is also a nonlinear function of temperature, solution (chemicals), pH, direct/reverse current or voltage waveforms magnitude, waveform pulse shapes (sinusoidal, rectangular, trapezoidal, etc.), duty ratio, plating area, etc. The simplified equations to calculate the thickness and electrodeposition time for the specified materials are

$$\text{Thickness}_{\text{material}} = \frac{\text{Time}_{\text{electroplating}} \times \text{Current}_{\text{density}} \times \text{Weight}_{\text{molecular}}}{\text{Faradey}_{\text{constant}} \times \text{Density}_{\text{material}} \times \text{Electron}_{\text{number}}},$$

$$\text{Time}_{\text{electroplating}} = \frac{\text{Thickness}_{\text{material}} \times \text{Faradey}_{\text{constant}} \times \text{Density}_{\text{material}} \times \text{Electron}_{\text{number}}}{\text{Current}_{\text{density}} \times \text{Weight}_{\text{molecular}}}.$$

It was emphasized that electroplating is used to deposit thin-film conductors and magnetic materials. However, microtransducers need the insulation layers, otherwise the core and coils as well as multilayer microcoils themselves will be short-circuited. Furthermore, the seed layers are embedded in microfabrication processes. As the ferromagnetic core is fabricated on top of the microcoils (or microcoils are made on the core), the seed layer is difficult to remove because it was placed at the bottom or at the center of the microstructure. The mesh seed layer can serve as the electroplating seed layer for the lower conductors, and as the microstructure is made, the edges of the mesh seed layer can be exposed and removed through plasma etching [6]. Thus, the microcoils are insulated. It should be emphasized that relatively high aspect ratio techniques must be used to fabricate the ferromagnetic core and microcoils, and patterning as well as surface planarization issues must be addressed.

Ferromagnetic cores in microstructures and microtransducers must be made. For example, the electroplated $Ni_{x\%}Fe_{100-x\%}$ thin films, such as *permalloy* $Ni_{80\%}Fe_{20\%}$, can be deposited to form the ferromagnetic core of microtransducers (actuators and sensors), inductors, transformers, switches, etc. The basic processes and sequential steps used are similar to the processes for the copper electrodeposition, and the electroplating is done in the electroplating bath. The windings (microcoils) must be insulated from cores, and therefore, the insulation layers must be deposited. The insulating materials used to insulate the windings from the core are benzocyclobutene, polyimide (PI-2611), etc. For example, the cyclotene 7200-35 is photosensitive and can be patterned through photolithography. The benzocyclobutene, used as the photoresist, offers good planarization and pattern properties, stability at low temperatures, and exhibits negligible hydrophilic properties.

The sketched fabrication process with sequential steps to make the electromagnetic microtransducer with movable and stationary members is illustrated in Figure 8.8. On the silicon substrate, the chromium-copper-chromium (Cr-Cu-Cr) mesh seed layer is deposited (through electron-beam evaporation) forming a seed layer for electroplating. The insulation layer (polyimide Dupont PI-2611) is spun on the top of the mesh seed layer to form the electroplating molds. Several coats can be done to obtain the desired thickness of the polyimide molds (one coat results in 8-12 μm insulation layer thickness). After coating, the polyimide is cured (at 280-310$^0$C) in nitrogen for one hour. A thin aluminum layer is deposited on top of the cured polyimide to form a hard mask for dry etching. Molds for the lower conductors are patterned and plasma etched until the seed layer is exposed. After etching the aluminum (hard mask) and chromium (top Cr-Cu-Cr seed layer), the molds are filled with the electroplated copper through the described copper electroplating process. One coat of polyimide insulates the lower conductors and the core (thus, the insulation is achieved). The seed layer is deposited, mesh-patterned, coated with polyimide, and hard-cured. The aluminum thin layers (hard mask for dry etching) are deposited, and the mold for the ferromagnetic cores is patterned and etched until the seed layer is exposed. After etching the aluminum (hard mask) and the chromium (top Cr-Cu-Cr seed layer), the mold is filled with the electroplated $Ni_{x\%}Fe_{100-x\%}$ thin films (electroplating process). The desired composition and thickness of the $Ni_{x\%}Fe_{100-x\%}$ thin films can be achieved as will be described later. One coat of the insulation layer (polyimide) is spin-cast and cured to insulate the core and upper conductors. The via holes are patterned in the sputtered aluminum layer (hard mask) and etched through the polyimide layer using oxygen plasma. The, vias are filled with the electroplated copper (electroplating process). The copper-chromium seed layer is deposited, and the molds for the upper conductors are formed using thick photoresist. The molds are filled with the electroplated copper and removed. Then, the gap for the movable member is made using the conventional processes. After removing the seed layer, the passivation layer (polyimide) is coated and cured to protect the top conductors. The polyimide is masked and etched to the silicon substrate. The bottom mesh seed layer is wet etched, and the microtransducer (with the ICs to control it) is diced and sealed.

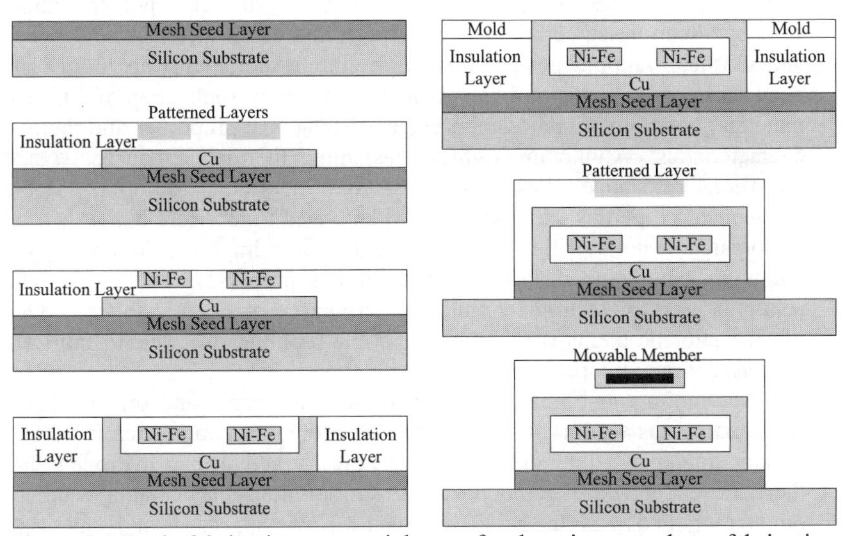

Figure 8.8 Basic fabrication sequential steps for the microtransducer fabrication.

As we emphasized, the electrodeposition process must be optimized to attain uniformity with ultimate goal to fabricate high-performance MEMS. As an example, Figure 8.9 illustrates the electrodeposited copper on the silicon substrate with the seed layer. It is evident that in most cases one cannot guarantee the ideal microstructures geometry and dimensions. To deposit windings, one can form deep grooves (25 – 75 μm) in silicon by anisotropic etching in KOH, etch the trenches (10-20 μm deep), electroplating conductors (Au, Cu or other metals) to form the windings, and finally electroplate Cu and PbSn to form the bumps (if needed), see Figure 8.9.

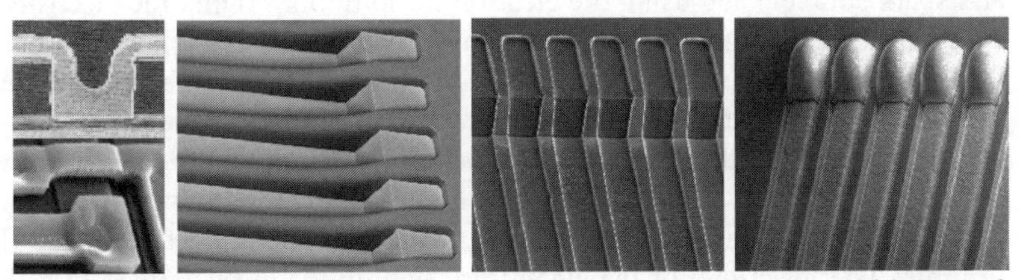

Figure 8.9 Electrodeposited copper, resist pattern used as an etching mask for etching the trenches, etched trenches (10 μm width), and windings with PbSn bumps.

Electroplated aluminum is the needed material to fabricate microstructures and microdevices. In particular, aluminum can be used as the conductor to fabricate microcoils as well as mechanical microstructures (gears, bearing, pins, reflecting surfaces, etc.). Advanced techniques and processes for the electrodeposition of aluminum are documented in [9]. Using the micromolding processes, high aspect ratio can be achieved. In CMOS and surface micromachining, single-crystal and polycrystalline silicon, silicon-based compounds, polymers, as well as various electroplated and physical vapor deposited metals (including aluminum) are used. The need for integrity and compliance of materials lead to application of different materials based upon their electro-chemo-mechanical properties and thermal characteristics. Aluminum (which has high thermal conductivity and corrosion resistance, low resistivity and neutron absorption, stable mechanical properties, etc.) has been widely used to fabricate ICs (it is used as conductors and sacrificial layers), and thus, it is important to develop the aluminum-compliant surface micromachining processes. The electrolytic solutions used to electroplate aluminum are based on organic solvents. One of the major problems is to find compatible molding materials to fabricate high aspect ratio microstructures.

Aluminum can be electrodeposited from inorganic and organic fused salt mixtures as well as from solutions of aluminum compounds in certain organic solvents (aluminum is more chemically active than hydrogen, and therefore cannot be electrodeposited from solutions that contain water or other compounds with the acidic hydrogen). However, fused salt baths result in inherent thermal distortion due to residual stresses in thin films as well as other drawbacks [9]. In contrast, the aluminum chloride-lithium aluminum – hydride ethereal baths provide low-stress thin films. The National Bureau of Standards' hydride process for the aluminum electroplating was commercialized. The aluminum electroplating solution composition is [9]:
$AlCl_3$ (aluminum chloride, 400 g/l) and $LiAlH_4$ (lithium aluminum hydride, 15 g/l) – electroplating is done at the temperature $10^0$C - $60^0$C.

Due to the nature of the bath, safety issues during electroplating must be strictly obeyed.

The electroplating must be carried out in an inert atmosphere (for example in the sealed glove box with dry nitrogen as the ambient gas). The care must be taken mixing the electrolytic solution (protecting from ignition and spark sources) and using the electrolytic solution. To mix the electrolytic solution in the nitrogen-filled sealed glove box, the $AlCl_3$ is slowly added to diethyl ether (this is exothermic reaction which produces heat, and thus, cooling is recommended during mixing to minimize evaporation and to permit rapid addition of the aluminum chloride). Then, the $LiAlH_4$ is slowly added mixing the solution [9]. For aluminum electroplating, the typical current density 10–15 mA/cm$^2$ results in the electroplating rate of 0.8–1.2 μm/min. The resistivity of the electrolytic solution is in the range of 95–110 ohm-cm. To fabricate micromolded electroplated aluminum microstructures, the material (including aluminum) must have desirable properties (high aspect ratio molds and removal) associated with conventional electroforming materials, as well as properties desirable for nonconventional electroplating processes (e.g., the ability to withstand solvent-based electrolytic solutions and chemical resistance to perform preprocessing required for electroplating). Thus, the standard photoresist electroplating mold processes cannot be utilized to make aluminum-based microstructures. The polyimide materials have the properties necessary to withstand aluminum electroplating conditions. The micromolding processes used to make the aluminum microstructures use photosensitive (photo-crosslinked upon exposure) and non-photosensitive polyimides [9]. The non-photosensitive polyimide process involves plasma or reactive ion etching to produce molds with the desired side-wall profiles and geometries. The major process for fabrication of electroplated microstructures using photosensitive polyimide is similar to the LIGA processes [2, 4, 10]. However, the photosensitive polyimide is used as the electroplating mask instead of polymethylmethacrylate, and the UV exposure source is used instead of the *x*-ray synchrotron radiation. The electroplating system consisting of the adhesion–seed–protective layers is deposited on the silicon substrate, see Figure 8.10.

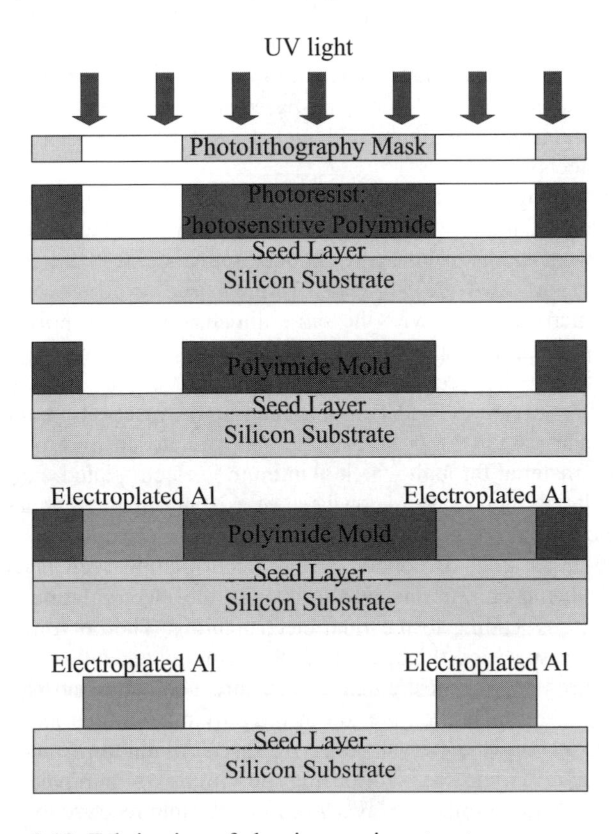


Figure 8.10 Fabrication of aluminum microstructures.

After the metal is deposited on the substrate, the antireflective coating is spun in the vacuum spinning station. The photosensitive polyimide is spun on top of the antireflective coating. The photosensitive polyimide is soft-baked in the oven and patterned (conventional microelectronic alignment and exposure stations can be used). The polyimide is developed and rinsed in solvent-based solutions to create the patterned molds in the thick polyimide films. After the polyimide molds are made, the antireactive coating is removed in oxygen plasma. The protective metal layer overlying the electroplating seed layer is removed (chromium is etched using the $HCl:H_2O$ 1:1 solution). Two additional steps are required to prepare the seed layer for aluminum electroplating. In particular, step 1: the sample is rinsed in isopropyl alcohol or methanol to remove residual water vapor from the surface; step 2: the sample is dipped in a solution of salicylic acid (100 g/l solution) in ether (this solution prepares the surface of the metal seed layer for electrodeposition removing residual metal oxides). Then, the sample is transferred to the electrolytic cell after removal from the salicylic acid solution. Electroplated aluminum is deposited to the surface of the mold. After completion of the electroplating process, the substrate is rinsed in alcohol to remove residues from the plating bath. The polyimide mold can be removed through wet etching of the polyimide using strippers or dry etching in oxygen-based plasmas [9]. Gold, copper, aluminum, and nickel can be used as seed layers for aluminum-based electroplated microstructures (electroplating using aluminum and nickel seed layers is sensitive to the oxidation effects).

Achieving high aspect ratio for aluminum-based microstructures is of great importance, and electroplating of aluminum using polymer micromolds can be viewed as the primary technique to make these microstructures. Article [9] reports that the developed refined process allows one to make aluminum microstructures with the same dimensions as the polyimide used as the mold. High aspect ratio is achieved utilizing the inversion characteristic of the process. In general, high aspect ratio polyimide microstructures are simpler to fabricate than high-aspect-ratio trenches. The fabrication starts with the basic process using photosensitive polyimide as the molding material through which aluminum is electroplated. For example, the electroplating metal system consisted of a 30 nm titanium adhesion layer between the silicon (or ceramic) substrate and the electroplating seed layer, 100 nm of either gold or copper as the electroplating seed layer, and an overlying 100 nm chromium layer to protect the electroplating seed layer during processes leading to the final electroplating. The polyimide mold is spun on at the specified thickness and photolithographically defined in the pattern required for the desired microstructure geometry and shape. After aluminum is electroplated on the top of the polyimide molds, the polyimide is removed, leaving the free-standing (released) aluminum microstructure. The seed layer is exposed (after the polyimide is removed), and the overlying chromium protective layer is removed using reactive ion etching.

### 8.6.2 $Ni_{x\%}Fe_{100-x\%}$ Thin Films Electrodeposition

As was reported, the ferromagnetic core of microstructures and microtransducers must be fabricated. Two major challenges in fabrication of high-performance microstructures are to make electroplated magnetic thin films with good magnetic properties as well as planarize microstructures (with primary emphasis on stationary and movable members). Electroplating and micromolding techniques and processes are used to deposit NiFe alloys ($Ni_{x\%}Fe_{100-x\%}$ thin films), and in conventional electromechanical motion devices, the $Ni_{80\%}Fe_{20\%}$ alloy is called *permalloy*, while $Ni_{50\%}Fe_{50\%}$ is called *orthonol*.

Let us document the deposition process. To deposit $Ni_{x\%}Fe_{100-x\%}$ thin films, the silicon wafer is covered with a seed layer (for example, for Cr-Cu-Cr seed layer, one may use 15-25 nm chromium, 100-200 nm copper, and 25-50 nm chromium) deposited using electron beam evaporation. The photoresist layer (e.g., 10-20 μm Shipley STR-1110) is deposited on the seed layer and patterned. Then, the electrodeposition of the $Ni_{x\%}Fe_{100-x\%}$ is performed at particular temperature (usually 20-30$^0$C) using a two electrode system, and, in general, the current density is in the range from 1 to 30 mA/cm$^2$. The temperature and pH should be maintained within the recommended values. High pH causes highly stressed NiFe thin films, and the low pH reduces leveling and causes chemical dissolving of the iron anodes resulting in disruption of the bath equilibrium and nonuniformity. High temperature leads to hazy deposits, and low temperature causes high current density burning. Thus, as was emphasized, many trade-offs must be taking into account.

For deposition, the pulse-width-modulation (with varied waveforms, different forward and reverse magnitudes, and controlled duty cycle) can be used applying commercial or in-house made power supplies. Denoting the duty cycle length as $T$, the forward and reverse pulse lengths are denoted as $T_f$ and $T_r$. The pulse length $T$ is usually in the range 5-20 μsec, and the duty cycle (ratio $T_f/T_r$) can be varied from 1 to 0.1. The ratio $T_f/T_r$ influences the percentage of Ni in the $Ni_{x\%}Fe_{100-x\%}$ thin films, e.g., the composition of $Ni_{x\%}Fe_{100-x\%}$ can be regulated based upon the desired properties which will be discussed later. However, varying the ratio $T_f/T_r$, the changes of the Ni are relatively modest (from 85 to 79%), and therefore other parameters vary to attain the desired composition.

It must be emphasized that the nickel (and iron) composition is a function of the current density, and Figure 8.11 illustrates the nickel (iron) composition in the $Ni_{x\%}Fe_{100-x\%}$ thin films.

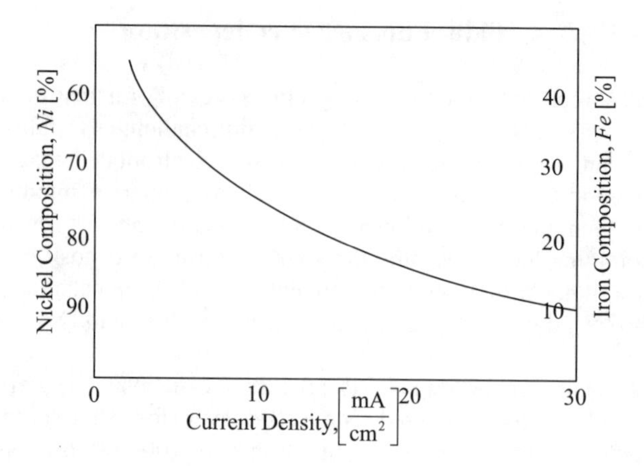


Figure 8.11 Nickel and iron compositions in $Ni_{x\%}Fe_{100-x\%}$ thin films as the functions of the current density.

The $Ni_{80\%}Fe_{20\%}$ thin films of different thickness (which is a function of the electrodeposition time) are usually made at the current density 14-16 mA/cm$^2$. This range of the current density can be used to fabricate a various thickness of *permalloy* thin films (from 500 nm to 50 μm). The rms value of the thin film roughness is 4-7 nm for the 30 μm thickness. It should be emphasized that to guarantee good surface quality, the current density should be kept at the specified range, and usually to change the composition of the $Ni_{x\%}Fe_{100-x\%}$ thin films, the reverse current is controlled.

To attain a good deposit of the *permalloy*, the electroplating bath may contain (other chemicals can be added and concentrations can be different): $NiSO_4$ (0.7 mol/l), $FeSO_4$ (0.03 mol/l), $NiCl_2$ (0.02 mol/l), saccharine (0.016 mol/l) as leveler (to reduce the residual stress allowing the fabrication of thicker films), and boric acid (0.4 mol/l).

The air agitation and saccharin are added to reduce internal stress and to keep the Fe composition stable. The deposition rate varies linearly as a function of the current density (the Faraday law is obeyed), and the electrodeposition slope is 100-150 nm-cm$^2$/min-mA. The *permalloy* thin films density is 9 g/cm$^3$ (as for the bulk *permalloy*).

The magnetic properties of the $Ni_{80\%}Fe_{20\%}$ (*permalloy*) thin films are studied, and the field coercivity ($H_c$) is a function of the thickness. For example, $H_c$=650 A/m for 150 nm thickness and $H_c$=30 A/m for 600 nm films.

Other possible solutions for electroplating the $Ni_{80\%}Fe_{20\%}$ (deposited at 25$^0$C) and $Ni_{50\%}Fe_{50\%}$ (deposited at 55$^0$C) thin films are:

- $Ni_{80\%}Fe_{20\%}$: $NiSO_4$–$6H_2O$ (200 g/l/), $FeSO_4$–$7H_2O$ (9 g/l), $NiCl_2$–$6H_2O$ (5 g/l), $H_3BO_3$ (27 g/l), saccharine (3 g/l), and pH (2.5-3.5);
- $Ni_{50\%}Fe_{50\%}$: $NiSO_4$–$6H_2O$ (170 g/l/), $FeSO_4$–$7H_2O$ (80 g/l), $NiCl_2$–$6H_2O$ (138 g/l), $H_3BO_3$ (50 g/l), saccharine (3 g/l), and pH (3.5-4.5).

To electroplate $Ni_{x\%}Fe_{100-x\%}$ thin films, various additives (chemicals) and components (available from M&T Chemicals and other suppliers) can be used to optimize (control) the thin film properties and characteristics. For example, in practice, the objective is to control the internal stress and ductility of the deposit, keep the iron content solublized, obtain bright film and leveling of the process, attain the desired surface roughness, and most importantly to guarantee the desired magnetic properties.

In general, *permalloy* thin films have optimal magnetic properties at the following composition of nickel and iron: 80.5% of Ni and 19.5% of Fe. Thin films with minimal magnetostriction usually have optimal coercivity and permeability properties. For $Ni_{80.5\%}Fe_{19.5\%}$ thin films, the magnetostriction has zero crossing, the coercivity is 20 A/m or higher (coercivity is a nonlinear function of the film thickness), and the permeability is from 600 to 2000. Varying the composition of Fe and Ni, the characteristics of the $Ni_{x\%}Fe_{100-x\%}$ thin films can be changed. The composition of the $Ni_{x\%}Fe_{100-x\%}$ thin films is controlled by changing the current density, $T_f/T_r$ ratio (duty cycle), bath temperature (varying the temperature, the composition of Ni can be varied from 75 to 92%), reverse current (varying the reverse current in the range 0-1 A, the composition of Ni can be changed from 72 to 90%), air agitation of the solution, paddle frequency (0.1 – 1 Hz), forward and reverse pulses waveforms, etc. The *B-H* curves for three different $Ni_{x\%}Fe_{100-x\%}$ thin films are illustrated in Figure 8.12. The $Ni_{80.5\%}Fe_{19.5\%}$ thin films have the saturation flux density 1.2 T, remanence $B_r$=0.26 T-A/m, and the relative permeability 600–2000.

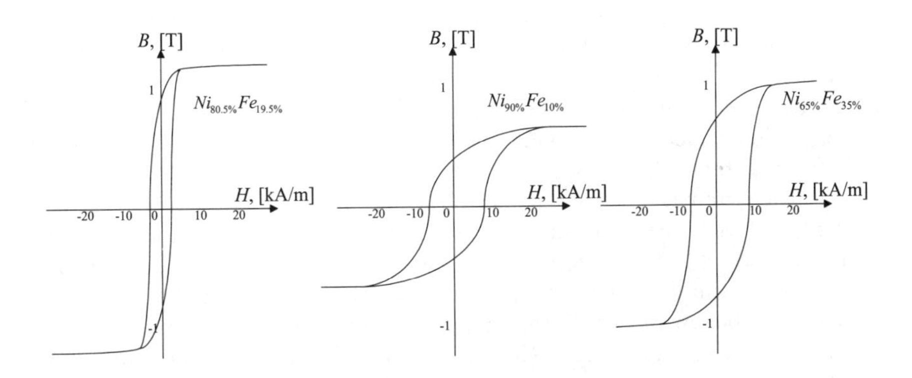


Figure 8.12 *B-H* curves for different $Ni_{x\%}Fe_{100-x\%}$ *permalloy* thin films.

It must be emphasized that other electroplated permanent magnets (NiFeMo, NiCo, CoNiMnP, and others) and micromachined polymer magnets exhibit good magnetic properties and can be used as the alternative solution to the $Ni_{x\%}Fe_{100-x\%}$ thin films widely used.

### 8.6.3 NiFeMo and NiCo Thin Films Electrodeposition

To attain the desired magnetic properties (flux density, coercivity, permeability, etc.) and thickness, different thin film alloys can be used. As was emphasized, magnetic materials are primarily used to fabricate stationary and movable members of microtransducers. Based upon the microstructures and microtransducers design, applications, and operating envelopes (temperature, shocks, radiation, humidity, etc.), different thin film films can be applied. As was discussed, the $Ni_{x\%}Fe_{100-x\%}$ thin films can be effectively used, and the desired magnetic properties can be readily achieved varying the composition of Ni. For sensors, the designer usually maximizes the flux density and permeability, and minimizes the coercivity. The $Ni_{x\%}Fe_{100-x\%}$ thin films have the flux density up to 1.2 T, coercivity 20 (for *permalloy*) to 500 A/m, and permeability 600 to 2000 (it was emphasized that the magnetic properties depend upon the thickness and other factors). Having emphasized the magnetic properties of the $Ni_{x\%}Fe_{100-x\%}$ thin films, let us perform the comparison.

It was reported in the literature that [11]:

- $Ni_{79\%}Fe_{17\%}Mo_{4\%}$ thin films have the flux density 0.7 T, coercivity 5 A/m and permeability 3400,
- $Ni_{85\%}Fe_{14\%}Mo_{1\%}$ thin films have the flux density 1-1.1 T, coercivity 8-300 A/m and permeability 3000–20000,
- $Ni_{50\%}Co_{50\%}$ thin films have the flux density 0.95-1.1 T, coercivity 1200-1500 A/m and permeability 100-150 ($Ni_{79\%}Co_{21\%}$. thin films have the permeability 20).

In general, high flux density, low coercivity, and high permeability lead to high-performance MEMS. However, other issues (affordability, compliance, integrity, operating envelope, fabrication processes, etc.) must be also addressed making the final choice. It must be emphasized that the magnetic characteristics, in addition to the film thickness, are significantly influenced by the fabrication processes and chemicals used.

The ferromagnetic core in microstructures and microtransducers must be made. Two major challenges in fabrication of high-performance microstructures and microtransducers are to make electroplated magnetic thin films with good magnetic properties as well as planarize the stationary and movable members. Electroplating and micromolding techniques and processes are used to deposit NiFe, NiFeMo, NiCo, and other thin films. For example, the $Ni_{80\%}Fe_{20\%}$, NiFeMo and NiCo (deposited at $25^0$C) electroplating solutions are:

- $Ni_{80\%}Fe_{20\%}$: $NiSO_4$–$6H_2O$ (200 g/l/), $FeSO_4$–$7H_2O$ (9 g/l), $NiCl_2$–$6H_2O$ (5 g/l), $H_3BO_3$ (27 g/l), and saccharine (3 g/l). The current density is 10–25 mA/cm$^2$ (nickel foil is used as the anode);
- NiFeMo: $NiSO_4$–$6H_2O$ (60 g/l/), $FeSO_4$–$7H_2O$ (4 g/l), $Na_2MoO_4$–$2H_2O$ (2 g/l), NaCl (10 g/l), citrid acid (66 g/l), and saccharine (3 g/l). The current density is 10–30 mA/cm$^2$ (nickel foil is used as the anode);

- $Ni_{50\%}Co_{50\%}$: $NiSO_4$–$6H_2O$ (300 g/l/), $NiCl_2$–$6H_2O$ (50 g/l), $CoSO_4$–$7H_2O$ (30 g/l), $H_3BO_3$ (30 g/l), sodium lauryl sulfate (0.1 g/l), and saccharine (1.5 g/l). The current density is 10–25 mA/cm$^2$ (nickel or cobalt can be used as the anode).

The most important feature is that the $Ni_{x\%}Fe_{100-x\%}$–NiFeMo–NiCo thin films (multilayer nanocomposites) can be fabricated shaping the magnetic properties of the resulting materials to attain the desired performance characteristics through design and fabrication processes.

## 8.6.4 Micromachined Polymer Magnets

Electromagnetic microtransducers can be devised, modeled, analyzed, designed, optimized, and then fabricated. Micromachined permanent magnet thin films, including polymer magnets (magnetically hard ceramic ferrite powder imbedded in epoxy resin), can be used. This section is focused on the application of the micromachined polymer magnets. Different forms and geometry of polymer magnets are available. Thin-film disks and plates are uniquely suitable for microtransducer applications. For example, to actuate the switches in microscale logic devices, to displace mirrors in optical devices and optical MEMS, etc. In fact, permanent magnets are used in rotational and translational (linear) microtransducers, microsensors, microswitches, etc. The polymer magnets have thicknesses ranging from hundreds of micrometers to several millimeters. Excellent magnetic properties can be achieved. For example, the micromachined polymer permanent-magnet disk with 80% strontium ferrite concentration (4 mm diameter and 90 μm thickness), magnetized normal to the thin-film plane (in the thickness direction), has the intrinsic coercivity $H_{ci}$=320000 A/m and a residual induction $B_r$=0.06 T [12]. Polymer magnets with thickness up to several millimeters can be fabricated by the low-temperature processes. To make the permanent magnets. The Hoosier Magnetics Co. strontium ferrite powder (1.1-1.5 μm grain size) and Shell epoxy resin (cured at 80$^0$C for two hours) can be used [12]. The polymer matrix contains a bisphenol-A-based epoxy resin diluted with cresylglycidyl ether, and the aliphatic amidoamine is used for curing. To prepare the polymer magnet composites, the strontium ferrite powder is mixed with the epoxy resin in the ball-mill rotating system (0.5 rad/sec for many hours). After the aliphatic amidoamine is added, the epoxy is deposited and patterned using screen-printing. Then, the magnet is cured at 80$^0$C for two hours and magnetized in the desired direction. It must be emphasized that magnets must be magnetized. That is, in addition to fabrication processes, one should study other issues, for example, the magnetization dynamics and permanent-magnet magnetic properties. The magnetic field in thin films are modeled, analyzed, and simulated solving differential equations, and the analytic and numerical results will be covered in this chapter.

### 8.6. 5 Planarization

The planarization of the movable and stationary members is a very important issue. For example, in fabrication of microtransducers, the gaps between the magnetic poles (permanent magnets) and teeth must be filled in order to eliminate the forces without disintegration and degradation of thin films. Copper can be applied to fill the gaps. Negative epoxy-based photoresistive SU-8, which has superior intrinsic adhesion characteristics (chemically resistant and thermally stable up to 250$^0$C), is widely used. The processing of the photoresistive epoxy resin starts with spin-coating in the dehumidified water at 150–250 rad/sec, resulting in the thickness of 15-25 μm. The softbrake can be made at 100$^0$C, following by cooling. After the relaxation exposure, the final hardbrake process is carried out. To deposit the copper in the desired area, photomasks are made which cover all regions except the regions where the deposition is needed (for example, between the teeth). The copper is deposited using the electroplating process described. It should be emphasized that the teeth gap can be filled with other deposited materials, e.g., aluminum (however, aluminum deposition is more challenging compared with the copper electroplating).

## 8.7 MAGNETIZATION DYNAMICS OF THIN FILMS

The magnetic field, including the magnetization distribution, in thin films is modeled, analyzed, and simulated solving differential equations. The dynamic variables are the magnetic field density and intensity, magnetization, magnetization direction, wall position domain, etc. The thin films must be magnetized. Therefore, let us study the magnetization dynamics in thin films. To attain high-fidelity modeling, the magnetization dynamics in the angular coordinates is described by the Landay-Lifschitz-Gilbert equations [13]

$$\frac{d\psi}{dt} = -\frac{\gamma}{M_s(1+\alpha^2)}\left(\sin^{-1}\psi\frac{\partial E(\theta,\psi)}{\partial\theta} + \alpha\frac{\partial E(\theta,\psi)}{\partial\psi}\right),$$

$$\frac{d\theta}{dt} = -\frac{\gamma\sin^{-1}\psi}{M_s(1+\alpha^2)}\left(\alpha\sin^{-1}\psi\frac{\partial E(\theta,\psi)}{\partial\theta} - \frac{\partial E(\theta,\psi)}{\partial\psi}\right),$$

where $M_s$ is the saturation magnetization; $E(\theta,\psi)$ is the total Gibb's thin film free energy density; $\gamma$ and $\alpha$ are the gyromagnetic and phenomenological constants.

The total energy consists the magnetocrystalline anisotropy energy, the exchange energy, and the magnetostatic self-energy (stray field energy) [14]. In particular, the Zeeman energy equation is

$$E = \int_v \left( \frac{k_{exh}}{J_s^2} \sum_{j=1}^{3} (\nabla J_j)^2 - \frac{k_J}{J_s^2} (\mathbf{a}_J \mathbf{J})^2 - \tfrac{1}{2} \mathbf{J} \cdot \mathbf{H}_D - \mathbf{J} \cdot \mathbf{H}_{ex} \right) dv ,$$

while the Gilbert equation is

$$\frac{\partial \mathbf{J}}{\partial t} = -|\gamma| \mathbf{J} \times \mathbf{H}_{eff} + \frac{\alpha}{J_s} \mathbf{J} \times \frac{\partial \mathbf{J}}{\partial t} .$$

Here, $\mathbf{J}$ is the magnetic polarization vector; $\mathbf{H}_D$ and $\mathbf{H}_{ex}$ are the demagnetizing and external magnetic fields; $\mathbf{H}_{eff}$ is the effective magnetic field (sum of the applied, demagnetization, and anisotropy fields); $k_{exh}$ and $k_J$ are the exchange and magnetocrystalline anisotropy constants; $\mathbf{a}_J$ is the unit vector parallel to the uniaxial easy axis.

Using the vector notations, we have

$$\frac{d\mathbf{M}}{dt} = -\frac{\gamma}{1+\alpha^2} \left( \mathbf{M} \times \mathbf{H}_{eff} + \frac{\alpha}{M_s} \mathbf{M} \times \left( \mathbf{M} \times \mathbf{H}_{eff} \right) \right) .$$

Thus, using the nonlinear differential equations given, the high-fidelity modeling and analysis of nanostructured nanocomposite permanent magnets can be performed using field and material quantities, parameters, constants, etc. For example, using the available software (developed by the National Institute of Standard and Technologies), the results of the three-dimensional simulation of the magnetic field in the *permalloy* thin film micromagnet (50–100–5 nm) are illustrated in Figure 8.13.

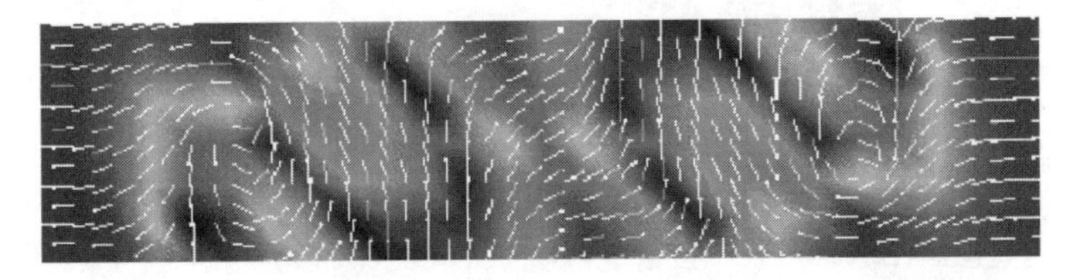

Figure 8.13 Magnetic field in the *permalloy* thin film micromagnet.

The characteristics of nano- and microscale structures (multilayer nanocomposed thin film layers) can be measured. The cryogenic microwave probe test-station for testing is illustrated in Figure 8.14.

Figure 8.14 Cryogenic probe test-station.

## 8.8 MICROSTRUCTURES AND MICROTRANSDUCERS WITH PERMANENT MAGNETS: MICROMIRROR ACTUATORS

The electromagnetic microactuator (permanent magnet on the cantilever flexible beam and spiral planar windings controlled by ICs fabricated using CMOS-MEMS technology) is illustrated in Figure 8.15. In addition, the microstructure with eight cantilever beams and eight-by-eight fiber switching matrix are illustrated.

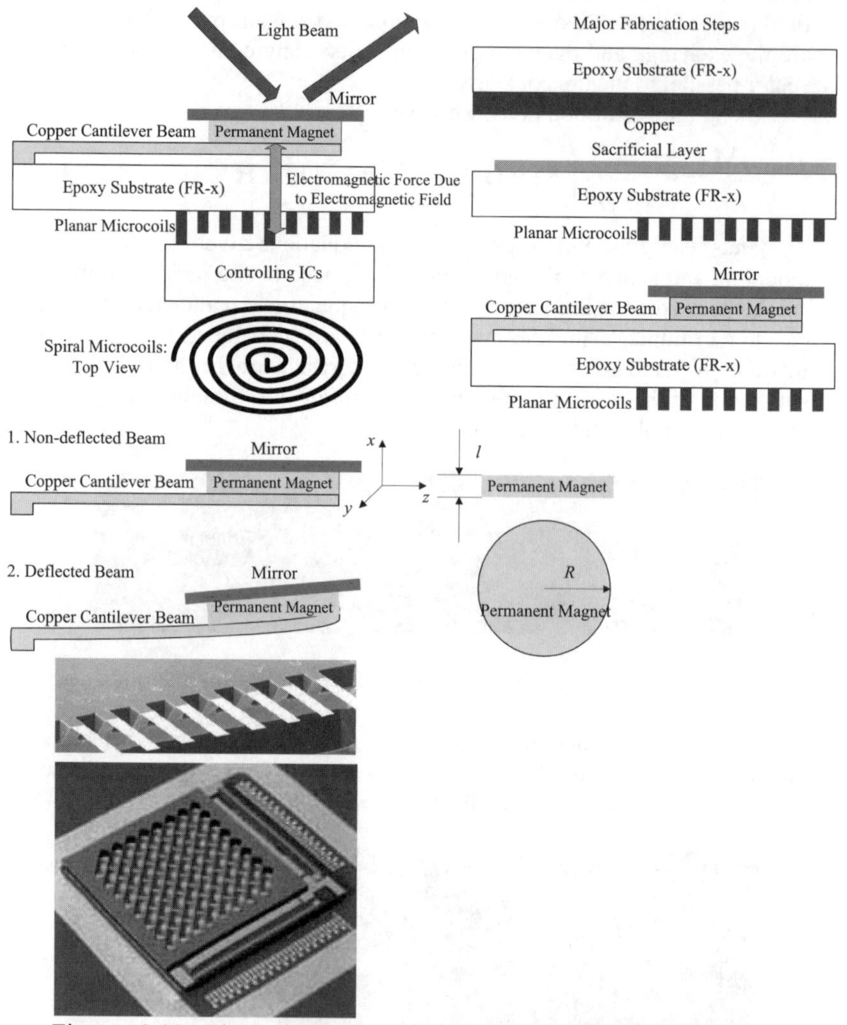


Figure 8.15 Electromagnetic microactuator with controlling ICs, microstructure with eight cantilever beams, and eight incoming–eight outgoing array of optical fiber (eight-by-eight fiber switching matrix).

The electromagnetic microactuators can be made using conventional surface micromachining and CMOS fabrication technologies through electroplating, screen printing, lamination processes, sacrificial layer techniques, photolithography, etching, etc. In particular, the electromagnetic microactuator studied can be made on the commercially available epoxy substrates (e.g., FR series) which have the one-sided laminated copper layer. It must be emphasized that the copper layer thickness (which can be from 10 μm and higher) is defined by the admissible current density and the current value needed to establish the desired magnetic field to attain the specified mirror deflection, deflection rate, settling time, and other steady-state and dynamic characteristics. The spiral planar microcoils can be made on the one-sided laminated copper layer using photolithography and wet etching in the ferric chloride solution. The resulting $x$-μm thick $N$-turn microwinding will establish the magnetic field (the number of turns is a function of the footprint area available, thickness, spacing, outer-inner radii, geometry, fabrication techniques and processes used, etc.). After fabrication of the planar microcoils, the cantilever beam with the permanent magnet and mirror is fabricated on the other side of the substrate. First, a photoresist sacrificial layer is spin-coated and patterned on the substrate. Then, the Cr-Cu-Cr or Ti-Cu-Cr seed layer is deposited to perform the copper electroplating (if copper is used to fabricate the flexible cantilever structure). The second photoresist layer is spun and patterned to serve as a mold for the electroplating of the copper-based cantilever beam. The copper cantilever beam is electroplated in the copper-sulfate-based plating bath. After the electroplating, the photoresist plating mold and the seed layer are removed releasing the cantilever beam structure. It must be emphasized that depending upon the permanent magnet used, the corresponding fabricated processes must be done before or after releasing the beam. The permanent-magnet disk is positioned on the cantilever beam's free end (for example, the polymer magnet can be screen-printed, and after curing the epoxy magnet, the magnet is magnetized by the external magnetic field). Then, the cantilever beam with the fabricated mirror is released by removing the sacrificial photoresist layer using acetone. The studied electromagnetic microactuator is fabricated using low-cost (affordable), high-yield micromachining–CMOS technology, processes, and materials. The most attractive feature is the application of the planar microcoils which can be easily made. The use of the polymer permanent magnets (which have good magnetic properties) allows one to design high-performance electromagnetic microactuators. It must be emphasized that the polysilicon can be used to fabricate the cantilever beam, and other permanent magnets can be applied.

In the article [12], the vertical electromagnetic force $F_{ze}$, acting on the permanent-magnet, is given by

$$F_{ze} = M_z \int_v \frac{dH_z}{dz} dv,$$

where $M_z$ is the magenetization; $H_z$ is the vertical component of the magnetic field intensity produced by the planar microwindings ($H_z$ is a nonlinear function of the current fed or voltage applied to the microwindings, number of turns, microcoils geometry, etc.; therefore, the thickness of the microcoils must be derived based on the maximum value of the current needed and the admissible current density).

The magnetically actuated cantilever microstructures were studied also in articles [15, 16], and the expressions for the electromagnetic torque are found as the functions of the magnetic field using assumptions and simplifications which, in general, limit the applicability of the results. The differential equations which model the electromagnetic and *torsional-mechanical* dynamics can be derived. In particular, the equations for the electromagnetic field are found using electromagnetic theory, and the electromagnetic filed intensity $H_z$ is controlled changing the current fed to the planar microwindings. The steady-state analysis, performed using the small-deflection theory [17], is also valuable. The static deflection of the cantilever beam $x$ can be straightforwardly found using the force and beam quantities. In particular, $x=\frac{l^3}{3EJ}F_n$, where, $l$ is the effective length of the beam; $E$ is the Young's (elasticity) modulus; $J$ is the equivalent moment of inertia of the beam with permanent magnet and mirror, and for the stand-alone cantilever beam with the rectangular cross section $J=\frac{1}{12}wh^3$; $w$ and $h$ are the width and thickness of the beam; $F_n$ is the net force which is normal to the cantilever beam.

Assuming that the magnetic flux is constant through the magnetic plane (loop), the torque on a planar current loop of any size and shape in the uniform magnetic field is

$$\mathbf{T}=i\mathbf{s}\times\mathbf{B}=\mathbf{m}\times\mathbf{B},$$

where $i$ is the current; $\mathbf{m}$ is the magnetic dipole moment [A-m$^2$].

Thus, the torque on the current loop always tends to turn the loop to align the magnetic field produced by the loop with permanent-magnet magnetic filed causing the resulting electromagnetic torque. For example, for the current loop shown in Figure 8.16, the torque is found to be

$$\mathbf{T}=i\mathbf{s}\times\mathbf{B}=\mathbf{m}\times\mathbf{B}$$
$$=1\times10^{-3}\left[(1\times10^{-3})(2\times10^{-3})\mathbf{a}_z\right]\times\left(-0.5\mathbf{a}_y+\mathbf{a}_z\right)=1\times10^{-9}\mathbf{a}_x\ \text{N-m}.$$

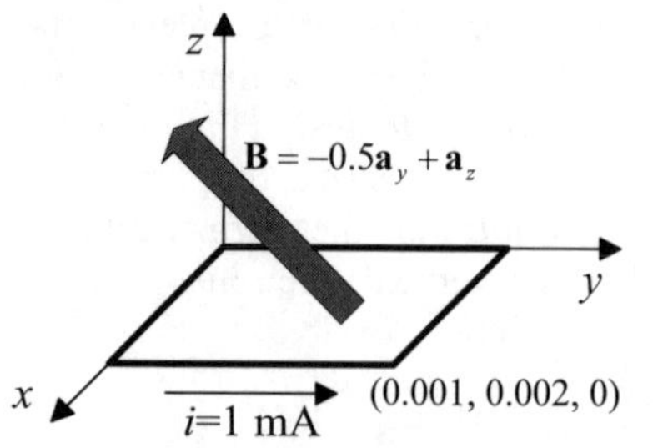

Figure 8.16 Rectangular planar loop in a uniform magnetic field with flux density $\mathbf{B}=-0.5\mathbf{a}_y+\mathbf{a}_z$.

The electromagnetic force is found as

$$\mathbf{F} = \oint_l i d\mathbf{l} \times \mathbf{B}.$$

In general, the magnetic field quantities are derived using

$$\mathbf{B} = \frac{\mu_0}{4\pi} i \oint_l \frac{d\mathbf{l} \times \mathrm{r}_0}{r^2} \text{ or } \mathbf{H} = \frac{1}{4\pi} i \oint_l \frac{d\mathbf{l} \times \mathrm{r}_0}{r^2},$$

and the Ampere circuital law gives $\oint_l \mathbf{H} \cdot d\mathbf{l} = i_{total}$ or $\oint_l \mathbf{H} \cdot d\mathbf{l} = Ni$.

Making use these expressions and taking note the variables defined in Figure 8.17, we have

$$\mathbf{H} = \frac{1}{4\pi} i \oint_l \frac{d\mathbf{l} \times \mathbf{r}_1}{r_1^3} \text{ and } \mathbf{B} = \frac{\mu_0}{4\pi} i \oint_l \frac{d\mathbf{l} \times \mathbf{r}_1}{r_1^3},$$

where $d\mathbf{l} = \mathbf{a}_\phi a d\phi = (-\mathbf{a}_x \sin\phi + \mathbf{a}_y \cos\phi) a d\phi$

and $\mathbf{r}_1 = \mathbf{a}_x (x - a\cos\phi) + \mathbf{a}_y (y - a\sin\phi) + \mathbf{a}_z z$.

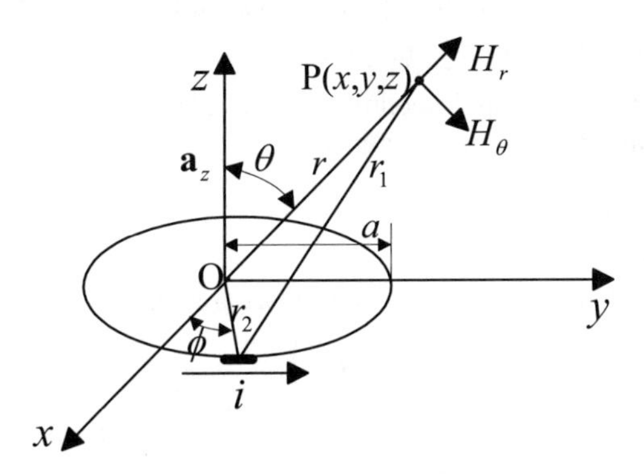


Figure 8.17 Planar current loop.

Hence,

$$d\mathbf{l} \times \mathbf{r}_1 = \left[\mathbf{a}_x z\cos\phi + \mathbf{a}_y z\sin\phi - \mathbf{a}_z(y\sin\phi + x\cos\phi - a)\right] a d\phi.$$

Then, neglecting the small quantities ($a^2 << r^2$), we have

$$r_1^3 = \left(x^2 + y^2 + z^2 + a^2 - 2ax\cos\phi - 2ay\sin\phi\right)^{3/2}$$

$$\approx r^3\left(1 - \frac{2ax}{r^2}\cos\phi - \frac{2ay}{r^2}\sin\phi\right)^{3/2}.$$

Therefore, one obtains

$$\frac{1}{r_1^3} = \frac{1}{r^3}\left(1 + \frac{3ax}{r^2}\cos\phi + \frac{3ay}{r^2}\sin\phi\right).$$

Thus, we have

$$\mathbf{B}=\frac{\mu_0 a}{4\pi}i\int_0^{2\pi}\left[\mathbf{a}_x z\cos\phi+\mathbf{a}_y z\sin\phi-\mathbf{a}_z\left(y\sin\phi+x\cos\phi-a\right)\right]a\frac{1}{r^3}\left(1+\frac{3ax}{r^2}\cos\phi+\frac{3ay}{r^2}\sin\phi\right)d\phi=$$

$$=\frac{\mu_0 a^2}{4\pi r^3}i\left[\mathbf{a}_x\frac{3xz}{r^2}+\mathbf{a}_y\frac{3yz}{r^2}-\mathbf{a}_z\left(\frac{3x^2}{r^2}+\frac{3y^2}{r^2}-2\right)\right].$$

Furthermore, using the coordinate transformation equations, in the spherical coordinate system one has

$$\mathbf{B}=\frac{\mu_0 a^2}{4\pi r^3}i(2\mathbf{a}_r\cos\theta+\mathbf{a}_\theta\sin\theta).$$

We have the expressions for the far-field components

$$B_r=\frac{\mu_0 a^2\cos\theta}{2\pi r^3}i,$$

$$B_\theta=\frac{\mu_0 a^2\sin\theta}{4\pi r^3}i,$$

$$B_\phi=0.$$

It is evident that due to the symmetry about the $z$ axis, the magnetic flux density does not have the $B_\phi$ component.

Using the documented technique, one can easily find the magnetic vector potential. In particular, in general

$$\mathbf{A}=\frac{\mu_0}{4\pi}i\oint_l\frac{d\mathbf{l}}{r_1}.$$

Assuming that $a^2 << r^2$, gives the following expression

$$\frac{1}{r_1}=\frac{1}{r}\left(1+\frac{ax}{r^2}\cos\phi+\frac{ay}{r^2}\sin\phi\right).$$

Therefore,

$$\mathbf{A}=\frac{\mu_0 a}{4\pi}i\int_0^{2\pi}(-\mathbf{a}_x\sin\phi+\mathbf{a}_y\cos\phi)\frac{1}{r}\left(1+\frac{ax}{r^2}\cos\phi+\frac{ay}{r^2}\sin\phi\right)d\phi$$

$$=\frac{\mu_0 a}{4\pi r^3}i(-\mathbf{a}_x y+\mathbf{a}_y x).$$

Hence, in the spherical coordinate system, we obtain

$$\mathbf{A}=(\mathbf{A}\cdot\mathbf{a}_r)\mathbf{a}_r+(\mathbf{A}\cdot\mathbf{a}_\phi)\mathbf{a}_\phi+(\mathbf{A}\cdot\mathbf{a}_\theta)\mathbf{a}_\theta=\frac{\mu_0 a}{4\pi r^2}i\mathbf{a}_\phi\sin\theta=A_\phi\mathbf{a}_\phi.$$

It should be emphasized that the equations derived can be expressed using the magnetic dipole.

However, in the microtransducer studied, high-fidelity analysis should be performed. Hence, let us perform the comprehensive analysis.

The vector potential is found to be

$$A_\phi(r,\theta)=\frac{\mu_0 ai}{4\pi}\int_0^{2\pi}\frac{\cos\phi d\phi}{\sqrt{a^2+r^2-2ar\sin\theta\cos\phi}},$$

and $B_r=\frac{1}{r\sin\theta}\frac{\partial(\sin\theta A_\phi)}{\partial\theta}$, $B_\theta=-\frac{1}{r}\frac{\partial(rA_\phi)}{\partial r}$, $B_\phi=0$.

Making use the following approximation

$$A_\phi(r,\theta)=\frac{\mu_0 ai}{4\pi}\int_0^{2\pi}\frac{\cos\phi d\phi}{\sqrt{a^2+r^2-2ar\sin\theta\cos\phi}}$$

$$\approx\frac{\mu_0 a^2 r\sin\theta i}{4(a^2+r^2)^{3/2}}\left(1+\frac{15a^2r^2\sin^2\theta}{8(a^2+r^2)^2}+\ldots\right),$$

one finds

$$B_r(r,\theta)=\frac{\mu_0 a^2\cos\theta i}{2(a^2+r^2)^{3/2}}\left(1+\frac{15a^2r^2\sin^2\theta}{4(a^2+r^2)^2}+\ldots\right),$$

$$B_\theta(r,\theta)=-\frac{\mu_0 a^2\sin\theta i}{4(a^2+r^2)^{5/2}}\left(2a^2-r^2+\frac{15a^2r^2\sin^2\theta(4a^2-3r^2)}{8(a^2+r^2)^2}+\ldots\right)$$

$B_\phi=0$.

We specify three regions:

- Near the axis $\theta<<1$;
- At the center $r<<a$;
- In far-field $r>>a$.

The electromagnetic torque and field depend upon the current in the microwindings and are nonlinear functions of the displacement.

Finally, it is important to emphasize that the magnetic dipole moment is

$$\mathbf{m}=\pi a^2 i\mathbf{a}_z,$$

where $\pi a^2$ is the planar area of the loop.

For sinusoidal field variation, using the equation for the vector magnetic potential $\nabla^2\mathbf{A}+\omega_f^2\mu\varepsilon\mathbf{A}=-\mu\mathbf{J}$ and the Lorentz condition $\nabla\cdot\mathbf{A}=-j\omega_f\mu\varepsilon V$, we have

$$\mathbf{A}=\frac{\mu}{4\pi}\int_v\frac{1}{r}\mathbf{J}e^{-j\beta r}dv,\ \beta=\omega_f\sqrt{\mu\varepsilon}.$$

Thus, for the current loop, the vector magnetic potential is given as

$$\mathbf{A}=\frac{\mu}{4\pi}\int_v\frac{1}{r_1}\mathbf{J}e^{-j\beta r_1}dv=\frac{\mu}{4\pi}\int_l\frac{1}{r_1}\mathbf{i}e^{-j\beta r_1}dl.$$

Taking note of $e^{-j\beta r_1} = e^{-j\beta r} e^{-j\beta(r_1 - r)} \approx e^{-j\beta r}\left[1 - j\beta(r_1 - r)\right]$, one has

$$\mathbf{A} = \frac{\mu a^2}{4r^2} i(1 + j\beta r)e^{-j\beta r} \sin\theta \mathbf{a}_\phi .$$

The expression for the electromagnetic forces and torques must be derived to model and analyze the *torsional-mechanical* dynamics. Newton's laws of motion can be applied to study the mechanical dynamics in the Cartesian or other coordinate systems (e.g., previously for the translational motion in the $x$-axis, we used $\frac{dv}{dt} = \frac{1}{m}\left(F_e - F_L\right)$ and $\frac{dx}{dt} = v$ to model the translational *torsional-mechanical* dynamics of the electromagnetic microactuators using the electromagnetic force $F_e$ and the load force $F_L$).

For the studied microactuator, the rotational motion can be studied, and the electromagnetic torque can be approximated as

$$T_e = 4R^2 t_{tf} M H_p \cos\theta ,$$

where $R$ and $t_{tf}$ are the radius and thickness of the permanent-magnet thin-film disk; $M$ is the permanent-magnet thin film magnetization; $H_p$ is the filed produced by the planar windings; $\theta$ is the displacement angle.

Then, the microactuator rotational dynamics is given by

$$\frac{d\omega}{dt} = \frac{1}{J}\left(T_e - T_L\right) \text{ and } \frac{d\theta}{dt} = \omega ,$$

where $T_L$ is the load torque which integrates the friction and disturbances torques.

It should be emphasized that more complex and comprehensive mathematical models can be developed and used integrating the nonlinear electromagnetic and six-degree-of-freedom rotational-translational motions (*torsional-mechanical* dynamics) of the cantilever beam. As an illustration we consider the high-fidelity modeling of the electromagnetic system.

### *Electromagnetic System Modeling in Microactuators With Permanent Magnets: High-Fidelity Modeling and Analysis*

In this section we focus our efforts to derive the expanded equations for the electromagnetic torque and force on cylindrical permanent-magnet thin films, see Figure 8.15. The permanent-magnet thin film is assumed to be uniformly magnetized, and the equations are developed for two orientations of the magnetization vector (the orientation is parallel to the axis of symmetry, and the orientation is perpendicular to this axis). Electromagnetic fields and gradients produced by the planar windings should be found at a point in inertial space which coincides with the origin of the permanent-magnet axis system in its initial alignment. Our ultimate goal is to control microactuators, and thus, high-fidelity mathematical models (which will result in viable analysis, control, and optimization) must be derived [18]. To attain our objective, the complete equations for the electromagnetic torque and force on a cylindrical permanent-magnet thin films are found.

The following notations are used: $A$, $R$ and $l$ are the area, radius, and length of the cylindrical permanent magnet; $\mathbf{B}$ is the magnetic flux density vector; $\mathbf{B}_e$ is the expanded magnetic flux density vector; $[\partial \mathbf{B}]$ is the matrix of field gradients [T/m]; $[\partial \mathbf{B}_e]$ is the matrix of expanded field gradients [T/m]; $\mathbf{F}$ and $\mathbf{T}$ are the total force and torque vectors on the permanent-magnet thin film; $i$ is the current in the planar microwinding; $\mathbf{m}$ is the magnetic moment vector [A-m$^2$]; $\mathbf{M}$ is the magnetization vector [A/m]; $\mathbf{r}$ is the position vector ($x$, $y$, $z$ are the coordinates in the Cartesian system), $\mathbf{r} = \begin{bmatrix} x \\ y \\ z \end{bmatrix}$; $T_r$ is the inertial coordinate vector-transformation matrix; $W$ and $\Pi$ are the work and potential energy; $\theta$ is the Euler orientation for the 3-2-1 rotation sequence; $\nabla$ is the gradient operator; $_{ij}$ is the partial derivative of $i$ component in $j$-direction; $_{(ij)k}$ is the partial derivative of $ij$ partial derivative in $k$-direction; $\bar{\ }$ (bar over a variable) indicates that it is referenced to the microactuator coordinates.

### *Electromagnetic Torques and Forces: Preliminaries*

The equations for the electromagnetic torque and force on a cylindrical permanent-magnet thin film are found by integrating the equations for torques and forces on an incremental volume of the permanent-magnet thin film with magnetic moment $\mathbf{M}dv$ over the volume. Figure 8.15 illustrates the microactuator with the cylindrical permanent-magnet thin film in the coordinate system which consists of a set of orthogonal body-fixed axes that are initially aligned with a set of orthogonal $x$-, $y$-, $z$-axes fixed in the inertial space.

The equations for the electromagnetic torque and force on an infinitesimal current can be derived using the fundamental relationship for the force on a current-carrying-conductor element in a uniform magnetic field. In particular, for a planar current loop (planar microwinding) with constant current $i$ in the uniform magnetic field $\mathbf{B}$ (vector $\mathbf{B}$ gives the magnitude and direction of the flux density of the external field), the force on an element $\mathbf{dl}$ of the conductor is found using the Lorentz force law

$$\mathbf{F} = \oint_l i d\mathbf{l} \times \mathbf{B}.$$

Assuming that the magnetic flux is constant through the magnetic loop, the torque on a planar current loop of any size and shape in the uniform magnetic field is

$$\mathbf{T}=i\oint_l \mathbf{r}\times(d\mathbf{l}\times\mathbf{B})=i\oint_l\left((\mathbf{r}\cdot\mathbf{B})d\mathbf{l}-\mathbf{B}\oint_l \mathbf{r}\cdot d\mathbf{l}\right).$$

Using Stokes's theorem, one has

$$\mathbf{T}=i\left(\int_s d\mathbf{A}\times\nabla(\mathbf{r}\cdot\mathbf{B})-\mathbf{B}\int_s(\nabla\times\mathbf{r})\cdot d\mathbf{A}\right)=i\int_s d\mathbf{A}\times\mathbf{B},$$

or $\mathbf{T}=i\mathbf{A}\times\mathbf{B}=\mathbf{m}\times\mathbf{B}$.

The electromagnetic torque **T** acts on the infinitesimal current loop in a direction to align the magnetic moment **m** with the external field **B**, and if **m** and **B** are misaligned by the angle $\theta$, we have

$$\mathbf{T}=\mathbf{mB}\sin\theta.$$

The incremental potential energy and work are found as

$$dW=d\Pi=\mathbf{T}d\theta=m\mathbf{B}\sin\theta d\theta,$$

and $W=\Pi=-m\mathbf{B}\cos\theta=-\mathbf{m}\cdot\mathbf{B}$.

Using the electromagnetic force, we have

$$dW=-d\Pi=\mathbf{F}\cdot d\mathbf{r}=-\nabla\Pi\cdot d\mathbf{r},$$

and

$$\mathbf{F}=-\nabla\Pi=\nabla(\mathbf{m}\cdot\mathbf{B})=(\mathbf{m}\cdot\nabla)\mathbf{B}$$

*Coordinate Systems and Electromagnetic Field*

The transformation from the inertial coordinates to the permanent-magnet coordinates is

$$\bar{\mathbf{r}}=T_r\mathbf{r}$$

$$=\begin{bmatrix}\cos\theta_y\cos\theta_z & \cos\theta_y\sin\theta_z & -\sin\theta_y\\ \sin\theta_x\sin\theta_y\cos\theta_z-\cos\theta_x\sin\theta_z & \sin\theta_x\sin\theta_y\sin\theta_z+\cos\theta_x\sin\theta_z & \sin\theta_x\cos\theta_y\\ \cos\theta_x\sin\theta_y\cos\theta_z+\sin\theta_x\sin\theta_z & \cos\theta_x\sin\theta_y\sin\theta_z-\sin\theta_x\cos\theta_z & \cos\theta_x\cos\theta_y\end{bmatrix}\begin{bmatrix}x\\y\\z\end{bmatrix},$$

$$\mathbf{r}=\begin{bmatrix}x\\y\\z\end{bmatrix},\ \bar{\mathbf{r}}=\begin{bmatrix}\bar{x}\\\bar{y}\\\bar{z}\end{bmatrix}$$

Using the transformation matrix

$$T_r=\begin{bmatrix}\cos\theta_y\cos\theta_z & \cos\theta_y\sin\theta_z & -\sin\theta_y\\ \sin\theta_x\sin\theta_y\cos\theta_z-\cos\theta_x\sin\theta_z & \sin\theta_x\sin\theta_y\sin\theta_z+\cos\theta_x\sin\theta_z & \sin\theta_x\cos\theta_y\\ \cos\theta_x\sin\theta_y\cos\theta_z+\sin\theta_x\sin\theta_z & \cos\theta_x\sin\theta_y\sin\theta_z-\sin\theta_x\cos\theta_z & \cos\theta_x\cos\theta_y\end{bmatrix},$$

if the deflections are small, we have $T_{rs}=\begin{bmatrix}1 & \theta_z & -\theta_y\\ -\theta_z & 1 & \theta_x\\ \theta_y & -\theta_x & 1\end{bmatrix}$.

It should be emphasized that we use the 3-2-1 orthogonal transformation matrix for the *z-y-x* Euler rotation sequence, and $\theta_x, \theta_y, \theta_z$ are the rotation Euler angles about the *x*, *y* and *z* axes.

The field **B** and gradients of **B** produced by the microcoils fixed in the inertial frame and expressed assuming that the electromagnetic fields can be described by the second-order Taylor series. Expanding **B** about the origin of the *x*, *y*, z system as a Taylor series, we have [18]

$$\mathbf{B}_e = \mathbf{B} + (\mathbf{r}\cdot\nabla)\mathbf{B} + \tfrac{1}{2}(\mathbf{r}\cdot\nabla)^2\mathbf{B},$$

or $B_{ei} = B_i + \dfrac{\partial B_i}{\partial \mathbf{r}}\mathbf{r} + \tfrac{1}{2}\mathbf{r}^T \dfrac{\partial^2 B_i}{\partial \mathbf{r}^2}\mathbf{r}$,

where $\dfrac{\partial B_i}{\partial \mathbf{r}} = \begin{bmatrix} \dfrac{\partial B_i}{\partial x} & \dfrac{\partial B_i}{\partial y} & \dfrac{\partial B_i}{\partial z} \end{bmatrix}$,

and $$\frac{\partial^2 B_i}{\partial \mathbf{r}^2} = \begin{bmatrix} \dfrac{\partial \frac{\partial B_i}{\partial x}}{\partial x} & \dfrac{\partial \frac{\partial B_i}{\partial x}}{\partial y} & \dfrac{\partial \frac{\partial B_i}{\partial x}}{\partial z} \\ \dfrac{\partial \frac{\partial B_i}{\partial y}}{\partial x} & \dfrac{\partial \frac{\partial B_i}{\partial y}}{\partial y} & \dfrac{\partial \frac{\partial B_i}{\partial y}}{\partial z} \\ \dfrac{\partial \frac{\partial B_i}{\partial z}}{\partial x} & \dfrac{\partial \frac{\partial B_i}{\partial z}}{\partial y} & \dfrac{\partial \frac{\partial B_i}{\partial z}}{\partial z} \end{bmatrix}.$$

We denote $B_{ij} = \dfrac{\partial B_i}{\partial j}$ and $B_{(ij)k} = \dfrac{\partial \frac{\partial B_i}{\partial j}}{\partial k}$.

Then,

$$\frac{\partial B_i}{\partial \mathbf{r}} = \begin{bmatrix} B_{ix} & B_{iy} & B_{iz} \end{bmatrix}$$

and $$\frac{\partial^2 B_i}{\partial \mathbf{r}^2} = \begin{bmatrix} B_{(ix)x} & B_{(ix)y} & B_{(ix)z} \\ B_{(iy)x} & B_{(iy)y} & B_{(iy)z} \\ B_{(iz)x} & B_{(iz)y} & B_{(iz)z} \end{bmatrix}.$$

Hence, the first-order gradients is given as

$$B_{eij} = B_{ij} + \frac{\partial \frac{\partial B_i}{\partial j}}{\partial \mathbf{r}}\mathbf{r} = B_{ij} + \begin{bmatrix} B_{(ij)x} & B_{(ij)y} & B_{(ij)z} \end{bmatrix}\mathbf{r}.$$

The expanded fields is expressed in the permanent-magnet coordinates as

$$\overline{\mathbf{B}}_e = \overline{\mathbf{B}} + (\overline{\mathbf{r}} \cdot \overline{\nabla})\overline{\mathbf{B}} + \tfrac{1}{2}(\overline{\mathbf{r}} \cdot \overline{\nabla})^2 \overline{\mathbf{B}},$$

where $\overline{\mathbf{B}} = T_r \mathbf{B}$ and $\overline{\nabla} = T_r \nabla$.

Using $\mathbf{r} = T_r^T \overline{\mathbf{r}}$, one has

$$B_{ei} = B_i + \frac{\partial B_i}{\partial \mathbf{r}} T_r^T \overline{\mathbf{r}} + \tfrac{1}{2} \overline{\mathbf{r}}^T T_r \frac{\partial^2 B_i}{\partial \mathbf{r}^2} T_r^T \overline{\mathbf{r}},$$

and $$\overline{\mathbf{B}}_e = T_r \begin{bmatrix} B_x + \frac{\partial B_x}{\partial \mathbf{r}} T_r^T \overline{\mathbf{r}} + \tfrac{1}{2} \overline{\mathbf{r}}^T T_r \frac{\partial^2 B_x}{\partial \mathbf{r}^2} T_r^T \overline{\mathbf{r}} \\ B_y + \frac{\partial B_y}{\partial \mathbf{r}} T_r^T \overline{\mathbf{r}} + \tfrac{1}{2} \overline{\mathbf{r}}^T T_r \frac{\partial^2 B_y}{\partial \mathbf{r}^2} T_r^T \overline{\mathbf{r}} \\ B_z + \frac{\partial B_z}{\partial \mathbf{r}} T_r^T \overline{\mathbf{r}} + \tfrac{1}{2} \overline{\mathbf{r}}^T T_r \frac{\partial^2 B_z}{\partial \mathbf{r}^2} T_r^T \overline{\mathbf{r}} \end{bmatrix}$$

*Electromagnetic Torques and Forces*

Now let us derive the fields and gradients at any point in the permanent magnet using the second-order Taylor series approximation. To eliminate the transformations between the inertial and permanent magnet coordinate systems and simplify the second-order negligible small components, we assume that the relative motion between the magnet and the reference inertial coordinate is zero and the $T_{rs}$ transformation matrix is used (otherwise, the second-order gradient terms will be very cumbersome).

The magnetization (the magnetic moment per unit volume) is constant over the volume of the permanent-magnet thin films, and **m**=**M**$v$.

Assuming that the magnetic flux is constant, the total electromagnetic torque and force on a planar current loop (microwinding) in the uniform magnetic field is

$$\overline{\mathbf{T}} = \int_v \left( \overline{\mathbf{M}} \times \mathbf{B}_e + \overline{\mathbf{r}} \times (\overline{\mathbf{M}} \cdot \nabla)\mathbf{B}_e \right) dv,$$

$$\overline{\mathbf{F}} = \int_v (\overline{\mathbf{M}} \cdot \nabla)\mathbf{B}_e dv,$$

where $$(\overline{\mathbf{M}} \cdot \nabla)\mathbf{B}_e = [\partial \mathbf{B}_e]\overline{\mathbf{M}} = \begin{bmatrix} B_{exx} & B_{exy} & B_{exz} \\ B_{eyx} & B_{eyy} & B_{eyz} \\ B_{ezx} & B_{ezy} & B_{ezz} \end{bmatrix} \begin{bmatrix} M_{\bar{x}} \\ M_{\bar{y}} \\ M_{\bar{z}} \end{bmatrix}.$$

*Case 1:* Magnetization Along the Axis of Symmetry

For orientation of the magnetization vector along the axis of symmetry ($x$-axis) of the permanent-magnet thin films, we have

$$(\overline{\mathbf{M}}\cdot\nabla)\mathbf{B}_e=[\partial\mathbf{B}_e]\overline{\mathbf{M}}=M_{\bar{x}}\begin{bmatrix}B_{exx}\\B_{exy}\\B_{exz}\end{bmatrix}.$$

Thus, in the expression $\overline{\mathbf{T}}=\int_v\left(\overline{\mathbf{M}}\times\mathbf{B}_e+\bar{\mathbf{r}}\times(\overline{\mathbf{M}}\cdot\nabla)\mathbf{B}_e\right)dv$,

the terms are

$$\bar{\mathbf{r}}\times(\overline{\mathbf{M}}\cdot\nabla)\mathbf{B}_e=M_{\bar{x}}\begin{bmatrix}-B_{exy}\bar{z}+B_{exz}\bar{y}\\B_{exx}\bar{z}-B_{exz}\bar{x}\\-B_{exx}\bar{y}+B_{exy}\bar{x}\end{bmatrix}\text{ and }\overline{\mathbf{M}}\times\mathbf{B}_e=M_{\bar{x}}\begin{bmatrix}0\\-B_{ez}\\B_{ey}\end{bmatrix}.$$

Therefore,

$$T_{\bar{x}}=M_{\bar{x}}\int_v\left(B_{exz}\bar{y}-B_{exy}\bar{z}\right)dv,$$

$$T_{\bar{y}}=-M_{\bar{x}}\int_v B_{ez}dv+M_{\bar{x}}\int_v\left(B_{exx}\bar{z}-B_{exz}\bar{x}\right)dv,$$

and $T_{\bar{z}}=M_{\bar{x}}\int_v B_{ey}dv+M_{\bar{x}}\int_v\left(B_{exy}\bar{x}-B_{exx}\bar{y}\right)dv$.

The terms in the derived equations must be evaluated.

Let us find the analytic expression for the electromagnetic torque $T_{\bar{x}}$.

In particular, we have

$$\int_v B_{exz}\bar{y}dv=B_{xz}\int_v\bar{y}dv+B_{(xx)z}\int_v\overline{xy}dv+B_{(xy)z}\int_v\bar{y}^2dv+B_{(xz)z}\int_v\overline{zy}dv,$$

where $\int_v\bar{y}dv=0$, $\int_v\overline{xy}dv=0$, $\int_v\overline{zy}dv=0$, and

$$\int_v\bar{y}^2dv=\int_{-\frac{1}{2}l}^{\frac{1}{2}l}\int_{-R}^{R}\int_{-\sqrt{R^2-\bar{z}^2}}^{\sqrt{R^2-\bar{z}^2}}\bar{y}^2d\bar{y}d\bar{z}d\bar{x}=\tfrac{1}{4}\pi lR^4=\tfrac{1}{4}vR^4.$$

Therefore, $M_{\bar{x}}\int_v B_{exz}\bar{y}dv=M_{\bar{x}}\tfrac{1}{4}B_{(xy)z}vR^4$.

Furthermore, $M_{\bar{x}}\int_v B_{exy}\bar{z}dv=M_{\bar{x}}\tfrac{1}{4}B_{(xy)z}vR^4$.

Thus, for $T_{\bar{x}}$, one has

$$T_{\bar{x}}=M_{\bar{x}}\int_v\left(B_{exz}\bar{y}-B_{exy}\bar{z}\right)dv=M_{\bar{x}}\left(\tfrac{1}{4}B_{(xy)z}vR^4-\tfrac{1}{4}B_{(xy)z}vR^4\right)=0.$$

Then, for $T_{\bar{y}}$, we obtain

$$T_{\bar{y}} = -M_{\bar{x}} \int_v B_{ez} dv + M_{\bar{x}} \int_v (B_{exx}\bar{z} - B_{exz}\bar{x}) dv$$

$$= M_{\bar{x}} \left[ -\left(B_z + B_{(zx)x} \tfrac{1}{24} l^2 + B_{(zy)y} \tfrac{1}{8} R^2 + B_{(zz)z} \tfrac{1}{8} R^2\right) v + B_{(xx)z} \left(\tfrac{1}{4} R^2 - \tfrac{1}{12} l^2\right) v \right]$$

$$= -v M_{\bar{x}} \left(B_z + B_{(xx)z} \left(\tfrac{1}{4} R^2 - \tfrac{1}{8} l^2\right) + B_{(yy)z} \tfrac{1}{8} R^2 + B_{(zz)z} \tfrac{1}{4} R^2\right).$$

Finally, we obtain the expression for $T_z$ as

$$T_{\bar{z}} = M_{\bar{x}} \int_v B_{ey} dv + M_{\bar{x}} \int_v (B_{exy}\bar{x} - B_{exx}\bar{y}) dv$$

$$= v M_{\bar{x}} \left(B_y + B_{(xx)y} \left(\tfrac{1}{8} l^2 - \tfrac{1}{4} R^2\right) - B_{(yy)y} \tfrac{1}{8} R^2 - B_{(yz)z} \tfrac{1}{8} R^2\right).$$

Thus, the following electromagnetic torque equations result

$$T_{\bar{x}} = 0,$$

$$T_{\bar{y}} = -v M_{\bar{x}} \left(B_z + B_{(xx)z} \left(\tfrac{1}{4} R^2 - \tfrac{1}{8} l^2\right) + B_{(yy)z} \tfrac{1}{8} R^2 + B_{(zz)z} \tfrac{1}{4} R^2\right),$$

$$T_{\bar{z}} = v M_{\bar{x}} \left(B_y + B_{(xx)y} \left(\tfrac{1}{8} l^2 - \tfrac{1}{4} R^2\right) - B_{(yy)y} \tfrac{1}{8} R^2 - B_{(yz)z} \tfrac{1}{8} R^2\right).$$

The electromagnetic forces are found as well. In particular, from

$$F_{\bar{x}} = M_{\bar{x}} \int_v B_{exx} dv,$$

$$F_{\bar{y}} = M_{\bar{x}} \int_v B_{exy} dv$$

and $F_{\bar{z}} = M_{\bar{x}} \int_v B_{exz} dv,$

using the expressions for the expanded magnetic fluxes, e.g., $\int_v B_{exx} dv = \int_v \left(B_{xx} + B_{(xx)x}\bar{x} + B_{(xx)y}\bar{y} + B_{(xx)z}\bar{z}\right) dv$, and performing the integration, one has the following expressions for the electromagnetic forces as the function of the magnetic field

$$F_{\bar{x}} = v M_{\bar{x}} B_{xx},$$

$$F_{\bar{y}} = v M_{\bar{x}} B_{xy},$$

$$F_{\bar{z}} = v M_{\bar{x}} B_{xz}.$$

The documented expressions for torques and forces, obtained as the nonlinear functions of the permanent-magnet parameters (magnetization, area, radius, and length) and magnetic field variables, allow one to model six-degree-of-freedom *torsional-mechanical* dynamics.

*Case 2:* Magnetization Perpendicular to the Axis of Symmetry

For orientation of the magnetization vector perpendicular to the axis of symmetry, the following equation is used to find the electromagnetic torque

$$\overline{\mathbf{T}} = \int_v \left(\overline{\mathbf{M}} \times \mathbf{B}_e + \overline{\mathbf{r}} \times (\overline{\mathbf{M}} \cdot \nabla)\mathbf{B}_e\right) dv,$$

where $(\overline{\mathbf{M}} \cdot \nabla)\mathbf{B}_e = [\partial \mathbf{B}_e]\overline{\mathbf{M}} = M_{\bar{z}} \begin{bmatrix} B_{exz} \\ B_{eyz} \\ B_{ezz} \end{bmatrix}$,

$$\overline{\mathbf{r}} \times (\overline{\mathbf{M}} \cdot \nabla)\mathbf{B}_e = M_{\bar{z}} \begin{bmatrix} -B_{eyz}\bar{z} + B_{ezz}\bar{y} \\ B_{exz}\bar{z} - B_{ezz}\bar{x} \\ -B_{exz}\bar{y} + B_{eyz}\bar{x} \end{bmatrix} \text{ and } \overline{\mathbf{M}} \times \mathbf{B}_e = M_{\bar{z}} \begin{bmatrix} -B_{ey} \\ B_{ex} \\ 0 \end{bmatrix}.$$

Thus,

$$T_{\bar{x}} = -M_{\bar{z}} \int_v B_{ey} dv + M_{\bar{z}} \int_v \left(B_{exz}\bar{y} - B_{eyz}\bar{z}\right) dv,$$

$$T_{\bar{y}} = M_{\bar{z}} \int_v B_{ex} dv + M_{\bar{z}} \int_v \left(B_{exz}\bar{z} - B_{ezz}\bar{x}\right) dv,$$

and $T_{\bar{z}} = M_{\bar{z}} \int_v \left(B_{eyz}\bar{x} - B_{exz}\bar{y}\right) dv$.

Expressing the fluxes and performing the integration, we have the following expressions for the torque components as the function of the magnetic field

$$T_{\bar{x}} = -vM_{\bar{z}}\left(B_y + B_{(xx)y}\tfrac{1}{24}l^2 + B_{(yy)y}\tfrac{1}{8}R^2 + B_{(yz)z}\tfrac{1}{8}R^2\right),$$

$$T_{\bar{y}} = vM_{\bar{z}}\left(B_x + B_{(xz)z}\left(\tfrac{3}{8}R^2 - \tfrac{1}{12}l^2\right) + B_{(xx)x}\tfrac{1}{24}l^2 + B_{(xy)y}\tfrac{1}{8}R^2\right),$$

$$T_{\bar{z}} = vM_{\bar{z}}B_{(xy)z}\left(\tfrac{1}{12}l^2 - \tfrac{1}{4}R^2\right).$$

The electromagnetic forces are found to be

$$F_{\bar{x}} = M_{\bar{z}} \int_v B_{exz} dv = vM_{\bar{z}}B_{xz},$$

$$F_{\bar{y}} = M_{\bar{z}} \int_v B_{eyz} dv = vM_{\bar{z}}B_{yz},$$

and $F_{\bar{z}} = M_{\bar{z}} \int_v B_{ezz} dv = vM_{\bar{z}}B_{zz}$.

Thus, the expressions for the electromagnetic force and torque components are derived. These equations provide one with the clear perspective how to model, analyze, and control the electromagnetic forces and torques changing the applied magnetic field because the terms $B_{ij} = \dfrac{\partial B_i}{\partial j}$ and $B_{(ij)k} = \dfrac{\partial \frac{\partial B_i}{\partial j}}{\partial k}$ can be viewed as the control variables. It must be emphasized that the electromagnetic field ($B_{ij}$ and $B_{(ij)k}$) is controlled by regulating the current in the planar microwindings and designing the microwindings (or other radiating energy microdevices). As was discussed, the derived forces and torques must be used in the torsional-mechanical equations of motion for the microactuator, and in general, the six-degree-of-freedom microactuator mechanical dynamics results. These mechanical equations of motion are easily integrated with the derived electromagnetic equations, and closed-loop systems can be designed to attain the desired microactuator performance. These equations guide us to the importance of electromagnetic features in the modeling, analysis, and design of microactuators.

*Some Other Aspects of Microactuator Design and Optimization*

In addition to the electromagnetic-mechanical (electromechanical) analysis and design, other design and optimization problems are involved. As an example, let us focus our attention on the planar windings. The ideal planar microwindings must produce the maximum electromagnetic field minimizing the footprint area taking into the consideration the material characteristics, operating conditions, applications, power requirements, and many other factors. Many planar winding parameters and characteristics can be optimized, for example, the dc resistance must be minimized to improve the efficiency, increase the flux, decrease the losses, etc. To attain good performance, in general, microwindings have the concentric circular current path and no interconnect resistances. For $N$-turn winding, the total dc resistance $r_t$ is found to be

$$r_t = \frac{2\pi\rho}{t_w} \sum_{k=1}^{N} \frac{1}{\ln\left( r_{Ok} / r_{Ik} \right)},$$

where $\rho$ is the winding material resistivity; $t_w$ is the winding thickness; $r_{Ok}$ and $r_{Ik}$ are the outer and inner radii of the $k$ turn winding.

To achieve the lowest resistance, the planar winding radii can be optimized by minimizing the resistance, and the minimum resistance is denoted as $r_{t\min}$. In particular, making use of first- and second-order necessary conditions for minimization, one has

$$\frac{dr_t}{dr_w} = 0 \text{ and } \frac{d^2 r_t}{dr_w^2} > 0,$$

where $r_w$ is the inner or outer radius of an arbitrary turn of the optimized planar windings from the standpoint of minimizing the resistance.

Then, the minimum value of the microcoil resistance is given by

$$r_{t\min} = \frac{2\pi\rho}{t_w} \frac{N}{\left( r_{OR} / r_{IR} \right)},$$

where $r_{OR}$ and $r_{IR}$ are the outer and inner radii of the windings (that is, $r_{O\ N\text{-th microcoil}}$ and $r_{I\ \text{1st microcoil}}$).

Thus, using the number of turns and turn-to-turn spacing, the outer and inner radii of the *k* turn winding are found as

$$\frac{r_{Ok}}{r_{Ik}} = \left(\frac{r_{OR}}{r_{IR}}\right)^{1/N}.$$

For spiral windings, the *averaging* (*equivalency*) concept should be used because the outer and inner radii are the functions of the planar angle, see Figure 8.18. Finally, it should be emphasized that the width of the *N*-th microcoil is specified by the rated voltage current density versus maximum current density needed, fabrication technologies used, material characteristics, etc.

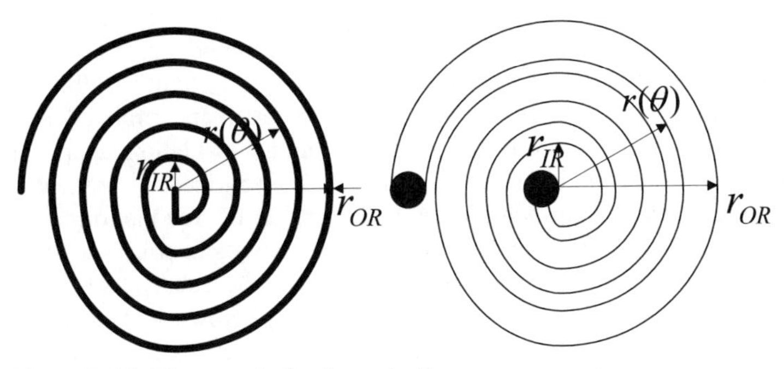


Figure 8.18 Planar spiral microwinding.

## 8.9 RELUCTANCE ELECTROMAGNETIC MICROMOTORS

Surface micromachining and CMOS technologies (photolithography, deposition, electroplating, molding, planarization, and other processes) are used to fabricate rotational and translational microtransducers. Microtransducers integrate magnetic materials (core), permanent magnets, windings, bearing, etc. It was emphasized that the NiFe thin films (as well other magnetic and nonmagnetic materials) and copper coils can be electroplated, and the fabrication processes were discussed. Manageable results are reported in [2, 4, 19, 20]. In general, the NiFe and copper electroplating baths, electroplating seed layers, and through-mask plating techniques are similar to those used to fabricate inductive thin-film heads. High-aspect-ratio optical lithography and x-ray lithography can be used to form the various resist layers needed. The stators and rotors can be fabricated separately or sequentially depending upon the microtransducers devised and fabrication technologies used (foe example, if stator and rotor are made separately, the microrotor is released from the substrate and installed on the shaft made as a part of the stator fabrication). High aspect ratio microstructure and electroplating-through-mask processes are the major steps to fabricate synchronous (including permanent-magnet and reluctance), induction, and other electromagnetic microtransducers. These microtransducers can be used as high-precision microscale drives and actuators in automotive, aerospace, power, manufacturing, biomedical, biotechnology, and other applications.

Shown in Figure 8.3, a synchronous reluctance micromotor has stationary and movable members. The stator and rotor are batch-fabricated microstructures made using the similar processes and materials because the same magnetic electroplated material can be used (however, the copper microwindings are electroplated on the stator). Microtransducers must be made through the developed batch-fabrication processes sequence (series of lithographic, electroplating, and planarization steps). As studied before, the process starts by the sputter-deposition of the Ti-Cu-Ti seed layer on an oxidized silicon wafer (Si-$SiO_2$ substrate), lithographically forming the plate-through (patterned) mask for the bottom part of the copper microwindings, removing the upper Ti layer, and electroplating the copper in the exposed seed layer regions. The photoresist and seed layer are removed from the regions between the electroplated structure, and the insulation layer (polymer dielectric) is deposited. Then, the surface is planarized. A thin patterned polymer dielectric is formed on top of the copper microcoils to insulate the windings from the ferromagnetic core (stator), leaving apertures at the end of each copper curl to form the microwinding in the subsequent steps. The second Ti-Cu-Ti seed layer is deposited, and the photoresist is patterned. The upper Ti is etched, and the NiFe thin film is electroplated. Removing the photoresist and seed layer, the NiFe thin film ferromagnetic core (stator) is covered by the insulation layer (dielectric). It should be emphasized that microcoils are connected and insulated to form the desired microwinding structure. The rotor is fabricated by electroplating the NiFe thin film on the lithographically patterned Si-$SiO_2$ substrate precoated with a copper layer (the adhesion layer under the copper is Ti or Ta). The rotor can be detached through the chemical dissolution of the copper layer and installed on the shaft located in the center of the stator cavity. The microshaft and support pins should be made during the photolithography and electroplating processes [20].

It was emphasized that the seed layers for the electrodeposited copper and NiFe thin films consist of the sputtered Cu (100-300 nm) on the adhesion layer (Ti or Ta). A layer of Ti is deposited on the Cu to enhance the resist adhesion. The fabrication processes are based upon high-aspect-ratio optical lithography using positive (novolac-type) and negative optical photoresists. Although the positive photoresists are attractive due to easy removal following electrodeposition, it is difficult to form large thickness and uniformity with respect to sidewall angle. The negative photoresist consists of solutions of different concentrations of Shell EPON Negative epoxy-based photoresist resin SU-8 and photosensitized with a commercially available triaryl sulfonium salt (Cyravure UVI, Union Carbide Co). Using negative photoresist, straight sidewall profiles (vertical sidewall with 85-90$^0$ angle) can be obtained for a photoresist thickness up to 130 μm for the single layers of resist on the copper and NiFe layers. In contrast, the positive resist (high-viscosity novolac, Shipley SJR 3740 or AZ P4620, Hoechst Celanese Co), can be applied using a multilayer spin method to achieve the thickness up to 70 μm. This photoresist can be exposed using UV light (patterning can be performed using contact printing on the UV Mask Aligner), and photoresist is removed without affecting the underlying structure. In particular, the hexamethyl disiloxane can be used to obtain good adhesion of the photoresist to the underlying electroplating copper seed layer. The baking process is necessary to stabilize the photoresist prior to UV light exposure.

The rotors and the stator are made using the NiFe thin films. The electrodeposition is carried out in the horizontal paddle cell connected to a plating solution reservoir. The solution should be continuously filtered (0.2 μm Millipore Co Filters can be used). To stabilize the temperature, the plating reservoir should have the automatic temperature-control system because the temperature must be 25$^0$C ± 0.1$^0$C. The solution pH (3±0.1) is controlled by adding the HCl. The ferrous ion concentration should be kept constant adding a ferrous sulfate solution (pH is 2) at the rate required to match the iron consumption rate. The nickel plate serves as the anode, and the stainless steel plate serves as an auxiliary electrode which is coplanar with the cathode. The auxiliary electrode reduces current nonuniformity at the edge of the cathode improving the deposit thickness uniformity across the cathode (particualrly at the edges). The current fed to the auxiliary electrode depends on the current density at the cathode. The current density commonly used is 10-15 mA/cm$^2$ to guarantee the uniform electrodeposition. It must be emphasized that the NiFe thin film layers can be planarized by polishing.

Electroplated copper is used to make microcoils, rotor support pins, and other microstructures (if needed). Electroplating processes, which guarantee the uniformity of the copper thickness, are commercially available and were described in this chapter.

It was emphasized that the insulation is very important in the microtransducer fabrication because the core must be insulated from the copper microcoils. Insulator layers must exhibit low stress and should be planarizable. Various dielectric materials (polyimide, hard-baked photoresist, and epoxy resin) have been studied from insulation and planarization viewpoints. The use of polyimide requires the barrier layer (electrolessly deposited material, e.g., NiP) to prevent a chemical reaction with copper during polyimide curing. In addition, voids formed due to the difficulty of removing solvent from the deep narrow regions, may result in short-circuits between the coils and core. The solventless epoxy resin eliminates the problem of voids, reduces stress, and does not require curing at high temperatures (temperature used for polyimides is 360-400$^0$C). The cured epoxy resin (dielectric material) is polished to planarize the surface and to expose copper and NiFe thin films. Through spin-coating, the planarized substrates with 5-7 μm epoxy resin layer can be made. The epoxy is patterned, covering microcoils to attain the insulation. After curing of the insulation layer, a seed layer is deposited, followed by photoresist. The insulation layer creates a nonplanar surface, however, this not a serious problem for the subsequent photoresist process. The difference in height between the core and the studs after NiFe thin film electrodeposition is removed through polishing following epoxy backfill.

The micromotor sizing depends upon the specifications and requirements (for example, the equation for the electromagnetic torque leads to the specified current, inductances, air gap, cross-sectional area, active magnetic length, etc.) as well as the processes and materials which significantly influence the magnetic and thermal characteristics, resistivity, etc. Processes and materials integration, as well as the integrity of the consequential steps, are key to fabricate micromotors. Microstators can be fabricated with 10-30 μm thick copper microconductors and 50 μm NiFe thin film using optical lithography (magnetic characteristics, resistivity, and inductance are the functions of the Cu and NiFe thickness). To fabricate microtransducers, different numbers of masks are needed. In particular, the studied microtransducer requires five mask levels as well as additional masks for reactive ion etching of the rotor cavity at the end of the fabrication to make the bearing. Also, electrodeposition processes are involved. The rotor cavity in the stator is opened using reactive ion etching with masking to ensure selectivity (epoxy resin dielectric removal rate in the range of 0.4-0.6 μm/min). It should be emphasized that the attention must be concentrated on the minimization of heat during sputter deposition of the electroplated seed layer. In fact, one must prevent the void formation at the Cu–insulator region because heat can cause cracks and voids in the thin insulation layer between the bottom Cu and NiFe. The effect of the thermal expansion mismatch must be minimized. During the reactive ion etching (to open the rotor cavity in the stator), the heating has to be minimized as well.

Thorough design can be performed optimizing the microtransducer characteristics by optimizing different parameters (sizing rotor and stator, shaping stator and rotor geometry, varying the air gap, deriving ferromagnetic core and permanent-magnet thickness, designing nanocomposite thin films, optimizing the magnetic properties of the magnetic materials such as permeability and hysteresis to minimize losses, minimizing the resistivity of the microwindings, maximizing the flux, reducing flux leakage and fringing effects, minimizing the friction, decreasing the torque ripple, attenuating viboacoustic phenomena, eliminating resonance effects, etc.). Many nonlinear phenomena cannot be modeled, analyzed, and even reasonably evaluated using the lumped-parameter mathematical models. Therefore, Maxwell's equations must be used integrating the basic nonlinear electromagnetic and torsional-mechanical phenomena, effects, and features.

Microtransducers can be fabricated using optical and *x*-ray lithography. Though the electromagnetic microtransducer design and operating principles are similar, different final design leads to many distinct features. As a result, fabrication processes and materials are different. For example, the size of microtransducers and microstructures is different, and high-aspect-ratio processes are usually involved. For optical lithography fabrication, the photoresist for electroplating the ferromagnetic core and conductor vias can be applied by spin coating. Using the LIGA process, the semirigid polymethylmethacrylate (PMMA) sheet can be glued to the substrate with PMMA, and the insulation of the core from the top and bottom copper conductors is needed. In fact, the majority of microtransducers have the bottom copper conductors level, the NiFe level, the top copper conductors level, and insulation layers. These three levels can be made using the Ti-Cu-Ti seed layer, spin-coating, and thermally curing the PMMA adhesion layer. Gluing a PMMA sheet (millimuter-thick Perspex CQ), fly-cutting the PMMA to the desired thickness (100-250 μm), and *x*-ray exposure are carried out. These processes are followed by resist-pattern development, pretreatment of the surface (remove Ti), electroplating Cu and/or NiFe thin films, and polishing the PMMA resist. Then, the AZ optical photoresist is applied, UV-exposing and developing of the photoresist are made, and copper is electroplated. Following removal of the AZ photoresist, PMMA and seed layers (through chemical etching or ion milling processes), an $SiO_2$-filled epoxy dielectric is applied and thermally cured. The excess dielectric is removed by polishing to expose the copper in preparation for the seed layer deposition and x-ray lithography as the needed steps to form next level.

The same sequential steps are performed to fabricate the NiFe thin film (second level). The connection of the top and bottom copper microconductors are made. After polishing the PMMA resist with respect to NiFe, the spacer pads (20-25 μm high) are electroplated using optically patterned photoresist. Following encapsulation of the NiFe with $SiO_2$-filled dielectric and polishing it to the NiFe thin film, the top copper conductor level is fabricated using the *x*-ray patterned PMMA resist. The *x*-ray lithography leads to the vertical sidewall profile leading the high aspect ratio microstructures. The encapsulated epoxy resin $SiO_2$-filled dielectric can be removed from the central rotor region through the reactive ion etching. The use of the epoxy resin $SiO_2$-filled dielectric complicates the etching process (compared with the optical lithography) because the etch rate of the silicon dioxide is slower compared with the epoxy resin. The rotor is electroplated on a separate substrate through the *x*-ray exposed PMMA, detached, and integrated with bearing following reactive ion etching. The rotor cavity on the stator is made.

In general, the x-ray lithography fabrication, which allows one to make thicker microstructures and larger microtransducers, leads to the following challenges [20]: (1) thicker layers lead to the thermal mismatch problems between copper microcoils, NiFe thin films, core, and insulator-dielectric (the thermal mismatch in the vertical direction is a serious issue – a perfectly planarized magnetic layer embedded in a polymer with high coefficient of thermal expansion can suffer delamination at the vertical metal-polymer interface when heated to $180^0$C during PMMA curing. This can cause cracking and delamination of the seed layer. This problem can be relaxed using a dielectric containing low-thermoexpansion composites, e.g., silicon dioxide. However, using the dielectric containing inorganic materials to reduce the thermal expansion mismatch results in changes of the reactive ion etching processes used to open the cavities to make the bearing for rotational microtransducers); (2) thick (100-400 μm) PMMA resist creates alignment problems due to the difficulty of accurately registering the mask to alignment marks under the thick resist (the processes for alignment through PMMA up to 400 μm thick with ±5 μm accuracy are developed by IBM [20]); (3) plating on the thick PMMA requires cautious design of the step sequences (surface pretreatment, PMMA development, and wetting initiation); (4) solvents and chemicals used in the PMMA processing must be compatible with underlying microstructure materials; (5) lamination of the PMMA sheet requires the minimization of the irregularities in the planarized wafer surface (planarization leads to the development of appropriate polishing techniques to deal with the various composite surfaces).

It should be emphasized that $SiO_2$ has been used as an insulating material in microelectronics. For example, $SiO_2$ is used as the gate dielectric of MOSFETs. Fabrication processes design, integration, optimization, and materials selection are the key in fabrication of electromagnetic microtransducers. The major fabrication problems that the designer faces are deposition of microconductors and magnetic thin films, planarization, core-insulator/dielectric-conductor thermal expansion mismatch, voids (including those caused by coefficient of thermal expansion mismatch), residue-free development in thick optically exposed photoresist and *x*-ray exposed PMMA resist, stress-free films, etc.

## 8.10 MICROMACHINED POLYCRISTALLINE SILICON CARBIDE MICROMOTORS

Multilayer fabrication processes at low temperature and micromolding techniques were developed to fabricate SiC microstructures and salient-pole micromotors which can be used at very high temperature ($400^0$C and higher) [21, 22]. This was done through the SiC surface micromachining.

Through the book it was emphasized that the MATLAB environment can be effectively used for CAD. The application of MATLAB is illustrated for the simples problem of the data fitting. The bulk modulus of SiC versus temperature is given as

$B=203_{T=20}, 200_{T=250}, 197_{T=500}, 194_{T=750}, 191_{T=1000}, 188_{T=1200}, 186_{T=1400}, 184_{T=1500}$ (the temperature is given in $^0$C).

The interpolation is performed using the `spline` MATLAB solver (spline fit).

The MATLAB file used is

```
T=[20 250 500 750 1000 1200 1400 1500]; % Temperature Data
B=[203 200 197 194 191 188 186 184];    % Bulk Modulus Data
Tinterpol=20:10:1500;
Binterpol=spline(T,B,Tinterpol);
plot(T,B,'o',Tinterpol,Binterpol,'-');
xlabel('Temperature, deg C');
ylabel('Bulk Modulus, GPa');
title('Temperature-Bulk Modulus Data and Spline Interpolation');
```

The resulting temperature – bulk modulus plot of the interpolated spline data (solid line) and the data values used are given in Figure 8.19.

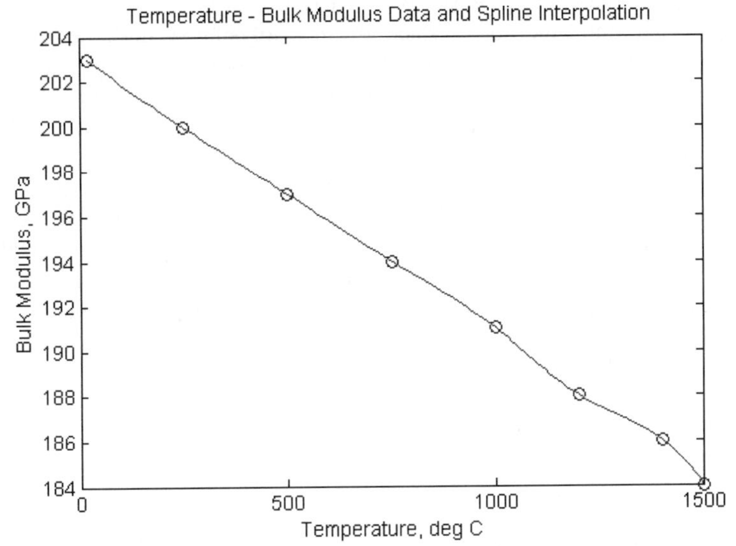


Figure 8.19 Temperature–bulk modulus data and its spline interpolation.

Advantages of the SiC micromachining and SiC technologies (high temperature and ruggedness) should be weighted against fabrication drawbacks because new processes must be designed and optimized. Reactive ion etching is used to pattern SiC thin films, however many problems such as masking, low etch rates, and poor etch selectivity, must be addressed and resolved. Articles [21, 22] report two single-layer reactive ion etching-based polycrystalline SiC surface micromachining processes using polysilicon or SiO as the sacrificial layer. In addition, the micromolding processes (used to fabricated polysilicon molds in conjunction with polycrystalline SiC film deposition and mechanical polishing to pattern polycrystaline SiC films) are introduced. The micromolding process can be used for single- and multi-layer SiC surface micromachining.

The micromotor fabrication processes are illustrated in Figure 8.20. A 5-10 μm thick sacrificial molding polysilicon is deposited through the LPCVD on a 3-5 μm sacrificial thermal oxide. The rotor-stator mold formation can be made on the polished (chemical-mechanical polishing) polysilicon surface, enabling the 2 μm fabrication features using standard lithography and reactive ion etching. After the mold formation and delineation, the SiC is deposited on the wafer using an atmospheric pressure chemical vapor deposition reactor. In particular, the phosphorus-doped (n-type) polycrystalline SiC films are deposited on the SiO sacrificial layers at $1050^0$C with 0.5–1 μm/hour rate (deposition is not selective, and SiC will be deposited on the surfaces of the polysilicon molds as well). Mechanical polishing of SiC is needed to expose the polysilicon and planarize the wafer surface. As reported in [21, 22], the polishing can be done with 3 μm diameter diamond suspension, 360 N normal force, and 15 rad/sec pad rotation (the removal rate of SiC is reported to be 100 nm/min). The wafers are polished until the top surface of the polysilicon mold is exposed (polishing must be stopped at once due to the fast polishing rate). The flange mold is fabricated through the polysilicon and the sacrificial oxide etching (using KOH and BHF, respectively). The 0.5 μm bearing clearance low-temperature oxide is deposited and annealed at $1000^0$C. Then, the 1 μm polycrystalline SiC film is deposited and patterned by reactive ion etching to make the bearing. The release begins with the etching (in BHF solution) to strip the left-over bearing clearance oxide. The sacrificial mold is removed by etching (KOH system) the polysilicon. It should be emphasized that the SiC and SiO are not etched during the mold removal step. Then, the moving parts of the micromotor are released. The micromotor is rinsed in water and methanol, and dried with the air jet.

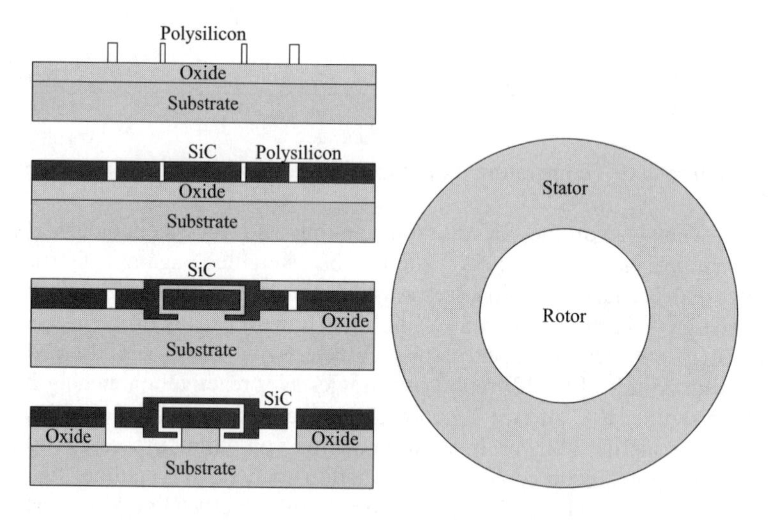


Figure 8.20 Fabrication of the SiC micromotors: cross-sectional schematics.

Using this fabrication process, the micromotor with the 100-150 μm rotor diameter, 2 μm airgap, and 21 μm bearing radius, was fabricated and tested in [21, 22]. The rated voltage was 100 V and the maximum angular velocity was 30 rad/sec. For silicon and polysilicon micromotors, one of the most critical problem are the bearing and ruggedness. The application of SiC reduces the friction and improves the ruggedness. These contributes to the reliability of the SiC-based-fabricated micromachines.

## 8.11 AXIAL ELECTROMAGNETIC MICROMOTORS

The major problem is to devise novel microtransducers in order to relax fabrication difficulties, guarantee affordability, efficiency, reliability, and controllability of MEMS. In fact, the electrostatic and planar micromotors fabricated and tested to date are found to be inadequate for a wide range of applications due to difficulties associated and cost. Therefore, this section is devoted to devising novel affordable rotational micromotors.

Figure 8.21 illustrates the devised axial topology micromotor which has the *closed-ended* electromagnetic system. The stator is made on the substrate with deposited microwindings (printed copper coils can be made using the fabrication processes described as well as using double-sided substrate with the one-sided deposited copper thin film through conventional photolithography process). The bearing post is fabricated on the stator substrate and the bearing hold is a part of the rotor microstructure. The rotor with permanent-magnet thin films rotates due to the electromagnetic torque developed. It is important to emphasize that stator and rotor are made using the conventional well-developed processes and materials.

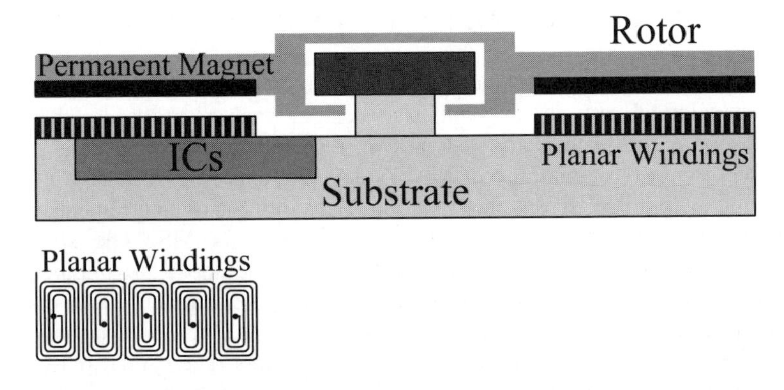


Figure 8.21 Slotless axial electromagnetic micromotor (cross-sectional schematics) with controlling ICs.

It is evident that conventional silicon and SiC technologies can be used. The documented micromotor has a great number of advantages. The most critical benefit is the fabrication simplicity. In fact, axial micromotors can be straightforwardly fabricated, and this will enable their wide applications as microactuators and microsensors. However, the axial micromotors must be designed and optimized to attain good performance. The optimization is based upon electromagnetic, mechanical, and thermal design. The micromotor optimization can be carried out using the steady-state concept (finite element analysis) and dynamic paradigms (lumped-parameter models or complete electromagnetic-mechanical-thermal high-fidelity mathematical models derived as a set of partial differential equations using Maxwell's, *torsional-mechanical* and heat equations). In general, the nonlinear optimization problems need to be addressed, formulated, and solved to guarantee the superior microtransducer performance. In addition to the microtransducer design, one must concentrate the attention on the ICs and controller design. In particular, the circuitry is designed based upon the converter and inverter topologies (e.g., hard- and soft-switching, one-, two-, or four-quadrant, etc.), filters and sensors used, rated voltage and current, etc. From the control prospective, the electromagnetic features must be thoroughly examined. For example, the electromagnetic micromotor studied is the synchronous micromachine. Therefore, to develop the electromagnetic torque, the voltages applied to the stator windings must be supplied as the functions of the rotor angular displacement. Therefore, the Hall-effect sensors must be used, or the so-called sensorless controllers (the rotor position is observed or estimated using the directly measured variables) must be designed and implemented using ICs. This brief discussion illustrates a wide spectrum of fundamental problems involved in the design of integrated microtransducers with controlling and signal processing ICs.

## 8.12 SYNERGETIC COMPUTER-AIDED DESIGN OF MEMS

The critical focus themes in MEMS technology development and implementation are rapid synthesis, design, and prototyping through synergetic multi-disciplinary system-level research in electromechanics [3]. Let us discuss the taxonomy of MEMS devising and optimization which is relevant to cognitive study, classification, and synthesis of electromechanical systems in general, including MEMS and many NEMS. The simplest example is devising microtransducers which compose a number of microstructures. Thus, the microtransducer can be defined in terms of microstructures needed to build it. Hence, one can use the Boolean theory to describe microtransducers in terms of microstructures. The similar reasoning can be made in describing the fabrication processes which can be represented as the sequential steps with specific materials and techniques applied. Different levels of description can be researched. For example, the electrodeposition processes are influenced by the chemicals used and the process parameters (current density, temperature, pH, etc.). Rather than emphasize the well-developed fabrication technologies and process (materials and processes characterization are very important problems which were studied and documented in [1-7, 9-12, 15, 16, 19-22]), our attention will be primarily concentrated on the device- and system-level synthesis and analysis. Devising MEMS is the closed evolutionary process to study possible system-level evolutions based upon synergetic integration of microscale structures, devices and other components in the unified functional core. The ability to devise and optimize MEMS to a large extent depends on the computational efficiency, adaptability, functionality, integrity, compliance, robustness, flexibility, prototypeability, visualability, interactability, decision-making, and intelligence of the computer-based design tools and environments developed. It is likely that the fundamental electromechanical theory and applied-experimental results in conjunction with high-performance interacting software (e.g., MATLAB and MATEMATICA) will allow the designer to device (synthesize), prototype, design, model, simulate, analyze, and optimize MEMS. Furthermore, the synergetic quantitative synthesis and symbolic descriptions can be efficiently used searching and evaluating possible organizations, architectures, configurations, topologies, geometries, and other descriptive features providing the evolutionary potential needed.

Although general, systematic, and straightforward approaches can be developed to optimize microtransducers, radiating energy and optical microdevices, as well as ICs, it is unlikely that general devising tools based upon abstract concepts cannot be effectively applied for MEMS. This is due to a great variety of possible MEMS solutions, architectures, subsystem-device-structure organizations, and electromagnetic-optical-mechanical and electromechanical features. However, restricting the number of the possible solutions based upon the specifications and requirements, the MEMS synthesis can be performed (using informatics theory, artificial intelligence, knowledge-based libraries, and expert techniques) emphasizing the electromagnetic-optical-mechanical-thermal phenomena, and the numerical optimization and design of complex MEMS can be accomplished through CAD. It was emphasized that the conceptual design of MEMS integrates devising (synthesis), prototyping and analysis with the consecutive design, optimization and verification tasks.

Currently, there does not exist a CAD environment which can perform functional synthesis, modeling, analysis, design, or optimization for even relatively simple MEMS and their components. CAD environments and tools, such as SPICE and VHDL, have been developed only to design and model application-specific integrated circuits which are the important MEMS components. The corresponding CAD packages allow one to design analog, digital, and hybrid ICs integrating the fabrication technologies using materials and processes databases, e.g., semiconductor and other materials, etchants with etching rates, electroplating and lithography processes, etc. Promising developments in the development of MEMS CAD tools have been reported [3, 23-25]. However, further joint synergetic efforts are needed to perform integrated electro-opto-mechanical analysis and design. The ultimate goal is to progress beyond the three-dimensional representation of MEMS (which has some degree of merit and is needed as the synthesis, design, and optimizations tasks are completed) and steady-state analysis, but devising and optimizing MEMS using data-intensive analysis and heterogeneous synthesis. The hierarchical synthesis, design, optimization, and fabrication flow as the MEMS-CAD diagram as illustrated in Figure 8.22.

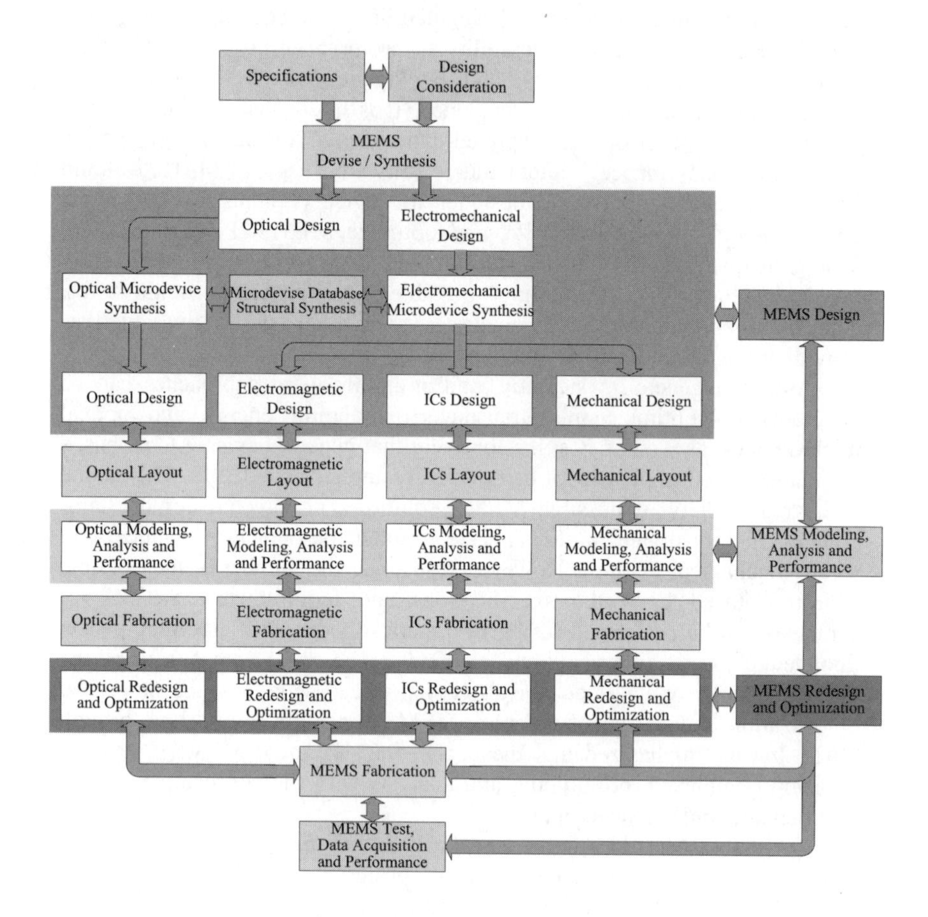


Figure 8.22 MEMS-CAD diagram.

As illustrated in Figure 8.22, the top-level specifications are used to devise (synthesize) novel MEMS, and then to devise electromechanical and electro-opto-mechanical (optical MEMS) subsystems, components, and microdevices. High-level integrated synthesis must be carried out using databases of microscale subsystems, devices, and structures. The current MEMS design tools allow one to perform the steady-state analysis and design of a limited number of microstructures and microdevices. Some CAD tools were extended to simulate and analyze electromechanical microdevices using linear and nonlinear differential equations applying the lumped-parameter mathematical modeling and simulation paradigms. For example, three-dimensional modeling and simulations of the cantilever beam were performed in [23, 24]. However, MEMS synthesis and classification, as well as analysis and optimization based upon the nonlinear electromagnetic features, to the best of the author's knowledge, have not been integrated in the design. The reusability and leverage of the extensive CAD developed for minitransducers are limited due to the critical need to devise novel microscale electromagnetic devices, microtransducers (for example, axial-flux/slotless/disk topologies), six-degree-of-freedom microactuators, sensorless control, etc. It is important to emphasize that the secondary phenomena and effects, usually neglected in conventional miniscale electromechanical motion devices (modeled using lumped-parameter models and analyzed using finite element analysis techniques) cannot be ignored. In addition, the microtransducer dynamics must be thoroughly studied applying high-fidelity mathematical models, and therefore the finite element analysis cannot be viewed as an enabling approach to be applied. It is the author's hope that analytic and numerical results, documented in this book, will allow us to progress toward the MEMS CAD developments.

In general, for microscale structures, devices, and systems, the evolutionary synthesis developments can be represented as the X-design-flow-map documented in Figure 8.23. This map illustrates the synthesis flow from devising (synthesis) to modeling–analysis–simulation–design, from modeling–analysis–simulation–design to optimization–refining, and finally from optimization–refining to fabrication–testing sequential evolutionary processes. The proposed X-map consists of four domains of MEMS representation. In particular,

- Devising (synthesis),
- Modeling–analysis–simulation–design,
- Optimization–refining,
- Fabrication–testing.

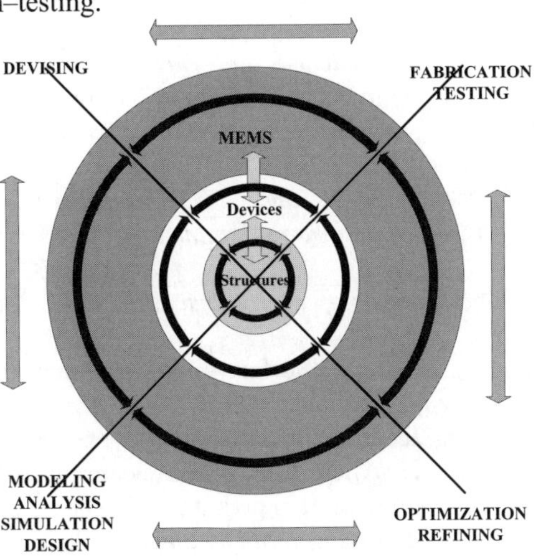


Figure 8.23 X-design-flow-map with four domains.

The desired degree of abstraction in the synthesis of new MEMS and NEMS requires one to apply this X-design-flow-map to devise novel MEMS and NEMS which integrate novel high-performance micro- and nanoscale structures and devices as components. The failure to verify the design of any MEMS component in the early phases at least causes the failure to design high-performance MEMS and leads to redesign.

The interaction between four domains (devising, modeling–analysis–simulation–design, optimization–refining and fabrication–testing) allows one to guarantee bi-directional top-down and bottom-up design features applying the low-level component data to high-level design, and using the high-level requirements to devise and design low-level components. The X-design-flow-map ensures hierarchy, modularity, locality, integrity, and other important features allowing one to design complex MEMS and NEMS. Using the results reported in this book, heterogeneous CAD can be developed. For example, to devise novel MEMS and NEMS, the Synthesis and Classification Solver is applied. The mathematical modeling problem is solved using micro- and nanoelectromechanics concepts reported. The developed mathematical models allow the designer to perform data-intensive analysis and heterogeneous simulations with outcome prediction. Design, control and optimization aspects are researched (these problems with application to MEMS were documented and illustrated in Chapter 7). Finally, as was illustrated in Chapter 3 and this chapter, using different fabrication technologies, techniques, processes and material, affordable high-performance MEMS can be made.

## References


1. S. A. Campbell, *The Science and Engineering of Microelectronic Fabrication*, Oxford University Press, New Yoork, 2001.
2. G. T. A. Kovacs, *Micromachined Transducers Sourcebook,* WCB McGraw-Hill, Boston, MA, 1998.
3. S. E. Lyshevski, *Nano- and Micro-Electromechanical Systems: Fundamental of Micro- and Nano- Engineering*, CRC Press, Boca Raton, FL, 1999.
4. M. Madou, *Fundamentals of Microfabrication*, CRC Press, Boca Raton, FL, 1997.
5. Y.-J. Kim and M. G. Allen, “Surface micromachined solenoid inductors for high frequency applications,” *IEEE Trans. Components, Packaging, and Manufacturing Technology*, part C, vol. 21, no. 1, pp. 26-33, 1998.
6. J. Y. Park and M. G. Allen, “Integrated electroplated micromachined magnetic devices using low temperature fabrication processes,” *IEEE Trans. Electronics Packaging Manufacturing*, vol. 23, no. 1, pp. 48-55, 2000.

7. D. J. Sadler, T. M. Liakopoulos and C. H. Ahn, "A universal electromagnetic microactuator using magnetic interconnection concepts," *Journal Microelectromechanical Systems*, vol. 9, no. 4, pp. 460-468, 2000.
8. S. E. Lyshevski, *Electromechanical Systems, Electric Machines, and Applied Mechatronics*, CRC Press, Boca Raton, FL, 1999.
9. A. B. Frazier and M. G. Allen, "Uses of electroplated aluminum for the development of microstructures and micromachining processes," *Journal Microelectromechanical Systems*, vol. 6, no. 2, pp. 91-98, 1997.
10. H. Guckel, T.R. Christenson, K.J. Skrobis, J. Klein, and M. Karnowsky, "Design and testing of planar magnetic micromotors fabricated by deep x-ray lithography and electroplating," *Technical Digest of International Conference on Solid-State Sensors and Actuators, Transducers 93*, Yokohama, Japan, pp. 60-64, 1993.
11. W. P. Taylor, M. Schneider, H. Baltes and M. G. Allen, "Electroplated soft magnetic materials for microsensors and microactuators," *Proc. Conf. Solid-State Sensors and Actuators, Transducers 97*, Chicago, IL, pp. 1445-1448, 1997.
12. L. K. Lagorce, O. Brand and M. G. Allen, "Magnetic microactuators based on polymer magnets," *Journal Microelectromechanical Systems*, vol. 8, no. 1, pp. 2-9, 1999.
13. D. O. Smith, "Static and dynamic behavior in thin permalloy films," *Journal of Applied Physics,* vol. 29, no. 2, pp. 264-273, 1958.
14. D. Suss, T. Schreft and J. Fidler, "Micromagnetics simulation of high energy density permanent magnets," *IEEE Trans. Magnetics*, vol. 36, no. 5, pp. 3282-3284, 2000.
15. J. W. Judy and R. S. Muller, "Magnetically actuated, addressable microstructures," *Journal Microelectromechanical Systems*, vol. 6, no. 3, pp. 249-256, 1997.
16. Y. W. Yi and C. Liu, "Magnetic actuation of hinged microstructures," *Journal Microelectromechanical Systems*, vol. 8, no. 1, pp. 10-17, 1999.
17. J. M. Gere and S. P. Timoshenko, *Mechanics of Materials*, PWS Press, 1997.
18. N. J. Groom and C. P. Britcher, "A description of a laboratory model magnetic suspension test fixture with large angular capability," *Proc. Conf. Control Applications, NASA Technical Paper* – 1997, vol. 1, pp. 454-459, 1992.
19. C. H. Ahn, Y. J. Kim and M. G. Allen, "A planar variable reluctance magnetic micromotor with fully integrated stator and coils," *Journal Microelectromechanical Systems,* vol. 2, no. 4, pp. 165-173, 1993.
20. E. J. O'Sullivan, E. I. Cooper, L. T. Romankiw, K. T. Kwietniak, P. L. Trouilloud, J. Horkans, C. V. Jahnes, I. V. Babich, S. Krongelb, S. G. Hegde, J. A. Tornello, N. C. LaBianca, J. M. Cotte and T. J. Chainer, "Integrated, variable-reluctance magnetic minimotor," *IBM Journal Research and Development,* vol. 42, no. 5, 1998.

21. A. A. Yasseen, C. H. Wu, C. A. Zorman and M. Mehregany, "Fabrication and testing of surface micromachined polycrystalline SiC micromotors," *IEEE Trans. Electron Device Letters,*vol. 21, no. 4, pp. 164-166, 2000.
22. A. A. Yasseen, C. A. Zorman and M. Mehregany, "Surface micromachining of polycrystalline silicon carbide films microfabricated molds of SiO and polysilicon," *Journal Microelectromechaniical Systems,*vol. 8, no. 1, pp. 237-242, 1999.
23. Z. Bai, D. Bindel, J. V. Clark, J. Demmel, K. S. J. Pister and N. Zhou, "New numerical techniques and tools in Sugar for 3D MEMS simulation," *Proc. Conf. Modeling and Simulation of Microsystems*, Hilton Head Island, SC, pp. 31-34, 2001.
24. J. V. Clark, N. Zhou, D. Bindel, L. Schenato, W. Wu, J. Demmel and K. S. J. Pister, "3D MEMS simulation modeling using modified nodal analysis," *Proc. Microscale Systems: Mechanics and Measurements Symposium*, Orlando, FL, pp. 68-75, 2000.
25. T. Mukherjee, G. K. Fedder, D. Ramaswamy and J. White, "Emerging simulation approaches for micromachined devices," *IEEE Trans. Computer-Aided Design of Integrated Circuits and Sytems,* vol. 19, no. 12, pp. 1572-1589, 2000.

# INDEX

**H**



**I**



**L**



**M**



**N**



**O**

## P



## Q



## R



## S



## T



## V



## W